Sulphate-reducing Bacteria

The Sulphate-reducing Bacteria (SRB) are a large group of anaerobic organisms that play an important role in many biogeochemical processes. Not only are they of early origins in the development of the biosphere, but their mechanisms of energy metabolism shed light on the limits of life processes in the absence of oxygen. They are widely distributed in nature, and are regular components of engineered systems including, for example, petroleum reservoirs and oil production facilities. SRB are currently subject to extensive genomic studies, which are yielding new understanding of their basic biochemical mechanisms, and aiding in the development of novel techniques for the analyses of their environmental roles. This volume provides a timely update on these important microorganisms, from basic science to applications, and will therefore serve as a valuable resource for researchers and graduate students in the fields of microbial ecology, microbial physiology, bioengineering, biogeochemistry and related areas of environmental science.

LARRY L. BARTON is a Professor in the Department of Biology at the University of New Mexico.

W. ALLAN HAMILTON is Professor Emeritus of Microbiology at the University of Aberdeen.

Sulphate-reducing Bacteria

Environmental and Engineered Systems

EDITED BY

LARRY L. BARTON

University of New Mexico

AND

W. ALLAN HAMILTON

University of Aberdeen

CAMBRIDGE UNIVERSITY PRESS
Cambridge, New York, Melbourne, Madrid, Cape Town, Singapore,
São Paulo, Delhi, Dubai, Tokyo

Cambridge University Press
The Edinburgh Building, Cambridge CB2 8RU, UK

Published in the United States of America by Cambridge University Press, New York

www.cambridge.org
Information on this title: www.cambridge.org/9780521123990

© Cambridge University Press 2007

This publication is in copyright. Subject to statutory exception and to the provisions of relevant collective licensing agreements, no reproduction of any part may take place without the written permission of Cambridge University Press.

First published 2007
This digitally printed version 2009

A catalogue record for this publication is available from the British Library

Library of Congress Cataloguing in Publication data

Sulphate-reducing bacteria: environmental and engineered systems / edited by Larry L. Barton and W. Allan Hamilton. -- 1st ed.
p. cm.
Includes bibliographical references and index.
ISBN-13: 978-0-521-85485-6 (hardback)
ISBN-10: 0-521-85485-7 (hardback)
1. Sulphur bacteria. I. Barton, Larry. II. Hamilton, W. Allan (William Allan), 1936- III. Title.

QR92.S8S855 2007
579.3′8--dc22

2007000588

ISBN 978-0-521-85485-6 Hardback
ISBN 978-0-521-12399-0 Paperback

Additional resources for this publication at www.cambridge.org/9780521123990

Cambridge University Press has no responsibility for the persistence or accuracy of URLs for external or third-party internet websites referred to in this publication, and does not guarantee that any content on such websites is, or will remain, accurate or appropriate.

Contents

List of Contributors — *page* vii
Preface — xvii

1 Energy metabolism and phylogenetic diversity of sulphate-reducing bacteria — 1
Rudolf K. Thauer, Erko Stackebrandt and W. Allan Hamilton

2 Molecular strategies for studies of natural populations of sulphate-reducing microorganisms — 39
David A. Stahl, Alexander Loy and Michael Wagner

3 Functional genomics of sulphate-reducing prokaryotes — 117
Ralf Rabus and Axel Strittmatter

4 Evaluation of stress response in sulphate-reducing bacteria through genome analysis — 141
J. D. Wall, H. C. Bill Yen and E. C. Drury

5 Response of sulphate-reducing bacteria to oxygen — 167
Henrik Sass and Heribert Cypionka

6 Biochemical, proteomic and genetic characterization of oxygen survival mechanisms in sulphate-reducing bacteria of the genus *Desulfovibrio* — 185
Alain Dolla, Donald M. Kurtz, Jr., Miguel Teixeira and Gerrit Voordouw

7 Biochemical, genetic and genomic characterization of anaerobic electron transport pathways in sulphate-reducing *Delta proteobacteria* — 215
Inês A. C. Pereira, Shelley A. Haveman and Gerrit Voordouw

Jean-Luc Cayol
Laboratoire de Microbiologie IRD, UMR 180
Universités de Provence et de la Méditerranée
ESIL-GBMA
Case 925
163 avenue de Luminy
13288 Marseille Cedex 09
France

John H. Cummings
Gut Group
University of Dundee
Ninewells Hospital Medical School
Dundee DD1 9SY
UK

Heribert Cypionka
Institut für Chemie und Biologie des Meeres
Universität Oldenburg
Carl-von-Ossietzky-Str. 9-11
D-26129 Oldenburg
Germany

Alain Dolla
BIP-IBSM-CNRS
13402 Marseilles Cedex 20
France

E. C. Drury
Biochemistry Department
University of Missouri–Columbia
Columbia
Missouri 65211
USA

Giovanni Esposito
Sub-department of Environmental Technology
Agricultural University of Wageningen
"Biotechnion" – Bomenweg, 2
P.O. Box 8129
6700 EV Wageningen
The Netherlands

Guy Fauque
Laboratoire de Microbiologie IRD, UMR 180
Universités de Provence et de la Méditerranée
ESIL-GBMA
Case 925
163 avenue de Luminy
13288 Marseille Cedex 09
France

Geoffrey M. Gadd
Division of Environmental and Applied Biology
Biological Sciences Institute
School of Life Sciences
University of Dundee
Dundee
DD1 4HN
Scotland
UK

Alexander Galushko
Max Planck Institute for Marine Microbiology
Celsiusstraße 1
D-28359 Bremen
Germany

Pablo Gonzalez
REQUIMTE, Departamento de Quimica
Centro de Quimica
Fina e Biotechnologia
Faculdade de Ciências e Technologia
Universidade Nova de Lisboa, 2829-516
Monte de Caparica
Portugal

Florence Goulhen
Institut de Biologie Structurale et Microbiologie
Unité de Bioénergétique et Ingénierie des Protéines, CNRS
31 chemin Joseph Aiguier
13402 Marseille Cedex 20
France

W. Allan Hamilton
Molecular and Cell Biology
Institute of Medical Sciences
University of Aberdeen
Aberdeen AB25 2ZD
Scotland
UK

Shelley A. Haveman
Department of Microbiology
University of Massachusetts
Amherst
Massachusetts 01003
USA

Simon L. Hockin
URS Corporation Ltd
St James' Building
61–95 Oxford Street
Manchester, M1 6EJ
UK

Katrin Knittel
Max Planck Institute for Marine Microbiology
Celsiusstraße 1, D-28359, Bremen
Germany

Donald M. Kurtz, Jr.
University of Texas at San Antonio
San Antonio
Texas 78249
USA

Piet N. L. Lens
Sub-department of Enivironmental Technology
Agricultural University of Wageningen
"Biotechnion" – Bomenweg, 2
P.O. Box 8129
6700 EV Wageningen
The Netherlands

Alexander Loy
Department of Microbial Ecology
University of Vienna
Vienna
Austria

George T. Macfarlane
Gut Group, University of Dundee
Ninewells Hospital Medical School
Dundee DD1 9SY
UK

Sandra Macfarlane
Gut Group, University of Dundee
Ninewells Hospital Medical School
Dundee DD1 9SY
UK

Isabel Moura
REQUIMTE, Departamento de Quimica
Centro de Quimica Fina e Biotechnologia
Faculdade de Ciências e Technologia
Universidade Nova de Lisboa
2829-516 Monte de Caparica
Portugal

José J. G. Moura
REQUIMTE, Departmento de Quimica
Centro de Quimica Fina e Biotechnologia,
Faculdade de Ciências e Technologia
Universidade Nova de Lisboa
2829-516 Monte de Caparica
Portugal

Florin Musat
Max Planck Institute for Marine Microbiology
Celsiusstraße 1
D-28359 Bremen
Germany

Santoshi Okabe
Department of Urban and Environmental Engineering
Graduate School of Engineering
Hokkaido University
North-13, West-8, Kita-ku
Sapporo 060-8628
Japan

Bernard Ollivier
Laboratoire de Microbiologie IRD, UMR 180
Universités de Provence et de la Méditerranée
ESIL-GBMA, Case 925
163 avenue de Luminy
13288 Marseille Cedex 09
France

R. John Parkes
School of Earth, Ocean and Planetary Sciences
University of Cardiff Main Building
Cardiff CF10 3YE
UK

Inês A. C. Pereira
Instituto de Technologia Química e Biológica
Universidade Nova de Lisboa
Av. Da República
(EAN) 2784-505 Ociras
Portugal

Richard M. Plunkett
Department of Biology, MSC03 2020
Laboratory of Microbial Chemistry
University of New Mexico
Albuquerque
New Mexico 87131
USA

Ralf Rabus
Max Planck Institue for Marine Microbiology
Celsiusstr. 1
D-28359 Bremen
Germany

Henrik Sass
School of Earth, Ocean and Planetary Sciences
Cardiff University
Park Place
Cardiff CF10 3YE
UK

Erko Stackebrandt
Deutsche Sammlung von Mikroorganismen und Zellkulturen GmbH
Mascheroder Weg 1B
D-38124 Braunschweig
Germany

David A. Stahl
Department of Microbiology
University of Washington
Seattle
Washington 98195
USA

Axel Strittmatter
Göttingen Genomics Laboratory
Georg-August-University
Grisebachstr. 8, D-37077 Göttingen
Germany

Jan A. Sunner
Department of Chemistry and Biochemistry
Montana State University
Bozeman
Montana 59717
USA

Miguel Teixeira
Instituto de Tecnologia Quimica e Biologica
Universidade Nova de Lisboa
Av. da República
(EAN) 2784-505 Oeiras
Portugal

Rudolf K. Thauer
Max Planck Institute for Terrestrial Microbiology
Karl-von-Frisch-Strasse
D-35043 Marburg/Lahn
Germany

Marcus Vallero
Sub-department of Environmental Technology
Agricultural University of Wageningen
"Biotechnion" – Bomenweg, 2
P.O.Box 8129
6700 EV Wageningen
The Netherlands

Gerrit Voordouw
Department of Biological Sciences
University of Calgary
Calgary
Alberta, T2N 1N4
Canada

Michael Wagner
Department of Microbial Ecology
University of Vienna
Vienna
Austria

J. D. Wall
Biochemistry Department
University of Missouri–Columbia
Columbia
Missouri 65211
USA

Friedrich Widdel
Max Planck Institute for Marine Microbiology
Celsiusstraße 1
D-28359 Bremen
Germany

H. C. Bill Yen
Biochemistry Department
University of Missouri–Columbia
Columbia
Missouri 65211
USA

Preface

Recognition of the biological nature of sulphate reduction in natural environments, and identification of the bacterial species involved dates to the latter part of the nineteenth century, and the seminal work of such giants of the early days of microbiology as Beijerinck and Winogradsky. The central role of environmental studies in highlighting the issues to be addressed and the problems to be solved, has remained to this day a constant theme in microbiological analyses of the sulphate reducers.

The modern era of such analyses, however, can be said to date from the period around 1960 when the demonstrations by Postgate and Peck, respectively, of the presence of cytochromes and of phosphorylation linked to anaerobic respiration in sulphate-reducing bacteria (SRB), fundamentally altered our view of the biochemical nature of these organisms and, in particular, of their mechanisms of energy conservation.

There then followed a period of intense activity centred on: elucidation of the metabolic pathways of substrate utilisation and the mechanisms of energy generation; cultural techniques and the identification of an ever-increasing number of new species; and the appreciation of their significant role in maintaining, or disrupting, the biological balance of many natural and man-made ecosystems.

These themes of biochemistry and cell physiology, phylogeny, and ecology remain central to the understanding of SRB themselves, and of their interactions with other components of the biosphere. In recent years, however, their study has undergone a further paradigm shift with the introduction of the many powerful experimental techniques and analytical approaches of molecular biology. As a direct consequence of these developments, sulphate-reducing prokaryotes (bacteria and archaea, as we have come to appreciate) are now the chosen organisms of study in many of the major microbiological laboratories worldwide. Additionally, there is an extensive literature covering their several unique characteristics which, in many cases, may help to shed light on certain issues of fundamental importance to our understanding of the evolution and development of life processes. It is the purpose of this book, therefore, to draw together many of the major players in the field of biological sulphate reduction, and to present a clear and full picture of the current state of our knowledge. We thank our authors who have accepted this challenge, and given so willingly of their time, effort and insight.

It has been a conscious decision to include in the same volume studies of: genomic and proteomic analyses; phylogenetic diversity; molecular characterisation of enzymes and respiratory systems; thermodynamic analyses of metabolic processes, including anaerobic oxidation of hydrocarbons; response to stress, most particularly with regard to oxygen and other alternative electron acceptors; extreme and specialised (micro)environments, including biofilms; environmental impact in, for example, bioremediation and corrosion; and medical microbiology. These apparently disparate subject areas nevertheless form an intellectual continuum, within which it is possible to see the interdependence of the techniques and thought processes of one study area impacting directly on another. Perhaps uniquely, our current knowledge of the SRB is sufficiently extensive for us to be able to recognise and make practical use of this cross-fertilisation, and yet not so extensive and subject to technique-dependency that scientists working in one field have neither knowledge of, nor empathy with, the work in other areas. Thus it may be that this book will be seen in later years to stand at the crossroads before any such parting of the ways, as research into the sulphate-reducers continues at its present exhilarating pace. We would hope so.

W. Allan Hamilton
Larry L. Barton

CHAPTER 1

Energy metabolism and phylogenetic diversity of sulphate-reducing bacteria

Rudolf K. Thauer, Erko Stackebrandt and W. Allan Hamilton

1.1 INTRODUCTION

Sulphate-reducing bacteria (SRB) are those prokaryotic microorganisms, both bacteria and archaea, that can use sulphate as the terminal electron acceptor in their energy metabolism, i.e. that are capable of dissimilatory sulphate reduction. Most of the SRB described to date belong to one of the four following phylogenetic lineages (with some examples of genera): (i) the mesophilic δ-proteobacteria with the genera *Desulfovibrio, Desulfobacterium, Desulfobacter,* and *Desulfobulbus*; (ii) the thermophilic Gram-negative bacteria with the genus *Thermodesulfovibrio*; (iii) the Gram-positive bacteria with the genus *Desulfotomaculum*; and (iv) the *Euryarchaeota* with the genus *Archaeoglobus* (Castro *et al.*, 2000). A fifth lineage, the *Thermodesulfobiaceae*, has been described recently (Mori *et al.*, 2003).

Many SRB are versatile in that they can use electron acceptors other than sulphate for anaerobic respiration. These include elemental sulphur (Bottcher *et al.*, 2005; Finster *et al.*, 1998), fumarate (Tomei *et al.*, 1995), nitrate (Krekeler and Cypionka, 1995), dimethylsulfoxide (Jonkers *et al.*, 1996), Mn(IV) (Myers and Nealson, 1988) and Fe(III) (Lovley *et al.*, 1993; 2004). Some SRB are even capable of aerobic respiration (Dannenberg *et al.*, 1992; Lemos *et al.*, 2001) although this process appears not to sustain growth, and probably provides these organisms only with energy for maintenance. Since dissimilatory sulphate reduction is inhibited under oxic conditions, SRB can grow at the expense of sulphate reduction only in the complete absence of molecular oxygen. SRB are thus considered to be strictly anaerobic microorganisms and are mainly found in sulphate-rich anoxic habitats (Cypionka, 2000; Fareleira *et al.*, 2003; Sass *et al.*, 1992). These conditions apply in marine sediments since ocean water is rich in sulphate, its concentration being as high as 30 mM. SRB are also present, however, in freshwater sediments, where the sulphate concentration is generally well below 1 mM but is continuously maintained at this level by the re-oxidation of the H_2S to sulphate at the oxic/anoxic interface due to the action of chemolithotrophic and photolithotrophic bacteria (Holmer and Storkholm, 2001). Since most SRB may use electron acceptors other than sulphate, they can also be found in anoxic habitats depleted in sulphate such as the human intestinal tract (Chapter 18). SRB are, however, most abundant in habitats where the availability of sulphate is not limiting.

1.1.2 Thermodynamics

The discussion of the energetics of sulphate reduction is introduced with consideration of the thermodynamic parameters that determine the possible interactions between potential electron donors and sulphate as electron acceptor.

The redox potential, $E^{\circ\prime}$, of the sulphate/HS^- couple is -217 mV under standard conditions, which are 1M concentrations of sulphate and HS^- at pH 7.0 and 25°C. Under physiological conditions, however, where the concentrations of sulphate are generally <30 mM and of HS^- <1 mM, the redox potential, E', is a little more positive and of the order of -200 mV. Thus at all expected concentrations of sulphate and HS^-, almost any organic compound generated by plants or animals, including carbohydrates, fatty acids, alkanes and aromatic compouds, should, in theory, be able to be completely oxidized to CO_2 since the redox potentials of each of these possible electron donors is significantly more negative than the –200 mV of the sulphate/HS^- couple (Table 1.1). Indeed, each of these bio-organic materials has now been shown to be completely mineralized by individual SRB, either alone or in syntrophic association with other organisms. This is even true for methane with a redox potential $E^{\circ\prime}$ of the CO_2/methane couple of -244 mV and therefore a redox potential difference $\Delta E'$ $[E'(SO_4^{2-}/HS^-) - E^{\circ\prime}(CO_2/CH_4)]$ of only $+44$ mV. Remarkably, it has recently been shown that a $\Delta E'$ of $+25$ mV, equivalent to a free energy change $\Delta G'$of –20 kJ/mol ($\Delta G' = -nF\Delta E'$; $n = 8$) is sufficient to sustain growth of SRB (Hoehler *et al.*, 2001).

For the prediction of electron flow in natural environments, in most cases $E^{\circ\prime}$ rather than E' of the reductant can be used since the two generally differ by only around 20 mV. There are, however, important exceptions, e.g. the H^+/H_2 couple and the S°/HS^- couple. The redox potential at pH 7.0 (H^+ concentration constant at 10^{-7} M) of the H^+/H_2 couple increases

Table 1.1. *Redox potentials E°′ of electron donors thermodynamically capable and not capable of dissimilatory sulphate, thiosulphate or bisulphite reduction*

Redox couple	*n*	E°′ (Mv)
$2CO_2$ + 2acetate /hexose	8	−670
CO_2 + acetate /pyruvate	2	−660
FeS_2/FeS + H_2S[d]	2	−613
SO_4^{2-}/HSO_3^-	**2**	**−516**
CO_2/CO	2	−520
Fe^{2+}/Fe°	2	−447
CO_2 + acetate/lactate	4	−430
CO_2 /formate	2	−432
$2H^+$/H_2	2	−414
		(−270 to −300)[a]
$6CO_2$ /hexose	24	−410
$S_2O_3^{2-}$/HS^- + HSO_3^-	**2**	**−402**
CO_2 + acetate + NH_3/alanine	4	−400
Acetate/ethanol	4	−390
CO_2/methanol	6	−370
$4CO_2$ /succinate	12	−312
$7CO_2$ /benzoate[e]	30	−300
2Acetate/butyrate	4	−290
CO_2 + acetate /glycerol	6	−290
$2CO_2$ /acetate	8	−290
$4CO_2$/butyrate	20	−280
$3CO_2$ /propionate	7	−280
N_2/NH_3[f]	6	−276
S°/H_2S*	**2**	**−270 (−120)[b]**
$6CO_2$/hexane[g]	38	−250
CO_2/CH_4	8	−244
SO_4^{2-}/HS^-	**8**	**−217 (−200)[c]**
SO_3H^-/HS^-	**6**	**−116**
Glycine/acetate + NH_3	2	−10
Fumarate/succinate	2	+33
Trimethylamine N-oxide/trimethylamine	2	+130
Dimethylsulfoxide/dimethylsulphide	2	+160
$Fe(OH)_3$ + HCO_3^-/$FeCO_3$[h]	1	+200
NO_2^-/NH_3	6	+330
NO_3^-/NH_3	8	+360

Mn^{4+}/Mn^{2+}	2	+407
NO_3^-/NO_2^-	2	+430
$2NO_3^-/N_2$	10	+760
$O_2/2H_2O$	4	+818
$2NO/N_2O$	2	+1175
$H_2O_2/2H_2O$	2	+1350
N_2O/N_2	2	+1.360

[a]Calculated for a H_2 partial pressure of 1 Pa and 10 Pa, respectively.
[b]Calculated for a $[HS^-] = 0.1$ mM.
[c]Calculated for [sulphate] = 30 mM and $[HS^-] = 0.1$ mM.
[d](Wächtershäuser, 1992).
[e]Calculated from the free energy of formation. $\Delta G°f$ of benzoate was estimated from $\Delta G°f$ for benzoic acid (crystalline solid state) (−245 kJ/mol), from the solubility of benzoic acid at 25 °C (27.8 mM) and from the pK of benzoic acid (4.2) to be −212.3 kJ/mol (Thauer and Morris, 1984).
[f]N_2 cannot be used as electron acceptor for energy conservation because of the too high energy of activation required for its reduction.
[g](Zengler *et al.*, 1999)
[h](Ehrenreich and Widdel, 1994)
Notes: $E°'$ at pH 7.0 are given for H_2, CO_2, CO, CH_4 and O_2 in the gaseous state at 10^5 Pa, for S° in the solid state and for all other compounds in aqueous solution at 1M concentration. The values in brackets are E' values calculated for physiological substrate and product concentrations. $E°'$ values were calculated from $\Delta G°'$ values: $\Delta G°' = -nF\Delta E$, where n is the number of electrons and $F = 96\ 487$ J/mol/volt. Except were indicated, $\Delta G°'$ values were taken from (Thauer *et al.*, 1977).

from −414 mV at an H_2 partial pressure of 10^5 Pa (standard condition) to values between −270 mV and −300 mV at H_2 partial pressures between 1 and 10 Pa, which is the concentration range of H_2 prevailing in sediments. Thus the oxidation of acetate to CO_2 ($E°' = -290$ mV) with H^+ as electron acceptor ($E' = -270$ mV at 1 Pa:H_2) becomes thermodynamically feasible and there are organisms that appear to live at the expense of this equation (Galouchko and Rozanova, 1996; Lee and Zinder, 1988; Shigematsu *et al.*, 2004).

$$CH_3COO^- + H^+ + 2H_2O = 2CO_2 + 4H_2 \quad \Delta G°' = +95\ \text{kJ/mol} \quad (1.1)$$

The redox potential of the $S°/HS^-$ couple (S° in the solid state and therefore constant) increases from −270 mV under standard conditions to −120 mV under physiological conditions where the HS^- concentrations can be 0.1 mM or even lower. As a consequence, under *in situ* conditions the reduction of sulphate with H_2S to S° is endergonic. In agreement with this prediction, SRB have been found that can grow at the expense of S° disproportionation to sulphate and HS^- (Finster *et al.*, 1998):

$$4S^0 + 4H_2O = SO_4^{2-} + 3HS^- + 5H^+ \qquad \Delta G^{\circ\prime} = +41\,\text{kJ/mol} \qquad (1.2)$$

1.1.3 Energy coupling

Growth of SRB with dissimilatory sulphate reduction indicates that substrate oxidation is coupled with adenosine triphosphate (ATP) synthesis from adenosine diphosphate (ADP) and inorganic phosphate. It leaves open, however, whether coupling is by substrate level phosphorylation and/or electron transport-linked phosphorylation (Thauer *et al.*, 1977). In substrate level phosphorylation an "energy-rich" intermediate is formed from organic substrates during exergonic oxidation reactions. The "energy-rich" intermediate is generally an acyl phosphate or an acyl thioester, which have group transfer potentials equivalent to that of ATP, and are in an enzyme-catalyzed equilibrium with the ADP/ATP system. In electron transport-linked phosphorylation the redox potential difference between electron carriers and the terminal electron acceptor is conserved in a transmembrane electrochemical proton or sodium ion gradient, which drives the phosphorylation of ADP via a membrane-bound ATP synthase.

For many years it was thought that it was only possible to grow SRB on organic substrates as electron donors for dissimilatory sulphate reduction, and this led to the belief that in these organisms energy is conserved mainly or exclusively via substrate level phosphorylation. However, in 1978 it was unambiguously shown that *Desulfovibrio vulgaris* can grow with H_2 and sulphate as the sole energy source (Badziong and Thauer, 1978).

$$4H_2 + SO_4^{2-} + H^+ = HS^- + 4H_2O \quad \Delta G^{\circ\prime} = -151.8\,\text{kJ/mol}. \qquad (1.3)$$

During growth on H_2 and sulphate, energy must be conserved by electron transport-linked phosphorylation since substrate level phosphorylation is only possible when the substrate oxidized is organic. There is, however, an exception to this general rule. The oxidation of bisulphite to sulphate can be coupled by substrate level phosphorylation via an energy-rich adenosine phosphosulphate (APS) intermediate. Using this reaction some SRB can grow at the expense of bisulphite disproportionation to sulphate and hydrogen sulphide (Bak and Cypionka, 1987; Frederiksen and Finster, 2003; Kramer and Cypionka, 1989).

$$4HSO_3^- + 4H^+ = 3SO_4^{2-} + HS^- \quad \Delta G^{\circ\prime} = -235.6\,kJ/mol. \tag{1.4}$$

This chapter first summarizes what is presently known about the biochemistry of dissimilatory sulphate reduction with H_2. Thermodynamic problems associated with the sulphate-dependent oxidations of substrates other than H_2 are then outlined. Subsequently, also under energetic aspects, the trophic interactions of SRB with other microorganisms in their habitats are described. Finally, the phylogenetic diversity of SRB is discussed.

Some of the arguments that will be put forward are based on genome sequence information. Until now, only the genome sequences of three SRB, of *Archaeoglobus fulgidus* (Klenk *et al.*, 1997), of *Desulfovibrio vulgaris* (Hildenborough) (Heidelberg *et al.*, 2004; Hemme and Wall, 2004) and *Desulfotalea psychrophila* (Rabus *et al.*, 2004), have been published. The genomes of many other SRB with different metabolic capacities and from different phylogenetic origins are presently being sequenced. Only when we have these data will we have a complete picture of the energy metabolism of SRB (Chapter 3).

For a previous review on the bioenergetic strategies of sulphate-reducing bacteria see Peck (1993).

1.2 DISSIMILATORY SULPHATE REDUCTION WITH H_2

The equations and proteins involved in dissimilatory sulphate reduction with H_2 will be described for *Desulfovibrio vulgaris* (Hansen, 1994; Matias *et al.*, 2005). The biochemistry of this δ-proteobacterium and of closely related species has been studied in detail and the genome sequence of the Hildenborough strain has recently been published (Heidelberg *et al.*, 2004). Most of the results can probably be generalized to other SRB capable of growth on H_2 as sole energy source such as *D. desulfuricans, Thermodesulphobacterium commune, Desulfobacterium autotrophicus, Desulfotomaculum orientis* and *Archaeoglobus profundus.*

1.2.1 Sulphate activation

Dissimilatory sulphate reduction with H_2 to H_2S in *D. vulgaris* proceeds via HSO_3^- as intermediate. The redox potential, $E^{\circ\prime}$, of the SO_4^{2-}/HSO_3^- couple is −516 mV and thus more than 100 mV more negative that of the H^+/H_2 couple (Table 1.1). Reduction of sulphate to bisulphite with H_2 can therefore only proceed after some input of energy. It has been shown that sulphate is first activated with ATP to adenosine phosphosulphate (= adenylylsulphate) (APS), and that the redox potential of the resulting APS/HSO_3^- couple is −60 mV. The reaction is catalyzed by ATP sulphurylase (Sperling *et al.*, 1998; Taguchi *et al.*, 2004) and consumes the energy of up to two enegy-rich bonds.

$$SO_4^{2-} + ATP + 2H^+ = APS + PPi \quad \Delta G^{\circ\prime} = -46\,kJ/mol \tag{1.5}$$

$$PPi + H_2O = 2Pi \quad \Delta G^{\circ\prime} = -21.9\,kJ/mol. \tag{1.6}$$

Desulfovibrio contain an active cytoplasmic inorganic pyrophosphatase that probably catalyzes the hydrolysis of most of the inorganic pyrophosphate generated in reaction 1.5 (Kobayashi *et al.*, 1975; Liu and Legall, 1990; Ware and Postgate, 1971; see also Weinberg *et al.*, 2004). Based on the genome sequence of *D. vulgaris*, there is no evidence of a membrane-associated inorganic pyrophosphatase which would allow conservation of part of the energy released during pyrophosphate hydrolysis in the form of a trans-membrane electrochemical proton potential, thus reducing somewhat the energy cost of sulphate activation.

1.2.2 Cytoplasmic APS reduction

The sulphate activation to APS increases the redox potential of the first step in dissimilatory sulphate reduction from −516 mV to −60 mV. This is well above the redox potential of the H^+/H_2 couple and allows the reduction with H_2 to proceed even at low H_2 concentrations. *Desulfovibrio* contain a cytoplasmic APS reductase (= adenylylsulphate reductase) whose direct electron donor is not yet known (Fritz *et al.*, 2002; Kremer and Hansen, 1988; Lopez-Cortes *et al.*, 2005; Yagi and Ogata, 1996).

$$APS + 2e^- + 2H^+ = HSO_3^- + AMP \quad E^{\circ\prime} = -60\,mV \tag{1.7}$$

1.2.3 Cytoplasmic bisulphite reduction

The reduction of APS to HSO_3^- is followed by the reduction of HSO_3^- to HS^-, a reaction catalyzed by a cytoplasmic bisulphite reductase (Crane *et al.*, 1997; Friedrich, 2002; Kremer and Hansen, 1988; Larsen *et al.*, 1999; Steger *et al.*, 2002; Zverlov *et al.*, 2005), whose direct electron donor has also still to be identified. For assay of enzyme activity reduced viologen dyes are generally used.

$$HSO_3^- + 6e^- + 6H^+ = HS^- + 3H_2O \quad E^{\circ\prime} = -116\,mV \tag{1.8a}$$

With respect to the mechanism of this equation, there has been considerable controversy over the last 30 years, which is not yet completely resolved. One of the reasons for this is that bisulphite reductase also catalyzes reactions 1.8b and 1.8c when the HSO_3^- concentration is high and the reductant concentration is limiting (Akagi, 1995). SRB contain a thiosulphate reductase which catalyzes equation 1.9.

$$3HSO_3^- + 2e^- + 3H^+ = S_3O_6^{2-} + 3H_2O \quad E^{\circ\prime} = -173\,mV \tag{1.8b}$$

$$S_3O_6^{2-} + 2e^- + H^+ = S_2O_3^{2-} + HSO_3^- \quad E^{\circ\prime} = +225\,mV \tag{1.8c}$$

$$S_2O_3^{2-} + 2e^- + H^+ = HS^- + HSO_3^- \quad E^{\circ\prime} = -402\,mV \tag{1.9}$$

Thus bisulphite reduction could proceed in three two-electron steps rather than in one six-electron step, especially when SRB are grown with bisulphite or thiosulphate rather than sulphate as terminal electron acceptor (Fitz and Cypionka, 1990; Sass *et al.*, 1992). However, recently it has been shown that when *D. vulgaris* is genetically impaired in thiosulphate reduction, this does not affect its ability to grow on sulphate and H_2 (Broco *et al.*, 2005). This is interpreted as indicating that, at least under these growth conditions, bisulphite is reduced in a single step. In the following, it is therefore assumed that APS and HSO_3^- are the only intermediary electron acceptors involved in dissimilatory sulphate reduction with H_2.

1.2.4 Periplasmic H_2 oxidation

D. vulgaris contains four periplasmic hydrogenases, three [NiFe]-hydrogenases and one [FeFe]-hydrogenase: of these, three hydrogenases couple with the major periplasmic poly-heme cytochrome *c* (*TpI-c3*), and one, the [NiFe]-hydrogenases 2, most probably with a second poly-heme cytochrome *c* (*TpII-c3*) (Heidelberg *et al.*, 2004; Matias *et al.*, 2005). When the organism is grown on H_2 and sulphate in medium depleted in nickel, only the [FeFe]-hydrogenase is synthesized without this having a noticeable effect on the growth rate (R. K. Thauer, unpublished results). Growth of *D. vulgaris* is also not impaired when the genes for the [FeFe]-hydrogenase (Haveman *et al.*, 2003; Pohorelic *et al.*, 2002) or one of the NiFe hydrogenases (Goenka *et al.*, 2005) are deleted. These findings indicate that the four hydrogenases can fully functionally replace each other, at least under the growth conditions employed in the laboratory where the H_2 concentration in the fermenters is kept high.

1.2.5 Transmembrane electron transport

Coupling of periplasmic hydrogen oxidation with cytoplasmic APS and HSO_3^- reduction must involve electron transport through the cytoplasmic membrane. The electron transport is most probably catalyzed by the Hmc complex, which is associated on the periplasmic side with a poly-heme cytochrome *c*, and on the cytoplasmic side with an iron-sulphur protein with sequence similarity to heterodisulphide reductase (Keon and Voordouw, 1996; Matias *et al.*, 2005; Rossi *et al.*, 1993). *D. vulgaris* deleted in the *hmc* genes grew normally on lactate and sulphate, but growth on H_2 and sulphate was hampered (Dolla *et al.*, 2000; Haveman *et al.*, 2003; Keon and Voordouw, 1996). The genome of *D. vulgaris* harbours two other polycistronic transcription units predicted to encode for transmembrane protein complexes (TpII-c_3 and Hme), associated on the periplasmic side with a cytochrome *c* and on the cytoplasmic side with an iron-sulphur protein, again with sequence similarity to heterodisulphide reductase (Heidelberg *et al.*, 2004; Matias *et al.*, 2005). The *Hme* genes are part of a locus that includes the genes for bisulphite reductase. The three transmembrane complexes Hmc, Hme and TpII-c_3 could have overlapping functions, in a like manner to the four periplasmic hydrogenases. In the genome, a further gene cluster for a transmembrane protein complex is found which lacks the periplasmic cytochrome *c* (Qmo complex) (Pires *et al.*, 2003). There is indirect evidence that the Qmo complex is involved in APS reduction (Matias *et al.*, 2005).

Heterodisulphide reductase from methanogens catalyzes the reduction of the heterodisulphide CoM-S-S-CoB to coenzyme M (HS-CoM) and coenzyme B (HS-CoB) (Hedderich *et al.*, 1998). Both coenzymes are absent from SRB, and cell extracts of SRB neither catalyze the reduction of CoM-S-S-CoB nor the oxidation of CoM-SH plus CoB-SH (Mander *et al.*, 2002; 2004). The iron-sulphur proteins in SRB with sequence similarity to heterodisulphide reductase must therefore have a different substrate specificity and/or a different function. However, since in methanogenic archaea heterodisulfide reduction links H_2 oxidation with methyl-coenzyme M reduction to methane, it is tempting to speculate that in SRB a disulphide/–SH couple might also be involved in the electron transport from H_2 to HSO_3^-.

1.2.6 ATP synthesis and sulphate transport

As indicated above, up to two ATP equivalents are required to activate sulphate before it can be reduced to HSO_3^-. The reduction of HSO_3^- to HS^- must therefore be coupled with the phosphorylation of at least 2 mol ADP in order that the SRB can grow on H_2 and sulphate. Growth yield data, extrapolated to infinite growth rates, for *D. vulgaris* on H_2 and sulphate were 12.2 g/mol sulphate, and on H_2 and thiosulphate 33.5 g/mol, indicating that HSO_3^- reduction with H_2 to HS^- is coupled with the net synthesis of approximately three ATP (Badziong and Thauer, 1978). In the interpretation of these growth yield data it has to be considered that sulphate must be transported into the cells before it can be reduced and that this transport also requires energy. Available evidence indicates that sulphate is symported with three protons (Cypionka, 1987; Kreke and Cypionka, 1992), or three sodium ions (Kreke and Cypionka, 1994) which is probably equivalent to the consumption of one third or one fourth of an ATP, as discussed below.

1.2.7 Proton stoichiometries

D. vulgaris contains a F_0F_1-type proton-translocating ATPase/ATP synthase (Heidelberg *et al.*, 2004; Hemme and Wall, 2004) of as yet unknown H^+ to ATP stoichiometry. Recent structural analyses of F_0F_1 ATPases from different organisms indicates that the H^+ to ATP stoichiometry may differ from organism to organism, and may be as high as five or as low as three in some organisms (Dimroth and Cook, 2004; Meier *et al.*, 2005; Mueller, 2004; Murata *et al.*, 2005). If the enzyme in *D. vulgaris* has a stoichiometry of three protons per ATP, then at least nine electrogenic protons are required for the synthesis of the three ATP predicted from growth yields to be formed during bisulphite reduction to HS^-. If the stoichiometry is five protons per ATP, then fifteen protons are required. Of these, six are generated from H_2 in the periplasm in a scalar reaction catalyzed by the periplasmic hydrogenases. The other protons required for ATP synthesis must therefore be generated during bisulphite reduction with H_2 by electrogenic proton translocation from the cytoplasm to the periplasm. Although menaquinone is the major quinone found in the cytoplasmic membrane of all SRB, its involvement in this proton translocation is unlikely since the redox potential, $E^{\circ\prime}$, of the menaquinone ox/red couple is −74 mV and thus more positive than that of the HSO_3^-/HS^- couple (−116 mV).

1.2.8 Cytoplasmic H_2 oxidation

The genome of *D. vulgaris* indicates that besides the four periplasmic hydrogenases, the SRB contains two energy-conserving membrane-associated hydrogenase complexes, EchABCDEF and CooMKLXUHF, which are phylogenetically closely related (Heidelberg *et al.*, 2004; Rodrigues *et al.*, 2003). These are known to catalyze the reduction of ferredoxin (Fd) with H_2 driven by the electrochemical proton potential (energy-driven reversed electron transport), or the reduction of protons to H_2 by reduced ferredoxin coupled with the generation of an electrochemical proton potential ($\Delta\mu H^+$) (Forzi *et al.*, 2005; Hedderich, 2004; Sapra *et al.*, 2003).

$$H_2 + \text{Fdox} + \Delta\mu H^+ = \text{Fdred}^{2-} + 2H^+ \tag{1.10}$$

The subunits of the energy-conserving hydrogenases catalyzing H_2 oxidation and ferredoxin reduction have a cytoplasmic orientation within the membrane complex.

For an understanding of the function of these energy-conserving hydrogenases in *D. vulgaris* it has to be appreciated that growth of the SRB on H_2 and sulphate is dependent on the presence of acetate and CO_2 as carbon sources. These are assimilated by the cells via acetyl-phosphate, acetyl-CoA and pyruvate, the last being formed from acetyl-CoA by reductive carboxylation in a reaction catalyzed by pyruvate:ferredoxin oxidoreductase. The redox potential, $E^{\circ\prime}$, of the acetyl-CoA + CO_2/pyruvate couple is −500 mV and thus considerably more negative than that of the H^+/H_2 couple, especially when the H_2 partial pressure is very low (1 to 10 Pa) (–270 to –300 mV). For pyruvate synthesis from acetyl-CoA, CO_2 and H_2 to become exergonic, the electrons from H_2 must be elevated to a more negative potential which is achieved by the energy-driven reversed electron transport from H_2 to ferredoxin, as catalyzed by the energy-conserving hydrogenase complexes EchABCDEF and/or CooMKLXUHF (Fricke *et al.*, 2006; Hedderich, 2004; Meuer *et al.*, 2002). The same arguments hold true for other reductive reactions such as the reductive carboxylation of succinyl-CoA to 2-oxoglutarate (−500 mV) or the reduction of CO_2 to CO (–520 mV).

Energy-driven reversed electron transport is not restricted to the energy-conserving hydrogenases and is of considerable general significance in anaerobic energy metabolism where there is often the need for reducing equivalents at low redox potential/high energy level to drive particular reductive reactions. Further examples to be considered later include lactate oxidation to pyruvate ($E^{\circ\prime} = -190$ mV) with cytochrome c_3 ($E^{\circ\prime} = -400$ mV) (Pankhania *et al.*, 1988), and the oxidation of succinate to fumarate ($E°' = +33$ mV) with menaquinone ($E°' = -60$ mV) as electron acceptor (Paulsen *et al.*, 1986).

When *D. vulgaris* metabolize organic substrates such as pyruvate ($E°' = -500$ mV) or CO ($E°' = -520$ mV), the oxidation of which yields reduced ferredoxin, then the two energy-conserving hydrogenases are involved in H_2 formation, as will be outlined in the section "Lactate oxidation and intraspecies hydrogen transfer".

These equation schemes are not, however, found in all SRB. *D. fructosovorans*, for example, has been shown to harbour a cytoplasmic NADP-reducing hydrogenase. This enzyme, which is absent from *D. vulgaris*, does not appear to be energy-coupled. Deletion mutants are not lethal (Malki *et al.*, 1997).

1.3 DISSIMILATORY SULPHATE REDUCTION WITH ELECTRON DONORS OTHER THAN H_2

As outlined in the previous section, dissimilatory sulphate reduction with H_2 in *D. vulgaris* involves the oxidation of H_2 in the periplasm, electron transport through the cytoplasmic membrane, and reduction of APS and bisulphite in the cytoplasm. Of the many other electron donors used by SRB for dissimilatory sulphate reduction, probably only formate is also oxidized in the periplasm. Biochemical and genomic data show that the three formate dehydrogenases in *D. vulgaris* are localized in the periplasm and, like the periplasmic hydrogenases, use poly-heme cytochromes *c* as electron acceptors (Heidelberg *et al.*, 2004). The oxidation of all other electron donors appears to occur in the cytoplasm or at the inner aspect of the cytoplasmic membrane. The redox potentials of the intermediates involved are summarized in Table 1.2. The energetic problems involved are illustrated by consideration of: (i) the oxidation of lactate to acetate and CO_2 in *D. vulgaris* (Steger *et al.*, 2002); (ii) the oxidation of acetate (acetyl-CoA) to CO_2 in SRB capable of complete oxidations (Thauer, 1988; Thauer *et al.*, 1989); and (iii) propionate oxidation to acetate and CO_2 in *Desulfobulbus propionicus* (Houwen *et al.*, 1991; Kremer and Hansen, 1988; Widdel and Pfennig, 1982).

1.3.1 Lactate oxidation and intraspecies H_2 transfer

D. vulgaris can grow on sulphate with lactate as the sole energy source. The hydroxy acid is incompletely oxidized to acetate and CO_2,

Table 1.2. *Redox potentials (E°′) of intermediates involved in dissimilatory sulphate reduction*

Redox couple	E°′ (mv)
Acetate/acetaldehyde	−581
CO_2/CO	−520
CO_2 + MFR/formyl-MFR	−520
Succinyl-CoA + CO_2/2-oxogrutarate	−520
Acetyl-CoA + CO_2/pyruvate	−498
CO_2/formate	−432
H^+/H_2	**−414**
$S_2O_3^{2-}/H_2S + HSO_3^-$	−402
Acetyl-CoA/acetaldehyde	−396
2-Oxoglutarate + CO_2/isocitrate	−364
Methenyl-H_4MPT/methylene-H_4MPT	−360
Methenyl-H_4MPT/methylene-H_4MPT	−330
Methenyl-H_4F/methylene-H_4F	−295
Acetoacetyl-CoA/ß-hydroxybutyryl-CoA	−238
Methylene-H_4F/methyl-H_4F	−200
Acetaldehyde/ethanol	−197
Dihydroxyacetone phosphate/ glycerol phosphate	−190
Pyruvate/lactate	−190
$3HSO_3^-/S_3O_6^{2-}$	−173
Oxaloacetate/malate	−172
CoM-S-S-CoB/HS-CoM + HS-CoB	−145
SO_3H^-/HS^-	**−116**
APS/AMP + HSO_3^-	**−60**
Crotonyl-CoA/butyryl-CoA	−10
Fumarate/succinate	+33
Acrylyl-CoA/propionyl-CoA	+69
$S_3O_6^{2-}/S_2O_3^{2-} + HSO_3^-$	+225

Notes: E°′ at pH 7.0 are given for H_2, CO_2 and CO in the gaseous state at 10^5 Pa, for S° in the solid state and for all other compounds in aqueous solution at 1M concentration (see legend to Table 1.1).

with the intermediary formation of pyruvate, acetyl-CoA and acetyl phosphate.

$$2\text{Lactate}^- + SO_4^{2-} + H^+ = 2\text{acetate}^- + 2CO_2 + HS^- + 2H_2O$$

$$\Delta G^{\circ\prime} = -196.4\,\text{kJ/mol} \qquad (1.11)$$

Equation 1.11 probably involves the following oxidoreduction steps:

$$\text{Lactate}^- + 2\text{cyt}c_3(\text{ox}) + \Delta\mu H^+ = \text{Pyruvate}^- + 2\text{cyt}c_3(\text{red})^{-1} \qquad (1.12)$$

$$\text{Pyruvate}^- + \text{CoA} + \text{Fdox} = \text{acetyl-CoA} + CO_2 + \text{Fdred}^{2-} + 2H^+ \qquad (1.13)$$

$$\text{Fdred}^{2-} + 2H^+ = \text{Fdox} + H_2 + \Delta\mu H^+ \qquad (1.10)$$

$$H_2 + 2\text{cyc } c_3(\text{ox}) = 2\text{cyt}c_3(\text{red})^{-1} + 2H^+ \qquad (1.14)$$

$$4\text{Cyt}c_3(\text{red})^{-1} + 0.5SO_4^{2-} = 4\text{cyt}c_3(\text{ox}) + 0.5H_2S \qquad (1.15)$$

Equation 1.12 is catalyzed by a membrane-associated lactate dehydrogenase complex, with its active site facing the cytoplasm (Hansen, 1994; Ogata *et al.*, 1981; Reed and Hartzell, 1999). From this site, the electrons generated by lactate oxidation ($E^{\circ\prime}$= –190 mV) are transferred through the cytoplasmic membrane, driven by $\Delta\mu H^+$, to one of the periplasmic cytochromes c_3 which are in enzyme-catalyzed equilibrium with the H^+/H_2 couple ($E^{\circ\prime}$= – 414 mV) (equation 1.14). The electron flow from lactate to cytochrome c_3 is an example of energy-driven reversed electron transport with the $\Delta\mu H^+$ generated in equation 1.10 used to drive equation 1.12. Equation 1.13 is catalyzed by a cytoplasmic pyruvate:ferredoxin oxidoreductase. The redox potential $E^{\circ\prime}$(–500 mV) of the acetyl-CoA + $CO_{2/}$pyruvate couple is well below that of the H^+/H_2 couple. Equation 1.10 is catalyzed by one of the two membrane-associated energy-conserving hydrogenase complexes, EchABCDEF or CooMKLXUHF, which are ferredoxin-specific and face the cytoplasm. Equation 1.14 is catalyzed by one of the four periplasmic cytochrome c_3-specific hydrogenases. The reduced cytochrome c_3 in the periplasm is finally re-oxidized by sulphate, which is reduced in the cytoplasm via transmembrane electron transport (equation 1.15) (see above). The H_2 generated in equation 1.10 has to diffuse into the periplasm in order to react with cytochrome c_3. This formation of H_2 in the cytoplasm and its re-oxidation in the periplasm has been termed intraspecies hydrogen transfer, or hydrogen cycling (Odom and Peck, 1984); Chapter 7, this volume).

Evidence for the proposed sequence of equations comes from the topology and specificity of the enzymes and electron carriers involved, and from the finding that H_2 is formed and re-consumed during growth of *D. vulgaris* on lactate and sulphate. In the absence of sulphate, H_2 is formed without re-consumption (equation 1.16).

$$\text{Lactate}^- + H_2O = \text{acetate}^- + CO_2 + 2H_2$$

$$\Delta G^{\circ\prime} = -8.8\,\text{kJ/mol} \quad (1.16)$$

H_2 formation from lactate was found to be inhibited by protonophores and by arsenate, demonstrating the involvement of an energy-requiring equation (Pankhania *et al.*, 1988). Contrary to the situation with lactate, H_2 formation from pyruvate, which is oxidized to acetate and CO_2, is not inhibited by protonophores and arsenate, showing that this equation does not require energy input (Pankhania *et al.*, 1988). These data demonstrate the validity of the equation scheme proposed above, and in particular show that lactate oxidation to pyruvate is the energy-requiring step.

Intraspecies H_2 transfer is probably also involved in dissimilatory sulphate reduction with CO. The SRB contains a cytoplasmic carbon monoxide dehydrogenase, which catalyzes the reduction of ferredoxin with CO (Soboh *et al.*, 2002; Voordouw, 2002).

$$CO + H_2O + \text{Fdox} = CO_2 + \text{Fdred}^{2-} + 2H^+ \quad (1.17)$$

$$\text{Fdred}^{2-} + 2H^+ = \text{Fdox} + H_2 + \Delta\mu H^+ \quad (1.10)$$

Besides intraspecies hydrogen transfer, formate cycling is another possible mechanism of electron transfer from the cytoplasm to the periplasm. The genome of *D. vulgaris* harbours genes for the expression of an active pyruvate-formate lyase which catalyzes the formation of acetyl-CoA and formate from pyruvate and CoA (Heidelberg *et al.*, 2004).

$$\text{Pyruvate}^- + \text{CoA} = \text{acetyl-CoA} + \text{formate}^- \quad \Delta G^{\circ\prime} = -16.3\,\text{kJ/mol} \quad (1.18)$$

Since pyruvate-formate lyase is a cytoplasmic enzyme and in *D. vulgaris* the three formate dehydrogenases are all localized in the periplasm (Haynes *et al.*, 1995; Heidelberg *et al.*, 2004), the formate generated from pyruvate must pass through the cytoplasmic membrane, most probably catalyzed via a proton symport, before it can be used as electron donor for dissimilatory sulphate reduction, or the reduction of protons to H_2.

This is therefore considered to be a case of intraspecies formate transfer (see Chapter 7).

1.3.2 Acetate (acetyl-CoA) oxidation to CO_2

There are many SRB that can oxidize organic compounds such as acetate, lactate, longer chain fatty acids, alkanes, or benzoic acid (Table 1.1) completely to CO_2 using sulphate as the electron acceptor. In all these cases acetyl-CoA is an intermediate in the pathway to CO_2. Some SRB use the citric acid cycle to oxidize acetyl-CoA to CO_2 with citrate, aconitate, isocitrate, 2-oxoglutarate, succinyl-CoA, succinate, fumarate, malate and oxaloacetate as intermediates. Amongst these is *Desulfobacter postgatei* (Thauer *et al.*, 1977). However, most SRB including *Archaeoglobus fulgidus*, use the oxidative acetyl-CoA synthase (decarbonylase)/carbon monoxide dehydrogenase pathway, either with tetrahydrofolate (H_4F) or tetrahydromethanopterin (H_4MPT) as C_1-carrier. In this pathway acetyl-CoA is oxidized via carbon monoxide, methyl-H_4F (methyl-H_4MPT), methylene-H_4F (methylene-H_4MPT), methenyl-H_4F (methenyl-H_4MPT), N^{10}-formyl-H_4F (N^5-formyl-H_4MPT) and formate (formylmethanofuran) as intermediates (Leaphart *et al.*, 2003; Thauer *et al.*, 1989; Thauer and Kunow, 1995) (for redox potentials see Table 1.2).

With respect to dissimilatory sulphate reduction, the oxidative acetyl-CoA synthase (decarbonylase)/carbon monoxide dehydrogenase pathway has the advantage that all oxidation steps involved proceed at redox potentials more negative than that of the APS/HSO_3^- couple (−60 mV) and that of the HSO_3^-/HS couple (−116 mV). On the contrary, the citric cycle involves one step, the oxidation of succinate to fumarate, with a redox potential of +33 mV. Oxidation of succinate to fumarate with sulphate as terminal electron acceptor is therefore an endergonic reaction which again requires energy-driven reversed electron transport to proceed.

1.3.3 Propionate oxidation

In SRB, propionyl-CoA, generated from propionate or by oxidation of uneven numbered fatty acids, is oxidized via methylmalonyl-CoA, succinyl-CoA, succinate, fumarate, malate, oxaloacetate, pyruvate and acetyl-CoA as intermediates. Biochemical evidence for this pathway has been obtained from *Desulfobulbus propionicus* growing on propionate and sulphate (Houwen *et al.*, 1991; Kremer and Hansen, 1988). The methylmalonyl-CoA pathway appears to be energetically more favourable than the oxidation of propionyl-CoA via acrylyl-CoA, since the redox potential of the acrylyl-CoA/propionyl-CoA couple is +69 mV and thus more problematical than that of the fumarate/succinate couple (+33 mV), and requiring an even greater input of energy before the reaction can proceed (Sato *et al.*, 1999).

1.4 TROPHIC INTERACTIONS OF SRB WITH OTHER MICROORGANISMS

In natural environments, SRB invariably coexist with a plethora of anaerobic microorganisms which either use other electron donors and acceptors, or compete with SRB through their utilization of the same substrates. The commensals frequently form catabolic end-products which can serve as good electron donors for the SRB. This is particularly important in the case of hydrogen. Oxidation by the SRB keeps the hydrogen partial pressure low such that the primary organisms may use H^+ as an electron acceptor and so be able to produce more oxidized carbon end-products in reactions which would otherwise be endergonic. The SRB and commensals thus derive mutual benefit from living together and generally form stable syntrophic associations in which cellular energetics become community energetics, and the rates of metabolism are controlled by the rates of interspecies transfer of reducing equivalents. In open non-mixed ecosystems, the rates of diffusion of extracellular electron donors and acceptors become dominant.

Specific examples of the metabolic activities associated with sulphate reduction in different parts of a marine sediment will be discussed with reference to two regions of such an ecosystem; (i) the oxic/anoxic interface which are the upper layers, including the overlying water column, where electron acceptors more positive than sulphate/HS^- are also present such as O_2, nitrate, Mn(IV) and Fe(III); and (ii) the deeper anoxic regions where sulphate is the electron acceptor with the most positive potential.

1.4.1 Oxic/anoxic interface

The first benefit SRB derive from the activity of other organisms is essentially physicochemical. As a direct consequence of the oxygenated character of the present day biosphere, anaerobic ecosystems arise from the exclusion or removal of oxygen. In a biologically active environment, this most commonly occurs through the rate of oxygen utilization by the metabolism of aerobic and facultative species exceeding the rate of its replenishment by diffusion, mixing or bioturbation. Such ecosystems are characterized by an oxic/anoxic interface where oxygen is replaced by a progressively less oxidizing terminal electron acceptor with increasing depth (see Chapters 5 and 6, this volume). Within a water column, the transition from oxic to anoxic conditions may occur over a depth of some centimetres, or even metres, and be subject to diurnal variation with the rise and fall of phototrophic species (Jorgensen, 1982). In sediments, and more particularly within biofilms, the interface may be very sharply defined, and measured in micrometres rather than centimetres or metres (Nielsen *et al.*, 1993); Chapter 12, this volume).

It is generally the case in oxic/anoxic ecosystems that maximal activities of the anaerobic SRB are found close to the interface with the oxic region. This arises from a number of possible causes. In addition to reducing the oxygen within the biosystem as a whole, the activities of aerobic and facultative species in the oxic zone are often responsible for the production of many of the compounds that form the carbon and electron donors utilized by the SRB. The main reservoir of sulphate is also found in oxic environments. Since the end-product of sulphate reduction, sulphide, is highly toxic to all life forms, including the SRB themselves, it is essential for the ongoing maintenance of biological activity that the sulphide be removed, or in some way neutralized. Most commonly, this is achieved by recycling across the interface with re-oxidation, biotic or abiotic, in the oxic region (Jorgensen, 1982).

The presence of other electron acceptors with more positive redox potentials than sulphate can give rise to both cooperative and competitive effects. Oxygen is generally considered to be inhibitory to SRB, although defence mechanisms have been demonstrated and species identified which can use oxygen as the electron acceptor in certain equations (see Chapter 5). However the rates of metabolism of, for example, sugars, alkanes and fatty acids by aerobic organisms greatly outcompete any potential activity found with the SRB using sulphate.

Many sulphate reducers can also use nitrate as terminal electron acceptor (see Chapter 8). This is the basis, for example, of the procedure in which nitrate is introduced to oil production facilities as a means of controlling sulphide generation in the reservoir, which give rise to so-called sour oil. It remains unclear, however, whether the observed reduction in sulphide is due to: nitrate competing as electron acceptor; the resulting nitrite acting as a non-competitive inhibitor of sulphidogenesis; or anaerobic oxidation of produced sulphide with nitrate as electron acceptor.

Recent work on the mechanisms and ecological significance of iron (FeIII) and manganese (MnIV) reduction has shed new light on the relative complexity of anaerobic respiratory processes in nature (Hamilton, 2003; Lovley *et al.*, 2004). Iron and manganese are amongst the most abundant elements found in the biosphere. Their redox potentials are in the biological range (Table 1.1), and many sulphate- and sulphur-reducing species have been shown also to have the capability of using them as terminal electron acceptors. The oxidations of Fe(II) and of Mn(II), and the reductions of Fe(III) and of Mn(IV) are, however, mediated by a wide range of microorganisms (Ehrlich, 1999; Ghiorse, 1984; Lovley *et al.*, 2004). This has been of great significance over geological time periods with the formation of iron and of manganese dioxide deposits. These same processes are now seen to be integral to the overall dynamics of anaerobic respiratory ecosystems, where the redox cycling of Fe(III)/Fe(II) and Mn(IV)/Mn(II) may play significant roles in coupling the re-oxidation of sulphide to oxygen, which thus serves as the ultimate electron acceptor (Thamdrup and Canfield, 1996). By this means, electron flux can be mediated over considerable distances between sulphidogenic and oxic zones in, for example, deepwater column biosystems (Nealson and Saffarini, 1994).

The largely insoluble character of both Fe(III) and Mn(IV) poses particular mechanistic problems for the organisms involved in their reduction. MnO_2, however, is extremely reactive and it has been found that considerable proportions of the reduced sulphide and Fe(II) in anaerobic ecosystems are re-oxidized abiotically in contact with MnO_2, without direct microbial involvement (Schiffers and Jorgensen, 2002; Thamdrup *et al.*, 1994); Chapter 5 this volume). This biotic/chemical redox cycling of Mn(IV)/Mn(II) is directly paralleled by the biotic/electrochemical redox cycling which has been shown to give rise to microbially influenced corrosion of stainless steels (Hamilton, 2003).

1.4.2 Anoxic regions where sulphate is the most positive electron acceptor

In the deeper anoxic regions of sediments, where the only electron acceptors present are sulphate (sulphidogenesis), CO_2 (methanogenesis and acetogenesis) and protons (hydrogen formation), the higher redox potential of the SO_4^{2-}/HS^- couple ensures that sulphidogenesis will be the dominant process, so long as there remains unreacted sulphate. Sulphidogenesis is also favoured by the more extensive range of substrates available to the sulphate reducers.

The major input of energy to anaerobic microbial communities is in the form of carbohydrates derived from the breakdown of plant and animal polysaccharides. Despite genomic evidence, however, for the presence of glycolytic enzymes (Heidelberg *et al.*, 2004), most SRB are not thought to degrade and grow on polysaccharides or their derived sugars in naturally occurring microbial ecosystems.

It appears rather that the many fermentative organisms in nature which are capable of sugar catabolism with high rates of activity outcompete any putative glycolytic activity that might be expressed by the SRB. On the other hand, in the anoxic conditions, sugar breakdown is incomplete and the reduced products derived from fermentative reactions are ideal substrates for sulphate reduction; for example, lactate, acetate formate or H_2. Being only partially degraded, these fermentation products retain redox energy which is available to anaerobic respiratory organisms such as the SRB. By this simple cooperation among species, complex polysaccharide materials can be completely degraded in anaerobic biosystems, with both fermentative and sulphate-respiring organisms each acquiring benefit from the throughput of carbon and energy donors.

The case of hydrogen is particularly significant. The essential character of fermentative mechanisms is that they maintain redox balance, without resource to an external acceptor as electron sink. Where the redox potential of the primary electron donor is more negative than that of the H^+/H_2 couple, however, H^+ can act as an electron acceptor with the formation of molecular hydrogen. Hydrogen is, of course, a highly favoured electron and energy donor for the SRB. Where the hydrogen partial pressure is maintained close to zero by this terminal respiratory mechanism, the primary fermentation is significantly altered with an increased hydrogen production allowing the formation of more oxidized carbon end-products. Where the fermentative and respiratory processes are coupled in this energetically mutually beneficial manner, the term used is "syntrophism" and the mechanism is referred to as "interspecies hydrogen transfer" (Schink and Stams, 2002). Syntrophism by hydrogen transfer is generally characterized by close physical association of the hydrogen-producing and the hydrogen-utilizing organisms (Conrad *et al.*, 1985). Hydrogen transfer between organisms is thus analogous to hydrogen cycling; or intraspecies hydrogen transfer, between cytoplasm and periplasm within a single organism.

As indicated by the redox potentials listed in Table 1.1, most of the compounds that can be oxidized by sulphate can also be oxidized by protons when the H_2 partial pressure is very low. Thus acetate can be first converted by a non-sulphate reducer to $2CO_2$ and $4H_2$, and then the hydrogen further oxidized by an SRB (Galouchko and Rozanova, 1996). Alternatively, the acetate can be oxidized directly by an SRB (Thauer *et al.*, 1989). Which of the two possibilities occurs in any one environment is not always evident. Only in the case of methane oxidation with sulphate does interspecies hydrogen transfer appear not to be possible, at least not unless H_2 partial pressures in marine sediments are lower than 1 Pa (Shima and Thauer, 2005). Here the intriguing possibility of extracellular electron transfer via microbial nanowires may have to be considered (Reguera *et al.*, 2005).

1.5 PHYLOGENETIC DIVERSITY

Sulphate-reducing bacteria have been described as early as the turn of the nineteenth century when Beijerinck (1895) reported the formation of hydrogen sulphide from sulphate by a species later reclassified as *Desulfovibrio desulfuricans*. This early period of microbiology saw the dawn of microbial systematics, microbial ecology and microbial physiology when Winogradsky (1890) discovered chemoautotrophy. Over 80 years the number of sulphate-reducing species remained small. The determination of the diversity and phylogenetic incoherence of sulphate- and sulphur-reducing bacteria coincided with the application of 16S rRNA oligonucleotide sequencing in the 1980s, reverse transcriptase sequencing of 16S rRNA genes in the late 1980s, and from 1990, the PCR-mediated amplification and sequencing techniques. Today, more than 120 species of 35 genera, belonging to 3 bacterial phyla and 1 archaeal phylum have been described and their metabolism elucidated. At the turn of the twenty-first century, a sufficient body of information on sequences of 16S rRNA had been accumulated in scientific journals to develop genus-specific oligonucleotide primers used in sequencing protocols on pure cultures and, as dye-labelled DNA fragments, in *in situ* detection of single cells in their natural habitat. It became possible to explore anoxic environments by using cultivation-independent techniques, and the enumeration of sulphate-reducing bacteria demonstrated the crucial role of sulphate reducers in the global sulphur cycle and in the food web of anoxic and mixed oxic/anoxic environments (Venter *et al.*, 2004; Hines *et al.*, 1999; Voordouw *et al.*, 1996). These studies were not exclusively based on the analysis of the 16S rRNA gene but also concentrated on the analysis of key proteins of sulphate reduction. Results of these studies confirm the dominant role of Gram-negative forms. Complete oxidizers were found in eutrophied environments, whereas incomplete oxidizers dominated pristine environments (Castro *et al.*, 2002). Among the gram-negative organisms members of *Desulfobacteriaceae* thrive on aromatic compounds (Koizumi *et al.*, 2002), while members of *Desulfovibrionaceae* did not use aromatic components of crude oil (Rabus *et al.*, 1996).

The chemical analysis of inorganic molecules of geologically ancient rocks indicated that the 3.8 Ga-old Isua sedimentary rocks did not show discrimation of ^{34}S over the lighter ^{32}S in sulphur compounds (Monster *et al.*, 1979). This finding was interpreted as the lack of biological reduction of sulphate. Enrichment of ^{32}S was, however, detected in rock dated 3.47 Ga indicating the antiquity of microbial sulphate reduction (Shen and Buick, 2004; Johnston *et al.*, 2005). The sulphate present on early Earth (rocks from the moon do not contain sulphate) was probably generated photochemically from volcanic SO_2 and H_2S in the atmosphere (Farquhar and Wing, 2003). All the available evidence indicates that the first SRB evolved before the cyanobacteria, and therefore dissimilatory sulphate reduction arose before oxygenic photosynthesis. New estimates date the origins of oxygenic photosynthesis back to 2.3–2.2 Ga (Kopp *et al.*, 2005).

Plotting the 16S rRNA similarity values against geological time, Stackebrandt (1995) concluded that the main radiation of sulphate-reducing bacteria evolved at the beginning of the formation of marine-banded iron; later than the origin of carbon fixation (photosynthesis, methanogenesis) but significantly earlier than the occurrence of O_2-dependent bacterial species. These data are in contrast to more recent findings on sequence similarities of genes coding for the alpha and beta subunits of the dissimilatory sulphite reductase, a key enzyme in the sulphur metabolism of sulphate reducers (Larsen *et al.*, 1999). These later data point to a common origin of these genes in *Archaeoglobus* (domain Archaea) and *Desulfotomaculum* (domain Bacteria), two groups of organisms that evolved from a common ancestor significantly earlier than the isotope fractionation data seem to indicate. Genes encoding for the subunits of a dissimilatory sulphite reductase-type protein have also been isolated from the hyperthermophilic crenarchaeote *Pyrobaculum islandicum* which shows structural and sequence similarities to the above-mentioned sulphate reducers (Molitor *et al.*, 1998). This indicates that most likely this enzyme-type belonged to the genetic makeup of the progenote, present even before its evolutionary split into archaea and bacteria.

In contrast to the low phylogenetic diversity of archaeal sulphate reducers, restricted to the two species of *Archaeoglobus* (*A. fulgidus*, *A. profundus*), the diversity of bacterial sulphate reducers is found in three phyla. On the other hand, considering that cultivated organisms have been described in more than 40 phyla, the evolution of sulphate reducers must be considered rather restricted. Though present in the ancestors of both domains, the ability to reduce sulphate was lost in most lineages, even in those descendents which still thrive today in anoxic environments. Only among the mesophilic Gram-negative sulphate reducers, members of the *Deltaproteobacteria*, has a phylogenetic diverse assemblage of forms evolved that have been described as novel taxa.

Members of *Thermodesulfovibrio* can be considered deep branching thermophilic descendants of the early Bacteria. This genus has been described for strictly anaerobic, non-sporeforming, thermophilic (optimum 65–75°C), sulphate-reducing bacteria from hot springs, hot oil reservoirs and hydrothermal vents. Species are either chemoorganotrophic or chemolithoautotrophic, in which sulphate is reduced to sulphide. Hydrogen or C1–C3 acids serve as electron donors. Chemotaxonomically they are unique as they contain non-phytanyl ether-linked lipids. Recently, *Thermodesulfobium* has been described, representing a new phylum, adjacent to candidate division OP9 (Mori *et al.*, 2003) (not shown). This organism, together with *Thermodesulfovibrio*, is placed within the *Desulfovibrio* branch of *Deltaproteobacteria* in the *apsA* genes tree. This indicates that enzymes involved in sulphate respiration may evolve in an independent manner. The finding of horizontal gene transfer among sulphate-reducing bacteria was first reported by Friedrich (2002).

The publications by Widdel and Bak (1992), Widdel and Pfennig (1984) and Rabus *et al.* (2001) summarize recent developments of the taxonomy of Gram-negative sulphate reducers, including a description of families and genera, along with an extensive coverage of metabolic properties. The sulphur- and sulphate-reducing organisms of this phylum are classified in four orders which contain a few genera for which sulphur-dependent pathways have not been detected: (a) the obligately intracellular *Lawsoni intracellularis* (McOrist *et al.*, 1995) and *Bilophila wadsworthia*, isolated from patients with gangrenous and perforated appendicitis (Baron *et al.*, 1989) within the family *Desulfovibrionaceae*; (b) *Malonomonas rubra* (Dehning and Schink, 1989) and *Pelobacter*, which ferment either acetoin or trihydroxybenzenoids to acetate and CO_2, though some species ferment either polyethylene glycol, 2,3-butandiol, or acetylene to acetate and alcohols (Schink, 1992) within the family *Desulfuromonadaceae*. Gram-positive sulphate reducers are located within the phylum Firmicutes in two phylogenetically neighbouring genera within the radiation of the many clostridial lineages.

Taxonomically the combination of phylogenetic position, morphology and physiology facilitate the affiliation of a novel isolate to taxa and decide whether a new taxon can be described (Table 1.3). The relatedness of phyla

Table 1.3. *Main taxonomic differences among genera of sulphate reducers*

Family	Genus	Morphology	Desulfoviridin	Oxidation of organic electron donors	Electron acceptors for growth (other than SO_4^{2-})	mol% G+C of DNA
Desulfovibrionaceae	*Desulfovibrio*	Vibrio	+	incomplete (i)	SO_3^{2-}, $S_2O_3^{2-}$, Fumarate	46–61
Desulfomicrobiaceae	*Desulfomicrobium*	Oval to rod	–	i	SO_3^{2-}, $S_2O_3^{2-}$,	52–60
Desulfohalobiaceae	*Desulfohalobium*	Rod	–	i	SO_3^{2-}, $S_2O_3^{2-}$, S^0	57
	Desulfonatronovibrio	Vibrio	–	nr	SO_3^{2-}, $S_2O_3^{2-}$	49
	Desulfonauticus	Curved rod	–	nr	SO_3^{2-}, $S_2O_3^{2-}$, S^0	34
	Desulfothermus	Rod	–	Complete (c)	$S_2O_3^{2-}$	37
Desulfonatronumaceae	*Desulfonatronum*	Vibrio	–	i	SO_3^{2-}, $S_2O_3^{2-}$	56–57
Desulfobacteraceae	*Desulfobacter*	Oval or vibrio	–	c	SO_3^{2-}, $S_2O_3^{2-}$	45–49
	Desulfobacterium	Oval	–	c	$S_2O_3^{2-}$	45–48
	Desulfobacula	Oval to curved rods	nd	c	$S_2O_3^{2-}$	41–42
	Desulfobotulus	Vibrio	–	i	SO_3^{2-}	53
	Desulfocella	Vibrio	–	i	–	35
	Desulfococcus	Sphere	+	c	SO_3^{2-}, $S_2O_3^{2-}$	56–57
	Desulfofaba (=*Desulfomusa*)	Rod	–	i	SO_3^{2-}, $S_2O_3^{2-}$	52
	Desulfofrigus	Rod	–	c	SO_3^{2-}, $S_2O_3^{2-}$, Fe(iii)-citrate	52–53

	Desulfonema	Multicellular filaments	v	c	SO_3^{2-}, $S_2O_3^{2-}$,	35–55
	Desulforegula	Rods	+	i	–	nd
	Desulfosarcina	Oval, aggregates	–	c	SO_3^{2-}, $S_2O_3^{2-}$	51–59
	Desulfospira	Vibrio	–	c	SO_3^{2-}, $S_2O_3^{2-}$	50
	Desulfotignum	Curved rods	–	c	SO_3^{2-}, $S_2O_3^{2-}$	62
	Desulfatibacillum	Rods	–	c	SO_3^{2-}, $S_2O_3^{2-}$	41
Desulfobulbaceae	*Desulfobulbus*	Lemon to rod	–	i	SO_3^{2-}, $S_2O_3^{2-}$, NO_3^-	50–60
	Desulfocapsa	Rod	–	i	–	47–51
	Desulfofustis	Rod	–	i	SO_3^{2-}, S^0	56
	Desulforhopalus	Oval	–	i	SO_3^{2-}, $S_2O_3^{2-}$	48–51
	Desulfotalea	Rod	–	i	SO_3^{2-}, $S_2O_3^{2-}$, Fe(iii)-citrate	42–47
	Desulfomonile	Rod	+	nd	$S_2O_3^{2-}$, 3-Cl-benzoate	49
	Desulfarculus	Vibrio	–	c	SO_3^{2-}	66
	Desulfacinum	Oval	–	nd	SO_3^{2-}, $S_2O_3^{2-}$	60–64
	Desulforhabdus	Rod	–	Complete	SO_3^{2-}, $S_2O_3^{2-}$	53

Table 1.3. (*cont.*)

Family	Genus	Morphology	Desulfoviridin	Oxidation of organic electron donors	Electron acceptors for growth (other than SO_4^{2-})	mol% G+C of DNA
	Desulfovirga	Rod	–	Complete, depending on the substrate	SO_3^{2-}, $S_2O_3^{2-}$, S^0	60
	Desulfobacca	Oval	–	c	SO_3^{2-}, $S_2O_3^{2-}$	51
	Desulfospira	Curved rods	–	c	SO_3^{2-}, $S_2O_3^{2-}$, S^0	49
	Thermodesulforhabdus	Rod	–	c	SO_3^{2-}	51
	Thermodesulfobium	Rod	nr	c	$S_2O_3^{2-}$, NO_3^-, NO_2^-	35
	Thermodesulfobacterium	Rod	–	i	$S_2O_3^{2-}$	28–40
Thermodesulfovibrionaceae	*Thermodesulfovibrio*	Vibrio	–	i	SO_3^{2-}, $S_2O_3^{2-}$	38
Firmicutes	*Desulfotomaculum*	Straight or curved rod	Sporulating Gram-positive	i or c	$S_2O_3^{2-}$, fumarate	48–52
	Desulfosporosinus	Straight or curved rod	Sporulating Gram-positive	i	$S_2O_3^{2-}$	45–46
Archaea	*Archaeoglobus*	Sphere	–	c	–	41–46

among each other with an emphasis of anaerobic bacteria has been summarized by Stackebrandt (2004). It should be noted that the order of phyla within the 16S rRNA gene sequences tree is not well resolved and phylogenetic trees, no matter which algorithm used for their generation, have a bush- or fan-like appearance. The hierarchic outline of sulphate reducers has been devised by Garrity *et al.* (2003). The reader interested in the original description of taxa at the genus and species level is referred to the homepage of Jean Euzeby (http://www.bacterio.cict.fr) which provides a nomenclatural database.

ACKNOWLEDGEMENT

This work was supported by the Max Planck Society and by the Fonds der Chemischen Industrie.

REFERENCES

Akagi, J. M. (1995). Respiratory sulphate reduction. In L. L. Barton (ed.), *Sulphate-Reducing Bacteria*, Vol. 8. New York: Plenum Press. pp. 89–111.

Badziong, W. and Thauer, R. K. (1978). Growth yields and growth rates of *Desulfovibrio vulgaris* (Marburg) growing on hydrogen plus sulphate and hydrogen plus thiosulphate as the sole energy sources. *Arch Microbiol*, **117**, 209–14.

Bak, F. and Cypionka, H. (1987). A novel type of energy-metabolism involving fermentation of inorganic sulfur-compounds. *Nature*, **326**, 891–2.

Baron, E. J., Summanen, P., Downes, J. *et al.* (1989). *Bilophila wadsworthia*, gen. nov. and sp. nov., a unique gram-negative anaerobic rod recovered from appendicitis specimens and human faeces. *J Gen Microbiol*, **135**, 3405–11.

Beijerinck, M. W. (1895). Über *Spirillum desulfuricans* als Ursache von Sulfatreduktion. *Zentralbl Bakteriol Parasitkd Infekt Abt II*, **1**, 49–59.

Bottcher, M. E., Thamdrup, B., Gehre, M. and Theune, A. (2005). S-34/S-32 and O-18/O-16 fractionation during sulfur disproportionation by *Desulfobulbus propionicus*. *Geomicrobiol J*, **22**, 219–26.

Broco, M., Rousset, M., Oliveira, S. and Rodrigues-Pousada, C. (2005). Deletion of flavoredoxin gene in *Desulfovibrio gigas* reveals its participation in thiosulphate reduction. *FEBS Lett*, **579**, 4803–7.

Castro, H., Reddy, K. R. and Ogram, A. (2002). Composition and function of sulphate-reducing prokaryotes in eutrophic and pristine areas of the Florida Everglades. *Appl Environ Microbiol*, **68**, 6129–37.

Castro, H. F., Williams, N. H. and Ogram, A. (2000). Phylogeny of sulphate-reducing bacteria. *FEMS Microbiol Ecol*, **31**, 1–9.

Conrad, R., Phelps, T. J. and Zeikus, J. G. (1985). Gas metabolism evidence in support of the juxtaposition of hydrogen-producing and methanogenic bacteria in sewage sludge and lake sediments. *Appl Environ Microbiol*, **50**, 595–601.

Crane, B. R., Siegel, L. M. and Getzoff, E. D. (1997). Structures of the siroheme- and Fe4S4-containing active center of sulfite reductase in different states of oxidation: Heme activation via reduction-gated exogenous ligand exchange. *Biochemistry*, **36**, 12101–19.

Cypionka, H. (1987). Uptake of sulphate, sulfite and thiosulphate by proton-anion symport in *Desulfovibrio desulfuricans*. *Arch Microbiol*, **148**, 144–9.

Cypionka, H. (2000). Oxygen respiration by *Desulfovibrio* species. *Annu Rev Microbiol*, **54**, 827–48.

Dannenberg, S., Kroder, M., Dilling, W. and Cypionka, H. (1992). Oxidation of H_2, organic-compounds and inorganic sulfur-compounds coupled to reduction of O_2 or nitrate by sulphate-reducing bacteria. *Arch Microbiol*, **158**, 93–9.

Dehning, I. and Schink, B. (1989). *Malonomonas rubra* gen. nov. sp. nov., a microaerotolerant anaerobic bacterium growing by decarboxylation of malonate. *Arch Microbiol*, **151**, 427–33.

Dimroth, P. and Cook, G. M. (2004). Bacterial Na^+- or H^+-coupled ATP synthases operating at low electrochemical potential. *Adv Microb Physiol*, **49**, 175–218.

Dolla, A., Pohorelic, B. K. J., Voordouw, J. K. and Voordouw, G. (2000). Deletion of the *hmc* operon of *Desulfovibrio vulgaris* subsp *vulgaris* Hildenborough hampers hydrogen metabolism and low-redox-potential niche establishment. *Arch Microbiol*, **174**, 143–51.

Ehrenreich, A. and Widdel, F. (1994). Anaerobic oxidation of ferrous iron by purple bacteria, a new-type of phototrophic metabolism. *Appl Environ Microb*, **60**, 4517–26.

Ehrlich, H. L. (1999). Microbes as geologic agents: their role in mineral formation. *Geomicrobiol J*, **16**, 135–53.

Fareleira, P., Santos, B. S., Antonio, C. *et al.* (2003). Response of a strict anaerobe to oxygen: survival strategies in *Desulfovibrio gigas*. *Microbiolog-SGM*, **149**, 1513–22.

Farquhar, J. and Wing, B. A. (2003). Multiple sulfur isotopes and the evolution of the atmosphere. *Earth and Planetary Science Letters*, **213**, 1–13.

Finster, K., Liesack, W. and Thamdrup, B. (1998). Elemental sulfur and thiosulphate disproportionation by *Desulfocapsa sulfoexigens* sp nov, a new anaerobic bacterium isolated from marine surface sediment. *Appl Environ Microbiol*, **64**, 119–25.

Fitz, R. M. and Cypionka, H. (1990). Formation of thiosulphate and trithionate during sulfite reduction by washed cells of *Desulfovibrio desulfuricans*. *Arch Microbiol*, **154**, 400–6.

Forzi, L., Koch, J., Guss, A. M. *et al.* (2005). Assignment of the 4Fe-4S clusters of Ech hydrogenase from *Methanosarcina barkeri* to individual subunits via the characterization of site-directed mutants. *FEBS J*, **272**, 4741–53.

Frederiksen, T. M. and Finster, K. (2003). Sulfite-oxido-reductase is involved in the oxidation of sulfite in *Desulfocapsa sulfoexigens* during disproportionation of thiosulphate and elemental sulfur. *Biodegradation*, **14**, 189–98.

Fricke, W. F., Seedorf, H., Henne, A. *et al.* (2006). The genome sequence of *Methanosphaera stadtmanae* reveals why this human intestinal archaeon is restricted to methanol and H_2 for methane formation and ATP synthesis. *J Bacteriol*, **188**, 642–58.

Friedrich, M. W. (2002). Phylogenetic analysis reveals multiple lateral transfers of adenosine-5′-phosphosulphate reductase genes among sulphate-reducing microorganisms. *J Bacteriol*, **184**, 278–89.

Fritz, G., Roth, A., Schiffer, A. *et al.* (2002). Structure of adenylylsulphate reductase from the hyperthermophilic *Archaeoglobus fulgidus* at 1.6-A resolution. *Proc Natl Acad Sci USA*, **99**, 1836–41.

Galouchko, A. S. and Rozanova, E. P. (1996). Sulfidogenic oxidation of acetate by a syntrophic association of anaerobic mesophilic bacteria. *Microbiology*, **65**, 134–9.

Garrity, G. M., Bell, J. A. and Lilburn, T. G. (2003). Taxonomic outline of the procaryotes. *Bergey's Manual of Systematic Bacteriology*. Second Edition. Release 5.0, New York: Springer Verlag 401 pages. DOI: 10.1007/bergeysoutline200405 (http://dx.doi.org/10.1007/bergeysoutline200405) New York: Springer-Verlag.

Ghiorse, W. C. (1984). Biology of iron- and manganese-depositing bacteria. *Annu Rev Microbiol*, **38**, 515–50.

Goenka, A., Voordouw, J. K., Lubitz, W., Gartner, W. and Voordouw, G. (2005). Construction of a NiFe-hydrogenase deletion mutant of *Desulfovibrio vulgaris* Hildenborough. *Biochem Soc Trans*, **33**, 59–60.

Hamilton, W. A. (2003). Microbially influenced corrosion as a model system for the study of metal microbe interactions: a unifying electron transfer hypothesis. *Biofouling*, **19**, 65–76.

Hansen, T. A. (1994). Metabolism of sulphate-reducing prokaryotes. *Antonie Van Leeuwenhoek International Journal of General and Molecular Microbiology*, **66**, 165–85.

Haveman, S. A., Brunelle, V., Voordouw, J. K. *et al.* (2003). Gene expression analysis of energy metabolism mutants of *Desulfovibrio vulgaris* Hildenborough indicates an important role for alcohol dehydrogenase. *J Bacteriol*, **185**, 4345–53.

Haynes, T. S., Klemm, D. J., Ruocco, J. J. and Barton, L. L. (1995). Formate dehydrogenase activity in cells and outer-membrane blebs of *Desulfovibrio gigas*. *Anaerobe*, **1**, 175–82.

Hedderich, R. (2004). Energy-converting NiFe hydrogenases from archaea and extremophiles: ancestors of complex I. *J Bioenerg Biomembr*, **36**, 65–75.

Hedderich, R., Klimmek, O., Kroeger, A. *et al.* (1998). Anaerobic respiration with elemental sulfur and with disulfides. *FEMS Microbiol Rev*, **22**, 353–81.

Heidelberg, J. F., Seshadri, R., Haveman, S. A. *et al.* (2004). The genome sequence of the anaerobic, sulphate-reducing bacterium *Desulfovibrio vulgaris* Hildenborough. *Nat Biotechnol*, **22**, 554–9.

Hemme, C. L. and Wall, J. D. (2004). Genomic insights into gene regulation of *Desulfovibrio vulgaris* Hildenborough. *Omics*, **8**, 43–55.

Hines, M. E., Evans, R. S., Sharak Genthner, B. R. *et al.* (1999). Molecular phylogenetic and biogeochemical studies of sulphate-reducing bacteria in the rhizosphere of *Spartina alterniflora*. *Appl Environ Microbiol*, **65**, 2209–16.

Hoehler, T., Alperin, M. J., Albert, D. B. and Martens, C. S. (2001). Apparent minimum free energy requirements for methanogenic Archaea and sulphate-reducing bacteria in an anoxic marine sediment. *FEMS Microbiol Ecol*, **38**, 33–41.

Holmer, M. and Storkholm, P. (2001). Sulphate reduction and sulphur cycling in lake sediments: a review. *Freshwater Biol*, **46**, 431–51.

Houwen, F. P., Dijkema, C., Stams, A. J. M. and Zehnder, A. J. B. (1991). Propionate metabolism in anaerobic bacteria – determination of carboxylation reactions with C-13-NMR spectroscopy. *Biochim Biophys Acta*, **1056**, 126–32.

Johnston, D. T., Wing, B. A., Farquhar, J. *et al.* (2005). Active microbial sulfur disproportionation in the Mesoproterozoic. *Science*, **310**, 1477–9.

Jonkers, H. M., van der Maarel, M. J. E. C., van Gemerden, H. and Hansen, T. A. (1996). Dimethylsulfoxide reduction by marine sulphate-reducing bacteria. *FEMS Microbiol Lett*, **136**, 283–7.

Jorgensen, B. B. (1982). Ecology of the bacteria of the sulfur cycle with special reference to anoxic oxic interface environments. *Philo Trans Roy Soc Ser B*, **298**, 543–61.

Keon, R. G. and Voordouw, G. (1996). Identification of the HmcF and topology of the HmcB subunit of the Hmc complex of *Desulfovibrio vulgaris*. *Anaerobe*, **2**, 231–8.

Klenk, H. P., Clayton, R. A., Tomb, J. F. *et al.* (1997). The complete genome sequence of the hyperthermophilic, sulphate-reducing archaeon *Archaeoglobus fulgidus*. *Nature*, **390**, 364–70.

Kobayashi, K., Morisawa, Y., Ishituka, T. and Ishimoto, M. (1975). Biochemical studies on sulphate-reducing bacteria. 14. Enzyme levels of adenylylsulphate reductase, inorganic pyrophosphatase, sulfite reductase, hydrogenase, and adenosine-triphosphatase in cells grown on sulphate, sulfite, and thiosulphate. *J Biochem (Tokyo)*, **78**, 1079–85.

Kopp, R. E., Kirschvink, J. L., Hilburn, I. A. and Nash, C. Z. (2005). The paleoproterozoic snowball Earth: a climate disaster triggered by the evolution of oxygenic photosynthesis. *Proc Natl Acad Sci USA*, **102**, 11131–6.

Koizumi, Y., Kelly, J. J., Nakagawa, T. *et al.* (2002). Parallel characterization of anaerobic toluene- and ethylbenzene-degrading microbial consortia by PCR-denaturing gradient gel electrophoresis, RNA-DNA membrane hybridization, and DNA microarray technology. *Appl Environ Microbiol*, **68**, 3215–25.

Kramer, M. and Cypionka, H. (1989). Sulphate formation via ATP sulfurylase in thiosulphate-disproportionating and sulfite-disproportionating bacteria. *Arch Microbiol*, **151**, 232–7.

Kreke, B. and Cypionka, H. (1992). Proton motive force in fresh-water sulphate-reducing bacteria, and its role in sulphate accumulation in *Desulfobulbus propionicus*. *Arch Microbiol*, **158**, 183–7.

Kreke, B. and Cypionka, H. (1994). Role of sodium-ions for sulphate transport and energy-metabolism in *Desulfovibrio salexigens*. *Arch Microbiol*, **161**, 55–61.

Krekeler, D. and Cypionka, H. (1995). The preferred electron-acceptor of *Desulfovibrio desulfuricans* Csn. *FEMS Microbiol Ecol*, **17**, 271–7.

Kremer, D. R. and Hansen, T. A. (1988). Pathway of propionate degradation in *Desulfobulbus propionicus*. *FEMS Microbiol Lett*, **49**, 273–7.

Larsen, O., Lien, T. and Birkeland, N. K. (1999). Dissimilatory sulfite reductase from *Archaeoglobus profundus* and *Desulfotomaculum thermocisternum*: phylogenetic and structural implications from gene sequences. *Extremophiles*, **3**, 63–70.

Leaphart, A. B., Friez, M. J. and Lovell, C. R. (2003). Formyltetrahydrofolate synthetase sequences from salt marsh plant roots reveal a diversity of acetogenic bacteria and other bacterial functional groups. *Appl Environ Microbiol*, **69**, 693–6.

Le, M. J. and Zinder, S. H. (1988). Isolation and characterization of a thermophilic bacterium which oxidizes acetate in syntrophic association with a methanogen and which grows acetogenically on H_2-CO_2. *Appl Environ Microbiol*, **54**, 124–9.

Lemos, R. S., Gomes, C. M., Santana, M. *et al.* (2001). The "strict" anaerobe *Desulfovibrio gigas* contains a membrane-bound oxygen respiratory chain. *J Inorg Biochem*, **86**, 314.

Liu, M. Y. and Legall, J. (1990). Purification and characterization of 2 proteins with inorganic pyrophosphatase activity from *Desulfovibrio vulgaris* – rubrerythrin and a new, highly-active,enzyme. *Biochem Biophys Res Commun*, **171**, 313–18.

Lopez-Cortes, A., Bursakov, S., Figueiredo, A. *et al.* (2005). Purification and preliminary characterization of tetraheme cytochrome c(3) and adenylylsulphate reductase from the peptidolytic sulphate-reducing bacterium *Desulfovibrio aminophilus* DSM 12254. *Bioinorg Chem Appl*, **3**, 81–91.

Lovley, D. R., Holmes, D. E. and Nevin, K. P. (2004). Dissimilatory Fe(III) and Mn(IV) reduction. *Adv Microb Physiol*, **49**, 221–86.

Lovley, D. R., Roden, E. E., Phillips, E. J. P. and Woodward, J. C. (1993). Enzymatic iron and uranium reduction by sulphate-reducing bacteria. *Mar Geol*, **113**, 41–53.

Malki, S., DeLuca, G., Fardeau, M. L. *et al.* (1997). Physiological characteristics and growth behavior of single and double hydrogenase mutants of *Desulfovibrio fructosovorans*. *Arch Microbiol*, **167**, 38–45.

Mander, G. J., Duin, E. C., Linder, D., Stetter, K. O. and Hedderich, R. (2002). Purification and characterization of a membrane-bound enzyme complex from the sulphate-reducing archaeon *Archaeoglobus fulgidus* related to heterodisulfide reductase from methanogenic archaea. *Eur J Biochem*, **269**, 1895–904.

Mander, G. J., Pierik, A. J., Huber, H. and Hedderich, R. (2004). Two distinct heterodisulfide reductase-like enzymes in the sulphate-reducing archaeon *Archaeoglobus profundus*. *Eur J Biochem*, **271**, 1106–16.

Matias, P. M., Pereira, I. A. C., Soares, C. M. and Carrondo, M. A. (2005). Sulphate respiration from hydrogen in *Desulfovibrio* bacteria: a structural biology overview. *Prog Biophys Mol Biol*, **89**, 292–329.

McOrist, S., Gebhart, C. J., Boid, R. and Barns, S. M. (1995). Characterization of *Lawsonia intracellularis* gen. nov., sp. nov., the obligately intracellular bacterium of porcine proliferative enteropathy. *Int J Syst Bacteriol*, **45**, 820–5.

Meier, T., Polzer, P., Diederichs, K., Welte, W. and Dimroth, P. (2005). Structure of the rotor ring of F-type Na^+-ATPase from *Ilyobacter tartaricus*. *Science*, **308**, 659–62.

Meuer, J., Kuettner, H. C., Zhang, J. K., Hedderich, R. and Metcalf, W. W. (2002). Genetic analysis of the archaeon *Methanosarcina barkeri* Fusaro reveals a central role for *Ech* hydrogenase and ferredoxin in methanogenesis and carbon fixation. *Proc Natl Acad Sci USA*, **99**, 5632–7.

Molitor, M., Dahl, C., Molitor, I. *et al.* (1998). A dissimilatory sirohaem-sulfite-reductase-type protein from the hyperthermophilic archaeon *Pyrobaculum islandicum*. *Microbiology-SGM*, **144**, 529–41.

Monster, J., Appel, P. W. U., Thode, H. G. *et al.* (1979). Sulfur isotope studies in early archaean sediments from Isua, West Greenland – implications for the antiquity of bacterial sulphate reduction. *Geochim Cosmochim Acta*, **43**, 405–13.

Mori, K., Kim, H., Kakegawa, T. and Hanada, S. (2003). A novel lineage of sulphate-reducing microorganisms: *Thermodesulfobiaceae* fam. nov., *Thermodesulfobium narugense*, gen. nov., sp nov., a new thermophilic isolate from a hot spring. *Extremophiles*, **7**, 283–90.

Mueller, V. (2004). An exceptional variability in the motor of archaeal A(1)A(0) ATPases: from multimeric to monomeric rotors comprising 6–13 ion binding sites. *J Bioenerg Biomembr*, **36**, 115–25.

Murata, T., Yamato, I., Kakinuma, Y., Leslie, A. G. W. and Walker, J. E. (2005). Structure of the rotor of the V-type Na^+-ATPase from *Enterococcus hirae*. *Science*, **308**, 654–9.

Myers, C. R. and Nealson, K. H. (1988). Bacterial manganese reduction and growth with manganese oxide as the sole electron-acceptor. *Science*, **240**, 1319–21.

Nealson, K. H. and Saffarini, D. (1994). Iron and manganese in anaerobic respiration – environmental significance, physiology, and regulation. *Annu Rev Microbiol*, **48**, 311–43.

Nielsen, P. H., Lee, W., Lewandowski, Z., Morrison, M. and Characklis, W. G. (1993). Corrosion of mild steel in an alternating oxic and anoxic biofilm system. *Biofouling*, **7**, 267–84.

Odom, J. M. and Peck, H. D. (1984). Hydrogenase, electron-transfer proteins, and energy coupling in the sulphate-reducing bacteria *Desulfovibrio*. *Annu Rev Microbiol*, **38**, 551–92.

Ogata, M., Arihara, K. and Yagi, T. (1981). D-Lactate dehydrogenase of *Desulfovibrio vulgaris. J Biochem (Tokyo)*, **89**, 1423–31.

Pankhania, I. P., Spormann, A. M., Hamilton, W. A. and Thauer, R. K. (1988). Lactate conversion to acetate, CO_2 and H_2 in cell suspensions of *Desulfovibrio vulgaris* (Marburg): indications for the involvement of an energy driven reaction. *Arch Microbiol*, **150**, 26–31.

Paulsen, J., Kröger, A. and Thauer, R. K. (1986). ATP-driven succinate oxidation in the catabolism of *Desulfuromonas acetoxidans. Arch Microbiol*, **144**, 78–83.

Peck, H. D. (1993). Bioenergetic strategies of the sulphate-reducing bacteria. In J. M. Odom and J. Rivers Singleton (eds.), *The Sulphate-Reducing Bacteria: Contemporary Perspectives.* New York, London: Springer-Verlag. pp. 41–76.

Pires, R. H., Lourenco, A. I., Morais, F. *et al.* (2003). A novel membrane-bound respiratory complex from *Desulfovibrio desulfuricans* ATCC 27774. *Biochim Biophys Acta-Bioenergetics*, **1605**, 67–82.

Pohorelic, B. K. J., Voordouw, J. K., Lojou, E. *et al.* (2002). Effects of deletion of genes encoding Fe-only hydrogenase of *Desulfovibrio vulgaris* Hildenborough on hydrogen and lactate metabolism. *J Bacteriol*, **184**, 679–86.

Rabus, R., Fukui, M., Wilkes, H. and Widdel, F. (1996). Degradative capacities and 16S rRNA-targeted whole-cell hybridization of sulphate-reducing bacteria in an anaerobic enrichment culture utilizing alkylbenzenes from crude oil. *Appl Environ Microbiol*, **62**, 3605–13.

Rabus, R., Hansen, T., and Widdel, F. (2001). An evolving electronic resource for the microbiological community. In S. Dworkin, M. Falkow, E. Rosenberg, K.-H. Schleifer and E. Stackebrandt (eds.), *The Prokaryotes.* New York: Springer-Verlag. pp. release 3.3, http://link.springer-ny.com/link/service/books/10125.

Rabus, R., Ruepp, A., Frickey, T. *et al.* (2004). The genome of *Desulfotalea psychrophila*, a sulphate-reducing bacterium from permanently cold Arctic sediments. *Environ Microbiol*, **6**, 887–902.

Reed, D. W. and Hartzell, P. L. (1999). The *Archaeoglobus fulgidus* D-lactate dehydrogenase is a Zn^{2+} flavoprotein. *J Bacteriol*, **181**, 7580–7.

Reguera, G., McCarthy, K. D., Mehta, T. *et al.* (2005). Extracellular electron transfer via microbial nanowires. *Nature*, **435**, 1098–101.

Rodrigues, R., Valente, F. M. A., Pereira, I. A. C., Oliveira, S. and Rodrigues-Pousada, C. (2003). A novel membrane-bound Ech NiFe hydrogenase in *Desulfovibrio gigas. Biochem Biophys Res Commun*, **306**, 366–75.

Rossi, M., Pollock, W. B. R., Reij, M. W. *et al.* (1993). The Hmc operon of *Desulfovibrio vulgaris* Subsp *vulgaris* Hildenborough encodes a potential transmembrane redox protein complex. *J Bacteriol*, **175**, 4699–711.

Sapra, R., Bagramyan, K. and Adams, M. W. W. (2003). A simple energy-conserving system: proton reduction coupled to proton translocation. *Proc Natl Acad Sci USA*, **100**, 7545–50.

Sass, H., Steuber, J., Kroder, M., Kroneck, P. M. H. and Cypionka, H. (1992). Formation of thionates by fresh-water and marine strains of sulphate-reducing bacteria. *Arch Microbiol*, **158**, 418–21.

Sato, K., Nishina, Y., Setoyama, C., Miura, R. and Shiga, K. (1999). Unusually high standard redox potential of acrylyl-CoA/propionyl-CoA couple among enoyl-CoA/acyl-CoA couples: a reason for the distinct metabolic pathway of propionyl-CoA from longer acyl-CoAs. *J Biochem (Tokyo)*, **126**, 668–75.

Schiffers, A. and Jorgensen, B. B. (2002). Biogeochemistry of pyrite and iron sulfide oxidation in marine sediments. *Geochim Cosmochim Acta*, **66**, 85–92.

Schink, B. (1992). The genus Pelobacter. In A. Balows, H. G. Trüper, M. Dworkin, W. Harder and K.-H. Schleifer (eds.), *The Prokaryotes*. New York: Springer-Verlag. pp. 3393–9.

Schink, B. and Stams, A. J. (2002). Syntrophism among Prokaryotes. In M. Dworkin (ed.), *The Prokaryotes (electronic version)*. New York: Springer Verlag. pp. 309–35.

Shen, Y. N. and Buick, R. (2004). The antiquity of microbial sulphate reduction. *Earth-Science Reviews*, **64**, 243–72.

Shigematsu, T., Tang, Y. Q. Kobayashi, T. *et al.* (2004). Effect of dilution rate on metabolic pathway shift between aceticlastic and nonaceticlastic methanogenesis in chemostat cultivation. *Appl Environ Microbiol*, **70**, 4048–52.

Shima, S. and Thauer, R. K. (2005). Methyl-coenzyme M reductase (MCR) and the anaerobic oxidation of methane (AOM) in methanotrophic archaea. *Curr Opin Microbiol*, **8**, 643–8.

Soboh, B., Linder, D. and Hedderich, R. (2002). Purification and catalytic properties of a CO-oxidizing: H_2-evolving enzyme complex from *Carboxydothermus hydrogenoformans*. *Eur J Biochem*, **269**, 5712–21.

Sperling, D., Kappler, U., Wynen, A., Dahl, C. and Truper, H. G. (1998). Dissimilatory ATP sulfurylase from the hyperthermophilic sulphate reducer *Archaeoglobus fulgidus* belongs to the group of homo-oligomeric ATP sulfurylases. *FEMS Microbiol Lett*, **162**, 257–64.

Stackebrandt, E. (1995). Origin and evolution of prokaryotes. In A. J. Gibbs, C. H. Calisher and F. Garcia-Arenal (eds.), *Molecular Basis of Virus Evolution*. Cambridge: Cambridge University Press, pp. 224–52.

Stackebrandt, E. (2004). The phylogeny and classification of anaerobic bacteria. In M. M. Nakano and P. Zuber (eds.), *Strict and Facultative Anaerobes. Medical and Environmental Aspects*. Wymondham, UK: Horizon Bioscience. pp. 1–25.

Steger, J. L., Vincent, C., Ballard, J. D. and Krumholz, L. R. (2002). *Desulfovibrio* sp genes involved in the respiration of sulphate during metabolism of hydrogen and lactate. *Appl Environ Microbiol*, **68**, 1932–7.

Taguchi, Y., Sugishima, M. and Fukuyama, K. (2004). Crystal structure of a novel zinc-binding ATP sulfurylase from *Thermus thermophilus* HB8. *Biochemistry*, **43**, 4111–18.

Thamdrup, B. and Canfield, D. E. (1996). Pathways of carbon oxidation in continental margin sediments off central Chile. *Limnol Oceanogr*, **41**, 1629–50.

Thamdrup, B., Fossing, H. and Jorgensen, B. B. (1994). Manganese, iron and sulfur cycling in a coastal marine sediment, Aarhus Bay, Denmark. *Geochim Cosmochim Acta*, **58**, 5115–29.

Thauer, R. K. (1988). Citric acid cycle, 50 years on: modifications and an alternative pathway in anaerobic bacteria. *Eur. J. Biochem.*, **176**, 497–508.

Thauer, R. K., Jungermann, K. and Decker, K. (1977). Energy conservation in chemotrophic anaerobic bacteria. *Bacteriol. Rev.* **41**, 100–80.

Thauer, R. K. and Kunow, J. (1995). Sulphate reducing Archaea. In N. Clark (ed.), *Biotechnology Handbook*. London: Plenum Publishing. pp. 33–48.

Thauer, R. K., Möller-Zinkhan, D. and Spormann, A. (1989). Biochemistry of acetate catabolism in anaerobic chemotrophic bacteria. *Annu. Rev. Microbiol.*, **43**, 43–67.

Thauer, R. K. and Morris, J. G. (1984). Metabolism of chemotrophic anaerobes: old views and new aspects. In D. P. Kelly and N. G. Carr (eds.), *The Microbe: 1984 Part II: Prokaryotes and Eukaryotes. Society for General Microbiology Symposium 36*. Cambridge: Cambridge University Press. pp. 123–68.

Tomei, F. A., Barton, L. L., Lemanski, C. L. *et al.* (1995). Transformation of selenate and selenite to elemental selenium by *Desulfovibrio desulfuricans*. *J Ind Microbiol*, **14**, 329–36.

Venter, J. C., Remington, K., Heidelberg, J. F. *et al.* (2004). Environmental genome shotgun sequencing of the Sargasso Sea. *Science*, **304**, 66–74.

Voordouw, G. (2002). Carbon monoxide cycling by *Desulfovibrio vulgaris* Hildenborough. *J Bacteriol*, **184**, 5903–11.

Voordouw, G., Armstrong, S. M., Reimer, M. F. *et al.* (1996). Characterization of 16S rRNA genes from oil field microbial communities indicates the presence of a variety of sulphate-reducing, fermentative, and sulfide-oxidizing bacteria. *Appl Environ Microb*, **62**, 1623–9.

Wächtershäuser, G. (1992). Groundworks for an evolutionary biochemistry: the iron-sulphur world. *Prog Biophys Mol Biol*, **58**, 85–201.

Ware, D. A. and Postgate, J. R. (1971). Physiological and chemical properties of a reductant-activated inorganic pyrophosphatase from *Desulfovibrio desulfuricans*. *J Gen Microbiol*, **67**, 145–60.

Weinberg, M. V., Jenney, F. E., Cui, X. Y. and Adams, M. W. W. (2004). Rubrerythrin from the hyperthermophilic archaeon *Pyrococcus furiosus* is a rubredoxin-dependent, iron-containing peroxidase. *J Bacteriol*, **186**, 7888–95.

Widdel, F. and Bak, F. (1992). Gram-negative mesophilic sulphate-reducing bacteria. In A. Balows, H. G. Trüper, M. Dworkin, W. Harder and K.-H. Schleifer (eds.), *The Prokaryotes*. New York: Springer-Verlag. pp. 3352–78.

Widdel, F. and Pfennig, N. (1982). Studies on dissimilatory sulphate-reducing bacteria that decompose fatty-acids. 2. Incomplete oxidation of propionate by *Desulfobulbus propionicus* Gen-Nov, Sp-Nov. *Arch Microbiol*, **131**, 360–5.

Widdel, F. and Pfennig, N. (1984). Dissimilatory sulphate- and sulfur-reducing bacteria. In N. R. Krieg and J. G. Holt (eds.), *Bergey's Manual of Systematic Bacteriology*. Baltimore, MD: Williams and Wilkins. pp. 663–79.

Winogradsky, S. (1890). Recherches sur les organismes de la nitrification. *Compt Rendue*, **110**, 1013–16. In T. D. Brock (ed.), *Milestones in Microbiology: 1556 to 1940*. ASM Press: Washington DC (1998). pp. 231–33.

Yagi, T. and Ogata, M. (1996). Catalytic properties of adenylylsulphate reductase from *Desulfovibrio vulgaris* Miyazaki. *Biochimie*, **78**, 838–46.

Zengler, K., Richnow, H. H., Rossello-Mora, R., Michaelis, W. and Widdel, F. (1999). Methane formation from long-chain alkanes by anaerobic microorganisms. *Nature*, **401**, 266–9.

Zverlov, V., Klein, M., Lucker, S. *et al.* (2005). Lateral gene transfer of dissimilatory (bi)sulfite reductase revisited. *J Bacteriol*, **187**, 2203–8.

CHAPTER 2

Molecular strategies for studies of natural populations of sulphate-reducing microorganisms

David A. Stahl, Alexander Loy and Michael Wagner

2.1 INTRODUCTION

An early focus on the use of molecular techniques to characterize natural populations of sulphate-reducing microorganisms (SRM) derived from the close relationship between their phylogenetic affiliation and their capability to anaerobically respire with sulphate. In other words, all so-far characterized SRM associate with lineages in the tree of life that predominantly consist of sulphate reducers. Known SRM are affiliated with two divisions (phyla) within the *Archaea* (the euryarchaeotal genus *Archaeoglobus* species and the crenarchaeotal genera *Caldivirga* and *Thermocladium*, affiliated with the *Thermoproteales*) and five divisions within the *Bacteria* (the *Deltaproteobacteria*, endospore-forming *Desulfotomaculum*, *Desulfosporosinus*, and *Desulfosporomusa* species within the Firmicutes division, *Thermodesulfovibrio* species within the Nitrospira division, and two divisions represented by *Thermodesulfobacterium* species and the recently isolated *Thermodesulfobium narugense*, the exact phylogenetic position of the latter is still ambiguous). Most described SRM are either Gram-positive bacteria with a low G+C content or Gram-negative *Deltaproteobacteria*. However, it is important to note that almost all major physiological properties of cultured and uncultured SRM, such as substrate usage patterns, the ability to completely oxidize a substrate to CO_2, and alternative ways of anaerobic energy generation cannot be unambiguously determined from comparative analysis of their 16S rRNA genes.

The generally tight association between phylogenetic affiliation and sulphate-reducing physiology offered a foundation to directly associate the population structure determined by 16S rRNA sequence type and process. These studies have now been complemented by the use of highly conserved genes in the pathway for sulphate respiration. As detailed elsewhere in this chapter, studies that examined both 16S rRNA and functional gene sequence diversity in samples recovered from the same habitat have often shown good correspondence. However, these studies have also revealed that we have as yet a rather incomplete understanding of the major phylogenetic groups of SRM.

2.2 16S rRNA SEQUENCING AND FINGERPRINTING

2.2.1 Technical considerations

One of the most unpredictable parameters of nucleic acid-based measures of diversity is the influence of the environmental matrix on recovery of nucleic acids. Although methods based on DNA probe hybridization to an RNA or DNA target are less sensitive to copurified impurities than enzymatic methods such as polymerase chain reaction (PCR), all nucleic acid-based methods are subject to interference from impurities such as humic acids which commonly contaminate soil and sediment extractions (Alm and Stahl, 2000). Since this chapter contribution is intended to provide an overview of insights gained via application of molecular measures, we do not address in detail the many protocols and methods now employed to recover nucleic acid from different habitats and the reader is referred to recent publications that address alternative methods of extraction and criteria to evaluate efficiency and biases associated with extraction (e.g. Alm and Stahl, 2000; Martin-Laurent *et al.*, 2001). Because of the many uncertainties now associated with the interpretation of different molecular data types, there is a need to employ complementary techniques to establish the relationship between sequence diversity and the abundance and activity of the contributing SRM populations. For example, the study by Llobet-Brossa *et al.* (2002) combined chemical, radiotracer, molecular, and microbiological methods to demonstrate that members of the *Desulfovibrio* and *Desulfosarcina-Desulfococcus-Desulfofrigus* groups are among the most active populations in coastal sediments, receiving very high inputs of organic matter. In our opinion there are far too many studies that are "snapshots" of diversity with limited replication.

There are two general methods for the recovery of nucleic acid sequence information: hybridization or sequence analysis of clonally propagated DNA fragments. The technical barriers to DNA sequence determination have virtually vanished in the last decade. As a consequence of these rapid changes in technology, high throughput sequence analysis is an increasingly attractive option for initial characterization of community structure and a number of primer sets have been developed for selective recovery of 16S rRNA gene sequences affiliated with recognized SRM (Table 2.1). Even so, as yet, there have been relatively few examples of targeted amplification to characterize SRM by standard cloning and sequencing methods. For example, Ravenschlag *et al.* used selective amplification to demonstrate that sequences affiliated with the *Desulfosarcina-Desulfococcus* group were dominant in a marine Arctic sediment (Ravenschlag *et al.*, 2000). Scheid and Stubner used primer sets selective for the *Desulfovibrionaceae*, *Desulfobacteriaceae*, and *Desulfobulbus* species to assess their presence on rice roots (Scheid and Stubner, 2001) and Loy *et al.* applied primers selective for the *Syntrophobacteraceae* and the genus *Desulfomonile* to monitor these organisms in acidic fens (Loy *et al.*, 2004). An extensive set of selective primers was designed by Daly *et al.* to examine SRM diversity in landfill leachates (Daly *et al.*, 2000). Selective amplification has more commonly been employed for denaturing gradient gel electrophoresis (DGGE) analyses, as is discussed in the following sections.

2.2.2 Fingerprinting methods

There are several popular fingerprinting methods for characterizing complex communities. Most of these rely on an initial amplification step using either general or specific primers for recovery of highly conserved genes, followed by fractionation based on size or conformation. The genes encoding the 16S rRNAs are most commonly employed. In addition, certain so-called functional genes encoding enzymes involved in metabolism relevant to SRM have been used to selectively identify SRM populations using fingerprinting techniques. These genes include reasonably well-conserved hydrogenases (Wawer and Muyzer, 1995) and highly conserved genes in the pathway of sulphate reduction. The primary advantage of fingerprinting methods is the provision of a format that offers a very rapid and comprehensive intersample comparison, for example, the characterization of changing population structure with depth in a marine or freshwater sediment, or water column.

2.2.3 Denaturing gradient gel electrophoresis (DGGE)

Analysis of environmental SRM by DGGE has used several diagnostic genes – the 16S rRNA genes (e.g. Maukonen *et al.*, 2006; Overmann *et al.*, 1999), genes encoding hydrogenases characteristic of *Desulfovibrio* species (Wawer and Muyzer, 1995), and genes encoding the dissimilatory

Table 2.1. *Selected 16S rRNA gene-targeted primer (pairs) for different taxa of sulphate-reducing microorganisms (SRM)*

Short name	Full name[a]	Suggested annealing temperature [°C]	Calculated melting temperature T_m [°C][b]	Sequence 5′–3′	Coverage according to RDP II probe match (perfectly matched hits/total searched)[c]	Target group	Reference
PROKA1492R[d]	S-*-Proka-1492-a-A-19	52, 60[e]	47–49	GGY TAC CTT GTT ACG ACT T	domain *Bacteria* (15959/18931) *Archaea* are not represented in RDP II	Most *Bacteria* and *Archaea*	Modified from Kane *et al.*, 1993
BACT11F[f]	S-D-Bact-0011-a-S-17	67	47	GTT TGA TCC TGG CTC AG	domain *Bacteria* (24218/37316)	Most *Bacteria*	Kane *et al.*, 1993
DELTA493F	S-C-dProt-0493-a-S-18	–	48–53	RRA GGA AGC ACC GGC TAA	domain *Bacteria* (3566/151831) class *Deltaproteobacteria* (3009/4269) order *Desulfurellales* (8/12)	Most *Deltaproteobacteria*	This chapter

					order *Desulfovibrionales* (544/660) order *Desulfobacterales* (1336/1572) order *Desulfuromonales* (426/497) order *Syntrophobacterales* (229/289) order *Myxococcales* (50/572)		
DSBAC355F[d,g]	S-*-Dsb-0355-a-S-18	60	48	CAG TGA GGA ATT TTG CGC	domain *Bacteria* (693/156970) family *Desulfobacteraceae* (471/607) family *Syntrophobacteraceae* (99/159)	Most *Desulfobacteraceae* and *Syntrophobacteraceae*	Scheid and Stubner, 2001

Table 2.1 (*cont.*)

Short name	Full name[a]	Suggested annealing temperature [°C]	Calculated melting temperature T_m [°C][b]	Sequence 5′–3′	Coverage according to RDP II probe match (perfectly matched hits/total searched)[c]	Target group	Reference
SRB385F[h]	S-*-Srb-0385-a-S-18	–	59	CCT GAC GCA GCG ACG CCG	domain *Bacteria* (4355/161606) class *Deltaproteobacteria* (1032/3822) family *Desulfovibrionaceae* (314/457) family *Desulfobulbaceae* (450/511)	Some *Deltaproteobacteria* and other *Bacteria*	Amann *et al.*, 1990
SRB385DbF[h]	S-*-Srb-0385-b-S-18	–	57	CCT GAC GCA GCA ACG CCG	domain *Bacteria* (4558/161606) class *Deltaproteobacteria* (1583/3822)	Some *Deltaproteobacteria* and other *Bacteria*	Rabus *et al.*, 1996

					family *Desulfobacteraceae* (561/683) family *Syntrophobacteraceae* (47/174)		
DCC305F	S-*-Dsb-0305-a-S-23	65	57–59	GAT CAG CCA CAC TGG RAC TGA CA	domain *Bacteria* (341/154572) family *Desulfobacteraceae* (315/575) genus *Desulfococcus* (6/7) genus *Desulfofaba* (3/4) genus *Desulfofrigus* (11/12) genus *Desulfonema* (1/12) genus *Desulfosarcina* (39/44)	Some *Desulfobacteraceae*	Daly *et al.*, 2000
DCC1165R	S-*-Dsb-1165-a-A-21	65	52–56	GGG GCA GTA	domain *Bacteria* (62/141333)	Few *Desulfobacteraceae*	Daly *et al.*, 2000

Table 2.1 (*cont.*)

Short name	Full name[a]	Suggested annealing temperature [°C]	Calculated melting temperature T_m [°C][b]	Sequence 5′–3′	Coverage according to RDP II probe match (perfectly matched hits/total searched)[c]	Target group	Reference
				TCT TYA GAG TYC	family *Desulfobacteraceae* (60/540) genus *Desulfococcus* (4/7) genus *Desulfosarcina* (13/42)		
DSB+57F	S-G-Dsb-0057-a-S-21	64	54	GCA AGT CGA ACG AGA AAG GGA	domain *Bacteria* (116/116499) genus *Desulfobacter* (15/20)	*Desulfobacter* and some other *Desulfobacteraceae*	This chapter
DSB1243R	S-G-Dsb-1243-a-A-21	64	54	AGT CGC TGC CCT TTG	domain *Bacteria* (18/100800) genus *Desulfobacter* (18/21)	*Desulfobacter*	This chapter

				TAC CTA			
DSB127F	S-G-Dsb-0127-a-S-22	60	55	GAT AAT CTG CCT TCA AGC CTG G	domain *Bacteria* (14/142899) genus *Desulfobacter* (14/20)	*Desulfobacter*	Daly *et al.*, 2000
DSB1273R	S-G-Dsb-1273-a-A-22	60	55–68	CYY YYY GCR RAG TCG STG CCC T	domain *Bacteria* (19/98854) genus *Desulfobacter* (17/21)	*Desulfobacter*	Daly *et al.*, 2000
DSN61F	S-*-Dsn-0061-a-S-17	52	52	GTC GCA CGA GAA CAC CC	domain *Bacteria* (3/121306) genus *Desulfonema* (3/12)	*Desulfonema limicola, D. ishimotonii*	Loy *et al.*, 2002
DSN+1201R	S-*-Dsn-1201-a-A-17	52	45	GAC ATA AAG GCC ATG AG	domain *Bacteria* (1415/101827) genus *Desulfonema* (7/8)	*Desulfonema* and some other *Bacteria*	Loy *et al.*, 2002

Table 2.1 (*cont.*)

Short name	Full name[a]	Suggested annealing temperature [°C]	Calculated melting temperature T_m [°C][b]	Sequence 5′–3′	Coverage according to RDP II probe match (perfectly matched hits/total searched)[c]	Target group	Reference
DBM169F	S-*-Dbm-0169-a-S-20	64	48–52	CTA ATR CCG GAT RAA GTC AG	domain *Bacteria* (6/146943) family *Desulfobacteraceae* (6/556)	*Desulfobacterium vacuolatum, D. niacini, D. autotrophicum*	Daly *et al.*, 2000
DBM1006R	S-*-Dbm-1006-a-A-21	64	49–50	ATT CTC ARG ATG TCA AGT CTG	domain *Bacteria* (13/107661) family *Desulfobacteraceae* (13/509)	*Desulfobacterium vacuolatum, D. niacini, D. autotrophicum*	Daly *et al.*, 2000
DSSC140F	S-*-Dssc-0140-a-S-20	60	52	GAA TTG GGG ATA ACG	domain *Bacteria* (33/144246) genus *Desulfosarcina* (31/43)	*Desulfosarcina*	This chapter

				TTG CG			
DSSC1438R	S-*-Dssc-1438-a-A-18	60	55	CCG AAG GGT TAG CCC GAC	domain *Bacteria* (7/57404) genus *Desulfosarcina* (4/16)	*Desulfosarcina*	This chapter
DSMON85F	S-G-Dsmon-0085-a-S-20	62	54–56	CGG GGT RTG GAG TAA AGT GG	domain *Bacteria* (9/130960) genus *Desulfomonile* (not classified in this RDP II release)	*Desulfomonile*	Loy *et al.*, 2004
DSMON1419R	S-G-Dsmon-1419-a-A-20	62	54–56	CGA CTT CTG GTG CAG TCA RC	domain *Bacteria* (4/61171) genus *Desulfomonile* (not classified in this RDP II release)	*Desulfomonile*	Loy *et al.*, 2004
DBACCA65F	S-S-Dbacca-0065-a-S-18	58	50	TAC GAG AAA GCC CGG CTT	domain *Bacteria* (1/121306) genus *Desulfobacca* (not classified in this RDP II release)	*Desulfobacca acetoxidans*	Loy *et al.*, 2004

Table 2.1 (*cont.*)

Short name	Full name[a]	Suggested annealing temperature [°C]	Calculated melting temperature T_m [°C][b]	Sequence 5′–3′	Coverage according to RDP II probe match (perfectly matched hits/total searched)[c]	Target group	Reference
DBACCA1430R	S-S-Dbacca-1430-a-A-18	58	50	TTA GGC CAG CGA CAT CTG	domain *Bacteria* (1/59323) genus *Desulfobacca* (not classified in this RDP II release)	*Desulfobacca acetoxidans*	Loy *et al.*, 2004
SYBAC+282F	S-*-Sybac-0282-a-S-18	60	53	ACG GGT AGC TGG TCT GAG	domain *Bacteria* (3066/154603) family *Syntropho-bacteraceae* (90/136)	*Syntropho-bacteraceae* and some other *Bacteria*	Loy *et al.*, 2004
SYBAC1427R	S-*-Sybac-1427-a-A-18	60	53	GCC CAC GCA CTT CTG GTA	domain *Bacteria* (47/59323) family *Syntropho-bacteraceae* (45/76)	*Syntropho-bacteraceae*	Loy *et al.*, 2004
DSBB280F	S-*-Dsbb-0280-a-S-18	58	53	CGA TGG TTA	domain *Bacteria* (454/154555)	*Desulfo-bulbaceae*	Kjeldsen *et al.*, 2007

				GCG GGT CTG	family *Desulfobulbaceae* (417/490)		
DSBB+1297R	S-*-Dsbb-1297-a-A-19	58	51	AGA CTC CAA TCC GGA CTG A	domain *Bacteria* (1023/97744) family *Desulfobulbaceae* (246/301)	*Desulfobulbaceae* and some other *Proteobacteria*	Kjeldsen *et al.*, 2007
DBB121F	S-*-Dbb-0121-a-S-23	66	55–57	CGC GTA GAT AAC CTG TCY TCA TG	domain *Bacteria* (5/142205) genus *Desulfobulbus* (5/85)	Few *Desulfobulbus*	Daly *et al.*, 2000
DBB1237R	S-*-Dbb-1237-a-A-23	66	59–61	GTA GKA CGT GTG TAG CCC TGG TC	domain *Bacteria* (563/100372) family *Desulfobulbaceae* (207/313)	*Desulfobulbaceae* and other *Bacteria*	Daly *et al.*, 2000

Table 2.1 (*cont.*)

Short name	Full name[a]	Suggested annealing temperature [°C]	Calculated melting temperature T_m [°C][b]	Sequence 5′–3′	Coverage according to RDP II probe match (perfectly matched hits/total searched)[c]	Target group	Reference
DSV682F	S-*-Dsv-0682-a-S-19	58	51	GGT GTA GGA GTG AAA TCC G	domain *Bacteria* (667/139919) order *Desulfovibrionales* (583/679)	*Desulfo-vibrionales* and some other *Deltaproteo-bacteria*	Devereux *et al.*, 1992
DSV+1402R	S-*-Dsv-1402-a-A-18	58	53	CTT TCG TGG TGT GAC GGG	domain *Bacteria* (2330/62983) order *Desulfovibrionales* (276/298)	*Desulfo-vibrionales* and some other *Bacteria*	Kjeldsen *et al.*, 2007
DSV230F	S-*-Dsv-0230-a-S-19	61	49–60	GRG YCY GCG TYY CAT TAG C	domain *Bacteria* (1714/152836) order *Desulfovibrionales* (382/472)	*Desulfo-vibrionales* and other *Bacteria*	Daly *et al.*, 2000

DSV838R	S-*-Dsv-0838-a-A-21	61	50–62	SYC CGR CAY CTA GYR TYC ATC	domain *Bacteria* (2474/124442) order *Desulfovibrionales* (327/647)	*Desulfo-vibrionales* and other *Bacteria*	Daly *et al.*, 2000
DVHO130F	S-*-Dvho-0130-a-S-18	58	50	ATC TAC CCG ACA GAT CGG	domain *Bacteria* (5/142899) genus *Desulfovibrio* (5/264)	*Desulfovibrio halophilus, D. oxyclinae*	This chapter
DVHO1424R	S-*-Dvho-1424-a-A-18	58	53	TGC CGA CGT CGG GTA AGA	domain *Bacteria* (5/60160) genus *Desulfovibrio* (5/186)	*Desulfovibrio halophilus, D. oxyclinae*	This chapter
DSM172F	S-G-Dsm-0172-a-S-19	56	49	AAT ACC GGA TAG TCT GGC T	domain *Bacteria* (57/147425) genus *Desulfomicrobium* (56/72)	*Desulfo-microbium*	Loy *et al.*, 2002

Table 2.1 (*cont.*)

Short name	Full name[a]	Suggested annealing temperature [°C]	Calculated melting temperature T_m [°C][b]	Sequence 5′–3′	Coverage according to RDP II probe match (perfectly matched hits/total searched)[c]	Target group	Reference
DSM1469R	S-G-Dsm-1469-a-A-18	56	50	CAA TTA CCA GCC CTA CCG	domain *Bacteria* (36/38000) genus *Desulfomicrobium* (23/31)	*Desulfo-microbium*	Loy *et al.*, 2002
DEM116F	S-*-DfmI-0117-a-S-18	63	48	GTA ACG CGT GGA TAA CCT	domain *Bacteria* (892/141333) genus *Desulfotomaculum* (69/77) genus *Pelotomaculum* (54/71) genus *Sporotomaculum* (3/3)	*Desulfo-tomaculum* cluster I[i] and some other *Firmicutes*	Stubner and Meuser, 2000
DEM1164R	S-*-DfmI-1165-a-A-18	63	48	CCT TCC TCC GTT	domain *Bacteria* (3455/102634) genus	*Desulfo-tomaculum* cluster I[i] and	Stubner and Meuser, 2000

				TTG TCA	*Desulfotomaculum* (75/78) genus *Pelotomaculum* (48/65) genus *Sporotomaculum* (3/3)	some other *Firmicutes*	
DFM140F	S-*-Dfm-0140-a-S-19	58	47–58	TAG MCY GGG ATA ACR SYK G	domain *Bacteria* (96/144246) genus *Desulfotomaculum* (56/78) genus *Pelotomaculum* (0/81) genus *Sporotomaculum* (2/3)	Most *Desulfotomaculum* cluster I[i] and some other *Bacteria*	Daly *et al.*, 2000
DFM842R	S-*-Dfm-0842-a-A-20	58	54	ATA CCC SCW WCW CCT AGC AC	domain *Bacteria* (115/123706) genus *Desulfotomaculum* (42/84)	Some *Desulfotomaculum* cluster I[i] and some other *Firmicutes*	Daly *et al.*, 2000

Table 2.1 (*cont.*)

Short name	Full name[a]	Suggested annealing temperature [°C]	Calculated melting temperature T_m [°C][b]	Sequence 5′–3′	Coverage according to RDP II probe match (perfectly matched hits/total searched)[c]	Target group	Reference
					genus *Pelotomaculum* (9/104) genus *Sporotomaculum* (5/5)		
DFSPOS219F[d]	S-G-Dfspos-0219-a-S-18	60	50	CGA TTA TGG ATG GAC CCG	domain *Bacteria* (48/152299) genus *Desulfosporosinus* (47/58)	*Desulfosporosinus*	Kjeldsen *et al.*, 2007
TDSV1329R[e]	S-G-Tdsv-1329-a-A-17	67	49	AGC GAT TCC GGG TTC AC	domain *Bacteria* (25/95191) genus *Thermodesulfovibrio* (24/27)	*Thermodesulfovibrio*	This chapter

TDSBM1361R[e]	S-G-Tdsbm-1361-a-A-16	67	49	ATT CAC GGC GGC ATG C	domain *Bacteria* (42/87018) genus *Thermodesulfobacterium* (37/40)	*Thermodesulfobacterium*	This chapter
ARGLO36F[d]	S-G-Arglo-0036-a-S-17	52	49	CTA TCC GGC TGG GAC TA	*Archaea* are not represented in RDP II	*Archaeoglobus*	Loy *et al.*, 2002

[a]Name of 16S rRNA gene-targeted oligonucleotide primer based on the nomenclature of Alm *et al.*, 1996.

[b]Melting temperature was calculated with the Oligonucleotide Properties Calculator (http://www.basic.northwestern.edu/biotools/OligoCalc.html) using the following equation: $T_m = 64.9 + 41*(yG+zC-16.4)/(wA+xT+yG+zC)$; under the assumption that annealing occurs under the standard conditions of 50 nM primer, 50 mM Na^+, and pH 7.0.

[c]RDP II probe match was performed with database release 9.34 (1 December 2005) containing 194 696 bacterial 16S rRNA sequences. The search for each probe was restricted to sequences with data in the respective probe binding region.

[d]Forward primers DSBAC355F, DFSPOS219F, and ARGLO36F are used together with the general prokaryotic reverse primer PROKA1492R.

[e]The recommended annealing temperature is 52°C when the primer is used with forward primer ARGLO36F and the annealing temperature is 60°C when the primer is used with forward primer DSBAC355F.

[f]Reverse primers TDSV1329R and TDSBM1361R are used together with the general bacterial forward primer BACT11F.

[g]Forward primer DSBAC355F should be applied with a competitive primer (5′-CAG TGG GGA ATT TTG CGC-3′ with a modified 3′ end e.g. didesoxynucleotide to prevent elongation) to ensure specificity (Scheid and Stubner, 2001).

[h]To ensure specific detection of deltaproteobacterial groups, forward primers SRB385R and SRB385DbR should only be applied with a specific reverse primer and not with a general bacterial reverse primer.

[i]Cluster designation of Gram-positive, spore-forming SRM according to Stackebrandt *et al.*, 1997.

(bi)sulphite reductase (Geets *et al.*, 2006). The principle of the technique has been well reviewed and is not elaborated here. Briefly, the method is based on the separation of different DNA fragments of approximately the same length but having different sequence composition (e.g. as derived from general amplification of 16S rRNA genes from an environmental sample) using gel electrophoresis. The combination of DGGE and population-specific probing has been used in multiple studies to resolve SRM diversity in both natural and engineered habitats (Matsui *et al.*, 2004; Okabe *et al.*, 2002; Santegoeds *et al.*, 1998). A slight variation of the technique is the use of nested PCR-DGGE for improved sensitivity of detection (Dar *et al.*, 2005).

2.2.4 Terminal restriction fragment length polymorphism

Terminal restriction fragment length polymorphism (T-RFLP) is a method based on PCR amplification of a conserved target sequence (most commonly a region of the 16S rRNA gene). One of the two PCR primers is fluorescent-labelled at the 5′ end, resulting in PCR amplification products which are tagged with a fluorescent dye at only one terminus. Following restriction enzyme digestion the fragments are resolved on a standard DNA sequencer, detecting only the terminal fragments that retain the fluorescent tag. On average 30–50 predominant terminal-restricted fragments are observed within a microbial ecosystem, and terminal fragment abundance might provide some information about relative abundance of the different sequence types. Primers specific for the archaeal or bacterial domains are most commonly used in the initial PCR reaction (Liu *et al.*, 1997; van der Maarel *et al.*, 1998). The general approach has also been used to assess the relative metabolic activity of populations by comparing T-RFLP fingerprints derived using either the rRNA gene or its transcript as initial template (Lümann *et al.*, 2000). Another approach to identifying metabolically active population combines T-RFLP with stable-isotope probing (Lueders *et al.*, 2004). As described for the DGGE and temperature gradient gel electrophoresis (TGGE) methods, T-RFLP has been combined with selective amplification of specific populations, including SRM (e.g. Wieland *et al.*, 2003). Software has been developed to infer the possible phylogenetic affiliation of predominant terminal restriction fragments within a sample, including TAP T-RFLP (http://35.8.164.52/html/TAP-trflp.html), T-RFLP Phylogenetic Assignment Tool (PAT) (http://trflp.limnology.wisc.edu/index.jsp), Microbial Community Analysis (MiCA) (http://mica.ibest.uidaho.edu/), tRFLP fragment sorter version 4.0 (http://www.oardc.ohio-state.edu/trflpfragsort/default.htm), and TRF-CUT (Ricke *et al.*, 2005).

In addition to the selective amplification of 16S rRNA genes affiliated with SRM, T-RFLP has also been used to selectively survey SRM by amplification of the highly conserved genes encoding the dissimilatory (bi)sulphite reductase. However, since this approach is dependent upon degenerated primers for amplification, it is subject to multiple biases. Most notably, the primers cannot achieve uniform amplification and have a tendency to amplify non-target sequences (Wagner *et al.*, 2005). Conclusions from studies that rely solely on the supposed fidelity of the amplification must be carefully interpreted (e.g. Watras *et al.*, 2005).

2.3 QUANTITATIVE MEMBRANE HYBRIDIZATION

The design of probes was, and continues to be, directed by phylogenetic inference based on comparison of near-complete 16S rRNA sequences and certain general rules of probe design that we do not address in this chapter. Since phylogenetic affiliation is determined by nucleotide sequence common within groups, short unique sequence motifs (probe targets of approximately 15–25 nucleotides) can be identified for most discrete clades. The first relatively complete set of probes for SRB was developed by Devereux *et al.* (1992) using the limited collection of 16S rRNA sequences available at that time (Devereux *et al.*, 1989). This initial set of probes continues to receive application, but has been modified and expanded with the availability of a more extensive collection of sequences for SRM (Table 2.2). These probes have been used for both DNA- and RNA-targeted hybridization, but have received greatest application in two formats for hybridizing native rRNA, either fluorescence *in situ* hybridization (FISH) or membrane hybridization. The accessibility of different sites within the ribosome is not as important with membrane hybridization as for application in FISH (discussed below), since the rRNA is essentially devoid of associated protein and denatured before immobilization on the membrane. Today a proposed standard for naming rRNA-targeted oligonucleotide probes is available (Alm *et al.*, 1996) together with an online resource for archiving probe sequences and associated performance information (Loy *et al.*, 2003).

2.3.1 Probe labelling and membrane hybridization

There are three general formats for labelling probes: incorporation of a radioactive label (generally phosphate); direct attachment of a fluorescent dye; or a ligand recognized by a reporter system (e.g. biotin or digoxigenin labelled). Although there are greater biosafety and disposal concerns

Table 2.2. *16S rRNA gene-targeted oligonucleotide probes for the detection of different taxa of SRM by fluorescence* in situ *hybridization*[a]

Probe name	Binding position[b]	FA [%] for standard FISH[c]	FA [%] for CARD-FISH[c]	Sequence 5′–3′	Coverage according to RDP II probe match (perfectly matched hits/total searched)[d]	Target group	Reference
SRB385 (SRB)[e]	385–402	35	N.D.	CGG CGT CGC TGC GTC AGG	domain *Bacteria* (4355/161606) class *Deltaproteobacteria* (1032/3822) family *Desulfovibrionaceae* (314/457) family *Desulfobulbaceae* (450/511)	Some *Deltaproteobacteria* and other *Bacteria*	Amann *et al.*, 1990; Manz *et al.*, 1998
SRB385Db[e]	385–402	30	N.D.	CGG CGT TGC TGC GTC	domain *Bacteria* (4558/161606) class *Deltaproteobacteria* (1583/3822)	Some *Deltaproteobacteria* and other *Bacteria*	Rabus *et al.*, 1996

				AGG	family *Desulfobacteraceae* (561/683) family *Syntrophobacteraceae* (47/174)		
DSS658	658–678	60	80	TCC ACT TCC CTC TCC CAT	domain *Bacteria* (576/141774) family *Desulfobacteraceae* (472/921) genus *Desulfococcus* (9/16) genus *Desulfofaba* (2/4) genus *Desulfofrigus* (14/14) genus *Desulfomusa* (2/2) genus *Desulfonema* (1/14)	Some *Desulfobacteraceae* and other *Bacteria*	Manz *et al.*, 1998; Mussmann *et al.*, 2005

Table 2.2 (*cont.*)

Probe name	Binding position[b]	FA [%] for standard FISH[c]	FA [%] for CARD-FISH[c]	Sequence 5′–3′	Coverage according to RDP II probe match (perfectly matched hits/total searched)[d]	Target group	Reference
					genus *Desulfosarcina* (142/154)		
804 (Dsb804)	804–821	No signal[f]	N.D.	CAA CGT TTA CTG CGT GGA	domain *Bacteria* (277/129089) family *Desulfobacteraceae* (264/887) genus *Desulfobacter* (28/31) genus *Desulfocella* (3/5) genus *Desulfococcus* (4/15) genus *Desulfofaba* (1/3)	Some *Desulfobacteraceae*	Devereux *et al.*, 1992; Rabus *et al.*, 1996

					genus *Desulfofrigus* (13/14) genus *Desulfonema* (6/14) genus *Desulforegula* (2/2) genus *Desulfosarcina* (94/146)		
DSB985	985–1004	20	N.D.	CACAGGATGTCAAACCCAG	domain *Bacteria* (76/108300) family *Desulfobacteraceae* (73/510) genus *Desulfobacter* (20/21) genus *Desulfobacula* (29/34) genus *Desulfospira* (1/1) genus *Desulfotignum* (10/10)	Most *Desulfobacter, Desulfobacula, Desulfospira* and *Desulfotignum*	Manz *et al.*, 1998

Table 2.2 (*cont.*)

Probe name	Binding position[b]	FA [%] for standard FISH[c]	FA [%] for CARD-FISH[c]	Sequence 5′–3′	Coverage according to RDP II probe match (perfectly matched hits/total searched)[d]	Target group	Reference
129 (DSB129)	129–146	15	N.D.	CAG GCT TGA AGG CAG ATT	domain *Bacteria* (15/142899) genus *Desulfobacter* (14/20)	Most *Desulfobacter*	Devereux *et al.*, 1992; Ramsing *et al.*, 1996
Dsb220 (desulfo-bacter)	220–239	N.D.	N.D.	TMC GCA RAC TCA TCC CCA AA	domain *Bacteria* (15/152335) family *Desulfobacteraceae* (15/577) genus *Desulfobacter* (14/20)	Most *Desulfobacter*	Amann *et al.*, 1990
221(DBM221)	221–240	35	N.D.	TGC GCG GAC TCA TCT TCA AA	domain *Bacteria* (8/152577) family *Desulfo-bacteraceae* (8/577)	*Desulfo-bacterium vacuolatum, D. niacini, D. auto-trophicum*	Devereux *et al.*, 1992; Manz *et al.*, 1998

DSS225	225–242	40	80	TGG TAC GCG GGC TCA TCT	domain *Bacteria* (164/152577) class *Deltaproteobacteria* (162/3383) family *Desulfobacteraceae* (155/577) genus *Desulfosarcina* (28/44)	Some *Desulfosarcina* and other *Desulfo-bacteraceae*	Mussmann *et al.*, 2005; Ravenschlag *et al.*, 2000
814 (Dscoc814)	814–883	No signal[e]	N.D.	ACC TAG TGA TCA ACG TTT	domain *Bacteria* (69/128568) family *Desulfobacteraceae* (66/879) genus *Desulfococcus* (4/15) genus *Desulfonema* (6/14)	Few *Desulfo-bacteraceae*	Devereux *et al.*, 1992; Ramsing *et al.*, 1996

Table 2.2 (*cont.*)

Probe name	Binding position[b]	FA [%] for standard FISH[c]	FA [%] for CARD-FISH[c]	Sequence 5′–3′	Coverage according to RDP II probe match (perfectly matched hits/total searched)[d]	Target group	Reference
					genus *Desulfosarcina* (26/146)		
cl81-644	644–661	25	55	CCC ATA CTC AAG TCC CTT	domain *Bacteria* (71/144572) family *Desulfobacteraceae* (69/936) genus *Desulfosarcina* (11/162)	Few *Desulfobacteraceae*	Mussmann *et al.*, 2005; Ravenschlag *et al.*, 2000
DSS449	449–466	N.D.	45	TTA GCA TAC TGC AGG TTC	domain *Bacteria* (30/160520) family *Desulfobacteraceae* (30/918) genus *Desulfosarcina* (29/155)	Few *Desulfosarcina*	Mussmann *et al.*, 2005

DSS138	138–155	N.D.	50	CGG GTT ATC CCC GAT TCG	domain *Bacteria* (15/144246) family *Desulfobacteraceae* (15/551)	Few *Desulfobacteraceae*	Mussmann *et al.*, 2005
DSC193	193–210	35	N.D.	AGG CCA CCC TTG ATC CAA	domain *Bacteria* (21/150278) genus *Desulfosarcina* (21/43)	*Desulfosarcina variabilis*	Ravenschlag *et al.*, 2000
DSF672	672–690	45	70	CCT CTA CAC CTG GAA TTC C	domain *Bacteria* (47/140736) order *Desulfobacterales* (30/1618) family *Desulfobacteraceae* (20/919) genus *Desulfofaba* (2/4) genus *Desulfofrigus* (14/14)	*Desulfofrigus, Desulfofaba gelida, Desulfomusa hansenii*	Mussmann *et al.*, 2005, Ravenschlag *et al.*, 2000

Table 2.2 (*cont.*)

Probe name	Binding position[b]	FA [%] for standard FISH[c]	FA [%] for CARD-FISH[c]	Sequence 5′–3′	Coverage according to RDP II probe match (perfectly matched hits/total searched)[d]	Target group	Reference
					genus *Desulfomusa* (2/2) family *Desulfobulbaceae* (8/629) genus *Desulfotalea* (5/22)		
DCC209	209–226	25	N.D.	CCC AAA CGG TAG CTT CCT	domain *Bacteria* (3/152178) genus *Desulfococcus* (3/7)	*Desulfococcus multivorans*	Ravenschlag *et al.*, 2000

DNMA657	657–676	30	N.D.	TTC CGC TTC CCT CTC CCA TA	domain *Bacteria* (61/141774) class *Deltaproteobacteria* (42/4334) family *Desulfobacteraceae* (42/921) genus *Desulfonema* (10/14)	Many *Desulfonema* and other *Delta- and Gamma proteobacteria*	Fukui *et al.*, 1999
OalgDEL136	136–153	N.D.	N.D.	GTT ATC CCC GAC TCG GGG	domain *Bacteria* (1/144246) family *Desulfobacteraceae* (1/551)	Deltaproteo- bacterial symbiont of *Olavius algarvensis*	Dubilier *et al.*, 2001
OcraDEL1	467–484	20	N.D.	CGT CAG CAC CTG GTG ATA	domain *Bacteria* (1/157182) family *Desulfobacteraceae* (1/914)	Deltaproteo- bacterial symbiont 1 of *Olavius crassitu- nicatus*	Blazejak *et al.*, 2005

Table 2.2 (*cont.*)

Probe name	Binding position[b]	FA [%] for standard FISH[c]	FA [%] for CARD-FISH[c]	Sequence 5′–3′	Coverage according to RDP II probe match (perfectly matched hits/total searched)[d]	Target group	Reference
OcraDEL2	443–471	20	N.D.	CAT GCA GAT TCT TCC CAC	domain *Bacteria* (1/160770) family *Desulfobacteraceae* (1/919)	Deltaproteobacterial symbiont 2 of *Olavius crassitunicatus*	Blazejak *et al.*, 2005
DSMA488	488–507	60	N.D.	GCC GGT GCT TCC TTT GGC GG	domain *Bacteria* (46/152818) order *Syntrophobacterales* (33/291) family *Syntrophaceae* (32/113) genus *Syntrophus* (32/113)	*Desulfarculus baarsii*, *Desulfomonile tiedjei*, and some *Syntrophus*	Manz *et al.*, 1998

DsmA455	455–480	20	N.D.	AGT TCY CTG AGC TAT TTA CTC AAA GA	domain *Bacteria* (5/160038) class *Deltaproteobacteria* (5/4331) family *Syntrophobacteraceae* (4/175)	*Desulfomonile*-related Lake Cadagno clones 618, 624, 626 and 651	Tonolla *et al.*, 2005
DsmB455	455–480	20	N.D.	AGA TCC CTG AGC TAT TTA CTC AAG GA	domain *Bacteria* (1/160038) family *Syntrophobacteraceae* (1/175)	*Desulfomonile*-related Lake Cadagno clone 650	Tonolla *et al.*, 2005
DSR651	651–668	35	70	CCC CCT CCA GTA	domain *Bacteria* (262/143696) phylum *Proteobacteria* (255/56830)	Some *Desulfobulbaceae*	Manz *et al.*, 1998; Mussmann *et al.*, 2005

Table 2.2 (*cont.*)

Probe name	Binding position[b]	FA [%] for standard FISH[c]	FA [%] for CARD-FISH[c]	Sequence 5′–3′	Coverage according to RDP II probe match (perfectly matched hits/total searched)[d]	Target group	Reference
				CTC AAG	family *Desulfobulbaceae* (250/633) genus *Desulfocapsa* (21/62) genus *Desulfofustis* (10/11) genus *Desulforhopalus* (76/114) genus *Desulfotalea* (1/22)		
Sval428	428–446	25	N.D.	CCA TCT GAC AGG ATT	domain *Bacteria* (281/161436) family *Desulfobulbaceae* (279/587)	Some *Desulfobulbaceae* (excluding most *Desulfobulbus*)	Mussmann *et al.*, 2005; Sahm *et al.*, 1999

				TTAC	genus *Desulfocapsa* (19/55) genus *Desulfofustis* (11/13) genus *Desulforhopalus* (85/102) genus *Desulfotalea* (20/22)		
660 (DBB660)	660–679	60	N.D.	GAATTCCACTTTCCCCTCTG	domain *Bacteria* (62/141774) genus *Desulfobulbus* (56/114)	Some *Desulfobulbus*	Devereux *et al.*, 1992; Manz *et al.*, 1998
DSR1256	1256–1273	N.D.	10	ACAGGTCGCCCTGTCGCT	domain *Bacteria* (36/99849) phylum *Proteobacteria* (36/37858) class *Deltaproteobacteria* (12/2495)	Few *Desulfobulbaceae* and some *Alphaproteobacteria*	Mussmann *et al.*, 2005

Table 2.2 (*cont.*)

Probe name	Binding position[b]	FA [%] for standard FISH[c]	FA [%] for CARD-FISH[c]	Sequence 5′–3′	Coverage according to RDP II probe match (perfectly matched hits/total searched)[d]	Target group	Reference
					family *Desulfobulbaceae* (12/311)		
Dblb1243	1243–1260	10	N.D.	GCG TGC CCT CTG TCT ATG	domain *Bacteria* (0/100285)[i]	*Desulfobacterium catecholicum*-related strains LacK1, LacK4 and LacK9	Mussmann *et al.*, 2005
Dcap1031	1031–1048	10	N.D.	TGT CAC CAA GCT CCT CTA	domain *Bacteria* (6/106341) genus *Desulfocapsa* (6/47)	Few *Desulfocapsa* including strain LacK10	Mussmann *et al.*, 2005

Dblb1032	1032–1049	10	N.D.	ACC TGT CAC CGA GCT CCT	domain *Bacteria* (7/106341) phylum *Proteobacteria* (7/41102) class *Deltaproteobacteria* (5/2747) family *Desulfobulbaceae* (4/369) genus *Desulforhopalus* (2/78)	*Desulfobacterium catecholicum* and related strains LacK1 and LacK9	Mussmann *et al.*, 2005
DSR186	186–203	N.D.	50	GCC ACC TTT CCT GAT AAA	domain *Bacteria* (22/149506) family *Desulfobulbaceae* (22/486) genus *Desulforhopalus* (14/79) genus *Desulfotalea* (1/20)	*Desulfobacterium catecholicum* and related strains LacK1 and LacK9	Mussmann *et al.*, 2005

Table 2.2 (*cont.*)

Probe name	Binding position[b]	FA [%] for standard FISH[c]	FA [%] for CARD-FISH[c]	Sequence 5′–3′	Coverage according to RDP II probe match (perfectly matched hits/total searched)[d]	Target group	Reference
DSC213	213–230	30	N.D.	CCT CCC TGT ACG ATA GCT	domain *Bacteria* (18/152318) genus *Desulfocapsa* (18/50)	*Desulfocapsa thiozymogenes*	Tonolla *et al.*, 2000
DSC441	441–459	30	N.D.	ATT ACA CTT CTT CCC ATC C	domain *Bacteria* (9/160770) genus *Desulfocapsa* (9/55)	Some *Desulfocapsa*	Tonolla *et al.*, 2000
SRB441	441–459	5	N.D.	CAT GCA	domain *Bacteria* (4/160770)	Few *Desulfo-bulbaceae*	Tonolla *et al.*, 2000

				CTTCTTTCCACTT	family *Desulfobulbaceae* (4/587)		
OcraDEL3	999–1016	20	N.D.	TTTCATAGAGCTTCCCGG	domain *Bacteria* (2/108094) family *Desulfobulbaceae* (2/372) genus *Desulforhopalus* (1/77)	Deltaproteobacterial symbiont 3 of *Olavius crassitunicatus*	Blazejak *et al.*, 2005
687 (DSV687)	687–702	15	N.D.	TACGGATTTCACTCCT	domain *Bacteria* (1004/139671) order *Desulfovibrionales* (588/679) order *Desulfuromonales* (243/474)	Most *Desulfovibrionales* (excluding Lawsonia) and many *Desulfuromonales*	Devereux *et al.*, 1992; Ramsing *et al.*, 1996
DSV (DSV321)	321–336	N.D.	N.D.	TGGGCCGTGTTTN	domain *Bacteria* (474/154020) phylum *Proteobacteria* (315/63004)	Some *Desulfovibrionaceae*, *Desulfomicrobiaceae*,	Küsel *et al.*, 1999

Table 2.2 (*cont.*)

Probe name	Binding position[b]	FA [%] for standard FISH[c]	FA [%] for CARD-FISH[c]	Sequence 5′–3′	Coverage according to RDP II probe match (perfectly matched hits/total searched)[d]	Target group	Reference
				CAG T	class *Deltaproteobacteria* (264/3486) order *Desulfovibrionales* (241/521) family *Desulfovibrionaceae* (216/398) genus *Desulfovibrio* (166/296) genus *Bilophila* (2/18) family *Desulfomicrobiaceae* (11/91)	*Desulfohalo-biaceae*, and other *Bacteria*	

					genus *Desulfomicrobium* (11/91) family *Desulfohalobiaceae* (10/18) genus *Desulfohalobium* (3/3) genus *Desulfonatronovibrio* (5/11)		
DSV698[g]	698–717	35	40	GTT CCT CCA GAT ATC TAC GG	domain *Bacteria* (271/139311) order *Desulfovibrionales* (263/681) family *Desulfovibrionaceae* (259/489) genus *Desulfovibrio* (214/398)	Some *Desulfo-vibrio*, *Bilophila wadsworthia*, and *Lawsonia intracellularis*	Manz *et al.*, 1998; Mussmann *et al.*, 2005

Table 2.2 (*cont.*)

Probe name	Binding position[b]	FA [%] for standard FISH[c]	FA [%] for CARD-FISH[c]	Sequence 5′–3′	Coverage according to RDP II probe match (perfectly matched hits/total searched)[d]	Target group	Reference
					genus *Bilophila* (14/16) genus *Lawsonia* (7/7)		
DSV1292	1292–1309	35	N.D.	CAA TCC GGA CTG GGA CGC	domain *Bacteria* (109/98009) order *Desulfovibrionales* (108/472) family *Desulfovibrionaceae* (107/365) genus *Desulfovibrio* (86/294) genus *Bilophila* (12/14)	Some *Desulfovibrio* and *Bilophila wadsworthia*	Manz *et al.*, 1998

DSV407	407–424	50	N.D.	CCG AAG GCC TTC TTC CCT	domain *Bacteria* (280/162399) phylum *Proteobacteria* (53/66508) family *Desulfovibrionaceae* (42/473) genus *Desulfovibrio* (42/366)	Few *Desulfovibrio* and other *Bacteria*	Manz *et al.*, 1998
DSD131	131–148	20	N.D.	CCC GAT CGT CTG GGC AGG	domain *Bacteria* (1/143587) genus *Desulfovibrio* (1/264)	"*Desulfovibrio aestuarii*"	Manz *et al.*, 1998
DSV185	185–202	10	N.D.	GCC CCC TTT CCC GTT	domain *Bacteria* (14/148919) order *Desulfovibrionales* (13/456)	*Desulfovibrio acrylicus*	Mussmann *et al.*, 2005

Table 2.2 (*cont.*)

Probe name	Binding position[b]	FA [%] for standard FISH[c]	FA [%] for CARD-FISH[c]	Sequence 5′–3′	Coverage according to RDP II probe match (perfectly matched hits/total searched)[d]	Target group	Reference
				TCC	family *Desulfovibrionaceae* (12/363)		
DSV445	445–462	10	N.D.	GAA CCA CAG TTT CTT CCC	domain *Bacteria* (0/160770)[h]	*Desulfovibrio* strain *EtOHK3*	Mussmann *et al.*, 2005
DSV64	64–81	10	N.D.	AAG AGG CCG TTC TCG CTC	domain *Bacteria* (0/121306)[h]	*Desulfovibrio* strain *EtOHK2*	Mussmann *et al.*, 2005

DSBO224	224–242	60	N.D.	GGG ACG CGG ACT CAT CCT C	domain *Bacteria* (14/152577) family *Desulfovibrionaceae* (8/371) genus *Desulfovibrio* (8/278) family *Desulfobacteraceae* (2/577)	*Desulfobotulus sapovorans, Desulfovibrio fairfieldensis,* and other *Deltaproteobacteria*	Manz *et al.*, 1998
DSV214	214–230	10	N.D.	CAT CCT CGG ACG AAT GC	domain *Bacteria* (62/152318) genus *Desulfomicrobium* (62/76)	Most *Desulfomicrobium*	Manz *et al.*, 1998
Dtm229	229–246	15	N.D.	AAT GGG ACG CGG AYC CAT[i]	domain *Bacteria* (255/152849) phylum *Firmicutes* (247/35709) genus *Desulfotomaculum* (71/81)	*Desulfotomaculum cluster I*[i] *and other Firmicutes*	Hristova *et al.*, 2000

Table 2.2 (*cont.*)

Probe name	Binding position[b]	FA [%] for standard FISH[c]	FA [%] for CARD-FISH[c]	Sequence 5′–3′	Coverage according to RDP II probe match (perfectly matched hits/total searched)[d]	Target group	Reference
					genus *Pelotomaculum* (71/93) genus *Sporotomaculum* (2/3)		
Dtm(bcd)230	230–247	10	N.D.	TAA TGG GAC GCG GAC CCA	domain *Bacteria* (217/152849) phylum *Firmicutes* (209/35709) genus *Desulfotomaculum* (48/81) genus *Pelotomaculum* (69/93) genus *Sporotomaculum* (2/3)	Many *Desulfotomaculum* cluster I[i] and other *Firmicutes*	Hristova *et al.*, 2000

DEM1164r	1164–1181	10	N.D.	CCT TCC TCC GTT TTG TCA	domain *Bacteria* (3455/102634) phylum *Firmicutes* (3314/25558) class *Clostridia* (3290/14998) genus *Clostridium* (286/1017) genus *Acetivibrio* (7/217) genus *Sporobacter* (735/847) genus *Anaerofilum* (338/350) genus *Ruminococcus* (187/417) genus *Eubacterium* (17/262) genus *Desulfotomaculum* (75/78) genus *Pelotomaculum* (48/65)	*Desulfotomaculum* cluster I[i] and other *Firmicutes*	Imachi *et al.*, 2006; Stubner and Meuser, 2000

					genus *Sporotomaculum* (3/3)		
Ih820[j]	820–837	20	N.D.	ACC TCC TAC ACC TAG CAC	domain *Bacteria* (94/127903) phylum *Firmicutes* (93/30650) family *Peptococcaceae* (81/361) genus *Pelotomaculum* (81/104)	*Desulfotomaculum* subcluster Ih (*Pelotomaculum* and *Cryptanaerobacter*)	Imachi *et al.*, 2006
TGP690	690–709	15	N.D.	CTC AAG TCC CTC AGT TTC AA	domain *Bacteria* (11/139671) phylum *Firmicutes* (7/32174) family *Peptococcaceae* (4/366) genus *Pelotomaculum* (4/104)	*Pelotomaculum thermopropionicum*	Imachi *et al.*, 2000
Tdesulfo848	848–866	20–30	N.D.	TTT CCC	domain *Bacteria* (18/122872)	Most *Thermodesulfovibrio*	Daims *et al.*, 2000

TTC GGC ACA GAG	phylum *Nitrospira* (18/349) family *Nitrospiraceae* (18/349) genus *Thermodesulfovibrio* (17/28)

[a] An up-to-date list of all SRM-specific probes can be viewed using the 'list probes by category' option of probeBase (http://www.microbialecology.net/probebase) (Loy *et al.*, 2003).

[b] Probe binding position according to *E. coli* 16S rRNA (Brosius *et al.*, 1981).

[c] FA [%]: formamide concentration in the hybridization buffer; N.D., not determined.

[d] RDP II probe match was performed with database release 9.34 (1 December 2005) containing 194 696 bacterial 16S rRNA sequences. The search for each probe was restricted to sequences with data in the respective probe binding region.

[e] To ensure specific detection of deltaproteobacterial groups, probe SRB385 or probe SRB385Db should only be applied simultaneously with another *Deltaproteobacteria* (sub)group-specific probe.

[f] Probe was tested for FISH but it showed no or a very weak fluorescence signal.

[g] Competitor probe for DSV698 (5′-GTT CCT CCA GAT ATC TAC GC-3′).

[h] Target sequence(s) not listed in RDP II.

[i] Cluster designation of Gram-positive, spore-forming SRM according to Stackebrandt *et al.*, 1997.

[j] Competitor probe for Ih820 (5′-ACC TCC TAC ACC TAG TAC-3′).

associated with the use of radioactive materials, this labelling format remains a preferred method for achieving good sensitivity, specificity, and linearity of response. The reader is referred to articles on alternative labelling formats (Davies *et al.*, 2000; Mansfield *et al.*, 1995; Stahl and Amann, 1991; Stahl *et al.*, 1988).

2.3.2 Probe characterization

Although the design of probes is relatively straightforward, confirming that hybridization to environmental nucleic acid of unknown composition derives from hybridization to the intended target remains one of the key challenges in the application of hybridization technology to the characterization of environmental systems. The first requirement is to adjust the hybridization conditions to the highest stringency possible that is compatible with achieving a useful signal. For example, quantification with species-specific probes should yield lower values than those obtained using a probe targeting a higher taxonomic (phylogenetic) level within a group of related organisms (Raskin *et al.*, 1996a). An additional level of characterization that is often no longer achieved, because of the required high investment in obtaining appropriate reference materials, is an empirical survey of the probe behaviour using an array of both target and non-target rRNAs immobilized on the same membrane which is challenged with the probe under optimized conditions of hybridization and washing (Devereux *et al.*, 1992). This level of characterization was achieved for the initial set of probes designed for major groups of SRM and a few other probe sets (e.g. for methanogens and *Fibrobacter* species) (Lin *et al.*, 1994; Raskin *et al.*, 1994).

2.3.3 Linearity of response

The membrane method relies upon an inference of target abundance in a mixture of environmental nucleic acid based on comparison of hybridization to the environmental RNA with a dilution series of an appropriate target rRNA immobilized on the same membrane. The general assumption of linearity of response with increasing application amounts has been demonstrated (Figure 2.1). However, a possibly confounding problem is the presence of inhibitory substances co-purifying with the RNA during extraction from an environmental matrix. These are often humic acids and have been shown to inhibit hybridization when present in significant amounts. The reader is referred to a paper by Alm *et al.* (2000) for a more complete description of this effect.

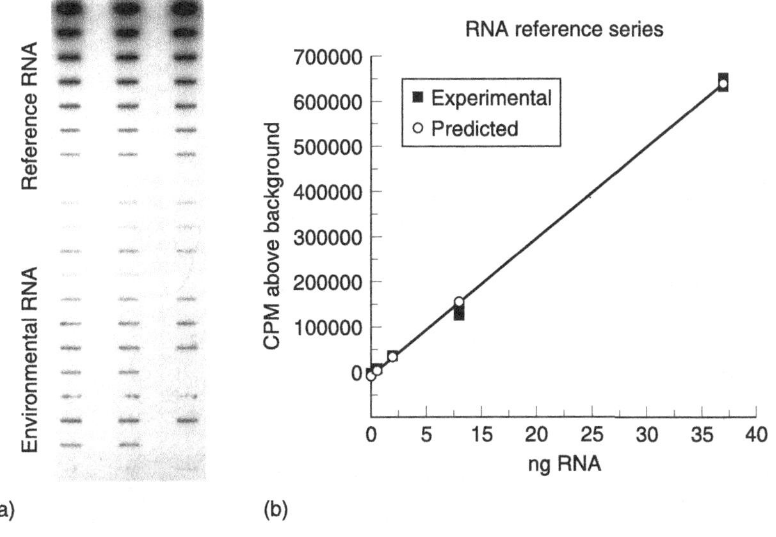

Figure 2.1. Example membrane hybridization of environmental RNA and a dilution series of RNA from a reference culture applied to the same membrane (a). The reference RNA dilution series was used to generate the standard curve (b) for inferring abundance of the target sequence in the environmental sample. All samples were applied in triplicate, and hybridized and washed under established conditions. The amount of radiolabelled probe retained in each application area (slot blot) was determined using a Phosphor Imager (Molecular Dynamics).

2.3.4 Analysis of natural systems

Membrane-based assessments of SRM abundance have been conducted for a variety of habitats in which sulphate respiration is known to be a major process. These have included the analysis of marine and freshwater sediments, stratified water column, gut compartments, and microbial mat communities (Devereux *et al.*, 1996; Koizumi *et al.*, 2004; Minz *et al.*, 1999a; Risatti *et al.*, 1994). One of the first examples of using group-specific probes to demonstrate a well-resolved stratification of SRM populations was a study of a saline microbial mat community (Risatti *et al.*, 1994). In that analysis, RNA extracted from approximately 1 mm depth interval slices of the mat was immobilized on membrane supports and hybridized with an early set of group-specific probes (Figure 2.2). Of particular note was the mostly non-overlapping distribution of different phylotypes (clades of SRM encompassed by individual probes) at different depth intervals. This was one of the

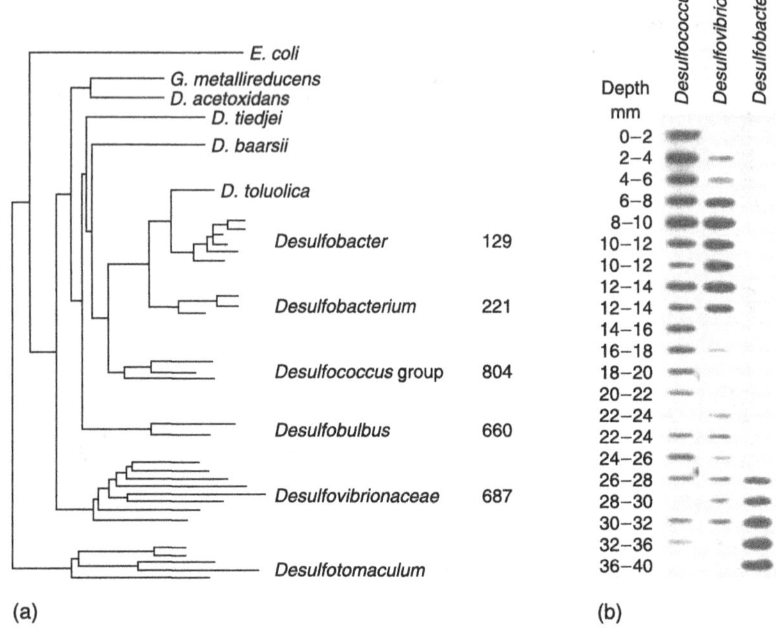

Figure 2.2. Depth profiles of major SRM phylotypes in a saline microbial mat. Group-specific probes (a) were hybridized to RNA extracted from millimetric depth intervals and applied to separate membrane panels (b). Each panel (labelled *Desulfococcus, Desulfovibrio,* and *Desulfobacterium*) was hybridized to the respective probe shown in panel A (adapted from Risatti *et al.*, 1994).

first clear demonstrations of phylogenetic affiliation ("phylotype") being correlated with habitat. This study and others since have consistently shown that SRM affiliated with the *Desulfococcus/Desufosarcina/Desulfonema* clade often dominate in regions of stratified marine environments near the oxic/anoxic boundary and receiving greatest organic input (Llobet-Brossa *et al.*, 2002; Minz *et al.*, 1999a; Ravenschlag *et al.*, 2000).

2.3.5 Engineered systems

One of the most detailed temporal studies of SRM community structure and response to system perturbation was conducted by Lutgarde Raskin as part of her doctoral studies (Raskin *et al.*, 1996b). This analysis employed a large set of group-specific probes for methanogens and SRM populations to examine their abundance in a biofilm reactor system under both sulphidogenic and sulphate-limited conditions. After establishing a stable community under these two conditions, the feed for the sulphate-limited reactor was amended with sulphate and sulphate was removed from the feed of the sulphidogenic reactor. Both reactors were monitored for changes in process and population structure over multiple months. Of particular note was the observation that SRM related to *Desulfovibrio* species sustained a large population under both sulphidogenic and sulphate-limiting conditions, suggesting that they could be sustained in the absence of sulphate via syntrophic association with hydrogenotrophic methanogens. Populations related to *Desulfococcus* species were more competitive under low-sulphate conditions and virtually eliminated within several weeks after sulphate addition to the medium feed.

2.4 FLUORESCENCE *IN SITU* HYBRIDIZATION (FISH) AND COMBINATIONS OF FISH WITH OTHER TECHNIQUES

Cultivation-independent identification and quantification of microorganisms by FISH with fluorescently-labelled rRNA-targeted oligonucleotide probes and subsequent microscopic or flow-cytometric analysis has become a standard method in environmental microbiology (see Amann and Schleifer, 2001; Daims *et al.*, 2005; Stahl and Amann, 1991; Wagner *et al.*, 2003 for some selected reviews). A suite of rRNA-targeted probes is available for FISH-based detection of a few recognized or yet uncultivated SRM groups (Table 2.2) and has been used successfully to visualize and quantify selected SRM in various habitats including oceans, freshwater systems, and waste water as well as in host eukaryotes (Amann *et al.*, 1992; Blazejak *et al.*, 2005; Boetius *et al.*, 2000; Detmers *et al.*, 2004; Dubilier *et al.*, 2001; Kleikemper *et al.*, 2002; Manz *et al.*, 1998; Poulsen *et al.*, 1993; Ramsing *et al.*, 1996; Ravenschlag *et al.*, 2000; Tonolla *et al.*, 2000). Because the specificity of some of these probes (e.g. SRB385 and SRB385Db) is rather low, multiple probes labelled with different dyes should be applied simultaneously in order to improve the detection reliability. However, the detection of SRM via FISH is not straightforward because Gram-negative and Gram-positive SRM co-occur in several habitats (Karnachuk *et al.*, 2006) and thus might require the use of different fixation and pre-treatment protocols (Wagner *et al.*, 2003). Furthermore, fluorescent phylogenetic staining of SRM in samples with high background fluorescence, like cyanobacterial mats (Minz *et al.*, 1999b), is difficult. Another problem is that SRM in their natural habitat often possess low cellular ribosome contents and can therefore only be detected by FISH if signal amplification strategies such as catalyzed reporter deposition

(CARD)-FISH (Pernthaler *et al.*, 2002) are used. This strategy was, for example, used to monitor diversity and vertical distribution of some selected deltaproteobacterial SRM in an intertidal mud flat of the Wadden Sea (Mussmann *et al.*, 2005a). However, one has to consider that even after prolonged starvation SRM retain low but still detectable amounts of rRNA (Fukui *et al.*, 1996). Consequently, rRNA-targeted CARD-FISH is not suited to differentiate between active and inactive SRM in the environment. This limitation might be overcome if recently developed FISH protocols for mRNA detection (Pernthaler and Amann, 2004; Wagner *et al.*, 1998b) are adapted to SRM. Promising target molecules would be the mRNA of the subunits of the dissimilatory (bi)sulphite reductase (*dsrA/dsrB*) or the adenosine-5′-phospho-sulphate reductase (*apsA*). It has already been shown that the cellular *dsrB* mRNA concentration is much more tightly linked to growth rate than the 16S rRNA, and therefore use of *dsrA/dsrB* as molecular markers is superior to 16S rRNA-based approaches for studying the response of SRM to different environmental conditions at the cellular level (Neretin *et al.*, 2003). Alternatively, SRM-specific FISH assays can be combined with techniques that allow one to infer physiological traits of the detected cells or cell clusters, such as microautoradiography (MAR) (Ito *et al.*, 2002; Nielsen and Nielsen, 2002); microsensors (Okabe *et al.*, 1999; Ramsing *et al.*, 1993), and secondary ion mass spectrometry (Orphan *et al.*, 2001; 2002).

2.5 REVERSE SAMPLE GENOME PROBING

Reverse sample genome probing (RSGP) employs the entire genome of a microorganism as a probe for the detection of the corresponding "genotype" in environmental samples or enrichment cultures (reviewed by Greene and Voordouw, 2003). The RSGP method was first described by the Voordouw laboratory as an approach to evaluate the composition of culturable SRM in oil fields.

In initial application of RSGP, a membrane panel of genomes from 16 cultured sulphate-reducing bacteria and four non-SRM heterotrophs was used to characterize produced water and metal corrosion coupons in several Western Canadian oil fields. SRM have long plagued the oil industry, contributing to souring (microbial production of H_2S) and to metal corrosion. Souring problems are reported to be aggravated by the use of water injection for oil recovery, generating a mixture of "produced" water and oil. The oil field study suggested a direct contribution of SRM to corrosion. Consistent with this expectation, this study demonstrated that biofilms on metal surfaces were typically dominated by one to three populations of SRM from the *Desulfovibrionaceae*, with much lower representation by the heterotrophs.

The RSGP method has two principal limitations in its current format. The microbial community is described only in terms of its culturable component and, since detection sensitivity of RSGP is defined by the extent of cross-hybridization of the standard DNAs, low abundance populations that hybridize below this experimentally defined threshold may not be detected in the presence of related organisms. However, the general approach could be extended to the use of large DNA fragments recovered in libraries of environmental DNA. Greene and Voordouw (2003) have suggested a strategy of screening a bacterial artificial chromosome (BAC) library for clones containing 16S rRNA genes. Following sequence determination of the 16S rRNA genes, only those BAC clones would be immobilized on the reference panel. This would serve to characterize much greater genotypic variation among environmental populations than possible using only the 16S rRNA gene sequence and not be dependent upon cultured reference organisms. Since many SRM are easily identified by 16S rRNA sequence type, this modification of method has considerable potential utility for the study of environmental populations.

2.6 THE SRP-PHYLOCHIP: A MICROARRAY FOR DETECTION OF ALL CULTIVATED SRM

As a consequence of the wide and scattered distribution of the capability for sulphate respiration in the prokaryotic parts of the tree of life, many specific rRNA-targeted probes or primers are needed to identify all recognized SRM in an environmental sample. Thus, a census of SRM by the above-mentioned, low-throughput methods requires a high number of separate experiments, rendering the analysis extremely tedious and time-consuming. This limitation can be bypassed by exploiting the miniaturized format of a DNA microarray, which offers the possibility to test for the presence and (within some limits) abundance of many different organisms by simultaneous hybridization of environmentally retrieved nucleic acids to thousands of surface-immobilized gene probes (Bodrossy and Sessitsch, 2004; Loy *et al.*, 2006; Palmer *et al.*, 2006; Stahl, 2004; Zhou, 2003). For SRM monitoring an encompassing 16S rRNA (gene)-targeted oligonucleotide microarray, the SRP-PhyloChip, is available (Loy *et al.*, 2002). With this tool, which carries in its most recent version more than 200 18-mer probes, environmental and clinical microbiologists are able to detect almost all described SRM at different phylogenetic levels in complex samples.

Environmental sample
Microarray analysis

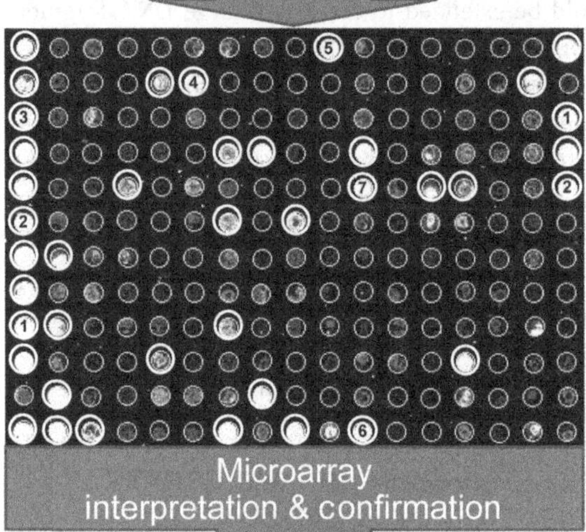
5
4
3
1
7
2
1
6
Microarray
interpretation & confirmation

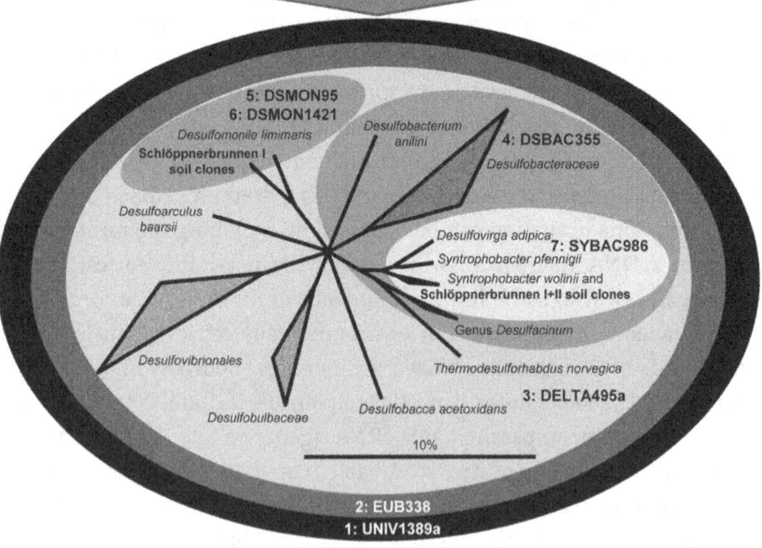
5: DSMON95
6: DSMON1421
Desulfomonile limimaris
Schlöppnerbrunnen I
soil clones
Desulfobacterium
anilini
4: DSBAC355
Desulfobacteraceae
Desulfoarculus
baarsii
Desulfovirga adipica
7: SYBAC986
Syntrophobacter pfennigii
Syntrophobacter wolinii and
Schlöppnerbrunnen I+II soil clones
Genus Desulfacinum
Desulfovibrionales
Thermodesulforhabdus norvegica
3: DELTA495a
Desulfobulbaceae
Desulfobacca acetoxidans
10%
2: EUB338
1: UNIV1389a

The SRP-PhyloChip carries probes with redundant or hierarchical specificity, a design strategy which partly compensates for a general limitation of microarrays, namely that at a defined stringency not all probes bind only to their fully complementary target nucleic acids (Loy and Bodrossy, 2006).

A very appealing feature of a microarray experiment is its low effort-to-output ratio compared with other, more traditional hybridization techniques such as FISH or membrane hybridization. In brief, a typical PhyloChip experiment starts with extraction of total community RNA or DNA from a sample of interest. In order to increase detection sensitivity, rRNA genes are usually amplified by PCR, using the environmental DNA extract as template, and are subsequently labelled with a fluorescent dye. Alternatively, total rRNA can be labelled directly with fluorescent molecules to avoid PCR biases. The labelled target nucleic acids are then hybridized with the PhyloChip, before the resulting hybridization pattern is recorded as a digital image with a fluorescence imager. Finally, image analysis programs are used to extract and interpret the wealth of data obtained in such experiments.

While the greatest benefit of DNA microarrays lies within their high parallelism, this feature also poses problems regarding the specificity, sensitivity, and quantification potential of this hybridization platform (Wagner *et al.*, 2007). However, these limitations are expected to be overcome by technical and data analysis innovations in the future. Currently, the application of diagnostic DNA microarrays in microbial ecology is most fruitful if these are combined with other well-established methods for microbial community analysis and used for rapid diversity pre-screening of environmental samples or for comparative studies involving large sample numbers, such as monitoring of community dynamics after perturbation or biogeographic surveys. For example, the diversity and distribution of SRM in different depths of two acidic fen soils receiving periodically fluctuating amounts of sulphate was analyzed with the SRP-PhyloChip and other complementary techniques (Loy *et al.*, 2004). Among other findings, this study revealed the presence of *Syntrophobacter*- and *Desulfomonile*-related bacteria in some of the fen samples (Figure 2.3). However, it currently

Figure 2.3. SRP-PhyloChip analysis of an acidic fen soil sample (site Schlöppnerbrunnen I, 22.5–30 cm depth; see Loy *et al.*, 2004 for details). Probe spots showing a positive signal on the microarray are indicated with bold white circles. SRP-PhyloChip results were confirmed and extended by selective PCR-based retrieval of 16S rRNA gene sequences from fen soil bacteria and subsequent phylogenetic analysis. Different grey shadings in the 16S rRNA tree and numbers indicate the specificities and the microarray positions, respectively, of relevant SRP-PhyloChip probes. Bar shows 10% estimated sequence divergence.

remains uncertain whether these organisms live/survive from respiration of minute amounts of sulphate or in syntrophy with hydrogenotrophic methanogens. In principle, such hypotheses on the ecophysiology of the identified SRM could also be tested by using the PhyloChip. In the so-called Isotope Array approach (Adamczyk *et al.*, 2003; Wagner *et al.*, 2006), an environmental sample is initially incubated with a ^{14}C-labelled substrate under the desired conditions (e.g. anoxic, with and without sulphate). In the next step, the environmental RNA pool, consisting of radioactive and non-radioactive RNA from active and inactive organisms, is extracted, labelled with fluorescent dyes, fragmented, and hybridized with a PhyloChip. Comparison of fluorescence and radioactivity readouts of the microarray finally allows deciphering of the substrate utilization performance of probe-defined microbial populations.

2.7 EVOLUTION AND ENVIRONMENTAL DIVERSITY OF FUNCTIONAL SRM GENES: IMPLICATIONS FOR SRM COMMUNITY ANALYSES

Initially, 16S rRNA-based approaches were of central importance for cultivation-independent analysis of SRM in the environment, but they restricted the view of microbial ecologists to those SRM which were at least moderately related to cultured SRM. Based on 16S rRNA sequence retrieval it is still impossible to discover novel SRM which are not closely related to known SRM and until recently no cultivation-independent tools were available to hunt for such organisms. This shortcoming of rRNA-based methods (if they are not combined with techniques that allow physiological properties of the identified microorganisms to be inferred (Wagner *et al.*, 2006a)) has forced microbial ecologists to search for alternative phylogenetic marker genes that are (more) specific for SRM. The first marker used for this purpose was the [NiFe] hydrogenase large-subunit gene and its transcript which were selected to monitor the diversity and activity of *Desulfovibrio* species in the environment (Wawer *et al.*, 1997; Wawer and Muyzer, 1995). However this approach cannot be extended to cover all SRM. Consequently, subsequent research focussed on genes encoding enzymes which are part of the unique pathway for dissimilatory reduction of sulphate and are thus common to all SRM. While the sulphate-activating enzyme ATP sulphurylase is also present in the assimilatory sulphate reduction metabolism of many non-SRM and plants (Rabus *et al.*, 2000), adenosine-5′-phosphosulphate (APS) reductase and dissimilatory (bi)sulphite reductase, catalyzing the two and six electron-involving reduction of APS and (bi)sulphite, respectively, are largely confined to SRM (exceptions are discussed below). The evolution of the latter two enzymes and the suitability of the genes encoding them as targets for PCR-based SRM community analyses were thus intensively studied (Friedrich, 2002; Hipp *et al.*, 1997; Karkhoff-Schweizer *et al.*, 1995; Wagner *et al.*, 1998a).

Degenerated PCR primers were designed for *apsA* (or *aprA*), the gene encoding the alpha subunit of APS reductase, and successfully used for the recovery of partial *apsA* sequences from representatives of almost all recognized SRM lineages (Friedrich, 2002). Largely congruent topologies of 16S rRNA and ApsA trees have shown that evolution of *apsA* in most SRM is consistent with a vertical mode of transmission (Friedrich, 2002). However, some SRM (such as members of the family *Syntrophobacteraceae*, the species *Desulfomonile tiedjei*, *Desulfoarculus baarsii*, and *Desulfobacterium anilini*, and the thermophilic genera *Thermodesulfovibrio* and *Thermodesulfobacterium*) have discordant positions in the two trees and unusual insertions/deletions in their ApsA sequences, indicating that they most likely received *apsA* via lateral gene transfer. While this finding in itself complicates the interpretation of environmentally recovered *apsA* sequences, further limitations of using *apsA* as a specific genetic marker for environmental SRM diversity surveys have become evident from genome sequencing projects. Some chemotrophic and phototrophic microorganisms involved in the oxidation of reduced sulphur compounds (sulphur-oxidizing microorganisms, SOM) also contain *aps* genes encoding an APS reductase, which presumably operates in the reverse direction to the homologous enzyme in SRM (Schwenn and Biere, 1979). In contrast to previous assumptions and the proposed functional differences of this enzyme in SRM and SOM (Hipp *et al.*, 1997), SOM and SRM are not separated into two distinct groups but are scattered in the ApsA tree (Figure 2.4). Furthermore, the genome of the chemotrophic SOM *Thiobacillus denitrificans* (Beller *et al.*, 2006) contains two highly divergent copies of *apsA* (ApsA amino acid identity of 48%) (Figure 2.4). The resulting problems regarding the identification and functional assignment of environmental *apsA* sequences are nicely illustrated by the following example. Partial *apsA* sequences were recovered from a hydrothermal sand patch located in shallow marine waters off the island of Dominica (Caribbean Sea) and phylogenetically analyzed (Figure 2.4; unpublished data). While some hydrothermal sand patch clones unambiguously clustered within the families *Desulfobacteraceae* and *Desulfobulbaceae* and thus obviously originate from SRM, the picture is less clear for other clones. For instance, another *apsA* clone group formed an independent lineage between phototrophic SOM of the phylum *Chlorobi* and the

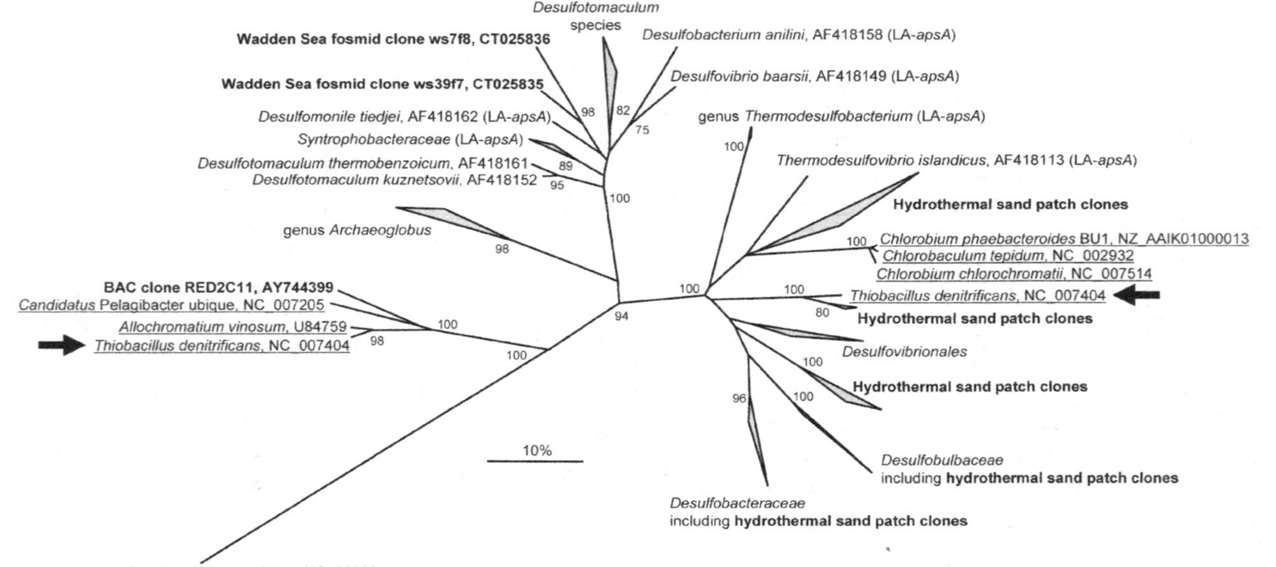

Figure 2.4. Genealogy of adenosine-5′-phosphosulphate reductases (ApsA). The phylogenetic tree was constructed based on 214 ApsA amino acid alignment positions using the neighbour-joining method with the Kimura model of amino acid substitution. Insertions and deletions were not included in the calculation. Parsimony bootstrap values (100 re-samplings) above 75% are shown. Environmental *apsA* clones, retrieved by PCR-based or metagenomic approaches, are depicted in bold. Bar indicates estimated sequence divergence. Prokaryotes known to oxidize reduced sulphur compounds are underlined. *Pyrobaculum aerophilum* contains an authentic frameshift in its *apsA*, reflecting the organism's inability to reduce sulphate (Fitz-Gibbon *et al.*, 2002). LA-*apsA*, sulphate-reducing prokaryotes with a laterally acquired *apsA* according to Friedrich (2002). Arrows highlight that *T. denitrificans* possesses two dissimilar *apsA* gene copies. BAC: bacterial artificial chromosome.

thermophilic SRM *Thermodesulfovibrio islandicus*, making it impossible to decide whether these novel types of ApsA have a function in the reductive or oxidative part of the sulphur cycle.

A similar picture emerges from comparative analyses of *dsrAB*, genes encoding the alpha and beta subunits of the dissimilatory (bi)sulphite reductase in SRM. The phylogenetic position of most organisms matches well in the 16S rRNA and DsrAB trees, with three exceptions (Figure 2.5) (Klein *et al.*, 2001; Wagner *et al.*, 1998a; Zverlov *et al.*, 2005). SRM of the euryarchaeotal genus *Archaeoglobus* (Larsen *et al.*, 1999), the genus *Thermodesulfobacterium*, and some members of the low G+C, Gram-positive phylum *Firmicutes* most likely have laterally acquired versions of *dsrAB* (Klein *et al.*, 2001). It was recently proposed that *Desulfobacterium anilini* is a representative of a deltaproteobacterial lineage whose members could have acted as *dsrAB* donors for the Gram-positive SRM (Figure 2.5) (Zverlov *et al.*, 2005). Our current perception of the evolutionary history of *dsrAB* of SRM is that it was dominated by vertical transmission but also influenced by a few lateral transfer events. The mode and mechanisms of such transfers are not well-understood. Non-matching distribution patterns of laterally acquired *apsA* and *dsrAB* (Figures 2.4 and 2.5) (Friedrich, 2002) and the distant location of *apsA* and *dsrAB* in sequenced SRM genomes suggest that lateral transfers of these genes occurred independent from each other, possibly by xenologous displacement of single genes *in situ* (Omelchenko *et al.*, 2003). However, this notion was recently challenged by the isolation of two large genomic fragments from yet uncultivated SRM thriving in Wadden Sea sediment (Mussmann *et al.*, 2005b). Each fragment contained a "metabolic island" with essential genes for dissimilatory sulphate reduction including *apsA* and *dsrAB* in close proximity. This finding suggests that, in addition to xenologous displacements of single genes, a whole gene set for sulphate reduction might migrate horizontally on a genetically mobile element, possibly transferring this metabolic trait in a single event. The adaptive advantage for a non-SRM, which receives a complete pathway for sulphate reduction *de novo*, is obvious. In contrast, the nature of the selective benefit conferred by the replacement of existing *apsA* or *dsrAB* by homologues from a phylogenetically distant SRM is not known.

Similar to *apsA*, the presence of *dsrAB* is not restricted to SRM, with these genes also present in bacteria which utilize sulphite (e.g. *Desulfitobacterium* species, *Carboxydothermus hydrogenoformans* (Wu *et al.*, 2005)) and/or organosulphonates (e.g. *Bilophila wadsworthia* (Laue *et al.*, 2001; Lie *et al.*, 1999)) but not sulphate as terminal electron acceptor(s). Furthermore, some *Pelotomaculum* and *Sporotomaculum* species, which are

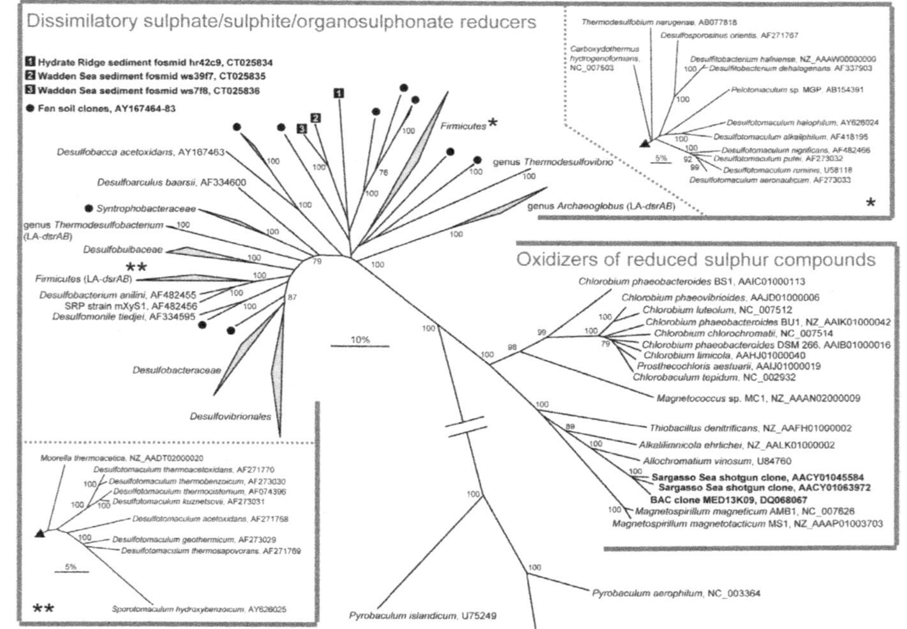

Figure 2.5. Genealogy of sulphite reductases (DsrAB). The phylogenetic tree was constructed based on 542 DsrAB amino acid alignment positions (see Figure 2.4 for further details). Insets indicated by one and two asterisks display the affiliation of *Firmicutes* with vertically and laterally acquired *dsrAB*, respectively. *Chlorobaculum tepidum* TLS contains two closely related copies of *dsrAB* in its genome (Eisen *et al.*, 2002), one of which contains an authentic frameshift in *dsrB*. LA-*dsrAB*, sulphate/sulphite/organosulphonate-reducing prokaryotes with a laterally acquired *dsrAB* according to Klein *et al.* (2001) and Zverlov *et al.* (2005). The long branch connecting *Pyrobaculum* species with the other organisms was interrupted and trimmed for display reasons. BAC: bacterial artificial chromosome.

related to SRM of the spore-forming genus *Desulfotomaculum*, have *dsrAB* but are apparently incapable of reducing sulphate, sulphite or organosulphonates (de Bok *et al.*, 2005; Imachi *et al.*, 2002). For members of the genus *Pelotomaculum* it has been hypothesized that they have lost their capability for sulphate/sulphite respiration as a consequence of their adaptation to a life in low-sulphate, methanogenic environments (Imachi *et al.*, 2006). The existence of *dsrAB* in these syntrophic bacteria could thus represent a genetic remnant of an ancient trait.

Some photo- and chemotrophic SOM also possess *dsrAB*, but these genes code for a reverse siroheme sulphite reductase (Schedel and Trüper, 1980). The involvement of this enzyme in the oxidative sulphur energy metabolism has been proven with mutants of the anoxygenic phototroph *Allochromatium vinosum* (Dahl *et al.*, 2005). It is important to note that not all SOM that have *apsA* contain *dsrAB* and vice versa, indicating that reverse APS reductase and reverse sulphite reductase are not (always) components of the same enzymatic pathway. In contrast to the patchy distribution of SOM and SRM in the ApsA tree, SOM and SRM are clearly separated into two independent groups based on their DsrAB phylogeny (Figure 2.5) (Sabehi *et al.*, 2005). A third deep-branching lineage is formed by DsrAB sequences of crenarcheotal *Pyrobaculum* species (Figure 2.5) (Dhillon *et al.*, 2005), which can grow on sulphite but not on sulphate (Fitz-Gibbon *et al.*, 2002; Molitor *et al.*, 1998). DsrAB from *Pyrobaculum* and from the other organisms depicted in Figure 2.5 might thus represent the archaeal and bacterial sulphite reductase versions, respectively, implying that the invention of DsrAB sulphite reductases occurred prior to the separation of the archaeal and bacterial domains. (N.B. *Archaea* of the genus *Archaeoglobus* acquired their *dsrAB* from the *Bacteria*.) Consequently, the clear dichotomy of the "bacterial" part of the DsrAB tree thus probably reflects a functional adaptation of this enzyme to the reductive and oxidative part of the sulphur cycle. It also renders *dsrAB* more appealing than *apsA* as a target for environmental diversity analyses because degenerated primers for *dsrAB* can be designed that selectively target either the dissimilatory sulphate/sulphite/organosulphonate reducers or SOM.

Novel primer variants and detailed protocols for using APS reductase and dissimilatory (bi)sulphite reductase as functional marker molecules for phylogenetic inventorying of SRM were recently published (Wagner *et al.*, 2005) and are thus not explicitly covered in this book chapter. While only two published SRM diversity studies used *apsA* (Deplancke *et al.*, 2000; Wagner *et al.*, 2005), numerous publications report recovery of *dsrAB* sequences from diverse environmental sources (Baker *et al.*, 2003;

Castro *et al.*, 2002; Chang *et al.*, 2001; Dhillon *et al.*, 2003; Dubilier *et al.*, 2001; Fishbain *et al.*, 2003; Loy *et al.*, 2004; Minz *et al.*, 1999b; Nakagawa *et al.*, 2004) and, recently, fingerprinting techniques for amplified *dsrAB* sequences have become available (Castro *et al.*, 2005; Geets *et al.*, 2006; Perez-Jimenez and Kerkhof, 2005; Wagner *et al.*, 2005). The most intriguing result of these studies was the discovery of many novel DsrAB lineages with only distant relationships to those of cultured SRM, demonstrating the existence of a previously not recognized SRM diversity in nature (including sulphite and organosulphonate reducers) (for an example see Figure 2.5). Since all SRM for which complete genome sequences are available contain only one *dsrAB* per genome, the most parsimonious explanation for the occurrence of these novel *dsrAB* sequences is that they either originate from microorganisms of yet undescribed phylogenetic clades, or from known phyla/classes whose members were thought to be incapable of anaerobically respiring with oxidized sulphur compounds. Deciphering the identities, physiologies, and environmental importance of these novel SRM is one of the big challenges in contemporary SRM research.

2.8 ENVIRONMENTAL GENOMICS AND TRANSCRIPTOMICS

Advances in DNA sequencing technology continue to challenge post-sequencing analysis of the genome sequence data. Current capacity for large genome sequencing facilities is extremely high, for example the US Department of Energy's Joint Genome Institute oversees the Community Sequencing Program (CSP). The CSP solicits proposals for genome sequencing (including specific organisms and DNA recovered from environmental samples) from an international user group. The current capacity of this one high throughput sequencing centre is approximately 20 Gb of DNA sequencing a year for the CSP program alone (http://www.jgi.doe.gov/CSP/). The program in Advanced Sequencing Technology sponsored by the National Human Genome Research Institute (NHGRI) continues to solicit proposals for new technologies that would dramatically reduce the cost of DNA sequencing. For example, 454 Life Sciences has developed an instrument with the capacity to sequence more than 20 million bases in 4.5 hours (www.454.com).

In comparison with these rapid changes in sequencing capacity, genomic studies of SRM remain relatively limited. At this time there are four completed genome sequences (*Archaeoglobus fulgidus, Desulfotalea psychrophila, Desulfovibrio vulgaris subsp. vulgaris* Hildenborough, and

Desulfovibrio desulfuricans G-20) (Heidelberg *et al.*, 2004; Klenk *et al.*, 1997; Rabus *et al.*, 2004). However, at least six genome sequencing projects are in progress, five bacterial (*Thermodesulfovibrio yellowstonii*, *Thermodesulfobacterium commune*, *Desulfotomaculum reducens* MI-1, and *Desulfovibrio vulgaris* DP4) and one archaeal (*Caldivirga maquilingensis*). The reader is directed to a comprehensive web site for continuously update progress on various genome projects (www.genomesonline.org).

The availability of completed genome sequences from pure cultures of this phylogenetically diverse set of SRM will serve as an initial framework to study the ecology and evolution of this key group of microorganisms. The origin(s) and diversification of SRM will certainly be informed by these data but there is an urgent need to complement these data with genomic information on SRM as yet not cultured. This is nicely illustrated by a recent study of Mussmann *et al.* (Mussmann *et al.*, 2005b). While all cultured SRM for which a genome sequence is available have their key genes for sulphate respiration scattered on the genome, this study recovered DNA fragments from a marine sediment which showed that the genome of some previously not recognized SRM contains a clustered set of genes for the dissimilatory reduction of sulphate. This finding has important implications for our perception of modes of horizontal gene transfer in SRM (as discussed above) and reminds us that environmental populations of SRM might differ substantially from those SRM available in culture collections. The recovery of genes associated with SRM from different environmental populations using shotgun or other cloning approaches will provide a broader perspective on their environmental diversity and also provide sequence data for direct characterization of activity and diversity in different environments. For example, this information would foster the development of more comprehensive DNA microarrays for functional genes and, as the technology develops, also serve for highly multiplexed expression analysis (Palumbo *et al.*, 2004; Schadt *et al.*, 2005).

Analysis of gene expression in environmental systems remains in its infancy, a limited number of studies having examined the transcription of highly expressed genes in abundant populations, for example using quantitative PCR (Q-PCR) to measure transcripts of genes encoding for glutamine synthetase (*glnA*) or ribulose-1,5-bisphosphate carboxylase/oxygenase (*rbcL*) in the open marine environment (e.g. Gibson *et al.*, 2006; Wawrik *et al.*, 2002). Transcriptional analysis of SRM has so far been limited to studies of pure cultures using either microarray or Q-PCR methods (e.g. Chhabra *et al.*, 2006; Neretin *et al.*, 2003). However, as techniques are developed that improve sensitivity and precision, we anticipate that the assessments of the environmental activities of SRM will become an attractive system for identifying environmental parameters that determine the activity of specific populations in complex communities such as marine sediments that sustain a diversity of different SRM.

ACKNOWLEDGEMENTS

We acknowledge support from (i) the US Department of Energy's Office of Biological and Environmental Sciences under the GTL-Genomics Program via the Virtual Institute for Microbial Stress and Survival (http://VIMSS.lbl.gov), NSF grant (DEB-0213186) and NASA NAI grant (NCC2-1273) support to D.A.S.; (ii) the European Community (Marie Curie Intra-European Fellowship within the 6th Framework Programme) and the Fonds zur Förderung der wissenschaftlichen Forschung (project P18836-B17) to A.L.; and (iii) the bmb+f (project 01 LC 0021A-TP2 in the framework of the BIOLOG II program) to M.W. D.A.S thanks Jennifer Becker and Liz Alm for providing the materials used for Figure 2.1 and to the many former lab members, including Rudi Amann, Richard Devereux, Lut Raskin, Matt Kane, Lars Poulsen, Rebecca Key, Bill Capman, Norman Fry, Jodi Flax, Greg Brusseau, Barbara MacGregor, Susan Fishbain, Brad Jackson, Dror Minz, Stefan Green, Heidi Gough, and Jesse Dillon, who have contributed in many ways to the material presented in this chapter. A.L. and M.W. are indebted to Michael Klein, Vladimir Zverlov, Angelika Lehner, Natuschka Lee, Stephan Duller, Doris Steger, Stephanie Füreder, Ivan Barisic, Sebastian Lücker, and Christian Baranyi (current and former members of our lab), who have greatly contributed to our work on SRM.

REFERENCES

Adamczyk, J., Hesselsoe, M., Iversen, N. *et al.* (2003). The isotope array, a new tool that employs substrate-mediated labeling of rRNA for determination of microbial community structure and function. *Appl. Environ. Microbiol.*, **69**, 6875–87.

Alm, E. W., Oerther, D. B., Larsen, N., Stahl, D. A. and Raskin, L. (1996). The oligonucleotide probe database. *Appl. Environ. Microbiol.*, **62**, 3557–9.

Alm, E. W. and Stahl, D. A. (2000). Critical factors influencing the recovery and integrity of rRNA extracted from environmental samples: use of an optimized protocol to measure depth-related biomass distribution in freshwater sediments. *J. Microbiol. Methods*, **40**, 153–62.

Alm, E. W., Zheng, D. and Raskin, L. (2000). The presence of humic substances and DNA in RNA extracts affects hybridization results. *Appl. Environ. Microbiol.*, **66**, 4547–54.

Amann, R. and Schleifer, K.-H. (2001). Nucleic acid probes and their application in environmental microbiology. In G. M. Garrity (ed.), *Bergey's Manual of Systematic Bacteriology*. 2nd edn. New York: Springer.

Amann, R. I., Binder, B. J., Olson, R. J. *et al.* (1990). Combination of 16S rRNA-targeted oligonucleotide probes with flow cytometry for analyzing mixed microbial populations. *Appl. Environ. Microbiol.*, **56**, 1919–25.

Amann, R. I., Stromley, J., Devereux, R., Key, R. and Stahl, D. A. (1992). Molecular and microscopic identification of sulphate-reducing bacteria in multispecies biofilms. *Appl. Environ. Microbiol.*, **58**, 614–23.

Baker, B. J., Moser, D. P., MacGregor, B. J. *et al.* (2003). Related assemblages of sulphate-reducing bacteria associated with ultradeep gold mines of South Africa and deep basalt aquifers of Washington State. *Environ. Microbiol.*, **5**, 267–77.

Beller, H. R., Chain, P. S., Letain, T. E. *et al.* (2006). The genome sequence of the obligately chemolithoautotrophic, facultatively anaerobic bacterium *Thiobacillus denitrificans*. *J. Bacteriol.*, **188**, 1473–88.

Blazejak, A., Erseus, C., Amann, R. and Dubilier, N. (2005). Coexistence of bacterial sulfide oxidizers, sulphate reducers, and spirochetes in a gutless worm (*Oligochaeta*) from the Peru margin. *Appl. Environ. Microbiol.*, **71**, 1553–61.

Bodrossy, L. and Sessitsch, A. (2004). Oligonucleotide microarrays in microbial diagnostics. *Curr. Opin. Microbiol.*, **7**, 245–54.

Boetius, A., Ravenschlag, K., Schubert, C. J. *et al.* (2000). A marine microbial consortium apparently mediating anaerobic oxidation of methane. *Nature*, **407**, 623–6.

Brosius, J., Dull, T. L., Sleeter, D. D. and Noller, H. F. (1981). Gene organization and primary structure of a ribosomal operon from *Escherichia coli*. *J. Mol. Biol.*, **148**, 107–27.

Castro, H., Newman, S., Reddy, K. R. and Ogram, A. (2005). Distribution and stability of sulphate-reducing prokaryotic and hydrogenotrophic methanogenic assemblages in nutrient-impacted regions of the Florida Everglades. *Appl. Environ. Microbiol.*, **71**, 2695–704.

Castro, H., Reddy, K. R. and Ogram, A. (2002). Composition and function of sulphate-reducing prokaryotes in eutrophic and pristine areas of the Florida Everglades. *Appl. Environ. Microbiol.*, **68**, 6129–37.

Chang, Y. J., Peacock, A. D., Long, P. E. *et al.* (2001). Diversity and characterization of sulphate-reducing bacteria in groundwater at a uranium mill tailings site. *Appl. Environ. Microbiol.*, **67**, 3149–60.

Chhabra, S. R., He, Q., Huang, K. H. *et al.* (2006). Global analysis of heat shock response in *Desulfovibrio vulgaris* Hildenborough. *J. Bacteriol.*, **188**, 1817–28.

Dahl, C., Engels, S., Pott-Sperling, A. S. *et al.* (2005). Novel genes of the *dsr* gene cluster and evidence for close interaction of Dsr proteins during sulfur oxidation in the phototrophic sulfur bacterium *Allochromatium vinosum. J. Bacteriol.*, **187**, 1392–404.

Daims, H., Nielsen, P. H., Nielsen, J. L., Juretschko, S. and Wagner, M. (2000). Novel *Nitrospira*-like bacteria as dominant nitrite-oxidizers in biofilms from wastewater treatment plants: diversity and *in situ* physiology. *Wat. Sci. Tech.*, **41**, 85–90.

Daims, H., Stoecker, K. and Wagner, M. (2005). Fluorescence in situ hybridization for the detection of prokaryotes. In Osborn, A. M. and Smith C. J. (eds) *Advanced Methods in Molecular Microbial Ecology.* Abingdon, UK: BIOS Scientific Publishers.

Daly, K., Sharp, R. J. and McCarthy, A. J. (2000). Development of oligonucleotide probes and PCR primers for detecting phylogenetic subgroups of sulphate-reducing bacteria. *Microbiology*, **146**, 1693–705.

Dar, S. A., Kuenen, J. G. and Muyzer, G. (2005). Nested PCR-denaturing gradient gel electrophoresis approach to determine the diversity of sulphate-reducing bacteria in complex microbial communities. *Appl. Environ. Microbiol.*, **71**, 2325–30.

Davies, M. J., Shah, A. and Bruce, I. J. (2000). Synthesis of fluorescently labelled oligonucleotides and nucleic acids. *Chem. Soc. Rev.*, **29**, 97–107.

de Bok, F. A., Harmsen, H. J., Plugge, C. M. *et al.* (2005). The first true obligately syntrophic propionate-oxidizing bacterium, *Pelotomaculum schinkii* sp. nov., co-cultured with *Methanospirillum hungatei*, and emended description of the genus *Pelotomaculum. Int. J. Syst. Evol. Microbiol.*, **55**, 1697–703.

Deplancke, B., Hristova, K. R., Oakley, H. A. *et al.* (2000). Molecular ecological analysis of the succession and diversity of sulphate-reducing bacteria in the mouse gastrointestinal tract. *Appl. Environ. Microbiol.*, **66**, 2166–74.

Detmers, J., Strauss, H., Schulte, U. *et al.* (2004). FISH shows that *Desulfotomaculum* spp. are the dominating sulphate-reducing bacteria in a pristine aquifer. *Microb. Ecol.*, **47**, 236–42.

Devereux, R., Delaney, M., Widdel, F. and Stahl, D. A. (1989). Natural relationships among sulphate-reducing eubacteria. *J. Bacteriol.*, **171**, 6689–95.

Devereux, R., Kane, M. D., Winfrey, J. and Stahl, D. A. (1992). Genus- and group-specific hybridization probes for determinative and environmental studies of sulphate-reducing bacteria. *Syst. Appl. Microbiol.*, **15**, 601–9.

Devereux, R., Winfrey, M. R., Winfrey, J. and Stahl, D. A. (1996). Depth profile of sulphate-reducing bacterial ribosomal RNA and mercury methylation in an estuarine sediment. *FEMS Microbiol. Ecol.*, **20**, 23–31.

Dhillon, A., Goswami, S., Riley, M., Teske, A. and Sogin, M. (2005). Domain evolution and functional diversification of sulfite reductases. *Astrobiology*, **5**, 18–29.

Dhillon, A., Teske, A., Dillon, J., Stahl, D. A. and Sogin, M. L. (2003). Molecular characterization of sulphate-reducing bacteria in the Guaymas Basin. *Appl. Environ. Microbiol.*, **69**, 2765–72.

Dubilier, N., Mulders, C., Ferdelman, T. *et al.* (2001). Endosymbiotic sulphate-reducing and sulphide-oxidizing bacteria in an oligochaete worm. *Nature*, **411**, 298–302.

Eisen, J. A., Nelson, K. E., Paulsen, I. T. *et al.* (2002). The complete genome sequence of *Chlorobium tepidum* TLS, a photosynthetic, anaerobic, green-sulfur bacterium. *Proc. Natl. Acad. Sci. USA*, **99**, 9509–14.

Fishbain, S., Dillon, J. G., Gough, H. L. and Stahl, D. A. (2003). Linkage of high rates of sulphate reduction in Yellowstone hot springs to unique sequence types in the dissimilatory sulphate respiration pathway. *Appl. Environ. Microbiol.*, **69**, 3663–7.

Fitz-Gibbon, S. T., Ladner, H., Kim, U. J. *et al.* (2002). Genome sequence of the hyperthermophilic crenarchaeon *Pyrobaculum aerophilum*. *Proc. Natl. Acad. Sci. USA*, **99**, 984–9.

Friedrich, M. W. (2002). Phylogenetic analysis reveals multiple lateral transfers of adenosine-5′-phosphosulphate reductase genes among sulphate-reducing microorganisms. *J. Bacteriol.*, **184**, 278–89.

Fukui, M., Suwa, Y. and Urushigawa, Y. (1996). High survival efficiency and ribosomal RNA decaying pattern of *Desulfobacter latus*, a highly specific acetate-utilizing organism, during starvation. *FEMS Microbiol. Ecol.*, **19**, 17–25.

Fukui, M., Teske, A., Assmus, B., Muyzer, G. and Widdel, F. (1999). Physiology, phylogenetic relationships, and ecology of filamentous sulphate-reducing bacteria (genus *Desulfonema*). *Arch. Microbiol.*, **172**, 193–203.

Geets, J., Borremans, B., Diels, L. *et al.* (2006). DsrB gene-based DGGE for community and diversity surveys of sulphate-reducing bacteria. *J. Microbiol. Methods*, **66**, 194–205.

Gibson, A. H., Jenkins, B. D., Wilkerson, F. P., Short, S. M. and Zehr, J. P. (2006). Characterization of cyanobacterial *glnA* gene diversity and gene expression in marine environments. *FEMS Microbiol. Ecol.*, **55**, 391–402.

Greene, E. A. and Voordouw, G. (2003). Analysis of environmental microbial communities by reverse sample genome probing. *J. Microbiol. Methods*, **53**, 211–19.

Heidelberg, J. F., Seshadri, R., Haveman, S. A. *et al.* (2004). The genome sequence of the anaerobic, sulphate-reducing bacterium *Desulfovibrio vulgaris* Hildenborough. *Nat. Biotechnol.*, **22**, 554–9.

Hipp, W. M., Pott, A. S., Thum-Schmitz, N. *et al.* (1997). Towards the phylogeny of APS reductases and sirohaem sulfite reductases in sulphate-reducing and sulfur-oxidizing prokaryotes. *Microbiology*, **143**, 2891–902.

Hristova, K. R., Mau, M., Zheng, D. *et al.* (2000). *Desulfotomaculum* genus- and subgenus-specific 16S rRNA hybridization probes for environmental studies. *Environ. Microbiol.*, **2**, 143–59.

Imachi, H., Sekiguchi, Y., Kamagata, Y. *et al.* (2002). *Pelotomaculum thermopropionicum* gen. nov., sp. nov., an anaerobic, thermophilic, syntrophic propionate-oxidizing bacterium. *Int. J. Syst. Evol. Microbiol.*, **52**, 1729–35.

Imachi, H., Sekiguchi, Y., Kamagata, Y. *et al.* (2006). Non-sulphate-reducing, syntrophic bacteria affiliated with *Desulfotomaculum* cluster I are widely distributed in methanogenic environments. *Appl. Environ. Microbiol.*, **72**, 2080–91.

Imachi, H., Sekiguchi, Y., Kamagata, Y., Ohashi, A. and Harada, H. (2000). Cultivation and in situ detection of a thermophilic bacterium capable of oxidizing propionate in syntrophic association with hydrogenotrophic methanogens in a thermophilic methanogenic granular sludge. *Appl. Environ. Microbiol.*, **66**, 3608–15.

Ito, T., Nielsen, J. L., Okabe, S., Watanabe, Y. and Nielsen, P. H. (2002). Phylogenetic identification and substrate uptake patterns of sulphate-reducing bacteria inhabiting an oxic-anoxic sewer biofilm determined by combining microautoradiography and fluorescent *in situ* hybridization. *Appl. Environ. Microbiol.*, **68**, 356–64.

Kane, M. D., Poulsen, L. K. and Stahl, D. A. (1993). Monitoring the enrichment and isolation of sulphate-reducing bacteria by using oligonucleotide hybridization probes designed from environmentally derived 16S rRNA sequences. *Appl. Environ. Microbiol.*, **59**, 682–6.

Karkhoff-Schweizer, R. R., Huber, D. P. and Voordouw, G. (1995). Conservation of the genes for dissimilatory sulfite reductase from

Desulfovibrio vulgaris and *Archaeoglobus fulgidus* allows their detection by PCR. *Appl. Environ. Microbiol.*, **61**, 290–6.

Karnachuk, O. V., Pimenov, N. V., Yusupov, S. K. *et al.* (2006). Distribution, diversity, and activity of sulphate-reducing bacteria in the water column in Gek-Gel Lake, Azerbaijan. *Microbiologiya*, **75**, 101–9.

Kjeldsen, K. U., Loy, A., Thomsen, T. R., *et al.* (2007). Diversity of sulfate-reducing bacteria from an extreme hypersaline sediment, Great Salt Lake (Utah, USA). *FEMS Microbiol. Ecol.*, in press.

Kleikemper, J., Schroth, M. H., Sigler, W. V. *et al.* (2002). Activity and diversity of sulphate-reducing bacteria in a petroleum hydrocarbon-contaminated aquifer. *Appl. Environ. Microbiol.*, **68**, 1516–23.

Klein, M., Friedrich, M., Roger, A. J. *et al.* (2001). Multiple lateral transfers of dissimilatory sulfite reductase genes between major lineages of sulphate-reducing prokaryotes. *J. Bacteriol.*, **183**, 6028–35.

Klenk, H.-P., Clayton, R. A., Tomb, J.-F. *et al.* (1997). The complete genome sequence of the hyperthermophilic, sulphate-reducing archaeon *Archaeoglobus fulgidus*. *Nature*, **390**, 364–70.

Koizumi, Y., Kojima, H. and Fukui, M. (2004). Dominant microbial composition and its vertical distribution in saline meromictic Lake Kaiike (Japan) as revealed by quantitative oligonucleotide probe membrane hybridization. *Appl. Environ. Microbiol.*, **70**, 4930–40.

Küsel, K., Pinkart, H. C., Drake, H. L. and Devereux, R. (1999). Acetogenic and sulphate-reducing bacteria inhabiting the rhizoplane and deep cortex cells of the sea grass *Halodule wrightii*. *Appl. Environ. Microbiol.*, **65**, 5117–23.

Larsen, O., Lien, T. and Birkeland, N. K. (1999). Dissimilatory sulfite reductase from *Archaeoglobus profundus* and *Desulfotomaculum thermocisternum*: phylogenetic and structural implications from gene sequences. *Extremophiles*, **3**, 63–70.

Laue, H., Friedrich, M., Ruff, J. and Cook, A. M. (2001). Dissimilatory sulfite reductase (desulfoviridin) of the taurine-degrading, non-sulphate-reducing bacterium *Bilophila wadsworthia* RZATAU contains a fused DsrB-DsrD subunit. *J. Bacteriol*, **183**, 1727–33.

Lie, T. J., Godchaux, W. and Leadbetter, E. R. (1999). Sulfonates as terminal electron acceptors for growth of sulfite-reducing bacteria (*Desulfitobacterium* spp.) and sulphate-reducing bacteria: effects of inhibitors of sulfidogenesis. *Appl. Environ. Microbiol.*, **65**, 4611–17.

Lin, C., Flesher, B., Capman, W. C., Amann, R. I. and Stahl, D. A. (1994). Taxon specific hybridization probes for fiber-digesting bacteria suggest novel gut-associated *Fibrobacter*. *Syst. Appl. Microbiol.*, **17**, 418–24.

Liu, W. T., Marsh, T. L., Cheng, H. and Forney, L. J. (1997). Characterization of microbial diversity by determining terminal restriction fragment length polymorphisms of genes encoding 16S rRNA. *Appl. Environ. Microbiol.*, **63**, 4516–22.

Llobet-Brossa, E., Rabus, R., Bottcher, M. E. *et al.* (2002). Community structure and activity of sulphate-reducing bacteria in an intertidal surface sediment: a multi-method approach. *Aquatic Microbial Ecology*, **29**, 211–26.

Loy, A. and Bodrossy, L. (2006). Highly parallel microbial diagnostics using oligonucleotide microarrays. *Clin. Chim. Acta*, **363**, 106–19.

Loy, A., Horn, M. and Wagner, M. (2003). probeBase: an online resource for rRNA-targeted oligonucleotide probes. *Nucleic Acids Res.*, **31**, 514–16.

Loy, A., Küsel, K., Lehner, A., Drake, H. L. and Wagner, M. (2004). Microarray and functional gene analyses of sulphate-reducing prokaryotes in low sulphate, acidic fens reveal co-occurence of recognized genera and novel lineages. *Appl. Environ. Microbiol.*, **70**, 6998–7009.

Loy, A., Lehner, A., Lee, N., Adamczyk, J. *et al.* (2002). Oligonucleotide microarray for 16S rRNA gene-based detection of all recognized lineages of sulphate-reducing prokaryotes in the environment. *Appl. Environ. Microbiol.*, **68**, 5064–81.

Loy, A., Taylor, M. W., Bodrossy, L. and Wagner, M. (2006). Applications of nucleic acid microarrays in soil microbial ecology. In J. E. Cooper and J. R. Rao (eds.), *Molecular approaches to soil, rhizosphere and plant microorganism analysis.* Wallingford, UK: CABI Publishing.

Lueders, T., Pommerenke, B. and Friedrich, M. W. (2004). Stable-isotope probing of microorganisms thriving at thermodynamic limits: syntrophic propionate oxidation in flooded soil. *Appl. Environ. Microbiol.*, **70**, 5778–86.

Lümann, H., Arth, I. and Liesack, W. (2000). Spatial changes in the bacterial community structure along a vertical oxygen gradient in flooded paddy soil cores. *Appl. Environ. Microbiol.*, **66**, 754–62.

Mansfield, E. S., Worley, J. M., McKenzie, S. E. *et al.* (1995). Nucleic-acid detection using nonradioactive labeling methods. *Mol. Cell. Probes*, **9**, 145–56.

Manz, W., Eisenbrecher, M., Neu, T. R. and Szewzyk, U. (1998). Abundance and spatial organization of Gram-negative sulphate-reducing bacteria in activated sludge investigated by *in situ* probing with specific 16S rRNA targeted oligonucleotides. *FEMS Microbiol. Ecol.*, **25**, 43–61.

Martin-Laurent, F., Philippot, L., Hallet, S. *et al.* (2001). DNA extraction from soils: old bias for new microbial diversity analysis methods. *Appl. Environ. Microbiol.*, **67**, 2354–9.

Matsui, G. Y., Ringelberg, D. B. and Lovell, C. R. (2004). Sulphate-reducing bacteria in tubes constructed by the marine infaunal polychaete *Diopatra cuprea*. *Appl. Environ. Microbiol.*, **70**, 7053–65.

Maukonen, J., Saarela, M. and Raaska, L. (2006). *Desulfovibrionales*-related bacteria in a paper mill environment as detected with molecular techniques and culture. *J. Ind. Microbiol. Biotechnol.*, **33**, 45–54.

Minz, D., Fishbain, S., Green, S. J. *et al.* (1999a). Unexpected population distribution in a microbial mat community: sulphate-reducing bacteria localized to the highly oxic chemocline in contrast to a eukaryotic preference for anoxia. *Appl. Environ. Microbiol.*, **65**, 4659–65.

Minz, D., Flax, J. L., Green, S. J. *et al.* (1999b). Diversity of sulphate-reducing bacteria in oxic and anoxic regions of a microbial mat characterized by comparative analysis of dissimilatory sulfite reductase genes. *Appl. Environ. Microbiol.*, **65**, 4666–71.

Molitor, M., Dahl, C., Molitor, I. *et al.* (1998). A dissimilatory sirohaem-sulfite-reductase-type protein from the hyperthermophilic archaeon *Pyrobaculum islandicum*. *Microbiology*, **144**, 529–41.

Mussmann, M., Ishii, K., Rabus, R. and Amann, R. (2005a). Diversity and vertical distribution of cultured and uncultured Deltaproteobacteria in an intertidal mud flat of the Wadden Sea. *Environ. Microbiol.*, **7**, 405–18.

Mussmann, M., Richter, M., Lombardot, T. *et al.* (2005b). Clustered genes related to sulphate respiration in uncultured prokaryotes support the theory of their concomitant horizontal transfer. *J. Bacteriol.*, **187**, 7126–37.

Nakagawa, T., Ishibashi, J., Maruyama, A. *et al.* (2004). Analysis of dissimilatory sulfite reductase and 16S rRNA gene fragments from deep-sea hydrothermal sites of the Suiyo Seamount, Izu-Bonin Arc, Western Pacific. *Appl. Environ. Microbiol.*, **70**, 393–403.

Neretin, L. N., Schippers, A., Pernthaler, A. *et al.* (2003). Quantification of dissimilatory (bi)sulphite reductase gene expression in *Desulfobacterium autotrophicum* using real-time RT-PCR. *Environ. Microbiol.*, **5**, 660–71.

Nielsen, J. L. and Nielsen, P. H. (2002). Quantification of functional groups in activated sludge by microautoradiography. *Water Sci. Technol.*, **46**, 389–95.

Okabe, S., Itoh, T., Satoh, H. and Watanabe, Y. (1999). Analyses of spatial distributions of sulphate-reducing bacteria and their activity in aerobic wastewater biofilms. *Appl. Environ. Microbiol.*, **65**, 5107–16.

Okabe, S., Santegoeds, C. M., Watanabe, Y. and de Beer, D. (2002). Successional development of sulphate-reducing bacterial populations and their activities in an activated sludge immobilized agar gel film. *Biotechnol. Bioengineering*, **78**, 119–30.

Omelchenko, M. V., Makarova, K. S., Wolf, Y. I., Rogozin, I. B. and Koonin, E. V. (2003). Evolution of mosaic operons by horizontal gene transfer and gene displacement in situ. *Genome Biol.*, **4**, R55.

Orphan, V. J., House, C. H., Hinrichs, K. U., McKeegan, K. D. and DeLong, E. F. (2001). Methane-consuming archaea revealed by directly coupled isotopic and phylogenetic analysis. *Science*, **293**, 484–7.

Orphan, V. J., House, C. H., Hinrichs, K. U., McKeegan, K. D. and DeLong, E. F. (2002). Multiple archaeal groups mediate methane oxidation in anoxic cold seep sediments. *Proc. Natl. Acad. Sci. USA*, **99**, 7663–8.

Overmann, J., Coolen, M. J. L. and Tuschak, C. (1999). Specific detection of different phylogenetic groups of chemocline bacteria based on PCR and denaturing gradient gel electrophoresis of 16S rRNA gene fragments. *Arch. Microbiol.*, **172**, 83–94.

Palmer, C., Bik, E. M., Eisen, M. B. *et al.* (2006). Rapid quantitative profiling of complex microbial populations. *Nucleic Acids Res.*, **34**, e5.

Palumbo, A. V., Schryver, J. C., Fields, M. W. *et al.* (2004). Coupling of functional gene diversity and geochemical data from environmental samples. *Appl. Environ. Microbiol.*, **70**, 6525–34.

Perez-Jimenez, J. R. and Kerkhof, L. J. (2005). Phylogeography of sulphate-reducing bacteria among disturbed sediments, disclosed by analysis of the dissimilatory sulfite reductase genes (*dsrAB*). *Appl. Environ. Microbiol.*, **71**, 1004–11.

Pernthaler, A. and Amann, R. (2004). Simultaneous fluorescence in situ hybridization of mRNA and rRNA in environmental bacteria. *Appl. Environ. Microbiol.*, **70**, 5426–33.

Pernthaler, A., Pernthaler, J. and Amann, R. (2002). Fluorescence in situ hybridization and catalyzed reporter deposition for the identification of marine bacteria. *Appl. Environ. Microbiol.*, **68**, 3094–101.

Poulsen, L. K., Ballard, G. and Stahl, D. A. (1993). Use of rRNA fluorescence in situ hybridization for measuring the activity of single cells in young and established biofilms. *Appl. Environ. Microbiol.*, **59**, 1354–60.

Rabus, R., Fukui, M., Wilkes, H. and Widdle, F. (1996). Degradative capacities and 16S rRNA-targeted whole-cell hybridization of sulphate-reducing bacteria in an anaerobic enrichment culture utilizing alkylbenzenes from crude oil. *Appl. Environ. Microbiol.*, **62**, 3605–13.

Rabus, R., Hansen, T. and Widdel, F. (2000). Dissimilatory sulphate- and sulfur-reducing prokaryotes. In M. Dworkin, S. Falkow, E. Rosenberg, K.-H. Schleifer and E. Stackebrandt (eds.), *The Prokaryotes: An evolving electronic resource for the microbiological community.* 3rd ed. New York: Springer-Verlag.

Rabus, R., Ruepp, A., Frickey, T. *et al.* (2004). The genome of *Desulfotalea psychrophila*, a sulphate-reducing bacterium from permanently cold Arctic sediments. *Environ. Microbiol.*, **6**, 887–902.

Ramsing, N. B., Fossing, H., Ferdelman, T. G., Andersen, F. and Thamdrup, B. (1996). Distribution of bacterial populations in a stratified fjord (Mariager Fjord, Denmark) quantified by in situ hybridization and related to chemical gradients in the water column. *Appl. Environ. Microbiol.*, **62**, 1391–404.

Ramsing, N. B., Kühl, M. and Jørgensen, B. B. (1993). Distribution of sulphate-reducing bacteria, O_2, and H_2S in photosynthetic biofilms determined by oligonucleotide probes and microelectrodes. *Appl. Environ. Microbiol.*, **59**, 3840–9.

Raskin, L., Capman, W. C., Kane, M. D., Rittmann, B. E. and Stahl, D. A. (1996a). Critical evaluation of membrane supports for use in quantitative hybridizations. *Appl. Environ. Microbiol.*, **62**, 300–3.

Raskin, L., Rittmann, B. E. and Stahl, D. A. (1996b). Competition and coexistence of sulphate-reducing and methanogenic populations in anaerobic biofilms. *Appl. Environ. Microbiol.*, **62**, 3847–57.

Raskin, L., Stromley, J. M., Rittmann, B. E. and Stahl, D. A. (1994). Group-specific 16S rRNA hybridization probes to describe natural communities of methanogens. *Appl. Environ. Microbiol.*, **60**, 1232–40.

Ravenschlag, K., Sahm, K., Knoblauch, C., Jørgensen, B. B. and Amann, R. (2000). Community structure, cellular rRNA content, and activity of sulphate-reducing bacteria in marine arctic sediments. *Appl. Environ. Microbiol.*, **66**, 3592–602.

Ricke, P., Kolb, S. and Braker, G. (2005). Application of a newly developed ARB software-integrated tool for in silico terminal restriction fragment length polymorphism analysis reveals the dominance of a novel *pmoA* cluster in a forest soil. *Appl. Environ. Microbiol.*, **71**, 1671–3.

Risatti, J. B., Capman, W. C. and Stahl, D. A. (1994). Community structure of a microbial mat: the phylogenetic dimension. *Proc. Natl. Acad. Sci.*, **91**, 10173–7.

Sabehi, G., Loy, A., Jung, K. H. *et al.* (2005). New insights into metabolic properties of marine bacteria encoding proteorhodopsins. *PLoS Biol.*, **3**, e273.

Sahm, K., Knoblauch, C. and Amann, R. (1999). Phylogenetic affiliation and quantification of psychrophilic sulphate-reducing isolates in marine arctic sediments. *Appl. Environ. Microbiol.*, **65**, 3976–81.

Santegoeds, C. M., Ferdelman, T. G., Muyzer, G. and de Beer, D. (1998). Structural and functional dynamics of sulphate-reducing populations in bacterial biofilms. *Appl. Environ. Microbiol.*, **64**, 3731–9.

Schadt, C. W., Liebich, J., Chong, S. C. *et al.* (2005). Design and use of functional gene microarrays (FGAs) for the characterization of microbial communities. *Methods Microbiol.*, **34**, 331–68.

Schedel, M. and Trüper, H. G. (1980). Anaerobic oxidation of thiosulphate and elemental sulfur in *Thiobacillus denitrificans. Arch. Microbiol.*, **124**, 205–10.

Scheid, D. and Stubner, S. (2001). Structure and diversity of Gram-negative sulphate-reducing bacteria on rice roots. *FEMS Microbiol. Ecol.*, **36**, 175–83.

Schwenn, J. D. and Biere, M. (1979). APS-reductase activity in the chromatophores of *Chromatium vinosum* strain D. *FEMS Microbiol. Lett.*, **6**, 19–22.

Stackebrandt, E., Sproer, C., Rainey, F. A. *et al.* (1997). Phylogenetic analysis of the genus *Desulfotomaculum*: evidence for the misclassification of *Desulfotomaculum guttoideum* and description of *Desulfotomaculum orientis* as *Desulfosporosinus orientis* gen. nov., comb. nov. *Int. J. Syst. Bacteriol.*, **47**, 1134–9.

Stahl, D. A. (2004). High-throughput techniques for analyzing complex bacterial communities. *Adv. Exp. Med. Biol.*, **547**, 5–17.

Stahl, D. A. and Amann, R. (1991). Development and application of nucleic acid probes. In E. Stackebrandt and M. Goodfellow (eds.), *Nucleic acid techniques in bacterial systematics.* Chichester, UK: John Wiley & Sons Ltd.

Stahl, D. A., Flesher, B., Mansfield, H. R. and Montgomery, L. (1988). Use of phylogenetically based hybridization probes for studies of ruminal microbial ecology. *Appl. Environ. Microbiol.*, **54**, 1079–84.

Stubner, S. and Meuser, K. (2000). Detection of *Desulfotomaculum* in an Italian rice paddy soil by 16S ribosomal nucleic acid analyses. *FEMS Microbiol. Ecol.*, **34**, 73–80.

Tonolla, M., Bottinelli, M., Demarta, A., Peduzzi, R. and Hahn, D. (2005). Molecular identification of an uncultured bacterium ("morphotype R") in meromictic Lake Cadagno, Switzerland. *FEMS Microbiol. Ecol.*, **53**, 235–44.

Tonolla, M., Demarta, A., Peduzzi, S., Hahn, D. and Peduzzi, R. (2000). *In situ* analysis of sulphate-reducing bacteria related to *Desulfocapsa thiozymogenes* in the chemocline of meromictic Lake Cadagno (Switzerland). *Appl. Environ. Microbiol.*, **66**, 820–4.

van der Maarel, M. J. E. C., Artz, R. R. E., Haanstra, R. and Forney, L. J. (1998). Association of marine Archaea with the digestive tracts of two marine fish species. *Appl. Environ. Microbiol.*, **64**, 2894–8.

Wagner, M., Horn, M. and Daims, H. (2003). Fluorescence in situ hybridisation for the identification and characterisation of prokaryotes. *Curr. Opin. Microbiol.*, **6**, 302–9.

Wagner, M., Loy, A., Klein, M. *et al.* (2005). Functional marker genes for identification of sulphate-reducing prokaryotes. *Methods Enzymol.*, **397**, 469–89.

Wagner, M., Nielsen, P. H., Loy, A., Nielsen, J. L. and Daims, H. (2006). Linking microbial community structure with function: fluorescence in situ hybridization-microautoradiography and isotope arrays. *Curr. Opin. Biotechnol.*, **17**, 1–9.

Wagner, M., Roger, A. J., Flax, J. L., Brusseau, G. A. and Stahl, D. A. (1998a). Phylogeny of dissimilatory sulfite reductases supports an early origin of sulphate respiration. *J. Bacteriol.*, **180**, 2975–82.

Wagner, M., Schmid, M., Juretschko, S. *et al.* (1998b). In situ detection of a virulence factor mRNA and 16S rRNA in *Listeria monocytogenes. FEMS Microbiol. Lett.*, **160**, 159–68.

Wagner, M., Smidt, H., Loy, A. and Jizhong, Z. (2007). Unravelling microbial communities with DNA-microarrays: challenges and future directions. *Microb. Ecol.*, in press.

Watras, C. J., Morrison, K. A., Kent, A. *et al.* (2005). Sources of methylmercury to a wetland-dominated lake in northern Wisconsin. *Environ. Sci. Technol.*, **39**, 4747–58.

Wawer, C., Jetten, M. S. and Muyzer, G. (1997). Genetic diversity and expression of the NiFe hydrogenase large-subunit gene of *Desulfovibrio* spp. in environmental samples. *Appl. Environ. Microbiol.*, **63**, 4360–9.

Wawer, C. and Muyzer, G. (1995). Genetic diversity of *Desulfovibrio* spp. in environmental samples analyzed by denaturing gradient gel electrophoresis of NiFe hydrogenase gene fragments. *Appl. Environ. Microbiol.*, **61**, 2203–10.

Wawrik, B., Paul, J. H. and Tabita, F. R. (2002). Real-time PCR quantification of *rbcL* (ribulose-1,5-bisphosphate carboxylase/oxygenase) *mRNA* in diatoms and pelagophytes. *Appl. Environ. Microbiol.*, **68**, 3771–9.

Wieland, A., Kuhl, M., McGowan, L. *et al.* (2003). Microbial mats on the Orkney Islands revisited: microenvironment and microbial community composition. *Microb. Ecol.*, **46**, 371–90.

Wu, M., Ren, Q., Durkin, A. S. *et al.* (2005). Life in hot carbon monoxide: the complete genome sequence of *Carboxydothermus hydrogenoformans* Z-2901. *PLoS Genet.*, **1**, e65.

Zhou, J. (2003). Microarrays for bacterial detection and microbial community analysis. *Curr. Opin. Microbiol.*, **6**, 288–94.

Zverlov, V., Klein, M., Lücker, S., Friedrich, M. W., Kellermann, J., Stahl, D. A., Loy, A. and Wagner, M. (2005). Lateral gene transfer of dissimilatory (bi)sulfite reductase revisited. *J. Bacteriol.*, **187**, 2203–8.

CHAPTER 3

Functional genomics of sulphate-reducing prokaryotes

Ralf Rabus and Axel Strittmatter

3.1 INTRODUCTION

Besides their challenging and ancient energy metabolism, and applied relevance, much of the interest in sulphate-reducing bacteria arises from their ecophysiological significance in marine environments (Widdel, 1998). In the biologically highly active shelf sediments they contribute to more than 50% of organic carbon remineralization (Jørgensen, 1982), which can only be explained by complete substrate oxidation (Fenchel and Jørgensen, 1977). While this capacity is not present among the frequently isolated and intensively studied *Desulfovibrio* spp., it could be demonstrated with e.g. the newly isolated *Desulfobacter postgatei* (Widdel and Pfennig, 1981) and *Desulfobacterium autotrophicum* (Brysch *et al.*, 1987). The latter employs the C1/CO-dehydrogenase pathway for complete oxidation of acetate to CO_2 as well as for CO_2-fixation (Schauder *et al.*, 1989). Most of the known sulphate-reducing bacteria can be grouped into the two deltaproteobacterial families *Desulfovibrionaceae* (Devereux *et al.*, 1990) or *Desulfobacteriaceae* (Widdel and Bak, 1992). This phylogenetic distinction is to a large extent paralleled by the capacities for incomplete (to acetate) and complete (to CO_2) oxidation of organic substrates, respectively.

At present, more than 450 prokaryotic genomes have been completely sequenced and about 1000 further prokaryotic genomes are in progress (http://www.genomesonline.org). While most genome projects primarily reflect biotechnological or biomedical research interests, environmentally relevant microorganisms have been selected for genome sequencing projects only during the last few years. This chapter provides an overview of the technologies involved and of the current status of genomic research with sulphate-reducing prokaryotes. Moreover, some major insights obtained from the current genome sequencing are summarized in respect of metabolism and regulatory potential.

3.2 BASIC PRINCIPLES OF GENOME SEQUENCING AND ANNOTATION

3.2.1 Genome sequencing

Our present day ability to rapidly and cost-efficiently determine complete prokaryotic as well as eukaryotic genome sequences relies on massive parallelism of applied technologies. Major components are: (i) fluorescence-based DNA sequencing with capillary-based instruments providing about 600–1000 bases of quality sequences per each of 16–384 parallel reactions, (ii) automated robotic systems for high throughput sample handling, (iii) automated software systems to analyze primary sequence data, and (iv) semi-automated software to perform sequence assembly. Two major strategies of genome sequencing can be distinguished: whole genome shotgun sequencing and hierarchical shotgun sequencing (Fraser and Fleischmann, 1997; Kaiser *et al.*, 2003).

Whole genome shotgun sequencing (Figure 3.1) was applied to determine the first complete genome sequence of a free-living organism, *Haemophilus influenzae* (Fleischmann *et al.*, 1995). The initial step in this approach is to fragment isolated total genomic DNA into small pieces of defined length (1–10 kb) by mechanical shearing or enzymatic cleavage. The fragments are then cloned into suitable plasmid vectors to generate a shotgun library covering the entire genome in the optimal case. Sequencing is then performed from both insert ends in a random, genome-wide and large-scale fashion. Typically, an 8- to 10-fold sequence coverage (redundancy of sequence reads relative to corresponding genomic segment) is generated to achieve highly accurate sequences (error rate <0.01%). After extended quality checks, all sequence reads (up to >100 000) are assembled into so-called contigs (overlapping series of sequence reads corresponding to contiguous genome segments). Progress of assembly is reflected by increasing size versus decreasing number of contigs. The whole genome shotgun sequencing approach relies on massive computational processing for linking and ordering of the high number of sequence reads and is best used in case of smaller prokaryotic genomes which are poor in repeats.

The hierarchical sequencing approach (Figure 3.1) on the other hand is more suitable for larger and more complex (e.g. repeat-rich) genomes, in particular those of eukaryotes. Here the genomic DNA is fragmented into

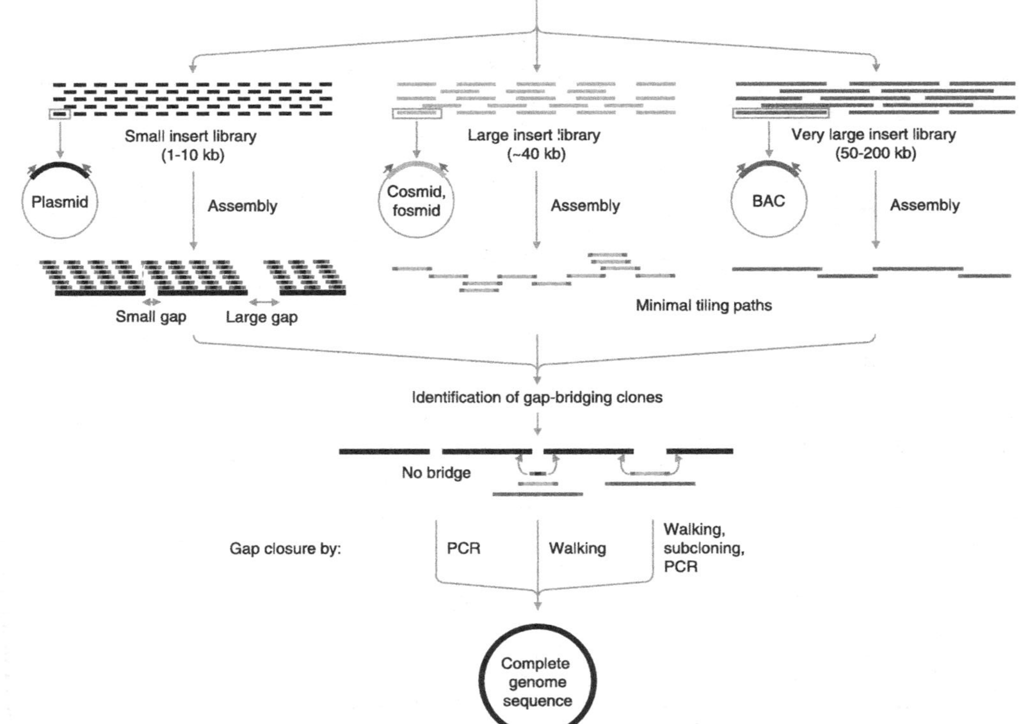

Figure 3.1. Determination of complete genome sequences by whole genome shotgun sequencing supported by large insert libraries for gap closure. Red arrows indicate pairwise end-sequencing and end-sequences are also marked in red. (For a colour version of this figure, please refer to colour plate section.)

larger pieces (50–200 kb) which are cloned into cosmid, fosmid or BAC-libraries (bacterial artificial chromosome). After ordering the large insert libraries into a minimal tiling path (minimal number of clones that together cover the genome in the optimal case) by end-sequencing, each of them can be subjected to an own shotgun sequencing. This approach makes large-scale misassembly less likely and provides finished contigs already at an early stage. Therefore, recent genome sequencing projects often apply hybrid strategies to benefit from the advantages of both approaches.

In parallel with the progress in sequencing (both strategies) the randomness of the constructed libraries and the even representation of the clones can be verified by comparison to the Lander–Waterman model (Lander and Waterman, 1988). Initial processing of raw sequence data comprises basecalling, as well as quality and vector-clipping, which are often performed with the PHRED programme (Ewing *et al.*, 1998). A typical PHRED quality is ≥20, meaning that the sequence is >99% correct. This processed sequence data can then be assembled with programmes such as PHRAP (http://www.phrap.org) and the assemblies can be visually inspected with e.g. the GAP4 STADEN package (Dear and Staden, 1991). Linking of contigs can be facilitated by screening large-insert libraries with the end-sequences of the contigs. Physical gaps are closed by either walking on bridging plasmids, direct sequencing of PCR products or walking on clones from large-insert libraries. Additionally, continuity of the genome sequence is assessed by long-range PCR across random or selected regions. It should be noted that several alternative technologies for ultrafast DNA sequencing (Metzker, 2005) are emerging that promise even more rapidly growing numbers of available genomes.

3.2.2 Genome annotation

The bioinformatical analysis of a complete genome sequence (or finished contig(s)) essentially involves three subsequent steps: (i) prediction of open reading frames (ORFs) and other DNA structures, (ii) assignment of gene functions, and (iii) data integration.

A combination of several gene-finding tools is currently applied to generate an optimized, non-redundant list of predicted ORFs. Automated prediction of start codons is often refined by manual inspection and/or manual adaptation of software (e.g. TICO; Tech *et al.*, 2005). For functional assignment, predicted ORFs are screened against nucleotide and protein databases (Galperin, 2006), by determining sequence similarities (e.g. BLAST; Altschul *et al.*, 1997) or conducting pattern and motive searches to assign the query sequence to a functionally characterized protein family (e.g. InterPro; Mulder *et al.*, 2005). An increasingly important aspect of efficient database searching is the controlled use of vocabularies and classifications, as promoted by the Gene Ontology (GO) Consortium (2004). The KEGG database for high-level complexity analysis allows the reconstruction of metabolic, cellular or regulatory networks (Kanehisa *et al.*, 2004).

Several software environments automatically integrate ORF prediction and functional assignment, e.g. MAGPIE (Gaasterland and Sensen, 1996), PEDANT (Riley *et al.*, 2005), YACOP (Tech and Merkl, 2003) and GenDB (Meyer *et al.*, 2003). Considering the available and continuously growing genome information, functional prediction will benefit from gene context analysis and comparative genomics, as implemented in systems like ERGO (Overbeek *et al.*, 2003), the TIGR Comprehensive Microbial Resources (Peterson *et al.*, 2001) and MicrobesOnline Web site (Alm *et al.*, 2005). An important prerequisite for comparative genomics is an equal annotation quality among published genomes. This issue will be addressed by the subsystem approach (SEED), where experts curate specific subsystems (e.g. histidine degradation) and project them to newly released genomes (Overbeek *et al.*, 2005).

3.3 BASIC PRINCIPLES OF TRANSCRIPTOMIC AND PROTEOMIC ANALYSIS

The availability of complete genome sequences has lead to a paradigm shift with respect to the analysis of gene regulation and function, moving from a single-gene to a genome-wide scale. The entirety of mRNAs (transcriptome) and proteins (proteome) formed by an organism under defined environmental conditions can now be semiquantitatively monitored with global technologies, e.g. DNA-microarray and gel-based proteomics. Condition-specific co-regulation of genes unlinked on the genome sequence, could decipher unpredicted facets of regulatory networks. In addition, expression data can provide first functional clues about genes of unknown function, which constitute about half of the genes predicted for a particular genome (Fraser *et al.*, 2000).

DNA-microarrays represent the state-of-the-art technology to analyze RNA profiles on the genome-wide level (Figure 3.2). Their solid glass surfaces typically carry oligonucleotides or PCR products representing all predicted genes of the studied genome. Condition-specific RNA/cDNA preparations are labelled with different fluorescent dyes and hybridized onto

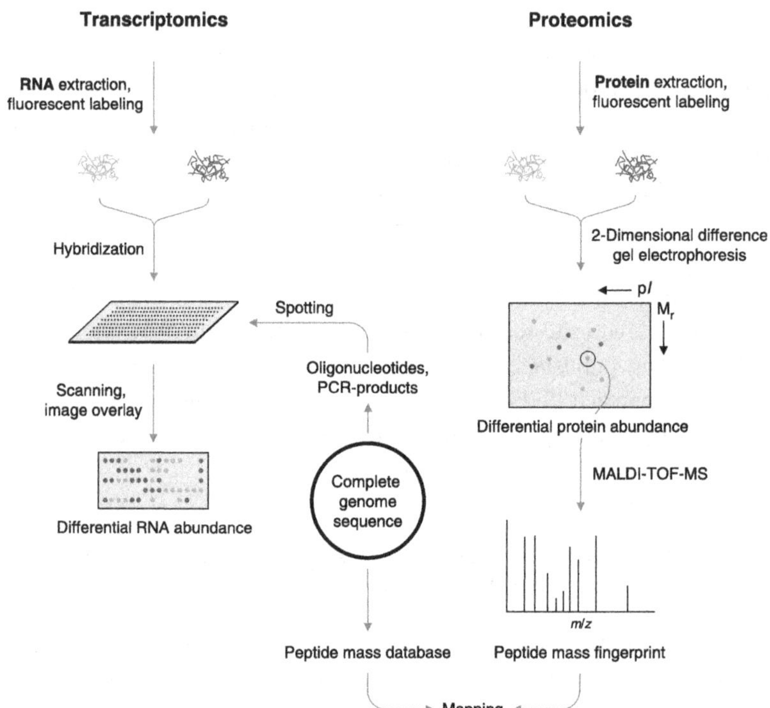

Figure 3.2. Genome-enabled workflow of transcriptomic and proteomic approaches to analyze regulatory processes. (For a colour version of this figure, please refer to colour plate section.)

the DNA-microarray. The resulting fluorescent signals correlate with the relative abundances of the individual mRNA species. Overviews on different DNA-microarray technologies were recently presented by others (e.g. Conway and Schoolnik, 2003)

Differences in protein profiles are most commonly determined with a gel-based approach (Figure 3.2). Quantitative analyses are enabled by the 2-dimensional difference gel electrophoresis (2D DIGE) system, which is based on co-separation of different, fluorescently labelled, condition-specific protein extracts on a single gel (Gade *et al.*, 2003). Identification of electrophoretically separated proteins is most commonly achieved by determining peptide mass fingerprints with MALDI-TOF-MS (matrix assisted laser desorption/ionization-time of flight-mass spectrometry) and matching them to the calculated masses of theoretical peptides generated by an *in silico* digest of each protein encoded by the studied genome (Mann *et al.*, 2001; Hufnagel and Rabus, 2006). An alternative approach for large-scale protein identification is LC-ESI-MS/MS, where peptides generated from complex protein mixtures are directly separated by multidimensional liquid chromatography (LC). The peptide fractions are then subjected to electron spray ionization (ESI) and the masses of the peptide ions are determined by mass spectrometry (MS). Identification occurs via mapping of peptide mass fingerprints or MS/MS data (Washburn *et al.*, 2001). Quantification can be achieved by e.g. tagging the intact proteins with a stable isotope prior to further processing and analysis (Schmidt *et al.*, 2005).

3.4 COMPLETE GENOMES OF SULPHATE-REDUCING PROKARYOTES

The complete genome sequences of the euryarchaeota *Archaeoglobus fulgidus* VC-16^T (Klenk *et al.*, 1997) and the three deltaproteobacteria *Desulfovibrio vulgaris* Hildenborough (Heidelberg *et al.*, 2004), *Desulfotalea psychrophila* LSv54^T (Rabus *et al.*, 2004) and *Desulfovibrio desulfuricans* strain G20 (www.jgi.doe.gov) are publicly available. It should be noted that *Dv. desulfuricans* G20 actually belongs to *Dv. alaskensis* and will be reclassified accordingly (Judy D. Wall, personal communication). Interest in the two *Desulfovibrio* spp. is to a large part based on their model character for basic biochemical and genetic investigations (Voordouw and Wall, 1993; Rabus *et al.*, 2000). The growth temperature ranges of *A. fulgidus* (60 to 95°C; Stetter, 1988) and *Dt. psychrophila* (−2 to 15°C; Knoblauch *et al.*, 1999) reflect the extreme temperature conditions prevailing in their respective natural habitats. The genome of *Db. autotrophicum* HRM2^T was only very recently completed (Strittmatter *et al.*, personal communication) as the first representative from the ecophysiologically important group of completely oxidizing sulphate-reducing bacteria.

An overview of the main physiological and general genome features of these five sulphate-reducing prokaryotes is presented in Table 3.1. While *A. fulgidus* (~2.2 Mb) and *Db. autotrophicum* (~5.6 Mb) differ most profoundly with respect to their genome sizes, they have almost identical G+C contents. The genomes of *Dv. vulgaris, Dv. desulfuricans* and *Dt. psychrophila* have very similar genomes sizes of ~3.5 to 3.7 Mb. The share

Table 3.1. *Comparison of the main physiological and general genome features of sulphate-reducing prokaryotes*

	A. fulgidus	*Dv. vulgaris*	*Dv. desulfuricans*	*Dt. psychrophila*	*Db. autotrophicum*
Main physiological features[a]					
Taxon affiliation above genus level	Euryarchaeota	*Desulfovibrionaceae*	*Desulfovibrionaceae*	*Desulfobulbaceae*	*Desulfobacteriaceae*
Habitat of isolation	Marine hydrothermal system	Freshwater sediment	Freshwater sediment	Arctic marine sediment	Mediterranean marine sediment
Optimal growth temperature [°C]	83	30–36	30–36	10	20–26
Electron acceptors other than SO_4^{2-}	SO_3^{2-}, $S_2O_3^{2-}$	SO_3^{2-}, $S_2O_3^{2-}$, fumarate	SO_3^{2-}, $S_2O_3^{2-}$, NO_3^-, fumarate	SO_3^{2-}, $S_2O_3^{2-}$	$S_2O_3^{2-}$, fumarate
Typical electron donors					
Hydrogen (autotrophically)	+	–	–	–	+
Hydrogen (+ acetate)	+	+	+	+	+
Acetate	+	–	–	–	+
Propionate	nr	–	–	–	+
Higher fatty acids (C_4–C_{16})	nr	–	–	–	+
Lactate	+	+	+	+	+

Ethanol	nr	+	+	+	+
Succinate, fumarate and/or malate	nr	+	(+)	+	+
Other substrates	Starch, gelatine	nr	Choline	Serine, alanine	Glutamate, isobutyrate
Fermentative growth	nr	+	+	+	+
Incomplete oxidation of organic compounds to acetate	–	+	+	+	–
Complete oxidation of organic compounds to CO_2	+	–	–	–	+
C1/CO-dehydrogen-ase pathway	+	–	–	–	+
General genome features[b]					
Chromosome size (bp)	2 178 400	3 570 858	3 730 232	3 523 383	~5 580 000
Plasmids (bp)	–	202 301	–	121 586 14 663	~63 000
G+C content (mol%)[c]	48.5	63.2	57.8	46.8	~49.0
Coding density (%)[c]	92.2	86.4	91.1	86.6	88
Predicted ORFs[c]	2436	3395	3784	3118	up to 5200
Average length (bp)[c]	822	908	892	968	≤ 1000

Table 3.1 (*cont.*)

	A. fulgidus	*Dv. vulgaris*	*Dv. desulfuricans*	*Dt. psychrophila*	*Db. autotrophicum*
ORFs with database match[c]	1797	2539	3430	2265	nd
ORFs with assigned putative function[c]	1096	1894	2369	1545	nd
ORFs with no database match[d]	43	302	354	110	nd
No. rRNA operons[c]	1	5	4	7	6
Encoded tRNAs[c]	46	68	66	64	50
Proteins with presumed regulatory function[c]	55	nr	200	31	>250
Transposase-related genes[c]	16	26	33	9	>150

[a] Data taken from Stetter *et al.* (1988) for *A. fulgidus*; Postgate and Campbell (1966) for *Dv. vulgaris* and *Dv. desulfuricans*; Knoblauch *et al.* (1999) for *Dt. psychrophila*; and Brysch *et al.* (1987) for *Db. autotrophicum*; and from Kuever *et al.* (2005) for general taxonomic information on sulphate-reducing deltaproteobacteria.

[b] Data from Klenk *et al.* (1997) for *A. fulgidus*; Heidelberg *et al.* (2004) for *Dv. vulgaris*; Rabus *et al.* (2004) for *Dt. psychrophila*, http://img.jgi.doe.gov/cgi-bin/pub/main.cgi and Judy D. Wall, personal communication for *Dv. desulfuricans* G20; Strittmatter *et al.* (personal communication) for preliminary analysis of the recently completed genome of *Db. autotrophicum*.

[c] Referring to chromosome only.

[d] Referring to complete genome; data taken from http://img.jgi.doe.gov/cgi-bin/pub/main.cgi?page=finalGenomes.

nr, not reported; nd, presently no final numbers available.

of ORFs displaying no database match is somewhat below 10%. The genome of *Db. autotrophicum* possesses an unusually large number (>150) of transposase-related genes, suggesting that lateral gene transfer might have played an important role in shaping the genome of this metabolically versatile sulphate reducer. Similarly high numbers of transposase-related genes have recently been reported for the denitrifying betaproteobacterium strain EbN1, featuring also a high degree of metabolic versatility (Rabus *et al.*, 2005). The other sulphate reducers have considerably lower numbers (9–33) of transposase-related genes. It will be interesting to learn whether the high number of transposase-related genes is a characteristic feature of genomes from members of the *Desulfobacteriaceae*, i.e. metabolic versatility is paralleled by pronounced genome plasticity.

3.5 METABOLISM

Several aspects of the *Desulfovibrio* spp. physiology have recently been investigated by genome-enabled studies and are described in detail in other chapters of this book: energy metabolism (Chapter 7 by Voordouw *et al.*); oxygen survival mechanisms (Chapter 6 by Dolla *et al.*); response to various environmental stress conditions (Chapter 4 by Wall *et al.*); and the reduction of mercury and metallic oxy-anions (Chapter 15 by Bruschi *et al.*). In the following, some principal metabolic aspects are briefly summarized.

3.5.1 Energy metabolism

Sulphate-reducing prokaryotes gain energy by dissimilatory reduction of sulphate to sulphide coupled to the oxidation of H_2 or organic substrates like lactate. Energetics and biochemistry of dissimilatory sulphate reduction are discussed in Chapter 1 by Thauer *et al.* Genomic analysis of *Desulfovibrio* spp. and *Dt. psychrophila* (Chapter 7 by Voordouw *et al.* and Table 7.1 therein) already revealed differences in hydrogenase contents and strategies for electron transfer from electron donors to the common enzymes of sulphate reduction.

Growth with H_2 as electron donor is a common property of sulphate-reducers and is also widespread among many other prokaryotes (Vignais *et al.*, 2001). Sulphate-reducing bacteria primarily oxidize H_2 in the periplasm via various types of hydrogenases. While *Dv. vulgaris* and *Dv. desulfuricans* possess periplasmic [Fe]-, [Ni/Fe]- or [Ni/Fe/Se]-type hydrogenases, *Dt. psychrophila* does not possess an [Fe]-type hydrogenase. In addition, the absence of energy-conserving cytoplasm-orientated

EchABCDEF and CooMLKXUHF hydrogenase complexes in *Dt. psychrophila* is particularly noteworthy, since these enzymes play a central role in the proposed model of hydrogen cycling during growth of *Dv. vulgaris* with lactate (Chapter 1 by Thauer *et al.* and Chapter 7 by Voordouw *et al.*). Interestingly, the deltaproteobacterial *Geobacter sulfurreducens* does not possess [Fe]-, periplasmic [Ni/Fe/Se]-type and hydrogen-evolving Ech-type hydrogenases, but several [Ni/Fe]-type hydrogenases, which are distinct from those of *Desulfovibrio* spp. (Coppi, 2005). Thus, great diversity of hydrogenase-encoding genes apparently exists among hydrogen-utilizing deltaproteobacteria.

Desulfovibrio spp. channel electrons generated during periplasmic oxidation of hydrogen or formate into a periplasmic cytochrome c_3 network, from where a direct transfer to periplasm-orientated cytochrome *c* subunits of transmembrane complexes (Hmc or TpIIc_3) occurs. An exception to this principle represents *Dt. psychrophila*, since this sulphate reducer apparently lacks a periplasmic cytochrome c_3 network and Hmc- or TpIIc_3-complexes. Therefore, one may speculate that *Dt. psychrophila* channels electrons generated during periplasmic oxidation of formate or hydrogen directly via membrane-anchoring subunits of periplasmic formate dehydrogenase and hydrogenase into the quinone pool or requires hitherto unknown periplasmic electron carriers.

The classical H^+-translocating NADH dehydrogenase (Ndh1–10) complex for regeneration of NADH was predicted for *Dt. psychrophila*, but is apparently absent in *Dv. vulgaris*, *Dv. desulfuricans* and *A. fulgidus*. However, all four sulphate reducers possess a Na^+-translocating NADH:quinone oxidoreductase (NqrA-E), known e.g. from marine *Vibrio alginolyticus* and raising the question about Na^+-dependent bioenergetics (Steuber, 2001).

Several membrane-bound protein complexes have been discovered in *Desulfovibrio* and *Archaeoglobus* spp. which could mediate electron transfer from the quinone pool or periplasmic formate dehydrogenase and hydrogenase to the cytoplasmic enzymes of sulphate reduction (Pires *et al.*, 2003; Haveman *et al.*, 2004; Matias *et al.*, 2005). The related Hmc and Hme complexes transfer electrons from the periplasm and are apparently specific for *Desulfovibrio* spp. and *A. fulgidus*, respectively. All sulphate-reducing bacteria with known genomes have the Qmo and DsrMKJOP complexes in common. The Qmo complex was suggested to deliver electrons from menaquinol to APS-reductase (Pires *et al.*, 2003; Haveman *et al.*, 2004), while the DsrMKJOP complex probably transfers electrons from the periplasm or the quinone pool to sulfite reductase (Pires *et al.*, 2006). At present it is unclear whether the electrons are directly transferred from these two complexes to APS- and sulfite reductase, or whether a cytoplasmic electron carrier is involved. The general similarity of the cytoplasmic DsrK subunits to the catalytic HdrD subunit could indicate the involvement of a sulphide/disulphide intermediate.

3.5.2 Genomic organization of metabolic genes

The organization of genes involved in carbon and energy metabolism in the genomes of *A. fulgidus, Dv. vulgaris, Dv. desulfuricans, Dt. psychrophila* and *Db. autotrophicum* is schematically shown in Figure 3.3. Overall, it can be concluded, that these genes are not clustered but rather scattered across the genomes. Interestingly, several genes and operon-like structures could at present not be detected in the genome of *Db. autotrophicum*, even though they are shared by the other four sulphate reducers.

3.6 REGULATORY CAPACITIES

Considering the low growth rates of sulphate-reducing prokaryotes and the low *in situ* concentrations of organic substrates, it has long been disputed to what extent these organisms are actually capable of regulating gene expression. Earlier investigations on the metabolic regulation have mostly been concerned with hydrogenase from *Desulfovibrio* spp. (Kremer *et al.*, 1988; van den Berg *et al.*, 1993). The availability of complete genome sequences now allows study of the regulatory potential by bioinformatical and global expression analysis. Several of the most recent studies were concerned with the adaptive responsiveness of Dv. *vulgaris* to environmental stress conditions. An overview of regulation-orientated studies is presented in Table 3.2. First results revealed unexpected and extensive regulatory capacities of sulphate-reducing prokaryotes.

Recent studies with *Db. autotrophicum* (Amann, 2004) revealed that expression of *adh* (alcohol dehydrogenase encoding) was induced by several alcohols, irrespective of the presence of lactate as alternative substrate. In contrast, propionate-dependent expression of *sbm* (methylmalonyl-CoA mutase encoding) was apparently repressed in the presence of lactate, indicating different hierarchical levels of regulation. The gene for the catalytic α-subunit of ACS/CODH (acetyl-CoA synthase/CO dehydrogenase) was not differentially regulated under autotrophic versus heterotrophic growth conditions. This finding suggests that the same enzymes function in oxidative and reductive directions of the C1/CO dehydrogenase pathway. Similar results were reported for homoacetogenic *Thermacetogenium phaeum*

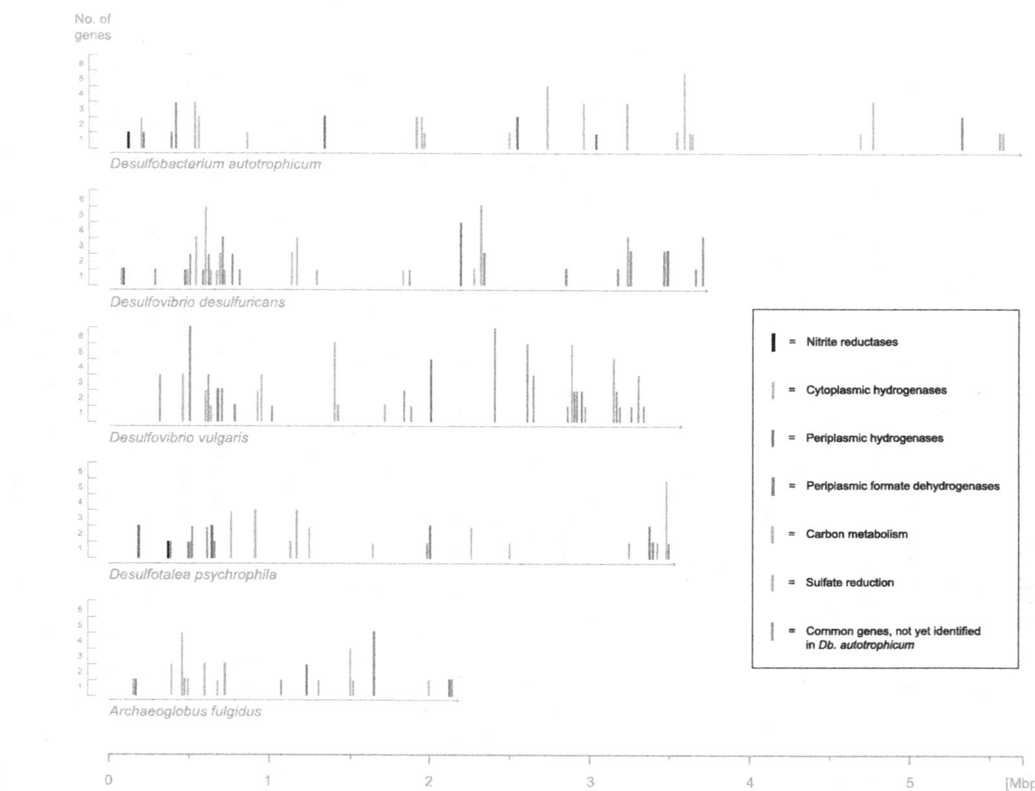

Figure 3.3. Genomic organization of genes involved in major metabolic functions in sulphate-reducing prokaryotes with known complete genome sequence. (For a colour version of this figure, please refer to colour plate section.)

Table 3.2. *Genome-enabled bioinformatical and experimental studies with sulphate-reducing prokaryotes*

Organism	Research question (approach)	Reference
Bioinformatical studies		
Dv. vulgaris	Conserved DNA motives within probable promoter regions (AlignACE)	Hemme and Wall, 2004
	Two-component signal transduction systems	Zhang *et al.*, 2006
	Influence of DNA sequence information on the synthesis of mRNA	Wu *et al.*, 2006
	Transcription factor binding sites for operons shared with *Geobacter* spp. (AlignACE)	Yan *et al.*, 2004
Dt. psychrophila	Two-component regulatory systems; detection of a new Ntr-subclass (phylogeny)	Rabus *et al.*, 2004
Deltaproteobacteria	Reconstruction of regulatory pathways; new CRP/FNR-like regulon controlling expression of *sat* and *apsAB* (GenomeExplorer)	Rodionov *et al.*, 2004
Experimental studies		
Dv. vulgaris	Function of oxygen resistance proteins (mutants, RT-PCR, proteomics)	Fournier *et al.*, 2003, 2006
	Correlation between mRNA and protein abundance (DNA-microarray, LC-MS/MS)	Nie *et al.*, 2006
	Response to Cu(II) and Hg(II) toxicity (RAP PCR, differential display, Q PCR)	Chang *et al.*, 2004
	Analysis of energy metabolism mutants (DNA-macroarray, proteomics)	Haveman *et al.*, 2003
	Analysis of inhibition by nitrite (DNA-macroarray)	Haveman *et al.*, 2004, 2005

Table 3.2 (*cont.*)

Organism	Research question (approach)	Reference
	Heat shock response (DNA-microarray)	Chhabra *et al.*, 2006
	Stress response to nitrite (DNA-microarray)	He *et al.*, 2006
	Response to salt stress (DNA-microarray, proteomics)	Mukhopadhyay *et al.*, 2006
Dv. desulfuricans G20	Signature-tagged mutagenesis (DNA-microarray)	Groh *et al.*, 2005
Db. autotrophicum	Psychrotolerant growth behaviour (proteomics)	Rabus *et al.*, 2002
	Metabolic regulation (proteomics, real-time RT-PCR)	Amann, 2004
A. fulgidus	Regulation of approx. 350 ORFs by heat shock (whole-genome DNA-microarray)	Rohlin *et al.*, 2005

(Hattori *et al.*, 2005) and methanogenic *Methanosarcina thermophila* (Grahame *et al.*, 2005). While only a single *acs/codh* operon is present in the latter, two have been reported for *M. acetivorans* (Galagan *et al.*, 2002) and *M. mazei* (Deppenmeier *et al.*, 2002) raising the question about a different functional/regulatory principle in these two *Methanosarcina* spp.

3.7 COMPARATIVE AND EVOLUTIONARY GENOMICS

Besides those described above, additional genomes of sulphate-reducing bacteria are in progress: *Dv. vulgaris* DP4 (http://www.jgi.doe.gov), *Dv. magneticus* RS-1 (http://www.bio.nite.go.jp) and the thermophilic, deeply-branching *Thermodesulfovibrio yellowstonii* and *Thermodesulfobacterium commune* (http://www.tigr.org). Moreover, complete genomes are also available from other deltaproteobacteria, e.g. metal-reducing *Geobacter sulfurreducens* (Methe *et al.*, 2003) and *G. metallireducens* (http://www.jgi.org), *Pelobacter carbinolicus* (http://www.jgi.org) and the predatory *Bdellovibrio bacteriovorus* HD100 (Rendulic *et al.*, 2004). Thus a rather comprehensive database for comparative and evolutionary analysis already exists.

Lateral gene transfer is generally regarded as a major mechanism driving the evolution of prokaryotic genomes (Boucher *et al.*, 2003).

A prominent example is represented by the genes involved in dissimilatory sulphate reduction, one of the most ancient modes of biological energy conservation (Shen *et al.*, 2001). Clusters of the coding genes – *sat* (ATP sulphurylase), *aps* (APS-reductase) and *dsrAB* (sulphite reductase) – are scattered across the known genomes of sulphate-reducing prokaryotes. Phylogenetic analysis of *dsrAB*, *aps* and 16S rDNA genes suggested a bacterial origin for *dsr* in *A. fulgidus* and acquisition of essential gene clusters via horizontal gene transfer (Friedrich, 2002; Klein *et al.*, 2001; Wagner *et al.*, 1998). A most relevant finding in this context emerged recently from a metagenomic investigation of marine sediments (Mussmann *et al.*, 2005). Surprisingly, the obtained DNA fragments from uncultured sulphate-reducing prokaryotes were characterized by tight clustering of all known genes related to sulphate reduction (including those for electron-channelling transmembrane complexes), supporting the hypothesis of their transfer among prokaryotes in a single event.

Using a phylogenetic approach Calteau and co-workers (2005) found evidence for the transfer of the complete *echABCDEF* operon from an archaeon belonging to the *Methanosarcina* clade to *Desulfovibrio* spp.. An example for a transfer event in opposite direction is the acquisition of the gene encoding 3-hydroxy-3-methylglutaryl-CoA reductase (isoprenoid biosynthesis) from a *Pseudomonas*-like bacterium by *A. fulgidus* and other members of the *Archaeoglobales* (Boucher *et al.*, 2001).

Comparison of the *Dt. psychrophila* and *A. fulgidus* genomes revealed a low number of shared homologues (Rabus *et al.*, 2004). Only 396 of the 3118 predicted proteins of *Dt. psychrophila* had homologues in *A. fulgidus*, with only 24 of them having the respective *A. fulgidus* protein as the closest homologue.

3.8 CONCLUSIONS

Most of our current molecular understanding about the energy and carbon metabolism of sulphate-reducing prokaryotes is based on biochemical and genetic studies with *Desulfovibrio* spp., leading to a certain conceptual bias. The recent genomic analysis of *Dt. psychrophila* and the ongoing analysis of *Db. autotrophicum* already indicate a wider diversity of energy and carbon metabolism-related genes, than inferred from *Desulfovibrio* spp. genomes alone. These genome-based findings/hypothesis present new challenges for detailed biochemical investigations to advance our understanding of principal metabolic processes in sulphate-reducing bacteria other than *Desulfovibrio* spp.. One may also expect additional insights from further genomes of the metabolically versatile members of the *Desulfobacteriaceae*. This would also advance our ecophysiological understanding, since members of the *Desulfococcus/Desulfosarcina* cluster apparently dominate in marine sediments (Llobet-Brossa *et al.*, 2002). In addition, functional genomics will yield many new insights into the thus far poorly understood regulatory capacities of sulphate reducers. Considering that changing environmental conditions (carbon flux or redox gradients) prevail in natural habitats, adaptability should represent a beneficial life strategy also for sulphate-reducing prokaryotes.

REFERENCES

Alm, E. J., Huang, K. H., Price, M. N. *et al.* (2005). The MicrobesOnline Web site for comparative genomics. *Genome Res*, **15**, 1015–22.

Altschul, S. F., Madden, T. L., Schäffer, A. A. *et al.* (1997). Gapped BLAST and PSI-BLAST: a new generation of protein database search programs. *Nucleic Acids Res*, **25**, 3389–402.

Amann, J. (2004). Metabolic regulation and reconstruction of *Desulfobacterium autotrophicum*. PhD thesis, University of Bremen.

Boucher, Y., Douady, C. J., Papke, R. T. *et al.* (2003). Lateral gene transfer and the origins of prokaryotic groups. *Annu Rev Genet*, **37**, 283–328.

Boucher, Y., Huber, H., L'Haridon, S., Stetter, K. O. and Doolittle, W. F. (2001). Bacterial origin for the isoprenoid biosynthesis enzyme HMG-CoA reductase of the archaeal orders *Thermoplasmatales* and *Archaeoglobales*. *Mol Biol Evol*, **18**, 1378–88.

Brysch, K., Schneider, C., Fuchs, G. and Widdel, F. (1987). Lithoautotrophic growth of sulphate-reducing bacteria, and description of *Desulfobacterium autotrophicum* gen. nov., sp. nov. *Arch Microbiol*, **148**, 264–74.

Calteau, A., Gouy, M. and Perrière, G. (2005). Horizontal transfer of two operons coding for hydrogenases between bacteria and archaea. *J Mol Evol*, **60**, 557–65.

Chang, I. S., Groh, J. L., Ramsey, M. M., Ballard, J. D. and Krumholz, L. R. (2004). Differential expression of *Desulfovibrio vulgaris* genes in response to Cu(II) and Hg(II) toxicity. *Appl Environ Microbiol*, **70**, 1847–51.

Chhabra, S. R., He, Q., Huang, K. H. *et al.* (2006). Global analysis of heat shock response in *Desulfovibrio vulgaris* Hildenborough. *J Bacteriol*, **188**, 1817–28.

Conway, T. and Schoolnik, G. K. (2003). Microarray expression profiling: capturing a genome-wide portrait of the transcriptome. *Mol Microbiol*, **47**, 879–89.

Coppi, M. V. (2005). The hydrogenases of *Geobacter sulfurreducens*: a comparative genomic perspective. *Microbiology*, **151**, 1239–54.

Dear, S. and Staden, R. (1991). A sequence assembly and editing program for efficient management of large projects. *Nucleic Acids Res*, **19**, 3907–11.

Deppenmeier, U., Johann, A., Hartsch, T. *et al.* (2002). The genome of *Methanosarcina mazei*: evidence for lateral gene transfer between bacteria and archaea. *J Mol Microbiol Biotechnol*, **4**, 453–61.

Devereux, R., He, S.-H., Doyle, C. L. *et al.* (1990). Diversity and origin of *Desulfovibrio* species: phylogenetic definition of a family. *J Bacteriol*, **172**, 3609–19.

Ewing, B., Hillier, L. D., Wendl, M. C. and Green, P. (1998). Base-calling of automated sequencer traces using PHRED. I. Accuracy assessment. *Genome Res*, **8**, 175–85.

Fenchel, T. M. and Jørgensen, B. B. (1977). Detritus food chains of aquatic ecosystems: the role of bacteria. In M. Alexander (ed.), *Advances in Microbial Ecology*, vol I. Plenum Press. pp. 1–58.

Fleischmann, R. D., Adams, M. D., White, O. *et al.* (1995). Whole-genome random sequencing and assembly of *Haemophilus influenzae* Rd. *Science*, **269**, 496–512.

Fournier, M., Aubert, C., Dermoun, Z. *et al.* (2006). Response of the anaerobe *Desulfovibrio vulgaris* Hildenborough to oxidative conditions: proteome and transcript analysis. *Biochimie*, **88**, 85–94.

Fournier, M., Zhang, Y., Wildschut, J. D. *et al.* (2003). Function of oxygen resistance proteins in the anaerobic sulphate-reducing bacterium *Desulfovibrio vulgaris* Hildenborough. *J Bacteriol*, **185**, 71–9.

Fraser, C. M., Eisen, J. A. and Salzberg, S. L. (2000). Microbial genome sequencing. *Nature*, **406**, 799–803.

Fraser, C. M. and Fleischmann, R. D. (1997). Strategies for whole microbial genome sequencing and analysis. *Electrophoresis*, **18**, 1207–16.

Friedrich, M. W. (2002). Phylogenetic analysis reveals multiple lateral transfers of adenosine-5′-phosphosulphate reductase genes among sulphate-reducing microorganisms. *J Bacteriol*, **184**, 278–89.

Gaasterland, T. and Sensen, C. W. (1996). Fully automated genome analysis that reflects user needs and preferences. A detailed introduction to the MAGPIE system architecture. *Biochimie*, **78**, 302–10.

Gade, D., Thiermann, J., Markowsky, D. and Rabus, R. (2003). Evaluation of two-dimensional difference gel electrophoresis for protein profiling. Soluble proteins of the marine bacterium *Pirellula* sp. strain 1. *J Mol Microbiol Biotechnol*, **5**, 240–51.

Galagan, J. E., Nusbaum, C., Roy, A. *et al.* (2002). The genome of *M. acetivorans* reveals extensive metabolic and physiological diversity. *Genome Res*, **12**, 532–42.

Galperin, M. Y. (2006). The molecular biology database collection: 2006 update. *Nucleic Acids Res*, **34**, D3–D5.

Gene Ontology Consortium. (2004). The Gene Ontology (GO) database and informatics resource. *Nucleic Acids Res*, **32**, D258–D261.

Grahame, D. A., Gencic, S. and DeMoll, E. (2005). A single operon-encoded form of the acetyl-CoA decarbonylase/synthase multienzyme complex responsible for synthesis and cleavage of acetyl-CoA in *Methanosarcina thermophila*. *Arch Microbiol*, **184**, 32–40.

Groh, J. L., Luo, Q., Ballard, J. D. and Krumholz, L. R. (2005). A method adapting microarray technology for signature-tagged mutagenesis of *Desulfovibrio desulfuricans* G20 and *Shewanella oneidensis* MR-1 in anaerobic sediment survival experiments. *Appl Environ Microbiol*, **71**, 7064–74.

Hattori, S., Galushko, A. S., Kamagata, Y. and Schink, B. (2005). Operation of the CO dehydrogenase/acetyl-coenzyme. A pathway in both acetate oxidation and acetate formation by the syntrophically acetate-oxidizing bacterium *Thermacetogenium phaeum*. *J Bacteriol*, **187**, 3471–6.

Haveman, S. A., Brunelle, V., Voordouw, J. K. *et al.* (2003). Gene expression analysis of energy metabolism mutants of *Desulfovibrio vulgaris* Hildenborough indicates an important role of alcohol dehydrogenase. *J Bacteriol*, **185**, 4345–53.

Haveman, S. A., Greene, E. A., Stilwell, C. P., Voordouw, J. K. and Voordouw, G. (2004). Physiological and gene expression analysis of inhibition of *Desulfovibrio vulgaris* Hildenborough by nitrite. *J Bacteriol*, **186**, 7944–50.

Haveman, S. A., Greene, E. A. and Voordouw, G. (2005). Gene expression analysis of the mechanism of inhibition of *Desulfovibrio vulgaris* Hildenborough by nitrate-reducing, sulfide-oxidizing bacteria. *Environ Microbiol*, **7**, 1461–5.

He, Q., Huang, K. H., He, Z. *et al.* (2006). Energetic consequences of nitrite stress in *Desulfovibrio vulgaris* Hildenborough inferred from global transcriptional analysis. *Appl Environ Microbiol*, in press.

Heidelberg, J. F., Seshadri, R., Haveman, S. A. *et al.* (2004). The genome sequence of the anaerobic, sulphate-reducing bacterium *Desulfovibrio vulgaris* Hildenborough. *Nature Biotechnol*, **22**, 554–9.

Hemme, C. L. and Wall, J. D. (2004). Genomic insights into gene regulation of *Desulfovibrio vulgaris* Hildenborough. *OMICS*, **8**, 43–55.

Hufnagel, P. and Rabus, R. (2006). Mass spectrometric identification of proteins in complex post-genomic projects. Soluble proteins of the metabolically versatile, denitrifying "*Aromatoleum*" sp. strain EbN1. *J Mol Microbiol Biotechnol*, **11**, 53–81.

Jørgensen, B. B. (1982). Mineralization of organic matter in the sea bed – the role of sulphate reduction. *Nature*, **296**, 643–5.

Kaiser, O., Bartels, D., Bekel, T. *et al.* (2003). Whole genome shotgun sequencing guided by bioinformatics pipelines – an optimized approach for an established technique. *J Biotechnol*, **106**, 121–33.

Kanehisa, M., Goto, S., Kawashima, S., Okuno, Y. and Hattori, M. (2004). The KEGG resource for deciphering the genome. *Nucleic Acids Res*, **32**, D277–D280.

Klein, M., Friedrich, M., Roger, A. J. *et al.* (2001). Multiple lateral transfers of dissimilatory sulfite reductase genes between major lineages of sulphate-reducing prokaryotes. *J Bacteriol*, **183**, 6028–35.

Klenk, H.-P., Clayton, R. A., Tomb, J.-F. *et al.* (1997). The complete genome sequence of the hyperthermophilic, sulphate-reducing archaeon *Archaeoglobus fulgidus*. *Nature*, **390**, 364–70.

Knoblauch, C., Sahm, K. and Jørgensen, B. B. (1999). Psychrophilic sulphate-reducing bacteria isolated from permanently cold Arctic marine sediments: description of *Desulfofrigus oceanense* gen. nov., sp. nov., *Desulfofrigus fragile* sp. nov., *Desulfofaba gelida* gen. nov., sp. nov., *Desulfotalea psychrophila* gen. nov., sp. nov. and *Desulfotalea arctica* sp. nov. *Inter J Syst Bacteriol*, **49**, 1631–43.

Kremer, D. R., Nienhuis-Kuiper, H. E. and Hansen, T. A. (1988). Ethanol dissimilation in *Desulfovibrio*. *Arch Microbiol*, **150**, 552–7.

Kuever, J., Rainey, F. A. and Widdel, F. (2005). Class IV. Deltaproteobacteria *class nov.* In D. J. Brenner, N. R. Krieg, J. T. Staley (eds.), Bergey's manual of systematic bacteriology, Vol 2, Part C. (2nd edn.). New York: Springer, p. 922.

Lander, E. S. and Waterman, M. S. (1988). Genomic mapping by fingerprinting random clones: a mathematical analysis. *Genomics*, **2**, 231–9.

Llobet-Brossa, E., Rabus, R., Böttcher, M. E. *et al.* (2002). Community structure and activity of sulphate-reducing bacteria in an intertidal surface-sediment: a multi-methods approach. *Aquat Microb Ecol*, **29**, 211–26.

Mann, M., Hendrickson, R. C. and Pandey, A. (2001). Analysis of proteins and proteomes by mass spectrometry. *Annu Rev Biochem*, **70**, 437–73.

Matias, P. M., Pereira, I. A. C., Soares, C. M. and Carrondo, M. A. (2005). Sulphate respiration from hydrogen in *Desulfovibrio* bacteria: a structural biology overview. *Prog Biophys Mol Biol*, **89**, 292–329.

Methé, B. A., Nelson, K. E., Eisen, J. A. *et al.* (2003). The genome of *Geobacter sulfurreducens*: metal reduction in subsurface environments. *Science*, **302**, 1967–9.

Metzker, M. L. (2005). Emerging technologies in DNA sequencing. *Genome Res*, **15**, 1767–76.

Meyer, F., Goesmann, A., McHardy, A. C. *et al.* (2003). GenDB – an open source genome annotation system for prokaryote genomes. *Nucleic Acids Res*, **31**, 2187–95.

Mukhopadhyay, A., He, Z., Yen, H.-C. *et al.* (2006). Salt stress in *Desulfovibrio vulgaris* Hildenborough: an integrated genomics approach. *Proc Natl Acad Sci USA*, (in press).

Mulder, N. J., Apweiler, R., Attwood, T. K. *et al.* (2005). InterPro, progress and status in 2005. *Nucleic Acids Res*, **33**, D201–D205.

Mussmann, M., Richter, M., Lombardot, T., *et al.* (2005). Clustered genes related to sulphate respiration in uncultured prokaryotes support the theory of their concomitant horizontal transfer. *J Bacteriol*, **187**, 7126–37.

Nie, L., Wu, G. and Zhang, W. (2006). Correlation between mRNA and protein abundance in *Desulfovibrio vulgaris*: a multiple regression to identify sources of variations. *Biochem Biophys Res Commun*, **339**, 603–10.

Overbeek, R., Begley, T., Butler, R. M. *et al.* (2005). The subsystems approach to genome annotation and its use in the project to annotate 1000 genomes. *Nucleic Acids Res*, **33**, 5691–702.

Overbeek, R., Larsen, N., Walunas, T. *et al.* (2003). The ERGO™ genome analysis and discovery system. *Nucleic Acids Res*, **31**, 164–71.

Peterson, J. D., Umayam, L. A., Dickinson, T., Hickey, E. K. and White, O. (2001). The Comprehensive Microbial Resource. *Nucleic Acids Res*, **29**, 123–5.

Pires, R. H., Lourenço, A. I., Morais, F. *et al.* (2003). A novel membrane-bound respiratory complex from *Desulfovibrio desulfuricans* ATCC 27774. *Biochim Biophys Acta*, **1605**, 67–82.

Pires, R. H., Venceslau, S. S., Morais, F. *et al.* (2006). Characterization of the *Desulfovibrio desulfuricans* ATCC 27774 DsrMKJOP complex – a membrane-bound redox complex involved in the sulphate respiratory pathway. *Biochemistry*, **45**, 249–62.

Postgate, J. R. and Campbell, L. L. (1966). Classification of *Desulfovibrio* species, the nonsporulating sulphate-reducing bacteria. *Bacteriol Rev*, **30**, 732–8.

Rabus, R., Brüchert, V., Amann, J. and Könneke, M. (2002). Physiological response to temperature changes of the marine, sulphate-reducing bacterium *Desulfobacterium autotrophicum*. *FEMS Microbiol Ecol*, **42**, 409–17.

Rabus, R., Hansen, T. A., and Widdel, F. (2000). Dissimilatory sulphate- and sulfur-reducing prokaryotes. In M. Dworkin, S. Falkow, E. Rosenberg, K.-H. Schleifer and E. Stackebrandt (eds.), *The prokaryotes: an evolving electronic resource for the microbiological community*. Heidelberg: Springer Science Online (http://www.prokaryotes.com).

Rabus, R., Kube, M., Heider, J. *et al.* (2005). The genome sequence of an anaerobic aromatic-degrading denitrifying bacterium, strain EbN1. *Arch Microbiol*, **183**, 27–36.

Rabus, R., Ruepp, A., Frickey, T. *et al.* (2004). The genome of *Desulfotalea psychrophila*, a sulphate-reducing bacterium from permanently cold Arctic sediments. *Environ Microbiol*, **6**, 887–902.

Rendulic, S., Jagtap, P., Rosinus, A. *et al.* (2004). A predator unmasked: life cycle of *Bdellovibrio bacteriovorus* from a genomic perspective. *Science*, **303**, 689–92.

Riley, M. L., Schmidt, T., Wagner, C., Mewes, H.-W. and Frishman, D. (2005). The PEDANT genome database in 2005. *Nucleic Acids Res*, **33**, D308–D310.

Rodionov, D. A., Dubchak, I., Arkin, A., Alm, E. and Gelfand, M. S. (2004). Reconstruction of regulatory and metabolic pathways in metal-reducing δ-proteobacteria. *Genome Biol*, **5**, R90.

Rohlin, L., Trent, J. D., Salmon, K. *et al.* (2005). Heat shock response of *Archaeoglobus fulgidus*. *J Bacteriol*, **187**, 6046–57.

Schauder, R., Preuß, A., Jetten, M. and Fuchs, G. (1989). Oxidative and reductive acetyl-CoA/carbon monoxide dehydrogenase pathway in *Desulfobacterium autotrophicum*. 2. Demonstration of the enzymes of the pathway and comparison of CO dehydrogenase. *Arch Microbiol*, **151**, 84–9.

Schmidt, A., Kellermann, J. and Lottspeich, F. (2005). A novel strategy for quantitative proteomics using isotope-coded protein labels. *Proteomics*, **5**, 4–15.

Shen, Y., Buick, R. and Canfield, D. E. (2001). Isotopic evidence for microbial sulphate reduction in the early Archaean era. *Nature*, **410**, 77–81.

Stetter, K. O. (1988). *Archaeoglobus fulgidus* gen. nov., sp. nov.: a new taxon of extremely thermophilic archaebacteria. *Syst Appl Microbiol*, **10**, 172–3.

Steuber, J. (2001). Na^+ translocation by bacterial NADH:quinone oxidoreductases: an extension to the complex-I family of primary redox pumps. *Biochim Biophys Acta*, **1505**, 45–56.

Tech, M. and Merkl, R. (2003). YACOP: enhanced gene prediction obtained by a combination of existing methods. *In Silico Biol*, **3**, 441–51.

Tech, M., Pfeifer, N., Morgenstern, B. and Meinicke, P. (2005). TICO: a tool for improving predictions of prokaryotic translation initiation sites. *Bioinformatics*, **21**, 3568–9.

Van den Berg, W. A. M., Stokkermans, J. P. W. G. and van Dongen, W. M. A. M. (1993). The operon for the Fe-hydrogenase in *Desulfovibrio vulgaris* (Hildenborough): Mapping of the transcript and regulation of expression. *FEMS Microbiol Lett*, **110**, 85–90.

Vignais, P. M., Billoud, B. and Meyer, J. (2001). Classification and phylogeny of hydrogenases. *FEMS Micobiol Rev*, **25**, 455–501.

Voordouw, G. and Wall, J. D. (1993). Genetics and molecular biology of sulphate-reducing bacteria. In M. Sebald (ed.), *Genetics and Molecular Biology of Anaerobic Bacteria*. New York: Springer-Verlag, pp. 456–73.

Wagner, M., Roger, A. J., Flax, J. L., Brusseau, G. A. and Stahl, D. A. (1998). Phylogeny of dissimilatory sulfite reductases supports an early origin of sulphate respiration. *J Bacteriol*, **180**, 2975–82.

Washburn, M. P., Wolters, D. and Yates 3rd, J. R. (2001). Large-scale analysis of the yeast proteome by multidimensional protein identification technology. *Nature Biotechnol*, **19**, 242–7.

Widdel, F. (1988). Microbiology and ecology of sulphate- and sulfur-reducing bacteria. In A. J. B. Zehnder (ed.), *Biology of Anaerobic Microorganisms*. New York: John Wiley & Sons, pp. 469–585.

Widdel, F., and Bak, F. (1992). Gram-negative mesophilic sulphate-reducing bacteria. In A. Balows, H. G. Trüper, M. Dworkin, W. Harder and K.-H. Schleifer (eds.), *The Prokaryotes*. Vol IV. (2nd edn.). New York: Springer-Verlag, pp. 3352–78.

Widdel, F. and Pfennig, N. (1981). Studies on dissimilatory sulphate-reducing bacteria that decompose fatty acids. I. Isolation of new sulphate-reducing bacteria enriched with acetate from saline environments. Description of *Desulfobacter postgatei* gen. nov., sp. nov. *Arch Microbiol*, **129**, 395–400.

Wu, G., Nie, L. and Zhang, W. (2006). Relation between mRNA expression and sequence information in *Desulfovibrio vulgaris*: combinatorial contributions of upstream regulatory motifs and coding sequence features to variations in mRNA abundance. *Biochem Biophys Res Commun*, **344**, 114–21.

Yan, B., Methé, B. A., Lovley, D. R. and Krushkal, J. (2004). Computational prediction of conserved operons and phylogenetic footprinting of transcription regulatory elements in the metal-reducing bacterial family *Geobacteraceae*. *J Theoret Biol*, **230**, 133–44.

Zhang, W., Culley, D. E., Wu, G. and Brockman, F. J. (2006). Two-component signal transduction systems of *Desulfovibrio vulgaris*: structural and phylogenetic analysis and deduction of putative cognate pairs. *J Mol Evol*, **62**, 473–87.

CHAPTER 4

Evaluation of stress response in sulphate-reducing bacteria through genome analysis

J. D. Wall, H. C. Bill Yen and E. C. Drury

4.1 INTRODUCTION

The unique ability of the anaerobic sulphate-reducing bacteria (SRB) to respire sulphate provides access to niches that may be restricted from other bacteria. However, environmental niches are by definition constantly in flux. Thus, for scientists to reach a level of understanding that will allow prediction and/or control of the activities of the SRB, it is necessary to learn how the bacteria respond to changes in environmental parameters; such as, nutrient availability, presence of toxic substances, altered salt concentrations, temperature fluctuations, and a myriad of other variables. With the recent sequencing of a number of SRB (Klenk *et al.*, 1997; Heidelberg *et al.*, 2004; Rabus *et al.*, 2004), the available proteins and regulatory sites of the bacteria have been revealed. Nevertheless, a significant percentage of the predicted open reading frames (ORFs) encode hypothetical or conserved hypothetical proteins for which functions remain obscure. Much work is yet to be done to elucidate the interplay of functions that allow the SRB to survive or even flourish in the changing conditions prevailing in their environment. Here we discuss preliminary transcriptional analyses of the responses of *Desulfovibrio vulgaris* Hildenborough to a number of environmental stressors.

A description of optimal growth conditions for *D. vulgaris* Hildenborough is derived from its early characterization. This strain is a mesophilic Gram-negative anaerobe which was isolated in 1946 from Wealden Clay near Hildenborough, Kent, in the United Kingdom (Postgate, 1984). The growth medium recommended by the National Collections of Industrial and Marine Bacteria (NCIMB, Aberdeen, Scotland) contains lactate, sulphate, and 0.1% (wt/vol) yeast extract and the growth conditions are 30°C and pH 7.4 in the absence of oxygen. Since this strain was isolated from clay soil, it is not a marine bacterium and was reported to be inhibited at sodium concentrations well below 1.0 M (Postgate, 1984).

Significant deviations from these optimal culture conditions certainly occur in environmental settings. How do the SRB alter their metabolism to tolerate or adapt to these non-optimal conditions? To explore this question, microarray analyses have been initiated by the Virtual Institute for Microbial Stress and Survival (http://vimss.lbl.gov/; Chhabra *et al.*, 2006; He *et al.*, 2006; Mukhopadhyay *et al.*, 2006). Previous studies have shown poor correlations between genes differentially expressed versus those needed for response to a given stress (Birrell *et al.*, 2002; Giaever *et al.*, 2002). However, an overall pattern of response may be informative. The reader should keep in mind that all inferences presented are from preliminary data and must be further explored to reach firm conclusions. Stimulating this research was the interest of the US Department of Energy in the potential use of these bacteria for bioremediation. That interest grew out of the observation that the SRB have the capacity to change the redox state, and thus the solubility, of a number of toxic metals and radionuclides (Lovley *et al.*, 1991; Gorby and Lovley, 1992). The conditions present at one of the contaminated test sites suggested the first stressors to be examined by microarrays (http://public.ornl.gov/nabirfrc/sitenarrative.cfm#Anchor11). These conditions included nitrate concentrations in groundwater from <1 to >150 mM, a pH between 3.25 and 6.5, and a dissolved oxygen content about 45 μM. The effects on the growth of *D. vulgaris* after the addition of various concentrations of a particular stressor to mid-exponential phase cells (OD_{600} of 0.3) were determined. A standard growth medium containing lactate and sulphate was used and, for most stressors, a 50% decrease in growth rate was targeted. For these preliminary experiments, no yeast extract was included to assure reproducible medium components. Samples of treated and untreated cultures were taken over a time course of 0 to 240 min for RNA examination. For the comparisons discussed here, a sample from each stress was chosen arbitrarily for examination. The stressors for which preliminary data have been gathered, the treatment time and the comparison samples are shown in Table 4.1. Data from oxygen exposure and from treatment with a sublethal concentration of chromate are included in tables but will not be discussed here since these treatments are discussed in detail in other chapters of this book.

Estimation of the differential gene expression in treatment versus the control samples from the microarray data was calculated as normalized $\log_2$ ratios ($\log_2 R$) using the following formula: $\log_2$ (treatment) $-\log_2$ (control). After further normalizations for signal

Table 4.1. *Stressors examined for transcriptional responses in* D. vulgaris *Hildenborough*

Stressor	Concentration or condition	Time of treatment (min)	Comparison culture
Cold[a]	8°C	240	30°C, 240 min
Heat[a,b]	50°C	15	37°C, 15 min
Oxygen[c]	0.1%	240	No O_2, 240 min
Alkaline pH[a,d]	pH 10	120	pH 7, 0 min
Acid pH[e]	pH 5.5	240	pH 7, 0 min
Nitrite[f]	2.5 mM	60	No NO_2^-, 60 min
Nitrate[c]	105 mM	240	No NO_3^-, 240 min
Sodium[c,g]	250 mM	120	No added Na^+, 120 min
Potassium[c,h]	250 mM	120	No added K^+, 120 min
Chromate[c]	0.55 mM	120	No$CrO_4^=$, 0 min
Stationary phase[i]	0.8 OD_{600}		Mid exponential phase 0.3 OD_{600}

[a] Growth was not observed at this level of stress.

[b] Extended incubation at 50°C resulted in death.

[c] Growth rate was inhibited by approximately 50%.

[d] pH was adjusted by addition of KOH.

[e] pH was adjusted by addition of H_2SO_4. Growth occurred only after a long lag and pH was increased at least one pH unit by the cells.

[f] Growth occurred after the nitrite concentration had been reduced below 0.5 mM.

[g] Total concentration of sodium in the treatment was about 462 mM because the lactate, sulphate, and other components were present as sodium salts.

[h] Total concentration of potassium in the treatment was 254 mM because the phosphate buffer was added as a potassium salt.

[i] Stationary phase was defined as the plateau in growth resulting from carbon limitation.

intensities (Colantuoni *et al.*, 2002) and sector-based artifacts, the significance of the ratios was calculated as a Z-score. Only ratios with an absolute Z value ≥2 were compared in the data considered here.

The stressors applied at the concentrations and times listed resulted in significant changes in expression of about 64% of the 3531 predicted protein-encoding genes. The numbers of genes responding with altered

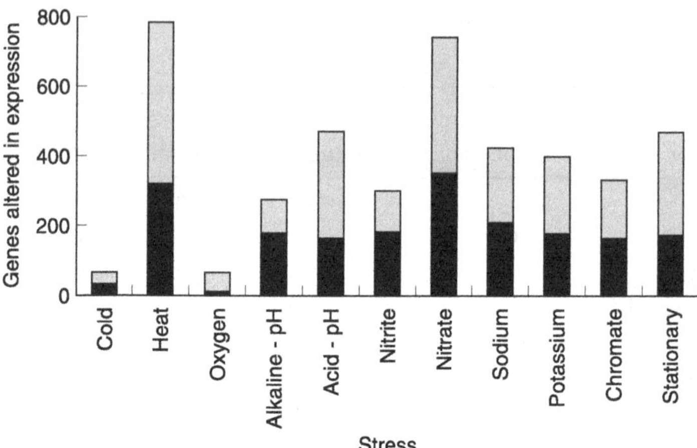

Figure 4.1. Numbers of differentially expressed genes (abs (Z) $\geq$2) are indicated for each of the treatments (details listed in Table 4.1). Grey bar indicates those increased in expression versus black for those decreased in expression. Bars are additive, not overlapping.

transcription in each treatment are shown in Figure 4.1. It can be readily seen that the cold treatment and the introduction of a low concentration of oxygen did not perturb the metabolism of *D. vulgaris* as dramatically as the other conditions. To determine whether the responses to the various treatments involved many of the same genes, all pairwise comparisons of the arrays of differentially expressed genes were made (Table 4.2). While definite conclusions cannot be made from this limited set of data, a few inferences can be made that deserve further exploration.

4.2 LOW TEMPERATURE STRESS

First, *D. vulgaris* subjected to the low temperature condition had only a few genes that changed significantly in transcription. In *Escherichia coli*, the response to cold shock is mediated through temperature induced changes in fluidity of the cytoplasmic membrane, nucleic acid structure, and ribosomes (Phadtare *et al.*, 2000). During adaptation to the cold by *E. coli*, translation is arrested by binding of Protein Y (YfiA) to the ribosome that prevents translation of all but the small set of cold shock proteins (Vila-Sanjurjo *et al.*, 2004). These proteins then alter the ribosomes so that protein biosynthesis resumes, albeit at a lower rate (Jones and Inouye, 1996). Whereas genes for the complement of cold shock proteins found in *E. coli* are not universal (Graumann and Marahiel, 1998) and are not recognized in the

Table 4.2. *Numbers of genes significantly changed in expression in response to a treatment and the number of differentially expressed genes found in two treatments*[a]

	Treatments[b]										
	Cold	Heat	O_2	pH10	pH5.5	NO_2^-	NO_3^-	NaCl	KCl	$CrO_4^=$	Stat[c]
Cold	68	*11*	*4*	*1*	*1*	*3*	*9*	*35*	*32*	*1*	*4*
Heat		785	*9*	*47*	*132*	*89*	*105*	*101*	*102*	*122*	*103*
O_2			70	*3*	*9*	*4*	*21*	*7*	*6*	*3*	*1*
pH10				280	*73*	*15*	*95*	*25*	*29*	*78*	*79*
pH5.5					471	*32*	*125*	*27*	*29*	*154*	*69*
NO_2^-						305	*36*	*19*	*15*	*22*	*39*
NO_3^-							742	*59*	*70*	*105*	*95*
NaCl								428	*254*	*33*	*24*
KCl									399	*37*	*28*
$CrO_4^=$										337	*57*
Stat											470

[a]Genes for which the $\log_2$ ratio met the abs(Z) ≥ 2 cut-off were counted. Those increased or decreased in expression were included. Genes that changed significantly but in opposite directions in the compared treatments were not included.

[b]Treatments are as described in Table 4.1.

[c]Stat is stationary phase.

genome of *D. vulgaris*, a *yfiA* homologue is found (DVU1629). This gene was not significantly altered in expression in response to the decrease in temperature applied here, but was increased upon heat shock. In addition to this putative translational regulator, *D. vulgaris* also appears to contain one ORF (DVU1858) annotated as a cold shock domain protein that responds transcriptionally to temperature changes like *yfiA*. Similar to studies for *E. coli* and *Bacillus subtilis* (Phadtare *et al.*, 2000), cold-shocked *D. vulgaris* did show significant increases in putative homologs for an RNA helicase (DVU3310), NusA involved in transcription termination/antitermination (DVU0510), translation initiation factor 2 (DVU0508), and ribosomal protein S-17 (DVU1312). These changes reflect the effects of lower temperature on nucleic acid structure and translation. Effects of the changes on membrane fluidity were revealed in that 15 of the 68 responding genes (22%) were annotated as cell envelope or transport functions, a slight increase in percentage versus their occurrence within the genome (16%). Putative genes for the UvrABC complex (DVU1987, DVU1605, and DVU0801, respectively) were all highly upregulated with corresponding $\log_2$ R of 2.0, 2.1, and 3.8. That this nucleotide excision repair pathway might be needed as the cells were cooled was unexpected. However, negative DNA supercoiling in *E. coli* has been shown to be transiently increased by cold shock (Phadtare *et al.*, 2000) which may change the DNA topology sufficiently for DNA repair systems to respond.

Interestingly, of these 68 genes, approximately half were also altered in expression in the salt-treated cells. This shared set also contained 27 genes which responded in both sodium and potassium treatments and 9 were annotated as involved in cell envelope biosynthesis or transporters. This observation is consistent with a major effect of these stressors on the cell membrane.

4.3 SALT STRESS

Both sodium- and potassium-treated cells had about 400 differentially expressed genes. Whereas some bacteria take up potassium transiently to offset the detrimental effects of osmotic shock (Csonka and Epstein, 1996), the presence of elevated sodium with incremental potassium was additive in decreasing growth rates of *D. vulgaris* (Mukhopadhyay *et al.*, 2006). This additivity was consistent with the overall similarity in the cellular response to increased concentrations of these cations, although the cells were inhibited in growth at lower concentrations of potassium than sodium. About 60% of the differentially expressed genes were shared in the sodium- and potassium-treated cells. In other two-way comparisons (Table 4.2), the numbers of genes changed in expression that were in common with each of the two cation treatments were also quite similar. Thus the cellular responses to these two salt treatments showed great overlap.

Osmotic stress in *E. coli* has been shown to be regulated in part by supercoiling of DNA that facilitates expression of *rpoS* (Cheung *et al.*, 2003). This gene encodes an alternative sigma factor known as the stress-induced sigma that is a key regulator for starvation, high osmolarity, high or low temperature, and acidic pH (Hengge-Aronis, 2000). *D. vulgaris* is among the majority of non-γ-proteobacteria lacking an RpoS ortholog (Nies, 2004). Therefore an alternative stress response paradigm is needed for this δ-proteobacterium that is likely to involve its many two component regulators.

A more detailed analysis of the responses of *D. vulgaris* to increased NaCl (Mukhopadhyay *et al.*, 2006) revealed that glycine betaine, a common microbial osmolyte (Bremer and Krämer, 2000), was accumulated by this bacterium. The genes for uptake of this compound were increased in expression in response to either salt (DVU2297, DVU2298, and DVU2299, minimum $\log_2 R = 1.9$). Further, isotope-coded affinity tagging coupled with liquid chromatography-mass spectrometry documented an increase in the expression of the ABC permease protein. Growth studies confirmed that micromolar concentrations of glycine betaine or ectoine could reverse the growth defects of these salts (Mukhopadhyay *et al.*, 2006). The related solute, carnitine, has also been shown to be effective in reversing the effects of high salt. No genes coding for biosynthetic functions for these osmolytes were recognized in the genome. In addition, providing biosynthetic intermediates for glycine betaine did not relieve salt stress (Mukhopadhyay *et al.*, 2006). *D. vulgaris* appears to scavenge osmolytes from its environment when faced with osmotically stressful conditions.

Table 4.3 shows that salt stress apparently increased the expression of genes for ATP synthase, sulphate adenylyltransferase, and the high molecular weight cytochrome operon. From these changes, we infer that *D. vulgaris* may increase energy production to support an additional protective strategy, active efflux of the offending cations. Incongruously, several lactate permeases and genes for acetate metabolism were down regulated suggesting the response is more complex than presently understood.

4.4 HEAT SHOCK

Heat shock is one of the best understood of environmental stressors. The major responses are the increased production of molecular chaperones that facilitate the refolding of denatured proteins and increased expression of a number of ATP-dependent proteases that degrade proteins which have been irretrievably damaged (Gross, 1996; Yura *et al.*, 2000). In *E. coli* these responses result from an increase in the alternative sigma factors, σ^{32} (encoded by *rpoH*) and σ^{24} (encoded by *rpoE*), the latter being activated under extreme heat stress (Raivio and Silhavy, 2000). Whereas *D. vulgaris* has a putative *rpoH* (DVU1584), it has no apparent *rpoE* orthologue. Interestingly, a gene annotated as *hrcA* (DVU0813), encoding a protein with homology to the *B. subtilis* heat-responsive repressor is present. This repressor binds to an operator with a highly conserved sequence, CIRCE (for controlling inverted repeat of chaperone expression, Zuber and Schumann, 1994). Derepression of the genes under control of this regulator occurs by heat inactivation of HrcA (Raivio and Silhavy, 2000). The HrcA-CIRCE repression

Table 4.3. D. vulgaris *Hildenborough transcriptional responses of representative genes in ATP synthesis, sulphate reduction and organic acid oxidation to stress treatments*[a]

ORF No.	Gene	Description	Treatment										
			Cold[b]	Heat	O_2	pH10	pH5.5	NO_2^-	NO_3^-	NaCl	KCl	$CrO_4^=$	Stat[c]
ATPase													
DVU0774	*atpC*	ATP synthase, F_1 epsilon subunit	–	–1.39[d]	–	2.47	–	–	–	1.40	1.33	–	–
DVU0775	*atpD*	ATP synthase, F_1 beta subunit	–	–2.07	–	–	–	–1.55	–	–	1.73	–	–1.30
DVU0776	*atpG*	ATP synthase, F_1 gamma subunit	–	–2.59	–	–	–	–1.55	–	–	1.62	–	–1.57
DVU0777	*atpA*	ATP synthase, F_1 alpha subunit	–	–2.06	–	–	–	–1.74	–	–	1.51	–	–
DVU0778	*atpH*	ATP synthase, F_1, delta subunit	–	–2.08	–	–	–	–1.67	–	1.44	1.52	–	–
DVU0779	*atpF2*	ATP synthase, F_0, B subunit, putative	–	–1.69	–	–	–	–1.43	–	1.67	1.51	–	–
DVU0780	*atpF1*	ATP synthase, F_0, B subunit, putative	–	–	–	–	–	–1.40	1.27	1.51	–	–	–
DVU0917	*atpE*	ATP synthase, F_0, C subunit	–	–	–	–	–	–1.27	–	–	–	–	–1.78
DVU0918	*atpB*	ATP synthase, F_0, A subunit	–	–1.91	–	–1.09	–	–1.34	1.10	–	–	–	–
$SO_4^=$ reduction													
DVU0053	NA[e]	Sulphate permease, putative	1.38	–	–	–	–1.15	–	–	–	–	–	–
DVU0279	NA	Sulphate permease family protein	–	–	1.09	–	–	–1.60	–	2.28	1.85	–	–
DVU1295	*sat*	Sulphate adenylyltransferase	–	–	–	–	–	–	1.10	–	–	–	–
DVU1566	*cysD*	PAPS reductase, putative	–	1.99	–	–	1.54	–	2.98	–	–	–	1.84
DVU1597	*sir*	Sulphite reductase, assimilatory-type	–	–	–	–	–	–	1.61	–	–	–	–
DVU0847	*apsA*	Adenylyl-sulphate reductase, alpha subunit	–	–1.71	–	–	–	–	1.73	–	–	–	–
DVU0846	*apsB*	Adenylyl-sulphate reductase, beta subunit	–	–2.11	–	–	–	–	1.21	–	–	–	–

DVU0404	*dsrD*	Dissimilatory sulphite reductase D	–	−2.96	–	–	1.78	–	–	–	–	1.81	–
DVU1286	*dsrP*	Integral membrane protein	–	–	–	–	–	−1.64	–	–	–	–	–
DVU1287	*dsrO*	Periplasmic (Tat), binds 2[4Fe-4S]	–	–	–	–	–	−2.40	–	–	–	–	−1.34
DVU1288	*dsrJ*	Periplasmic (Sec), trihaem cytochrome *c*	–	–	–	–	–	−2.37	–	–	–	–	–
DVU1289	*dsrK*	Cytoplasmic, binds 2[4Fe-4S]	–	−1.54	–	–	–	−2.20	–	–	–	–	–
DVU1290	*dsrM*	Inner membrane protein binds 2 haem b	–	–	–	–	–	−2.65	–	–	–	–	–
DVU1636	*ppaC*	Inorganic pyrophosphatase, Mn-dependent	–	–	−1.14	–	–	–	2.14	–	–	–	–
DVU0529	*rrf2*	Rrf2 protein	–	2.84	–	–	–	–	–	2.77	2.29	–	–
DVU0530	*rrf1*	Rrf1 protein	–	2.09	–	–	–	–	–	2.44	2.34	–	–
DVU0531	*hmcF*	HmcF, 52.7 kD protein	–	2.19	–	–	–	–	–	2.37	1.96	–	–
DVU0532	*hmcE*	HmcE, 25.3 kD protein	–	1.86	–	–	1.63	–	–	1.34	1.38	–	–
DVU0533	*hmcD*	HmcD, 5.8 kD protein	–	2.39	–	–	–	–	–	1.97	2.29	–	–
DVU0534	*hmcC*	HmcC, 43.2 kD protein	–	1.49	−1.07	–	–	–	1.86	–	–	–	–
DVU0535	*hmcB*	HmcB, 40.1 kD protein	–	–	–	–	–	–	3.24	2.67	1.53	–	–
DVU0536	*hmcA*	HmcA, high-molecular weight cytochrome *c*	–	–	–	–	–	–	–	2.96	2.91	1.35	–
Lactate/pyruvate oxidation													
DVU0600	*ldh*	L-lactate dehydrogenase	–	2.46	–	–	–	–	–	–	–	–	–
DVU2784	*lldD*	Dehydrogenase, FMN-dependent family	−1.33	–	–	–	–	–	−1.67	–	–	–	–
DVU2110	*b2975*	L-lactate permease	–	–	–	–	–	–	1.31	–	–	–	–
DVU2285	*NA*	L-lactate permease family protein	–	−1.87	–	–	−1.22	−1.52	–	–	–	–	–
DVU2451	*NA*	L-lactate permease family protein	−1.21	−2.46	–	−1.47	–	–	−1.27	−1.66	−2.29	–	–
DVU2683	*NA*	L-lactate permease family protein	–	–	–	–	–	−1.80	–	−1.32	−1.29	–	–

Table 4.3 (*cont.*)

ORF No.	Gene	Description	Treatment										
			Cold[b]	Heat	O_2	pH10	pH5.5	NO_2^-	NO_3^-	NaCl	KCl	$CrO_4^=$	Stat[c]
DVU3284	*b2975*	L-lactate permease	–	3.68	–	–	2.15	–	2.69	–	–	–	–
DVU1569	*porA*	Pyruvate ferredoxin oxidoreductase alpha subunit	–	–	–	1.49	–	–	−1.68	–	–	–	–
DVU1947	*oorC*	Pyruvate ferredoxin oxidoreductase gamma subunit putative	–	–	–	–	–	–	−1.87	–	–	−1.50	–
DVU3025	*por*	Pyruvate ferredoxin oxidoreductase	–	−3.53	–	–	–	–	1.48	−1.98	−1.43	–	–
DVU3026	NA	L-lactate permease family protein	–	−3.60	–	–	–	–	–	−1.48	−1.77	–	–
DVU3027	*glcD*	Glycolate oxidase, subunit GlcD	–	−2.31	–	–	–	–	1.20	−1.47	−1.66	–	–
DVU3028	NA	Iron-sulphur cluster binding protein	–	−1.61	–	2.23	–	–	1.55	–	–	1.21	–
DVU3029	*pta*	Phosphate acetyltransferase	–	−3.77	–	–	–	–	–	−1.63	−1.63	–	–
DVU3030	*ackA*	Acetate kinase	–	−3.55	–	–	–	–	–	−2.34	−2.27	–	–
DVU3031	NA	Conserved hypothetical protein	–	−2.92	–	–	–	–	–	−1.66	−1.84	–	–
DVU3032	NA	Conserved hypothetical protein	–	−3.05	–	–	–	–	–	−2.29	−2.23	–	–
DVU3033	NA	Iron-sulphur cluster binding protein	–	−2.98	–	–	–	–	–	−2.08	−1.71	–	–
Fur regulon													
DVU0303	NA	Hypothetical protein	–	2.86	1.49	2.02	1.35	2.79	–	–	–	2.19	–
DVU0304	NA	Hypothetical protein	–	–	–	–	–	2.77	–	–	–	2.34	–

DVU0763	*gdp*	GGDEF domain protein	–	1.50	–	1.10	–	–	–	1.74	–	1.97	–
DVU2378	*foxR*	Transcriptional regulator, AraC family	–	1.38	–	–	–	1.66	1.61	–	–	–	–
DVU2571	*feoB*	Fe^{2+} transport protein B	–	3.41	–	1.74	–	2.95	–	1.54	1.15	1.50	–
DVU2572	*feoA*	Fe^{2+} transport protein A	–	3.28	–	1.86	1.56	3.25	–	1.49	–	2.95	2.78
DVU2573	NA	Hypothetical protein	–	3.77	–	–	2.27	2.49	–	1.25	–	2.60	1.66
DVU2574	*feoA*	Fe^{2+} transporter component, feoA	–	3.54	–	–	1.40	2.49	–	1.18	–	1.97	–
DVU2680	*fld*	Flavodoxin, iron-repressed	–	1.27	1.22	–	–	–	−1.59	–	–	1.74	–
DVU3330	NA	Fe-regulated P-type ATPase, hypothetical	–	1.93	–	–	1.12	2.09	–	–	–	–	–
PerR regulon													
DVU0772	NA	Hypothetical protein	1.58	–	2.05	–	–	–	–	–	–	–	–
DVU2247	*ahpC*	Alkyl hydroperoxide reductase C	–	2.14	2.56	–	–	1.56	−1.52	1.45	–	–	–
DVU2318	*rbr2*	Rubrerythrin, putative	–	–	2.34	–	–	–	–	–	–	−2.49	–
DVU3093	*rdl*	Rubredoxin-like protein	–	–	2.28	–	–	–	–	–	–	–	–
DVU3094	*rbr*	Rubrerythrin	–	2.12	2.04	–	–	–	–	–	–	–	–
DVU3095	*perR*	Peroxide-responsive regulator PerR	1.33	–	1.87	−1.15	–	–	–	1.81	1.45	–	–

[a] Treatments are as described in Table 4.1.

[b] For genes and treatments for which there are no entries the data were either not obtained or did not achieve a sufficient Z-score.

[c] Stat is stationary phase.

[d] Numbers are $\log_2$ ratio of treatment versus control for which abs (Z) ≥ 2.

[e] NA, no gene acronym assigned.

system has now been found in a number of bacteria including *Caulobacter crescentus* (Roberts *et al.*, 1996) and *Bradyrhizobium japonicum* (Minder *et al.*, 2000).

In silico searches for binding motifs for σ^{32} and HrcA in the *D. vulgaris* genome led to the prediction of regulons for each of these regulators (Rodionov *et al.*, 2004). Interestingly none of the genes predicted to be part of these regulons (with the exception of DVU1337, Lon protease) was actually increased in expression by the heat treatment (cut-off of abs (Z) $\geq$2 for the dataset; Table 4.4). However, quite a number of predicted chaperones and proteases were observed to respond to this heat shock regime (Table 4.4). A recent report on the heat shock response in *D. vulgaris* attributed a portion of the regulation to the σ^{54} regulator (DVU1628) (Chhabra *et al.*, 2006). This sigma factor does not belong to the σ^{70} family; it is more eukaryotic-like, strictly requiring activator proteins for promoting gene transcription (Cases *et al.*, 2003). From a genome search of *D. vulgaris*, at least 99 genes were predicted to have σ^{54} promoter sites and 34 were significantly altered in expression (1.7 fold change or greater and abs (Z) $\geq$1.7) (Chhabra *et al.*, 2006). Interestingly, none of the heat-responsive genes predicted to be σ^{54}-regulated was found among the chaperones and proteases in Table 4.4. In addition to transcriptional analysis, differential protein content in response to heat shock was also examined by Ettan® DIGE (difference gel electrophoresis) (Amersham) (Chhabra *et al.*, 2006). A poor correlation was obtained between the proteins identified as differentially expressed and the genes with altered transcript levels; however, three proteins whose genes were putatively regulated by σ^{54} were found to be differentially expressed when heat shocked: FliM, flagellar motor switch protein (DVU0910); HtrA, peptidase (DVU1468); and AcpD, acyl carrier protein phosphodiesterase (DVU2548). Four chaperones, GroEL (DVU1976), GroES (DVU1977), DnaK (DVU0811), and HtpG (DVU2643), were identified as increased in expression at the protein level, but only the latter two had significant transcript changes in this study (Table 4.4). In conclusion, the classical model of heat shock regulation by alternative sigma factors does not yet fully explain the data from *D. vulgaris*. It appears that the major regulatory features are yet to be deciphered.

4.5 NITRITE STRESS

Nitrate and nitrite effects on *D. vulgaris* were examined because of the high levels of nitric acid in many of the sites where metal bioremediation might be implemented. *D. vulgaris* Hildenborough does not grow with

Table 4.4. *Expression of putative heat shock genes of* D. vulgaris

Gene No.	Name	Annotation	Log_2R[b]
Predicted HrcA regulated (CICRE site)[a]			
DVU1976	*groEL*	Chaperonin, 60 kDa	–[c]
DVU1977	*groES*	Chaperonin, 10 kDa	–
Predicted RpoH (σ^{54}) regulated[a]			
DVU 1001	b0965	CoA binding protein	–
DVU1002	NA	Conserved hypothetical protein	–
DVU1003	NA	dnaJ domain protein	–
DVU1334	*tig*	Trigger factor	−1.43
DVU1335	*clpP*	ATP-dependent Clp protease, proteolytic subunit	–
DVU1336	*ClpX*	ATP-dependent Clp protease, ATP-binding subunit ClpX	–
DVU1337	*lon*	ATP-dependent protease La	1.77
DVU1584-DVU1578	*rpoH*	Heat shock sigma factor (first gene of operon)	–
DVU1976	*groEL*	Chaperonin, 60 kDa	–
DVU1977	*groES*	Chaperonin, 10 kDa	–
Heat shock responsive protein repair genes[d]			
DVU0684	*hflK*	hflK protein, putative	1.52
DVU0811	*dnaK*	dnaK protein	2.94
DVU0812	*grpE*	Heat shock protein GrpE	1.83
DVU0813	*hrcA*	Heat-inducible transcription repressor HrcA	2.68
DVU0864	NA	Glycoprotease family protein, putative	3.83
DVU1191	*lon*	ATP-dependent protease La, putative	2.02
DVU1278	*ftsH*	Cell division protein FtsH	2.55
DVU1457	*trxB*	Thioredoxin reductase, putative	2.33
DVU1577	*hslV*	ATP-dependent protease hslV	1.88
DVU1602	*clpA*	ATP-dependent Clp protease, ATP-binding subunit ClpA	2.8
DVU1603	*aat*	leucyl/phenylalanyl-tRNA–protein transferase	2.52
DVU1838	*trxB*-2	Thioredoxin reductase	1.76
DVU1874	*clpB*	ATP-dependent Clp protease, ATP-binding subunit ClpB	4.24
DVU2442	NA	Heat shock protein, Hsp20 family	3.66

Table 4.4 (*cont.*)

Gene No.	Name	Annotation	Log_2R^b
DVU2470	b0786	Membrane protein, putative	2.19
DVU2494	NA	Peptidase, M48 family	3.44
DVU2643	*htpG*	Heat shock protein HtpG	2.63
DVU3243	*dnaJ*	dnaJ protein	1.57
DVU0576	*msrB*	Peptide methionine sulfoxide reductase MsrB	4.65

[a] Regulon predicted by motif searches in genome (Rodionov *et al.*, 2004).
[b] Log_2R is log_2 of the ratio of treated sample to the control as given in Table 4.1.
[c] Dash means no significant change in gene expression was observed.
[d] Only genes increased in transcription are listed.

nitrate respiration and has no nitrate reductase (Moura *et al.*, 1997). In contrast, it does possess a nitrite reductase, as do many eubacteria. That enzyme is thought to protect the cell from toxic levels of nitrite produced as an intermediate from nitrate respirers in the environmental vicinity (Pereira *et al.*, 2000; Greene *et al.*, 2003). Preliminary analyses of the transcriptional responses to nitrite stress of *D. vulgaris* have been carried out (He *et al.*, 2006). The growth of *D. vulgaris* was shown to be inhibited by nitrite at concentrations at or below 5.0 mM. Growth inhibition was reversed once the nitrite concentration had been reduced to the μM range. This result was consistent with the dramatic increase in sensitivity seen with a deletion in the nitrite reductase gene *nrfA* (DVU0625) (Haveman *et al.*, 2004). The microarray analysis of changes in transcription (He *et al.*, 2006) were in excellent agreement with the transcription changes reported from the more limited array of genes by Haveman and co-workers (Haveman *et al.*, 2004). Table 4.3 shows that the genes for ATP synthase were decreased in expression as were those encoding the membrane-bound electron transport complex, DsrMKJOP, believed to supply electrons to DsrAB, the sulfite reductase. Nitrite has been reported to directly inhibit the DsrAB complex (Wolfe *et al.*, 1994), thus blocking the use of sulphate as electron acceptor for *D. vulgaris.* These results have been interpreted to mean that the diversion of electrons from sulphate reduction to nitrite reduction reduces the respiratory production of ATP. He *et al.* (2006) also observed an approximately twofold increase in expression of lactate dehydrogenase (DVU0600) and an almost threefold increase for the fumarate reductase operon (DVU3261–3). These changes could indicate that fumarate was providing a sink for electrons from lactate oxidation and/or that substrate-level phosphorylation from lactate was supporting metabolism.

4.6 NITRATE STRESS

Nitrate is not often considered to be a common environmental stressor. However, point sources of high levels of nitric acid contamination can be found in sites where heavy metals and radionuclides were processed. For microbes such as *D. vulgaris*, that are unable to use nitrate as a nitrogen source or as a terminal electron acceptor, the effects of high concentrations of nitrate can not easily be predicted. Three factors could be interacting. First any non-physiological pH would be detrimental (discussed in subsequent sections). Second, high concentrations of any molecular species would manifest an osmotic shock. Third, the production of small amounts of nitrite from non-specific reduction of nitrate by the low potential reductases of the cell (for example the multiple formate dehydrogenases, the dimethyl sulfoxide reductase, or multihaem *c*-type cytochromes) could be inhibitory.

To identify any specific transcriptional responses to sodium nitrate, increasing concentrations of this salt were added to buffered batch cultures of *D. vulgaris* in defined medium. At a concentration of just over 100 mM, a significant effect on the growth rate of the bacterium was observed. Figure 4.2 presents a diagrammatic illustration of the overlap of genes differentially transcribed in response to stresses imposed by increased concentrations of sodium nitrate, sodium nitrite, or sodium chloride. The percentage of total ORFs responding to nitrate was 21%; to nitrite, 8.6%; and to NaCl, 12%. The ORFs found responsive to nitrate and to nitrite were 1.0% of the genome, quite near the 1.8% randomly expected. Similarly 1.7% of the total ORFs responded to both nitrate and NaCl, just below the random expectation of 2.5%.

In these preliminary experiments, we infer that the response to nitrate is not a sum of salt and nitrite responses, but that high nitrate concentrations elicit a unique profile of transcript changes. A comparison of the overlap of genes responding to nitrate and stressors other than nitrite or sodium (Table 4.2) reinforces the conclusion that the nitrate response is not a composite of salt and nitrite stresses.

A notable feature of the *D. vulgaris* response to nitrate was a strong downregulation of genes for transposases (Table 4.5). Interestingly, this response was exhibited by cells subjected to shifts in environmental pH

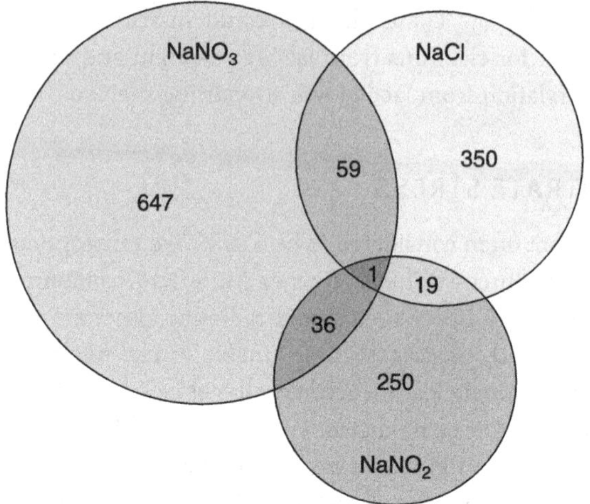

Figure 4.2. Illustration of the number of differentially expressed genes that were in common between and among the three stressors nitrate, nitrite, and sodium chloride in *D. vulgaris*.

as well, both high and low. In contrast, a number of ion transporters, in particular those involving sodium for symport or antiport, were upregulated by exposure to high nitrate (Table 4.5). This treatment also resulted in increases in transcription of a number of flagellar biosynthesis genes, a result that was consistent with microscopic observation of their motility (data not shown). The confirmation and meaning of these changes awaits further experimentation.

4.7 ACID pH STRESS

Exposure to environmental pHs outside of the physiologically acceptable range of a bacterium occurs frequently. For pathogens traversing the stomach, multiple strategies for surviving this low pH environment have evolved. A least three systems have been documented in *E. coli* (Foster, 2004). The first is a poorly understood adaptation occurring in cells grown at an intermediate acid pH that then allows the cells to survive exposure to what is considered a lethal pH, 2.5. This response has been shown to be RpoS regulated.

Although *D. vulgaris* has no recognizable ORF encoding RpoS, preliminary studies were carried out to explore the existence of an adaptive acid response. No indication of such a response was obtained. The second

Table 4.5. *Differentially transcribed genes for transposases and transporters in* D. vulgaris *cells stressed by high* $NaNO_3$ *concentrations*

Transposases			Transporters		
Gene No.	Annotation	$\log_2 R^a$	Gene No.	Annotation	$\log_2 R^a$
DVU0556	ISDvu3, transposase OrfA	−2.23	DVU0088	Na^+/pantothenate symporter	2.86
DVU0557	ISDvu3, transposase OrfB	−2.50	DVU0381	Na^+/H^+ antiporter NhaC-1	4.28
DVU0562	ISD1, transposase OrfA	−2.43	DVU0413	K^+ uptake, Trk1	1.43
DVU0564	ISDvu4, transposase truncation	−1.26	DVU0446	Na^+/solute symporter family	1.52
DVU2010	ISD1, transposase OrfB	−1.64	DVU3108	Na^+/H^+ antiporter NhaC-2	5.13
DVU2017	ISDvu5, transposase	−1.95	DVU3332	Heavy metal P-type ATPase	2.23
DVU2004	ISDvu4, transposase	−1.72	DVU0177	*modA*, periplasmic protein, Mo ABC transporter	1.56
D U2049	Transposase OrfA, IS3 family	−1.97	DVU0180	*modC*, ATP binding protein Mo ABC transporter	1.32
DVU2178	ISDvu2, transposase OrfB	−1.30			
DVU2179	ISDvu2, transposase, OrfA	−2.49			

[a] R is the ratio of experimental transcript to control.

and third systems of acid resistance in *E. coli* are based on intracellular proton consumption by amino acid decarboxylases coupled with antiporters that remove the decarboxylated product in exchange for another amino acid (Foster, 2004). Glutamate, arginine, and possibly lysine participate in such systems to maintain pH homeostasis.

When *D. vulgaris* is inoculated into growth medium with a pH below about 6.8, cells do not grow until the pH has been adjusted to the circumneutral range by metabolic processes. The acid stress condition used for transcript analysis, pH 5.5, resulted in an extended lag of >100 h. Although the cells did eventually grow, the cell yield was always lower than non-stressed cells. When the pH was determined after growth, it was increased over 0.5 pH unit from the uninoculated medium (unpublished). The genome sequence was explored to determine whether the potential for acid resistance systems similar to the amino acid dependent ones of *E. coli* could account for the pH adjustment by *D. vulgaris.* No apparent orthologues were present for any genes of the glutamate decarboxylation system or its known control elements. However, an operon encoding arginine decarboxylase was found and four of the five genes in this operon were significantly increased in expression by the acid treatment (Table 4.6). The apparent hallmark of the acid resistance system, the adjacent location of a corresponding antiporter, is not apparent in *D. vulgaris* although there are many ORFs elsewhere that are annotated as amino acid transporters. In contrast, the genes nearby that encoding arginine decarboxylase appear to be involved

Table 4.6. *Selected* D. vulgaris *genes increased in transcription in response to pH 5.5*

Gene No.	Annotation	Log_2R^a
DVU0417	Arginine decarboxylase	2.79
DVU0418	Saccharopine dehydrogenase	0.80^b
DVU0419	Carboxynorspermidine decarboxylase	4.69
DVU0420	Hypothetical protein	2.92
DVU0421	Agmatinase, putative	1.81
DVU3336	K^+ channel, histidine kinase domain	1.58
DVU3337	K^+-transporting ATPase, C subunit	3.43
DVU3338	K^+-transporting ATPase, B subunit	2.57
DVU3339	K^+-transporting ATPase, A subunit	4.72

[a] R is the ratio of the experimental transcript to the control.
[b] This ratio was not statistically significant.

in polyamine biosynthesis (Table 4.6). Work with *Vibrio cholerae* has recently shown that polyamine may serve as a signal for biofilm formation (Karatan *et al.*, 2005), a growth mode shown to provide protection against less than optimal environments. Interestingly, acid treatment of *D. vulgaris* causes a quite obvious increase in adherence to glassware (unpublished observations). Whether there is a causal relationship between this phenotype and the expression of polyamine biosynthesis has yet to be established.

The potassium transporting ATPase operon was considerably increased in acid treated cells as well (Table 4.6). If ATP were used to pump potassium from the cells, perhaps the potassium could subsequently re-enter through an antiporter in exchange for protons facilitating the maintenance of a compatible internal pH. Although a potassium/proton anitporter was not identified, a sodium/proton antiporter (DVU0381) was increased by exposure to acid (log_2R of 1.33).

Additional systems that responded by upregulation to the acid conditions included 14 genes involved in flagellum biosynthesis and DNA repair genes. As mentioned previously, genes coding for seven different transposases were decreased in expression. Curiously, the *hrcA* gene (DVU0813), annotated to code for the heat inducible transcription repressor, was upregulated (log_2R of 2.88). The collection of changes in gene expression in *D. vulgaris* in response to acidic pH requires additional physiological and biochemical studies for clarity.

4.8 ALKALINE pH STRESS

For most mesophylls, growth at an external pH of about 7.8 or greater results in an inverted pH gradient across the cytoplasmic membrane. *D. vulgaris* Hildenborough tolerates growth medium at slightly alkaline pHs (up to 8.5) better than pHs below 6.8; however, at pH 10, the treatment tested in these microarray studies (Table 4.1), growth was inhibited. Bacteria often shift the external pH by their own metabolism; for example, organic or inorganic acids can be produced or consumed. In particular, *D. vulgaris* is an incomplete oxidizer of organic acids producing acetate (Postgate, 1984); however, the pH actually becomes alkaline with growth likely because of the escape of the end products H_2S and CO_2. To determine how *D. vulgaris* begins to compensate for exposure to high pH, a preliminary analysis of the transcriptional response of this bacterium to a high pH shock was performed.

In *E. coli*, reactions that minimize the alkalization of the cytoplasm are identified among the base-responsive proteins. In rich medium, amino acid catabolism is apparently directed to produce acids (Stancik *et al.*, 2002). Tryptophanase that deaminates serine, cysteine, and tryptophan (McFall and Newman, 1996), becomes one of the most highly expressed proteins as pH rises (Blankenhorn *et al.*, 1999). A gene encoding that enzyme, *tnaA*, upstream of a tryptophan transport protein, *mtr*, is annotated in *D. vulgaris* (DVU2204 and DVU2205, respectively). These genes were not increased in base-shocked cells. If the tryptophanase gene in *D. vulgaris* were regulated similar to that in *E. coli*, then the absence of the inducer tryptophan and the presence of acetate that represses *tnaA* synthesis (Stancik *et al.*, 2002) could account for this lack of response.

Another amino acid biosynthesis enzyme, CysK (*o*-acetylserine sulfhydrylase A) was found among *E. coli* proteins induced by base treatment. The transcript of *cysK* was also confirmed to be increased by high pH exposure (Stancik *et al.*, 2002). It was proposed that CysK might act in a degradative mode, producing pyruvate to contribute to an acidification of the internal pH. Although the transcript for the *D. vulgaris* ortholog (DVU0663) was increased by high pH, it is not understandable that a cell growing in defined medium with lactate as the sole carbon source would degrade cysteine for pH control. Interestingly, amino acid biosynthetic genes were enriched among those genes most responsive to high pH in *D. vulgaris*.

The challenge of high pH in the environment is often coincident with other stressors such as sodium. Table 4.2 shows that the number of genes that responded to high pH which were also differentially expressed in response to sodium was not greater than expected by random. Following the shift up in pH, there were no obvious patterns in expression changes of the major genes coding for energy generation or substrate utilization (Table 4.3). The only hint at compensation for the inverted membrane pH gradient was an increase in the expression of the Na^+/H^+ transporter encoded by *nhaC-2* (DVU3108, $\log_2$R of 1.91). Because sulphate can be transported by symport with sodium (Cypionka, 1995), the accumulated sodium could, in turn, be used to drive the uptake of low concentrations of protons to moderate the alkalization of the cytoplasm. Clearly, much more must be learned before the ability of *D. vulgaris* to grow at alkaline pHs is understood.

4.9 UNIVERSAL STRESS RESPONSE

Analysis of stationary phase responses in *E. coli* led to the conceptual understanding of global regulators and allowed the identification of the master regulator, RpoS or σ^S. Not only does this sigma factor control many stationary phase-inducible genes, but it also induces genes in response to many other stresses (Hengge-Aronis, 2000). *D. vulgaris*, like many other non-γ-proteobacteria (Nies, 2004), does not have a recognized RpoS. Therefore, a question addressed with these preliminary data was whether *D. vulgaris* had a common core of genes responding to multiple stresses. At least nine ORFs of this bacterium were annotated as belonging to the "universal stress protein family". However, their expression was not coordinated or consistent and only three of the genes were significantly altered in expression in more than a single stress.

When the ORFs responding in four or more of the treatments were examined, genes that had been suggested to be part of the Fur (iron sensor) or PerR (peroxide sensor) regulon were notably over-represented (Table 4.3). These regulon members had been identified by whole genome searches for conserved binding motifs (Rodionov *et al.*, 2004). The Fe^{2+}-dependent repressor Fur regulates the uptake of iron in most Gram-negative bacteria and some of the genes for oxidative response (Hantke and Braun, 2000). Thus Fur is considered to be a global regulator responding to a number of stresses. In *D. vulgaris*, an orthologue for *fur* (DVU0942) and two additional paralogues, *perR* (DVU3095) and *zur* (DVU1340; annotated as encoding a zinc sensor) are identified. Table 4.3 shows the transcriptional responses of most of the genes predicted to be members of the Fur regulon in *D. vulgaris*. A number of these genes are annotated as acquisition systems for Fe(II) and are orthologues of genes found in aerobes (Kammler *et al.*, 1993; Robey and Cianciotto, 2002). In anaerobic environments where members of the *Desulfovibrio* genus are found, Fe(II) would be expected to be available. Consistent with the expectation that Fe(II) would be the source of iron, genes for production of siderophores for Fe(III) acquisition have not been recognised in the *D. vulgaris* genome. Interestingly, most of the Fur regulon was increased in expression in many of the stress treatments. From this observation, it is tempting to infer that Fe(II) was limiting during responses to the stresses. Perhaps repair of damage and possibly loss of Fe(II) from the many iron-containing proteins is a general stress response in this bacterium. This damage is distinguishable from the response to oxygen which, in these studies, was notable for derepression of the PerR regulon (Table 4.3).

In conclusion, an examination of differentially expressed genes in response to various stresses has allowed the beginnings of insights into the robust life style of the sulphate-reducing bacteria. Overlap between responses to cold or to salt and the uniqueness of a response to nitrate concentrations deserve further exploration. In addition, there are features of the heat shock response of *D. vulgaris* that may reveal the involvement of new regulators not yet suspected as players in this physiology. Finally, just how these anaerobes compensate for an apparent absence of the global regulator, RpoS, remains to be elucidated. Further experiments with selected mutants will begin to clarify some of these questions, along with the combined power of proteomic and metabolomic analyses. As these experiments progress, the ability to predict the metabolic capacity of these bacteria in environmental settings were bioremediation is desired will steadily improve.

ACKNOWLEDGEMENTS

This work was part of the Virtual Institute for Microbial Stress and Survival (http://vimss.lbl.gov) supported by the US Department of Energy, Office of Science, Office of Biological and Environmental Research, Genomics Program: GTL through contract DE-AC02–05CH11231 between Lawrence Berkeley National Laboratory and the US Department of Energy with a subcontract to the University of Missouri–Columbia.

REFERENCES

Birrell, G. W., Brown, J. A., Wu, H. I. *et al.* (2002). Transcription response of *Saccharomyces cerevisiae* to DNA-damaging agents does not identify the genes that protect against these agents. *PNAS*, **99**, 8778–83.

Blankenhorn, D., Phillips, J. and Slonczowski, J. L. (1999). Acid- and base-induced proteins during aerobic and anaerobic growth of *Escherichia coli* revealed by two-dimensional gel electrophoresis. *J. Bacteriol.*, **181**, 2209–16.

Bremer, E. and Kramer, R. (2000). Coping with osmotic challenges: osmoregulation through accumulation and release of compatible solutes in bacteria. In G. Storz and R. Hengge-Aronis (eds.), *Bacterial stress reponses*. Washington, DC: ASM Press. pp. 79–97.

Cases, I., Ussery, D. W. and de Lorenzo, V. (2003). The sigma54 regulon (sigmulon) of *Pseudomonas putida*. *Environ. Microbiol.*, **5**, 1281–93.

Cheung, K. J., Badarinarayana, V., Selinger, D. W., Janse, D. and Church, G. M. (2003). A microarray-based antibiotic screen identifies a regulatory role for supercoiling in the osmotic stress response of *Escherichia coli*. *Genome Res.*, **13**, 206–15.

Chhabra, S. R., He, Q., Huang, K. H. *et al.* (2006). Global analysis of heat shock response in *Desulfovibrio vulgaris* Hildenborough. *J. Bacteriol.*, **188**, 1817–28.

Colantuoni, C., Henry, G., Zeger, S. and Pevsner, J. (2002). Local mean normalization of microarray element signal intensities across an array surface: quality control and correction of spatially systematic artifacts. *Biotechniques*, **32**, 1316–20.

Csonka, L. N. and Epstein, W. (1996). Osmoregulation. In F. C. Neidhardt, R. Curtiss III, J. L. Ingraham *et al.* (eds.), *Escherichia coli and Salmonella: cellular and molecular biology*, 2nd edn. Washington, DC: ASM Press. pp. 1210–23.

Cypionka, H. (1995). Solute transport and cell energetics. In L. L. Barton (ed.), *Sulphate-reducing bacteria*. New York: Plenum Press. pp. 151–84.

Foster, J. W. (2004). *Escherichia coli* acid resistance: tales of an amateur acidophile. *Nat. Rev. Microbiol.*, **11**, 898–907.

Giaever, G., Chu, A. M., Ni, L. *et al.* (2002). Functional profiling of the *Saccharomyces cerevisiae* genome. *Nature*, **418**, 387–91.

Gorby, Y. A. and Lovley, D. R. (1992). Enzymatic uranium precipitation. *Environ. Sci. Technol.*, **26**, 205–7.

Graumann, P. L. and Marahiel, M. A. (1998). A superfamily of proteins that contain the cold-shock domain. *Trends Biochem. Sci.*, **23**, 286–90.

Greene, E. A., Hubert, C., Nemati, M., Jenneman, G. E. and Voordouw, G. (2003). Nitrite reductase activity of sulphate-reducing bacteria prevents their inhibition by nitrate-reducing sulfide-oxidizing bacteria. *Environ. Microbiol.*, **5**, 607–17.

Gross, C. A. (1996). Function and regulation of the heat shock proteins. In F. C. Neidhardt, R. Curtiss III, J. L. Ingraham, E. C.C. *et al.* (eds.), *Escherichia coli and Salmonella: cellular and molecular biology*, 2nd edn. Washington, D.C.: ASM Press. pp. 1382–99.

Hantke, K. and Braun, V. (2000). The art of keeping low and high iron concentrations in balance. In G. Storz and R. Hengge-Aronis (eds.), *Bacterial stress reponses*. Washington, DC: ASM Press. pp. 275–88.

Haveman, S. A., Greene, E. A., Stilwell, C. P., Voordouw, J. K. and Voordouw, G. (2004). Physiological and gene expression analysis of inhibition of *Desulfovibrio vulgaris* Hildenborough by nitrite. *J. Bacteriol.*, **186**, 7944–50.

He, Q., Huang, K. H., He, Z. *et al.* (2006). Energetic consequences of nitrite stress in *Desulfovibrio vulgaris* Hildenborough inferred from global transcriptional analysis. *Appl. Environ. Microbiol.*, **72**, 4370–81.

Heidelberg, J. F., Seshadri, R., Haveman, S. A. *et al.* (2004). The genomic sequence of the anaerobic, sulphate-reducing bacterium *Desulfovibrio vulgaris* Hildenborough. *Nat. Biotechnol.*, **22**, 554–9.

Hengge-Aronis, R. (2000). The general stress response in *Escherichia coli.* In G. Storz and R. Hengge-Aronis (eds.), *Bacterial stress reponses.* Washington, DC: ASM Press. pp. 161–78.

Jones, P. G. and Inouye, M. (1996). RbfA, a 30S ribosomal binding factor, is a cold-shock protein whose absence triggers the cold-shock response. *Mol. Microbiol.*, **21**, 1207–18.

Kammler, M., Schon, C. and Hantke, K. (1993). Characterization of the ferrous iron uptake system of *Escherichia coli. J. Bacteriol.*, **175**, 6212–9.

Karatan, E., Duncan, T. R. and Watnick, P. I. (2005). NspS, a predicted polyamine sensor, mediates activation of *Vibrio cholerae* biofilm formation by norspermidine. *J. Bacteriol.*, **187**, 7434–43.

Klenk, H. P., Clayton, R. A., Tomb, J. F. *et al.* (1997). The complete genome sequence of the hyperthermophilic, sulphate-reducing archaeon *Archaeoglobus fulgidus. Nature,* **390**, 364–70.

Lovley, D. R., Phillips, E. J. P., Gorby, Y. A. and Landa, E. (1991). Microbial reduction of uranium. *Nature,* **350**, 413–16.

McFall, E. and Newman, E. B. (1996). Amino acids as carbon sources. In F. C. Neidhardt, R. Curtiss III, J. L. Ingraham (eds.), *Escherichia coli and Salmonella: Cellular and Molecular Biology,* 2nd edn. Washington, DC: ASM Press. pp. 358–79.

Minder, A. C., Fischer, H.-M., Hennecke, H. and Narberhaus, F. (2000). Role of HrcA and CIRCE in the heat shock regulatory network of *Bradyrhizobium japonicum. J. Bacteriol.*, **182**, 14–22.

Moura, I., Bursakov, S., Costa, C. and Moura, J. J. G. (1997). Nitrate and nitrite utilization in sulphate-reducing bacteria. *Anaerobe,* **3**, 279–90.

Mukhopadhyay, A., He, Z., Yen, H.-C. (2006). Salt stress in *Desulfovibrio vulgaris* Hildenborough: an integrated genomics approach. *J. Bacteriol.*, **188**, 4068–78.

Nies, D. H. (2004). Incidence and function of sigma factors in *Ralstonia metallidurans* and other bacteria. *Arch. Microbiol.*, **181**, 255–68.

Pereira, I. A. C., LeGall, J., Zavier, A. V. and Teixeira, M. (2000). Characterization of heme c nitrite reductase from a non-ammonifying microorganism, *Desulfovibrio vulgaris* Hildenborough. *Biochim. Biophys. Acta,* **1481**, 119–30.

Phadtare, S., Yamanaka, K. and Inouye, M. (2000). The cold shock response. In G. Storz and R. Hengge-Aronis (eds.), *Bacterial stress responses.* Washington, DC: ASM Press. pp. 33–45.

Postgate, J. R. (1984). *The sulphate reducing bacteria* (2nd edn). Cambridge and London: Cambridge University Press.

Rabus, R., Ruepp, A., Frickey, T. *et al.* (2004). The genome *Desulfotalea psychrophila*, a sulphate-reducing bacterium from permanently cold Artic sediments. *Environ. Microbiol.*, **6**, 887–902.

Raivio, T. L. and Silhavy, T. J. (2000). Sensing and responding to envelope stress. In G. Storz and R. Hengge-Aronis (eds.), *Bacterial stress reponses.* Washington, DC: ASM Press. pp. 19–32.

Roberts, R. C., Toochinda, C., Avedissian, M. *et al.* (1996). Identification of a *Caulobacter crescentus* operon encoding *hrcA*, involved in negatively regulating heat-inducible transcription and the chaperone gene *grpE*. *J. Bacteriol.*, **178**, 1829–41.

Robey, M. and Cianciotto, N. P. (2002). *Legionella pneumophila feoAB* promotes ferrous iron uptake and intracellular infection. *Infect. Immun.*, **70**, 5659–69.

Rodionov, D. A., Dubchak, I., Arkin, A. P., Alm, E. J. and Gelfand, M. S. (2004). Reconstruction of regulatory and metabolic pathways in metal-reducing δ-proteobacteria. *Genome Biol.*, **5**, R90.

Stanik, L. M., Stanik, D. M., Schmid, B. *et al.* (2002). pH-Dependent expression of periplasmic proteins and amino acid catabolism in *Escherichia coli*. *J. Bacteriol.*, **184**, 4246–58.

Vila-Sanjurjo, A., Schuwirth, B. S., Hau, C. W. and Cate, J. H.D. (2004). Structural basis for the control of translation initiation during stress. *Nature Struc. Mol. Bio.*, **11**, 1054–9.

Wolfe, B. M., Lui, S. M. and Cowan, J. A. (1994). Desulfoviridin, a multimeric-dissimilatory sulfite reductase from *Desulfovibrio vulgaris* Hildenborough purification, characterization, kinetics and EPR studies. *Eur. J. Biochem.*, **223**, 79–89.

Yura, T. K., Kanemori, M. and Morita, M. T. (2000). The heat shock response: regulation and function. In G. Storz and R. Hengge-Aronis (eds.), *Bacterial stress reponses*. Washington, DC: ASM Press. pp. 3–18.

Zuber, U. and Schumann, W. (1994). CIRCE, a novel heat shock element involved in regulation of heat shock operon *dnaK* of *Bacillus subtilis*. *J. Bacteriol.*, **176**, 1359–63.

CHAPTER 5

Response of sulphate-reducing bacteria to oxygen

Henrik Sass and Heribert Cypionka

5.1 PRESENCE OF SULPHATE-REDUCING BACTERIA IN OXIDISED HABITATS

During the second half of the nineteenth century the formation of sulphide from sulphate was recognised as a biogenic process (Meyer, 1864). While it was initially suggested that algae were the catalysing organisms (Cohn, 1867), Hoppe-Seyler demonstrated in 1886 that the process required anoxic conditions and was chemotrophic, requiring external electron donors. In 1895, Beijerinck proved that sulphate reduction is catalysed by bacteria and described the first pure culture, *Spirillum desulfuricans*. This organism was described as strictly anaerobic and was irreversibly inhibited by oxygen.

The view that sulphate-reducing bacteria (SRB) are extremely sensitive to oxygen started to change in the late 1970s when sulphate reduction was demonstrated to occur also in oxidised sediment layers which showed no traces of FeS and were considered oxic (Jørgensen, 1977). Similarly, cultivation-based studies revealed the presence of viable sulphate reducers within these layers (Laanbroek and Pfennig, 1981; Battersby *et al.*, 1985; Jørgensen and Bak, 1991). However, it was found that oxygen did not penetrate as deep into sediments as previously assumed and that large parts of the oxidised, hence FeS-free layers, were in fact anoxic. In sediment layers that contain oxidised manganese or iron species, sulphide can be chemically reoxidised to elemental sulphur (Aller and Rude, 1988), or, in the case of manganese oxide, even to thiosulphate (Schippers and Jørgensen, 2001). Using these sulphur species as alternative electron acceptors is advantageous for sulphate reducers as it saves adenosine triphosphate (ATP) that is normally used for sulphate activation (Cypionka *et al.*, 1985). Another reason to live close to the sediment surface might be the better supply of organic material, either by sedimentation or by exudation from benthic photosynthetic organisms (Wind and Conrad, 1995; Sass *et al.*, 1997; Blaabjerg *et al.*, 1998). Several investigations based on cultivation or molecular methods have demonstrated that numbers of SRB peak in the chemoclines of sediments or microbial mats (Risatti *et al.*, 1994; Sass *et al.*, 1997; Minz *et al.*, 1999a; Wieringa *et al.*, 2000) with a second peak found deeper in the permanently anoxic layers. This depth distribution seems not to be restricted to sedimentary environments. Similar patterns were found in stratified water bodies of fjords (Ramsing *et al.*, 1996; Teske *et al.*, 1996) and lakes (Tonolla *et al.*, 2000). In biofilms and microbial mats, however, high numbers of sulphate reducers were observed even under fully oxygenated conditions (Visscher *et al.*, 1992; Ramsing *et al.*, 1993; Risatti *et al.*, 1994; Ito *et al.*, 2002; Okabe *et al.*, 2003).

In hypersaline microbial mats strong diurnal changes of oxygen concentrations occur. During daytime, photosynthesis leads to oxygen tensions above air saturation, while during the night oxygen often is depleted. However, sulphate reduction rates measured during daytime by the use of radiotracers were similar to those under anoxic conditions during the night (Canfield and Des Marais, 1991; Fründ and Cohen, 1992; Jørgensen, 1994; Teske *et al.*, 1998). So far this "aerobic sulphate reduction" remains elusive since it has only been observed in natural environments and has not been demonstrated in laboratory experiments with enrichments or pure cultures.

Microbiological and molecular studies have revealed that chemoclines and oxic layers harbour different types of SRB than permanently anoxic layers (Risatti *et al.*, 1994; Sass *et al.*, 1998b; Minz *et al.*, 1999b). In the oxic layers predominantly *Desulfococcus, Desulfonema* and *Desulfovibrio* species were detected, while *Desulfobulbus, Desulfobacter* and *Desulfotomaculum* species appear to be restricted to permanently anoxic horizons. Differences have also been found at the species level. In the sediments of an oligotrophic lake, the layers close to the oxic–anoxic interface were dominated by other *Desulfovibrio* species than the permanently anoxic layers (Sass *et al.*, 1998a; 1998b).

The presence of SRB in oxidized sediments was first explained by the presence of small anoxic microniches within the sediment. For marine sediments Jørgensen (1977) estimated a minimum niche diameter of 200 μm, even at oxygen concentrations as low as 3 to 5 $\mu mol{\cdot}l^{-1}$. However, the development of high-resolution microelectrodes made it possible to investigate oxygen profiles in detail. By these methods, anoxic microniches were neither detected in sediments, nor in microbial mats or biofilms (Ramsing *et al.*, 1993). In contrast, the existence of anoxic microniches was proven for sludge flocks (Schramm *et al.*, 1999) and marine detrital aggregates (Ploug *et al.*, 1997). However, the steep gradients observed in these aggregates, which are necessary to maintain anoxic conditions in the centre, also indicated that the organic material would not support this high oxygen consumption over periods longer than several hours. Such ephemeral anoxic conditions, however, do not allow the development of stable communities of sulphate-reducing bacteria.

5.2 OXYGEN TOLERANCE AND DETOXIFICATION

Sulphate-reducing bacteria possess various physiological adaptations for coping with oxygen. Hardy and Hamilton (1981) isolated some strains from oxic North Sea waters and found that they survived oxygen exposure of up to 72 h. Physiological experiments demonstrated strong differences in oxygen tolerance among different species (Sass *et al.*, 1996). The capacity to survive long periods of oxygen exposure can be seen as an adaptation for dispersion to new environments. For example, sulphate-reducing bacteria colonising drinking water biofilms were characterised as extremely oxygen-tolerant, surviving up to 72 days of aeration (Bade *et al.*, 2000).

The toxicity of molecular oxygen is largely due to the formation of partially reduced species (Imlay, 2002). The presence of reduced free iron, sulphide or sulphydryl groups leads to the formation of superoxide radicals and peroxides which strongly increase the toxicity of oxygen (Cypionka *et al.*, 1985). The ability to detoxify hydrogen peroxide or superoxide radicals seems to be a major prerequisite for thriving close to the oxic–anoxic interface. Abdollahi and Wimpenny (1990) found that the presence of even low oxygen concentrations leads to an increase in respiration and to elevated levels of superoxide dismutase and nicotinamide adenine dinucleotide (NADH)-oxidase activity. So far, several additional proteins involved in the response to oxidising conditions have been identified, for example catalase and rubredoxin oxygen oxidoreductase (Fareleira *et al.*, 2003; Fournier *et al.*, 2003). Sass *et al.* (1997) found that SRB isolated from increasing depths of lake sediment differed in respect of catalase activity. All isolates from the upper oxidised three centimetres were shown to possess catalase activity, while this feature was absent from most isolates from the deeper permanently anoxic zones. However, some strains exhibiting catalase activity were nevertheless oxygen-sensitive, indicating that coping with oxygen radicals is probably only one aspect amongst others involved in oxygen tolerance.

5.3 AEROBIC RESPIRATION

In 1990, Dilling and Cypionka discovered that sulphate-reducing bacteria can use molecular oxygen as a terminal electron acceptor and that they can even gain ATP from this process. Respiration rates in some sulphate-reducers can be extraordinarily high, exceeding those of some aerobes like *Escherichia coli* (Kuhnigk *et al.*, 1996). Several mechanisms for oxygen reduction in sulphate-reducing bacteria are known (reviewed in Cypionka, 2000). In *Desulfovibrio desulfuricans, D. salexigens* and *D. gigas,* cytoplasmic NADH oxidase activity was found (Abdollahi and Wimpenny, 1990; Chen *et al.*, 1993; van Niel and Gottschal, 1998). In *D. gigas* a NADH rubredoxin oxidoreductase and a FAD and haem-containing enzyme were found to reduce oxygen to hydrogen peroxide or water (Chen *et al.*, 1993; Gomes *et al.*, 1997). An additional membrane-bound oxygen-reducing respiratory chain involving a cytochrome of the *bd*-type was also found in *D. gigas* (Lemos *et al.*, 2001). Cytochrome *bd* has been detected not only in sulphate reducers, but also in other "strict" anaerobes like homoacetogens and *Bacteroides fragilis.* Here it was suggested to be involved in protection against oxidative stress but not in energy generation (Baughn and Malamy, 2004; Das *et al.*, 2005). In the sulphate reducers with the highest oxygen respiration rates, *Desulfovibrio vulgaris* and *D. termitidis,* periplasmic hydrogenase and cytochrome *c* were found to play a major role in oxygen reduction (Baumgarten *et al.*, 2001). A periplasmic *c*-type cytochrome was also found in the microaerotolerant fermenter *Malonomonas rubra* lacking respiratory activity (Kolb *et al.*, 1998) and was suggested to be involved in oxygen defence.

Several substrates serve as electron donors for aerobic respiration including organic substrates but also molecular hydrogen and even reduced sulphur compounds (Dannenberg *et al.*, 1992; Kuhnigk *et al.*, 1996; Sass *et al.*, 1997). While some strains readily oxidise a broad variety of substrates, other sulphate reducers show constant oxygen reduction activity which cannot be stimulated by the addition of external substrates. This might be explained by the oxygen sensitivity of dissimilatory enzymes like lactate dehydrogenase (Stams and Hansen, 1982) or hydrogenase (Baumgarten *et al.*, 2001) and by the use of endogenous substrates like polyglucose for the reduction of oxygen (van Niel *et al.*, 1996).

When molecular oxygen is present simultaneously with nitrate or oxidised sulphur species, sulphate-reducing bacteria will first reduce oxygen (Krekeler and Cypionka, 1995). As mentioned above, no pure culture reducing sulphate in the presence of oxygen has been reported

(Krekeler and Cypionka, 1995; Sass *et al.*, 1996; Kjeldsen *et al.*, 2004; Jonkers *et al.*, 2005).

5.4 BEHAVIOUR IN OXYGEN GRADIENTS

For sulphur-oxidizing bacteria both H_2S and O_2 are essential energy substrates, and it is not surprising that they have developed specific adaptations for life at oxygen–sulphide boundaries. Sulphur oxidizers are even able to establish conditions under which the chemical reaction between H_2S and O_2 is minimised and biotic oxidation prevails. Many of them form mats and some species (*Thiovulum*) floating veils at the oxygen–sulphide interface (Fenchel, 1994). Vacuolated species (*Thioploca* or *Beggiatoa*) migrate into sulphidic zones after having accumulated nitrate as an electron acceptor (Schulz and Jørgensen, 2001). If the sulphate reducers were all strict anaerobes, a comparable behaviour would not appear necessary. However, they show various behavioural responses to oxygen that appear as complex as those of their sulphur-oxidising counterparts. Obviously, for sulphate reducers oxygen is not just a poison that has to be avoided. As mentioned above, in many natural environments the presence of oxygen is accompanied by an increased availability of electron donors. Sulphate reducers can use a variety of electron acceptors including molecular oxygen. Why they did not evolve the capacity for sustainable aerobic growth (Marschall *et al.*, 1993; Johnson *et al.*, 1997; Sigalevich *et al.*, 2000) cannot be simply explained, but we propose a hypothesis below.

5.5 AGGREGATION

Among responses to oxygen, aggregation of cells or attachment to surfaces appears to be the simplest one. Sulphate-reducing bacteria tend to form aggregates when they are exposed to oxygen in homogeneously mixed suspensions (Krekeler *et al.*, 1998; Sass *et al.*, 1998a; Teske *et al.*, 1998; Mogensen *et al.*, 2005). It is not yet clear whether aggregation represents an active behaviour or results indirectly from chemical changes at the cell surfaces, e.g. oxidation of thiol groups. Aggregated bacteria might be able to keep the interior of the aggregate anoxic as has been observed for sludge flocks (Schramm *et al.*, 1999). But even just particle-associated sulphate reducers reveal higher survival rates upon oxygen exposure than free-living bacteria (Fukui and Takii, 1990; 1994). Since the environmental conditions at oxic–anoxic interfaces are often variable, there is a good chance to return to more favourable anoxic conditions after some time. By oxygen respiration the sulphate reducers themselves may contribute to this change.

5.6 MIGRATION

In microbial mats, sulphate reducers were found to migrate vertically in response to the oxygen regime. MPN counts of sulphate-reducing bacteria in the upper 3 mm of a cyanobacterial mat of Solar Lake (Sinai, Egypt) were 20-fold lower during the day when benthic photosynthesis leads to high oxygen concentrations, than at night in the absence of oxygen (Krekeler *et al.*, 1998). Thus, although sulphate reduction was measured even at oxygen saturation (Canfield and DesMarais, 1991), at least part of the sulphate-reducing populations appeared to avoid high oxygen concentrations. In the same mat, a *Desulfonema* species diurnally migrating within and below the oxic surface layers was detected by means of molecular techniques (Teske *et al.*, 1998).

5.7 AEROTAXIS

Mobile sulphate reducers show chemotactic responses. The complete genome analysis has shown that *Desulfovibrio vulgaris* harbours numerous chemotaxis genes (Heidelberg *et al.*, 2004). Lactate, sulphate, thiosulphate and even sulphide were found to be attractants (Sass *et al.*, 2002). Oxygen causes aerotactic band formation. A *c*-type haem-containing methyl-accepting protein, *DcrA*, functions as a sensor for oxygen concentrations or redox potentials (Fu *et al.*, 1994; Fu and Voordouw, 1997). Deletion of the corresponding gene led to a decrease in oxygen tolerance (Voordouw and Voordouw, 1998). The bands in oxygen gradients are formed at low oxygen concentrations within a few minutes. The behaviour of SRB resembles that of microaerophilic bacteria forming "Atmungsfiguren" [respiratory figures] as described by Beijerinck (1893). This includes a negative response to high oxygen concentrations, but no strict action of O_2 as a repellent, as the cells stay within the oxic environment (Eschemann *et al.*, 1999; Sass *et al.*, 2002). Recently, aerotactic band formation by *D. desulfuricans* DSM 9104 was studied in a stopped-flow diffusion chamber (Fischer and Cypionka, 2005). By means of this chamber, reproducible, steep oxygen gradients in a flat capillary could be created, allowing time-lapse video recordings and spatio-temporal analysis of band formation. The cells formed two types of bands. Bands of the first type evolved quickly after starting the experiment and were located near the oxic–anoxic interface. Bands of the second type appeared typically several minutes later and a few millimetres inside the initially anoxic volume of the capillary. Band formation was dependent on the presence of an electron donor and could be stimulated by lactate addition, thus appearing to be energy taxis. Mathematical modelling of oxygen diffusion and respiration within the chamber revealed that bands formed preferentially at oxygen concentrations close to 4% air saturation. The swimming speed of the cells was highest (up to $58\,\mu m \cdot s^{-1}$) close to the oxic–anoxic interfaces. These observations can be interpreted as elements of an active defence strategy; the cells try to remove oxygen from the environment to re-establish anoxic conditions for growth. As oxygen reduction is at least partially coupled to proton translocation (Fitz and Cypionka, 1991), the rising electrochemical potential could change the motility pattern and speed.

5.8 GROWTH IN OXYGEN-SULPHIDE GRADIENTS

Even though some SRB can respire with oxygen, the latter does not support sustainable growth. So far, only a slight increase in biomass at very low oxygen concentrations has been reported (Marschall *et al.*, 1993; Johnson *et al.*, 1997). In mixed cultures with aerobes, sulphate-reducing bacteria have been grown under continuous aeration. However, in these experiments steady growth was only achieved when oxygen was consumed by the syntrophic partners (Gottschal and Szewzyk, 1985; van den Ende *et al.*, 1997; Sigalevich *et al.*, 2000). In a comparative study on sulphate-reducing bacteria isolated from lake sediments, remarkable differences in oxygen tolerance were found (Sass *et al.*, 1996; 1997). Some strains were irreversibly inhibited by exposure to concentrations as low as $1\,\mu mol \cdot l^{-1}$, while others survived exposure to 20% air saturation. In some oxygen-exposed cultures cells were strongly elongated but obviously did not divide (Figure 5.1). These findings indicate that, at least in some strains, biomass may be formed in the presence of oxygen, but cell division is inhibited by molecular oxygen. This might be due to inactivation of ribonucleotide reductase, an enzyme involved in the replication of DNA (LeGall and Xavier, 1996). This rather simple inhibition mechanism might explain why sulphate reducers often resume growth as soon as oxygen is depleted (Figure 5.2) (Sass *et al.*, 1996; van Niel *et al.*, 1996; Kjeldsen *et al.*, 2004). In microbial mats, sulphate reducers might start growing immediately after establishing anoxic conditions during night, while growth ceases during daytime when oxygen is present (van Niel *et al.*, 1996). A comparable behaviour was observed

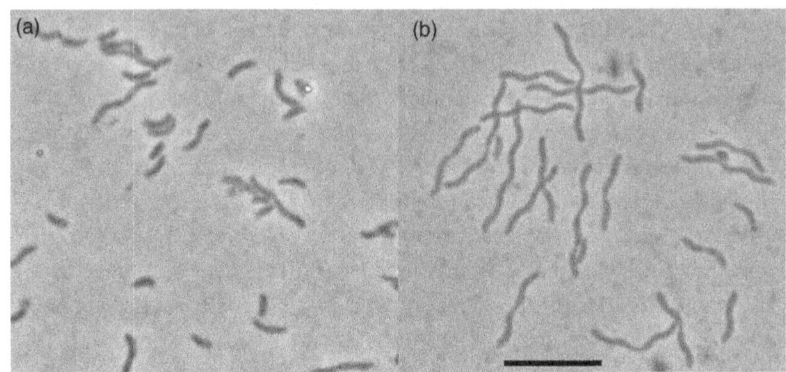

Figure 5.1. Exponential phase cells of *Desulfovibrio cuneatus* STL1 grown with lactate and sulphate. (a) under anoxic conditions. (b) exposed to 12.5 $\mu mol \cdot l^{-1}$ O_2.

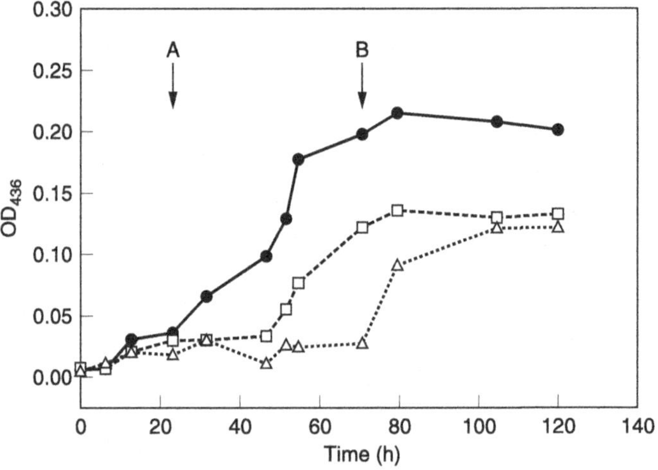

Figure 5.2. Influence of oxygen on growth of *Desulfomicrobium* sp. STL8. Cultures were grown under a N_2/CO_2 (80/20, v/v) atmosphere. Oxygen was added after 24 h to the headspace (A). Oxygen concentrations were calculated considering liquid culture volume and headspace. Culture 1 (closed circles) remained anoxic, while culture 2 (open squares) and 3 (open triangles) received 1% and 12.5% oxygen, respectively. In culture 2, oxygen was depleted by the bacteria and after reestablishing reducing conditions (as indicated by resazurin), growth started again. After 70 h, the headspace of culture 3 was flushed with N_2/CO_2 (B) and growth commenced. Note that growth yield as indicated by optical density was reduced in the oxygen-treated cultures (data from Sass, 1997).

in growth experiments employing sulphate reducers immobilised in oxygen–sulphide counter-gradients (Sass *et al.*, 2002). From the almost linear depth profile of cell numbers it was obvious that the cells were depleting oxygen first and that growth started after the conditions turned anoxic.

5.9 CONCLUSIONS

Sulphate-reducing bacteria employ a variety of mechanisms to respond to the presence of O_2, suggesting that these capacities are significant for this physiological group.

Although being a potential electron acceptor that produces oxic conditions, the presence of O_2 often also coincides with a supply of easily degradable organic matter. This might be the one of the key reasons for sulphate reducers to approach oxic environments.

Sulphate reducers have a very versatile respiratory metabolism. Many of them are able to utilise different sulphur compounds as electron acceptors. Some even grow coupled to the reduction of nitrate, nitrite (Widdel and Pfennig, 1982) or several metal ions (Coleman *et al.*, 1993; Tebo and Obraztsova, 1998). In many cases the growth yields with alternative electron acceptors are even higher than those with sulphate (Seitz and Cypionka, 1986). If sulphate reducers have evolved the capacity of reducing oxygen and to gain ATP from this process by chemiosmotic mechanisms, why did they not develop the ability to grow with oxygen like aerobic heterotrophs? Part of the answer might be that sulphate reducers are not only characterised by their capacity to reduce sulphate but also by the restricted spectrum of substrates they can use. Most of the electron donors consumed originate from fermentation processes (Widdel, 1988; Widdel and Hansen, 1992). In particular close to oxic–anoxic interfaces, many of the fermentation products are formed by facultative aerobes which would oxidise their substrates completely if oxygen was available. Life in oxic environments would force the sulphate reducers to use other types of electron donors. By contrast, oxygen consumption by sulphate-reducing bacteria at the oxic–anoxic interface would promote the production of fermentation products and avoid substrate competition with aerobes. In fact, in support of this the extremely high oxygen respiration rates detected in some SRB point to a defence rather than to an energy-conserving process. Comparably high oxygen consumption rates have been found in nitrogen-fixing bacteria that protect their nitrogenases, but usually not in other aerobic bacteria (Cypionka and Meyer, 1982).

An instructive example for the life of sulphate reducers at oxic-anoxic interfaces is their role in the gut of termites. *Desulfovibrio* species isolated from termite guts were found to have extremely high aerobic respiration rates, e.g. with hydrogen as an electron donor, and the capacity to fix nitrogen (Kuhnigk *et al.*, 1996; Fröhlich *et al.*, 1999). They are found generally at the inner surface of the hindgut wall (Berchtold *et al.*, 1999). It was suggested that the sulphate reducers consume oxygen and fermentative hydrogen, thus generating anoxic conditions and promoting fermentation. Since the *Desulfovibrio* belong to the incompletely oxidising species, they will not consume acetate and other fatty acids. These fermentation products can be utilised by the termite, while the sulphate reducers could feed on products like H_2, lactate, ethanol, and additionally fix nitrogen for their host.

Thus, while sulphur oxidisers try to keep oxygen away from sulphide in order to avoid an abiotic reaction, sulphate reducers appear to consume oxygen in order to avoid possible competition with aerobes and stimulate the release of fermentation products at the same time.

ACKNOWLEDGEMENTS

The authors wish to thank Derek Martin for critically reading the manuscript and for his valuable suggestions.

REFERENCES

Abdollahi, H. and Wimpenny, J. W. T. (1990). Effects of oxygen on the growth of *Desulfovibrio desulfuricans*. *J Gen Microbiol*, **136**, 1025–30.

Aller, R. C. and Rude, P. D. (1988). Complete oxidation of solid phase sulfides by manganese and bacteria in anoxic marine sediments. *Geochim Cosmochim Acta*, **52**, 751–65.

Bade, K., Manz, W. and Szewzyk, U. (2000). Behavior of sulphate-reducing bacteria under oligotrophic conditions and oxygen stress in particle-free systems related to drinking water. *FEMS Microbiol Ecol*, **32**, 215–23.

Battersby, N. S., Malcolm, S. J., Brown, C. M. and Stanley, S. O. (1985). Sulphate reduction in oxic and suboxic North East Atlantic sediments. *FEMS Microbiol Ecol*, **31**, 225–8.

Baughn, A. D. and Malamy, M. H. (2004). The strict anaerobe *Bacteroides fragilis* grows in and benefits from nanomolar concentrations of oxygen. *Nature*, **427**, 441–4.

Baumgarten, A., Redenius, I., Kranczoch, J. and Cypionka, H. (2001). Periplasmic oxygen reduction by *Desulfovibrio* species. *Arch Microbiol,* **176**, 306–9.

Beijerinck, M. W. (1893). Über Atmungsfiguren beweglicher Bakterien. *Zentralbl Bakteriol Parasitenkunde,* **14**, 827–45.

Beijerinck, M. W. (1895). Ueber *Spirillum desulfuricans* als Ursache von Sulfatreduction. *Centralbl Bakteriol II Abt,* **1**, 1–9, 49–59, 104–14.

Berchtold, M., Chatzinotas, A., Schönhuber, W., *et al.* (1999). Differential enumeration and in situ localization of microorganisms in the hindgut of the lower termite *Mastotermes darwiniensis* by hybridization with rRNA-targeted probes. *Arch Microbiol,* **172**, 407–16.

Blaabjerg, V., Mouritsen, K. N. and Finster, K. (1998). Diel cycles of sulphate reduction rates in sediments of a *Zostera marina* bed (Denmark). *Aquat Microb Ecol,* **15**, 97–102.

Canfield, D. E. and DesMarais, D. J. (1991). Aerobic sulphate reduction in microbial mats. *Science,* **251**, 1471–3.

Chen, L., Liu, M. Y., LeGall, J. *et al.* (1993). Purification and characterization of a NADH-rubredoxin oxidoreductase involved in the utilization of oxygen by *Desulfovibrio gigas. Eur J Biochem,* **216**, 443–8.

Cohn, F. (1867). Beiträge zur Physiologie der Phycochromaceen und Florideen. *Arch Mikroskopie Anatomie,* **3**, 1–60.

Coleman, M. L., Hedrick, D. B., Lovley, D. R., White, D. C. and Pye, K. (1993). Reduction of Fe(III) in sediments by sulphate-reducing bacteria. *Nature,* **361**, 436–8.

Cypionka, H. (2000). Oxygen respiration by *Desulfovibrio* species. *Annu Rev Microbiol,* **54**, 827–48.

Cypionka, H. and Meyer, O. (1982). Influence of carbon monoxide on growth and respiration of carboxydotrophic and other aerobic organisms. *FEMS Microbiol Lett,* **15**, 209–14.

Cypionka, H., Widdel, F. and Pfennig, N. (1985). Survival of sulphate-reducing bacteria after oxygen stress, and growth in sulphate-free oxygen-sulfide gradients. *FEMS Microbiol Ecol,* **27**, 189–93.

Dannenberg, S., Kroder, M., Dilling, W. and Cypionka, H. (1992). Oxidation of H_2, organic compounds and inorganic sulfur compounds coupled to reduction of O_2 or nitrate by sulphate-reducing bacteria. *Arch Microbiol,* **158**, 93–9.

Das, A., Silaghi-Dumitrescu, R., Ljungdahl, L. G. and Kurtz, D. M. Jr. (2005). Cytochrome *bd* oxidase, oxidative stress, and dioxygen tolerance of the strictly anaerobic bacterium *Moorella thermoacetica. J Bacteriol,* **187**, 2020–9.

Dilling, W. and Cypionka, H. (1990). Aerobic respiration in sulphate-reducing bacteria. *Arch Microbiol*, **71**, 123–8.

Eschemann, A., Kühl, M. and Cypionka, H. (1999). Aerotaxis in *Desulfovibrio*. *Environ Microbiol*, **1**, 489–94.

Fareleira, P., Santos, B. S., António, C. *et al.* (2003). Response of a strict anaerobe to oxygen: survival strategies in *Desulfovibrio gigas*. *Microbiology*, **149**, 1513–22.

Fenchel, T. (1994). Motility and chemosensory behaviour of the sulphur bacterium *Thiovulum majus*. *Microbiology*, **140**, 3109–16.

Fischer, J. P. and Cypionka, H. (2005). Analysis of aerotactic band formation by *Desulfovibrio desulfuricans* in a stopped-flow diffusion chamber. *FEMS Microbiol Ecol*, **55**, 186–94.

Fitz, R. M. and Cypionka, H. (1991). Generation of a proton gradient in *Desulfovibrio vulgaris*. *Arch Microbiol*, **155**, 444–8.

Fournier, M., Zhang, Y., Wildschut, J. D. *et al.* (2003). Function of oxygen resistance proteins in the anaerobic sulphate-reducing bacterium *Desulfovibrio vulgaris* Hildenborough. *J Bacteriol*, **185**, 71–9.

Fröhlich, J., Sass, H., Babenzien, H.-D. *et al.* (1999). Isolation of *Desulfovibrio intestinalis* sp. nov. from the hindgut of the lower termite *Mastotermes darwiniensis*. *Can J Microbiol*, **45**, 145–52.

Fründ, C. and Cohen, Y. (1992). Diurnal cycles of sulphate reduction under oxic conditions in cyanobacterial mats. *Appl Environ Microbiol*, **58**, 70–7.

Fu, R. and Voordouw, G. (1997). Targeted gene-replacement mutagenesis of *dcrA* encoding an oxygen sensor of the sulphate-reducing bacterium *Desulfovibrio vulgaris* Hildenborough. *Microbiology*, **143**, 1815–26.

Fu, R., Wall J. D. and Voordouw, G. (1994). *DcrA* a *c*-type heme-containing methyl-accepting protein from *Desulfovibrio vulgaris* Hildenborough, senses the oxygen concentration or redox potential of the environment. *J Bacteriol*, **176**, 344–50.

Fukui, M. and Takii, S. (1990). Colony formation of free-living and particle-associated sulphate-reducing bacteria. *FEMS Microbiol Ecol*, **73**, 85–90.

Fukui, M. and Takii, S. (1994). Kinetics of sulphate respiration by free-living and particle-associated sulphate-reducing bacteria. *FEMS Microbiol Ecol*, **13**, 241–7.

Gomes, C. M., Silva, G., Oliveira, S. *et al.* (1997). Studies on the redox centers of the terminal oxidase from *Desulfovibrio gigas* and evidence for its interaction with rubredoxin. *J Biol Chem*, **272**, 22502–8.

Gottschal, J. C. and Szewzyk, R. (1985). Growth of a facultative anaerobe under oxygen-limiting conditions in pure culture and in co-culture with a sulphate-reducing bacterium. *FEMS Microbiol Ecol*, **31**, 159–70.

Hardy, J. A. and Hamilton, W. A. (1981). The oxygen tolerance of sulphate-reducing bacteria isolated from North Sea waters. *Curr Microbiol*, **6**, 259–62.

Heidelberg, J. F., Seshadri, R., Haveman, S. A. *et al.* (2004). The genome sequence of the anaerobic, sulphate-reducing bacterium *Desulfovibrio vulgaris* Hildenborough. *Nature Biotechnology*, **22**, 554–9.

Hoppe-Seyler, F. (1886). Ueber die Gährung der Cellulose mit Bildung von Methan und Kohlensäure. II. Der Zerfall der Cellulose durch Gährung unter Bildung von Methan und Kohlensäure und die Erscheinungen, welche dieser Process veranlasst. *Z Physiol Chem*, **10**, 401–40.

Imlay, J. A. (2002). How oxygen damages microbes: oxygen tolerance and obligate anaerobiosis. *Adv Microb Physiol*, **46**, 111–53.

Ito, T., Nielsen, J. L., Okabe, S., Watanabe, Y. and Nielsen, P. H. (2002). Phylogenetic identification and substrate uptake patterns of sulphate-reducing bacteria inhabiting an oxic-anoxic sewer biofilm determined by combining microautoradiography and fluorescent *in situ* hybridization. *Appl Environ Microbiol*, **68**, 356–64.

Johnson, M. S., Zhulin, I. G., Gapuzan, M. E. R. and Taylor, B. L. (1997). Oxygen-dependent growth of the obligate anaerobe *Desulfovibrio vulgaris* Hildenborough. *J Bacteriol*, **179**, 5598–601.

Jonkers, H. M., Koh, I. O., Behrend, P., Muyzer, G. and de Beer, D. (2005). Aerobic organic carbon mineralization by sulphate-reducing bacteria in the oxygen-saturated photic zone of a hypersaline microbial mat. *Microb Ecol*, **49**, 291–300.

Jørgensen, B. B. (1977). Bacterial sulphate reduction within reduced microniches of oxidized marine sediments. *Mar Biol*, **41**, 7–17.

Jørgensen, B. B. (1994). Sulphate reduction and thiosulphate transformations in a cyanobacterial mat during a diel oxygen cycle. *FEMS Microbiol Ecol*, **13**, 303–12.

Jørgensen, B. B. and Bak, F. (1991). Pathways and microbiology of thiosulphate transformations and sulphate reduction in a marine sediment (Kattegatt, Denmark). *Appl Environ Microbiol*, **57**, 847–56.

Kjeldsen, K. U., Joulian, C. and Ingvorsen, K. (2004). Oxygen tolerance of sulphate-reducing bacteria in activated sludge. *Environ Sci Technol*, **38**, 2038–43.

Kolb, S., Seeliger, S., Springer, N., Ludwig, W. and Schink, B. (1998). The fermenting bacterium *Malonomonas rubra* is phylogenetically related to sulfur-reducing bacteria and contains a *c*-type cytochrome similar to those of sulfur and sulphate reducers. *System Appl Microbiol*, **21**, 340–5.

Krekeler, D. and Cypionka, H. (1995). The preferred electron acceptor of *Desulfovibrio desulfuricans* CSN. *FEMS Microbiol Ecol*, **17**, 271–8.

Krekeler, D., Teske, A. and Cypionka, H. (1998). Strategies of sulphate-reducing bacteria to escape oxygen stress in a cyanobacterial mat. *FEMS Microbiol Ecol*, **25**, 89–96.

Kuhnigk, T., Branke, J., Krekeler, D., Cypionka, H. and König, H. (1996). A feasible role of sulphate-reducing bacteria in the termite gut. *System Appl Microbiol*, **19**, 139–49.

Laanbroek, H. J. and Pfennig, N. (1981). Oxidation of short-chain fatty acids by sulphate-reducing bacteria in freshwater and marine sediments. *Arch Microbiol*, **128**, 330–5.

LeGall, J. and Xavier, A. V. (1996). Anaerobes response to oxygen: the sulphate-reducing bacteria. *Anaerobe*, **2**, 1–9.

Lemos, R. S., Gomes, C. M., Santana, M. *et al.* (2001). The 'strict' anaerobe *Desulfovibrio gigas* contains a membrane-bound oxygen-reducing respiratory chain. *FEBS Lett*, **496**, 40–3.

Marschall, C., Frenzel, P. and Cypionka, H. (1993). Influence of oxygen on sulphate reduction and growth of sulphate-reducing bacteria. *Arch Microbiol*, **159**, 168–73.

Meyer, L. (1864). Chemische Untersuchungen der Thermen zu Landeck in der Grafschaft Glatz. *J Prakt Chem*, **91**, 1–15.

Minz, D., Fishbain, S., Green, S. J. *et al.* (1999a). Unexpected population distribution in a microbial mat community: sulphate-reducing bacteria localized to the highly oxic chemocline in contrast to a eukaryotic preference for anoxia. *Appl Environ Microbiol*, **65**, 4659–65.

Minz, D., Flax, J. L., Green, S. J. *et al.* (1999b). Diversity of sulphate-reducing bacteria in oxic and anoxic regions of a microbial mat characterized by comparative analysis of dissimilatory sulfite reductase genes. *Appl Environ Microbiol*, **65**, 4666–71.

Mogensen, G. L., Kjeldsen, K. U. and Ingvorsen, K. (2005). *Desulfovibrio aerotolerans* sp. nov., an oxygen-tolerant sulphate-reducing bacterium isolated from activated sludge. *Anaerobe*, **11**, 339–49.

Okabe, S., Ito, T. and Satoh, H. (2003). Sulphate-reducing bacterial community structure and their contribution to carbon mineralization in a wastewater biofilm growing under microaerophilic conditions. *Appl Microbiol Biotechnol*, **63**, 322–34.

Ploug, H., Kühl, M., Buchholz-Cleven, B. and Jørgensen, B. B. (1997). Anoxic aggregates – an ephemeral phenomenon in the pelagic environment? *Aquat Microb Ecol*, **13**, 285–94.

Ramsing, N. B., Fossing, H., Ferdelmann, T. G., Andersen, F. and Thamdrup, B. (1996). Distribution of bacterial populations in a stratified fjord (Mariager Fjord, Denmark) quantified by in situ hybridization and related to chemical gradients in the water column. *Appl Environ Microbiol*, **62**, 1391–404.

Ramsing, N. B., Kühl, M. and Jørgensen, B. B. (1993). Distribution of sulphate-reducing bacteria, O_2, and H_2S in photosynthetic biofilms determined by oligonucleotide probes and microelectrodes. *Appl Environ Microbiol*, **59**, 3840–9.

Risatti, J. B., Capman, W. C. and Stahl, D. A. (1994). Community structure of a microbial mat: the phylogenetic dimension. *Proc Natl Acad Sci USA*, **91**, 10173–7.

Sass, A. M., Eschemann, A., Kühl, M. *et al.* (2002). Growth and chemosensory behavior of sulphate-reducing bacteria in oxygen-sulfide gradients. *FEMS Microbiol Ecol*, **40**, 47–54.

Sass, H. (1997). *Vorkommen und Aktivität sulfatreduzierender Bakterien in der Chemokline limnischer Sedimente.* PhD thesis, University of Oldenburg.

Sass, H., Berchtold, M., Branke, J. *et al.* (1998a). Psychrotolerant sulphate-reducing bacteria from an oxic freshwater sediment, description of *Desulfovibrio cuneatus* sp. nov. and *Desulfovibrio litoralis* sp. nov. *System Appl Microbiol*, **21**, 212–19.

Sass, H., Cypionka, H. and Babenzien, H.-D. (1996). Sulphate-reducing bacteria from the oxic layers of the oligotrophic Lake Stechlin. *Arch Hydrobiol – Spec Iss Adv Limnol*, **48**, 241–6.

Sass, H., Cypionka, H. and Babenzien, H.-D. (1997). Vertical distribution of sulphate-reducing bacteria at the oxic–anoxic interface in sediments of the oligotrophic Lake Stechlin. *FEMS Microbiol Ecol*, **22**, 245–55.

Sass, H., Wieringa, E., Cypionka, H., Babenzien, H.-D. and Overmann, J. (1998b). High genetic and physiological diversity of sulphate-reducing bacteria isolated from an oligotrophic lake sediment. *Arch Microbiol*, **170**, 243–51.

Schippers, A. and Jørgensen, B. B. (2001). Oxidation of pyrite and iron sulfide by manganese dioxide in marine sediments. *Geochim Cosmochim Acta*, **65**, 915–22.

Schramm, A., Santegoeds, C. M., Nielsen, H. K. *et al.* (1999). On the occurrence of anoxic microniches, denitrification, and sulphate reduction in aerated activated sludge. *Appl Environ Microbiol*, **65**, 4189–96.

Schulz, H. N. and Jørgensen, B. B. (2001). Big bacteria. *Annu Rev Microbiol*, **55**, 105–37.

Seitz, H. J. and Cypionka, H. (1986). Chemolithotrophic growth of *Desulfovibrio desulfuricans* with hydrogen coupled to ammonification of nitrate or nitrite. *Arch Microbiol*, **146**, 63–7.

Sigalevich, P., Meshorer, E., Helman, Y. and Cohen, Y. (2000). Transition from anaerobic to aerobic growth conditions for the sulphate-reducing bacterium *Desulfovibrio oxyclinae* results in flocculation. *Appl Environ Microbiol*, **66**, 5005–12.

Stams, A. J. M. and Hansen, T. A. (1982). Oxygen-labile L(+) lactate dehydrogenase activity in *Desulfovibrio desulfuricans*. *FEMS Microbiol Lett*, **13**, 389–94.

Tebo, B. M. and Obraztsova, A. Y. (1998). Sulphate-reducing bacterium grows with Cr(VI), U(VI), Mn(IV), and Fe(III) as electron acceptors. *FEMS Microbiol Lett*, **162**, 193–8.

Teske, A., Ramsing, N. B., Habicht, K. *et al.* (1998). Sulphate-reducing bacteria and their activities in cyanobacterial mats of Solar Lake (Sinai, Egypt). *Appl Environ Microbiol*, **64**, 2943–51.

Teske, A., Wawer, C., Muyzer, G. and Ramsing, N. B. (1996). Distribution of sulphate-reducing bacteria in a stratified fjord (Mariager Fjord, Denmark) as evaluated by most-probable-number counts and denaturing gradient gel electrophoresis of PCR-amplified ribosomal DNA fragments. *Appl Environ Microbiol*, **62**, 1405–15.

Tonolla, M., Demarta, A., Peduzzi, S., Hahn, D. and Peduzzi, R. (2000). In situ analysis of sulphate-reducing bacteria related to *Desulfocapsa thiozymogenes* in the chemocline of meromictic Lake Cadagno (Switzerland). *Appl Environ Microbiol*, **66**, 820–4.

Van den Ende, F. P., Meier, J. and van Gemerden, H. (1997). Syntrophic growth of sulphate-reducing bacteria and colorless sulfur bacteria during oxygen limitation. *FEMS Microbiol Ecol*, **23**, 65–80.

Van Niel, E. W. J. and Gottschal, J. C. (1998). Oxygen consumption by *Desulfovibrio* strains with and without polyglucose. *Appl Environ Microbiol*, **64**, 1034–9.

Van Niel, E. W. J., Pedro Gomez, T. M., Willems, A. *et al.* (1996). The role of polyglucose in oxygen-dependent respiration by a new strain of *Desulfovibrio salexigens*. *FEMS Microbiol Ecol*, **21**, 243–53.

Visscher, P. T., Prins, R. A. and van Gemerden, H. (1992). Rates of sulphate reduction and thiosulphate consumption in a marine microbial mat. *FEMS Microbiol Ecol*, **86**, 283–94.

Voordouw, J. K. and Voordouw, G. (1998). Deletion of the *rbo* gene increases the oxygen sensitivity of the sulphate-reducing bacterium *Desulfovibrio vulgaris* Hildenborough. *Appl Environ Microbiol*, **64**, 2882–7.

Widdel, F. (1988). Microbiology and ecology of sulphate-reducing and sulfur-reducing bacteria. In A. J. B. Zehnder (ed.), *Biology of anaerobic microorganisms*, New York, NY: John Wiley and Sons. pp. 469–585.

Widdel, F. and Hansen, T. A. (1992). Dissimilatory sulphate- and sulfur-reducing bacteria. In A. Balows, H. G. Trüper, M. Dworkin, W. Harder and K. H. Schleifer (eds.), *The prokaryotes*, vol. 1, 2nd edn. New York, NY: Springer. pp. 583–24.

Widdel, F. and Pfennig, N. (1982). Studies on dissimilatory sulphate-reducing bacteria that decompose fatty acids. II. Incomplete oxidation of propionate by *Desulfobulbus propionicus* gen. nov., sp. nov. *Arch Microbiol*, **131**, 360–5.

Wieringa, E. B. A., Overmann, J. and Cypionka, H. (2000). Detection of abundant sulphate-reducing bacteria in marine oxic sediment layers by a combined cultivation and molecular approach. *Environ Microbiol*, **2**, 417–27.

Wind, T. and Conrad, R. (1995). Sulfur compounds, potential turnover of sulphate and thiosulphate, and numbers of sulphate-reducing bacteria in planted and unplanted paddy soil. *FEMS Microbiol Ecol*, **18**, 257–66.

CHAPTER 6

Biochemical, proteomic and genetic characterization of oxygen survival mechanisms in sulphate-reducing bacteria of the genus *Desulfovibrio*

Alain Dolla, Donald M. Kurtz, Jr., Miguel Teixeira and Gerrit Voordouw

6.1 INTRODUCTION

Sulphate-reducing bacteria (SRB) are anaerobes, which derive energy for growth from anaerobic metabolism, coupling the oxidation of organic substrates with the dissimilatory reduction of sulphate to hydrogen sulphide (sulphate respiration). Although generally considered as strict anaerobes, more and more data indicate a higher abundance and metabolic activity in oxic zones of biotopes, such as marine and freshwater sediments, than in neighbouring anoxic zones (Ravenschlag *et al.*, 2000; Sass *et al.*, 1997; 1998). A well-documented example of sulphate reduction under oxic conditions is also provided by cyanobacterial mats. Here a zone of photosynthetic oxygen synthesis overlaps with a zone of sulphide production by SRB and a zone of oxygen-dependent microbial sulphide oxidation, creating steep, opposing gradients of oxygen and sulphide, which fluctuate with the rhythm of day and night (Canfield and des Marais, 1991; Teske *et al.*, 1998; Caumette *et al.*, 1994). In cyanobacterial mats from the saline evaporation pond in Baja California, *Desulfobacter* and *Desulfobacterium* are restricted to greater depths while the *Desulfococcus* and *Desulfovibrio* groups are predominant in the upper part of the photo-oxic zone (Risatti *et al.*, 1994). The high numbers of SRB found in these oxic environments indicate that these organisms are able to deal with temporal exposures to oxygen concentrations as high as 1.5 mM (Sigalevich and Cohen, 2000).

In these oxygen-exposed systems SRB of the genus *Desulfovibrio* are among the most oxygen-tolerant, e.g. *Desulfovibrio oxyclinae* was isolated from Solar Lake microbial mats (Krekeler *et al.*, 1997) and *Desulfovibrio desulfuricans* strain DvO1 was isolated from activated sludge aerated to atmospheric oxygen saturation (Kjeldsen *et al.*, 2005). Although this could simply reflect the general ease with which *Desulfovibrio* spp. are isolated in pure culture, the possibility that these organisms are genuinely more aerotolerant than many other SRB should not be ruled out. Indeed, some *Desulfovibrio* spp., such as *D. vulgaris* Hildenborough (referred to from now on as *D. vulgaris*), have been shown not only to survive exposure to an aerobic atmosphere (Lumppio *et al.*, 2001; Fournier *et al.*, 2003), but also to swim towards and grow optimally near the oxic/anoxic interfaces of sulphate-containing media exposed to air (Marschall *et al.*, 1993; Fu and Voordouw, 1997; Eschemann *et al.*, 1999). The availability of two complete genomic sequences for *D. vulgaris* (Heidelberg *et al.*, 2004) and for *D. desulfuricans* G20 (http://www.jgi.doe.gov) gives unprecedented possibilities to uncover their mechanisms of aerotolerance. This contribution will, therefore, focus primarily on the mechanisms involved in aerotolerance of members of this genus. Different systems of oxygen defence in aerated environments, including oxygen reduction and Reactive Oxygen Species (ROS) detoxification will be described. A global survey of oxygen defense systems in *Desulfovibrio* spp., based on bioinformatic and proteomic analyses is provided first, followed by a more detailed consideration of individual defence proteins.

6.2 PROTEOMICS OF THE OXYGEN STRESS RESPONSE

Influence of oxygen exposure on the proteome of *D. vulgaris* has been evaluated by 2D-gel electrophoresis and mass spectrometry (Fournier *et al.*, 2006). The profile of soluble proteins of cells grown anaerobically in lactate-sulphate medium was compared with that of cells exposed to 100% (vol/vol) oxygen for 1 hour. These severe oxygen exposure conditions may be comparable to what cells encounter in cyanobacterial mats (Sigalevich and Cohen, 2000) and resulted in significant changes in the intensities of 54 protein spots, 35 being decreased and 19 increased. Less abundant were several proteins involved in nucleic acid and protein synthesis, as well as proteins involved in essential cellular processes such as cell division or protein folding. Several enzymes involved in ROS detoxification also showed decreased abundance, including superoxide reductase (Sor) and rubrerythrins 1 and 2 (Rbr1 and Rbr2), which catalyze reduction of superoxide and hydrogen peroxide, respectively. Lower concentrations of ROS detoxifying enzymes may contribute to the loss of viability observed under these oxidative stress conditions. It is uncertain why concentrations of these enzymes are reduced under these conditions. Repression of gene expression is not intuitively logical: if anything, one would expect induction under these conditions! Cell viability is definitively compromised under these conditions and these enzymes may be considered the battered-to-death foot soldiers in a war that is about to be lost and that are, therefore, not being replaced. Alternatively, repression of gene expression could be part of a deliberate strategy by the cell to limit the amounts of these and other iron-containing proteins in an attempt to reduce the concentration of Fenton-reactive iron. Determining the extent to which these two alternatives are correct will require more extensive study of the gene expression response at varying oxygen doses.

Several proteins whose abundance increased under these severe oxidative conditions are directly linked to defence mechanisms. They include a thiol-peroxidase and a bacterioferritin co-migratory protein (Bcp), homologues of which have thiol-peroxidase activity, a glutaredoxin, which has disulphide reductase activity and, a NifU homologue that might function as an iron-sulphur cluster scaffold involved in repairing oxidative damage to thiols and iron-sulphur clusters. The abundance of some proteins of unknown function was also differentially affected, suggesting that these also play a role in oxygen defence (Fournier *et al.*, 2006). However, analysis of the transcriptome, i.e. through the use of microarrays, and of a wider range of oxygen exposure conditions is needed to gain a deeper appreciation of the response of *D. vulgaris* to oxidative conditions.

A list of oxygen reduction, ROS detoxification and repair enzymes suggested by bioinformatic analyses of the genomes of *D. vulgaris* and *D. desulfuricans* G20 and by differential analysis of the *D. vulgaris* proteome is provided in Table 6.1. A model indicating most of the components which participate in oxygen defence is shown in Figure 6.1. Wherever possible, the functions of these various components has been inferred from comparison with those of homologues in well-studied microorganisms, such as *Escherichia coli*, or by biochemical and biophysical characterization, as described below.

6.3 MEMBRANE-BOUND AND SOLUBLE OXYGEN REDUCTASES IN *DESULFOVIBRIO* SPP.

6.3.1 Oxygen reduction by membrane-bound oxygen reductases

Aerobic organisms reduce oxygen to water by membrane-bound oxidases, acting as the terminal enzymes of aerobic respiratory chains. These include the haem-copper oxygen reductases (also generally called cytochrome *c* oxidases, Cox), the *bd*-type oxygen reductases and the

Table 6.1. *Protein components involved in oxygen reduction, Reactive Oxygen Species (ROS) detoxification and oxygen damage repair in* D. vulgaris *Hildenborough and* D. desulfuricans *G20*[a]

Enzymatic activity	Name	Abbr.	Locus in *D. vulgaris*[b]	Locus in *D. desulfuricans*[c]
ROS detoxification				
Superoxide dismutation	Superoxide dismutase	Sod	DVU2410	Dde0882
Superoxide reduction	Superoxide reductase	Sor	DVU3183	Dde3193
Hydrogen peroxide dismutation	Catalase	Kat	DVUA0091[d]	*ni*[e]
NADH peroxidases	Rubrerythrin isoenzyme 1	Rbr1	DVU3094	Dde1222
	Rubrerythrin isoenzyme 2	Rbr2	DVU2310	Dde1320
	Nigerythrin	Ngr	DVU0019	Dde3337
Thiol-specific peroxidase	Thiol peroxidase		DVU1228	Dde2313
	Bacterioferritin comigratory protein	Bcp	DVU0814	Dde1027
Hydroperoxidase		AhpC	DVU3077	Dde0305
	AhpC/TSA family proteins		DVU2247	Dde1311
	AhpF family protein	AhpF	DVU0283	Dde0210

Oxygen reduction				
Oxygen reduction	Rubredoxin:oxygen oxidoreductase	Roo	DVU3185	Dde3195
	Roo homologue		DVU2014	
	Haem-copper oxygen reductase	Cox	DVU1811–1815	Dde1823–1827
	Cytochrome *bd* oxygen reductase	Cbd	DVU3270–3271	Dde3204–3205
	[Fe] hydrogenase	[Fe] Hyd	DVU1769–1770	Dde2280–2281
				Dde0081–0082
Damage repair				
Disulfide reduction	Glutaredoxin	NdrH (Glr)[f]	DVU0883	Dde2739
		TrxB-1	DVU0377–0378	Dde0464–0465
	Thioredoxins/thioredoxin reductases	TrxB-2	DVU1838–1839	Dde1202–1203
		Trx		Dde2066–2067
		(Thr)[f]		
	Thioredoxin reductase	TrxB	DVU1457	Dde2151
	Thioredoxin-related, disulfide isomerase	CcmG	DVU1586	Dde2114
				Dde2781
				Dde3416

Table 6.1. (*cont.*)

Enzymatic activity	Name	Abbr.	Locus in *D. vulgaris*[b]	Locus in *D. desulfuricans*[c]
Fe-S cluster synthesis	NifU homolog; Fe-S scaffold protein	NifU-like	DVU0662–0665	Dde3078–3081
Methionine sulfoxide reduction	Methionine sulfoxide reductase	MsrA	DVU1984	Dde1002;
		MsrB	DVU0576	Dde2482
				Dde0714
				Dde1003
Iron transport, scavenging				
	Ferritin		DVU1568	Dde1791
	Bacterioferritin	Bfr	DVU1397	Dde0133

[a]The list is based on bioinformatic and proteomic analyses. The function of several of the listed proteins has been determined through genetic and biochemical studies. Most are thought to be cytoplasmic, or membrane-bound (Cox, Cbd). Sod and [Fe] hydrogenase are periplasmic.

[b]Heidelberg *et al.*, 2004.

[c]www.jgi.doe.gov.

[d]plasmid-borne.

[e]ni: not identified.

[f]names in brackets correspond to those indicated in Figure 6.1.

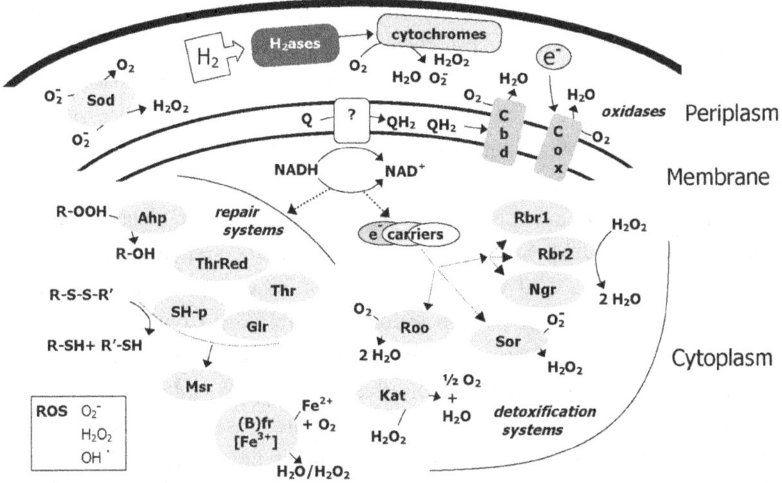

Figure 6.1. Model of oxygen defence in *Desulfovibrio* spp. The function of the various components is outlined in the text. The abbreviated names are explained in Table 6.1. Electrons for reduction of oxygen and ROS are also derived from cytoplasmic oxidation of lactate to acetate and CO_2, from periplasmic oxidation of formate and/or from oxidation of sulphide to sulphur. These reactions are not shown. Hence, although the figure covers key oxygen respiration and oxygen defence components, many more proteins are expected to be involved. The cellular location of the components shown is not proven in all cases. H_2ases: hydrogenases; e^-: electron; Q/QH_2: oxidized/reduced quinone pool.

alternative oxidases. The haem-copper oxygen reductases have been classified into families A (further divided into A1 and A2), B and C based on amino acid sequence similarities, catalytic properties and intraprotein proton channels (Pereira *et al.*, 2001; Pereira and Texeira, 2004). All have a catalytic subunit I with a low-spin and a high-spin haem, which together with a copper ion form the catalytic heterobimetallic centre. The A- and B-family enzymes have a second conserved subunit II, which may contain a di-copper centre on the periplasmic side of the membrane. The C-family enzymes, comprising the *cbb3* oxidases, have subunits with one and two *c*-type haems in addition to the catalytic subunit I. The *bd* type oxygen reductases are quinol oxidases with a catalytic site containing one *b*-type and one *d*-type haem, distinct from the haem-copper enzymes. The alternative oxidases are quinol oxidases with a diiron centre as the catalytic site. These were known to occur in plants and fungi, but have now been detected also in bacteria.

While mitochondria have only A1 type haem-copper enzymes, prokaryotes often contain several distinct oxygen reductases. This confers a robust system to cope with changes in oxygen availability: the C-family haem-copper oxygen reductases and the cytochrome *bd* reductases have the highest affinities for oxygen, and are maximally expressed under microaerophilic conditions.

The presence of oxygen reductases in SRB was first suggested by sequencing the *D. vulgaris* Miyazaki cytochrome c_{553} gene which appeared to contain a gene for a haem-copper oxygen reductase in close proximity (Kitamura *et al.*, 1995). Subsequently, the genome sequences of *D. vulgaris* and *D. desulfuricans* G20 were found to contain genes for both a haem-copper-type and a *bd*-type oxygen reductase (Table 6.1: Cox and Cbd). The strict conservation of the haem and metal-binding sites in the derived amino acid sequences showed undoubtedly that these genes code for fully functional enzymes. The haem-copper oxygen reductases of both *Desulfovibrio* species are of the A2 type and have, besides the ligands for the di-copper centre, two binding motifs for *c*-type haems in their subunit II, which constitutes a novelty in this enzyme family. This subunit II C-terminal extension does not appear to be related to the monohaemic cytochrome *c* domain of caa_3 oxygen reductases (Srinivasan *et al.*, 2005).

Neither of these oxygen reductases has been isolated from either *D. vulgaris* or *D. desulfuricans* G20. However, a canonical cytochrome *bd* quinol:oxygen oxidoreductase was purified from *D. gigas* cells grown on media with equal concentrations of sulphate and fumarate as the electron acceptors (Lemos *et al.*, 2001). Approximately 50% of the respiratory activity of membranes isolated from sulphate-fumarate grown cells was sensitive to potassium cyanide, a potent inhibitor of haem-copper oxygen reductases, but not of cytochromes *bd*. Therefore, a haem-copper enzyme appears also to be expressed in *D. gigas*. Although the presence of *a*-type haem, a characteristic of A- and B-family haem-copper oxygen reductases, was demonstrated the enzyme was not successfully isolated from *D. gigas*. Further proof for the presence of a haem-copper oxygen reductase is also provided by the fact that the *D. vulgaris* genome contains a gene for a protohaem IX farnesyl transferase. This enzyme inserts the farnesyl chain in haem *b*, yielding haem *o* en route to the synthesis of haem *a*.

Which enzymes donate electrons to the membrane-bound, oxygen reducing electron transport chain in *Desulfovibrio* spp.? Succinate and NADH are the canonical electron donors for many electron transport chains to oxygen (Saraste, 1999; Pereira *et al.*, 2004). Both genomes have genes for succinate:quinone oxidoreductases, which are homologous to fumarate:quinone oxidoreductases. Indeed, a fumarate:quinone oxidoreductase was isolated from *Desulfovibrio gigas* (Lemos *et al.*, 2002). The genomic data regarding NADH-oxidizing enzymes is much less clear: no genes for obvious homologues of either type I NADH:quinone oxidoreductase (NDH-I) or of sodium-dependent NDH are present. However, genes for a putative type II NDH, annotated as pyridine nucleotide disulphide oxidoreductase, are found. In addition, the possibility of an as yet unidentified NDH complex or the use of an unidentified redox-active metabolite in sulphate reducers cannot be ruled out. The high respiratory activity of *D. gigas* membranes using NADH as electron donor suggest the presence of such an activity. The genomes of *D. vulgaris* and *D. desulfuricans* also contain genes encoding the *rnf* operon, which have considerable homologies to sodium-translocating NDHs (Na^+-NQR), but whose function is still not fully understood (Kumagai *et al.*, 1997).

The presence of putative haem-copper oxygen reductases with a periplasmic electron accepting site on subunit II poses another problem, i.e. which are the periplasmic electron carriers and which enzyme serves as their reductase? In other aerobic chains, these carriers are reduced by quinol:electron carrier oxidoreductases, the prototype of which is the quinol:cytochrome *c* oxidoreductase (the bc_1 complex, or complex III). However, such a complex appears to be absent from *Desulfovibrio* (Lemos *et al.*, 2001) and there is at least one precedent for an organism lacking a bc_1 complex, but having a functional analogue, containing a multihaem cytochrome (Pereira *et al.*, 1999). Therefore, since SRB have many periplasmic multihaem cytochromes associated with the membrane as part of large membrane-bound complexes, it is possible that one of these functionally acts like the bc_1 complex in these bacteria. The monohaem c_{553} (DVU1817), the gene for which is located in close proximity to the oxidase genes (Table 6.1: DVU1811–1815), or its homologue (DVU3041) appear to be the most obvious candidates for electron carrier to the haem-copper oxygen reductases.

In summary, it appears that sulphate-reducing bacteria have the necessary membrane-bound enzymes to reduce oxygen to water. Their oxygen-reducing active site is in the membrane, located more closely to the peri- than to the cytoplasmic side. Although these enzymes themselves release essentially no ROS during oxygen reduction, ROS may form during transport of electrons to their oxygen-reducing active site. An as yet unanswered question is whether the presence of these enzymes lowers overall ROS production, i.e. we do not know whether oxygen defence is the primary function of these enzymes or whether the energy conserving nature of the oxygen reduction process is more important to the organism.

6.3.2 Oxygen reduction by soluble, cytoplasmic rubredoxin: oxygen oxidoreductase

During oxygen stress *Desulfovibrio gigas* can metabolize internal polyglucose reserves through the glycolytic pathway with lactate as the presumed end-product to produce adenosine triphosphate (ATP) through substrate level phosphorylation. The NADH generated in this process is reoxidized by the membrane-bound oxidases, as well as by a proposed cytoplasmic electron transfer chain, composed of NADH:rubredoxin oxidoreductase (Nro, a flavoprotein containing one FMN and one FAD), type I rubredoxin (Rd) and rubredoxin:oxygen oxidoreductase (Roo) (Chen *et al.*, 1993a; b). Nro oxidizes NADH transferring the electrons to Rd, which serves as the intermediate carrier in the Roo-mediated reduction of oxygen to water. Roo and Rd form a dicistronic transcriptional unit (Frazao *et al.*, 2000; Silva *et al.*, 2001), but this genetic organization is not conserved in all *Desulfovibrio* spp. Roo is functionally a homodimer of 86 kDa, each subunit harbouring one FMN and two iron atoms. This enzyme was the first example of a now well-established family of flavo-diiron proteins (initially referred to as the A-type flavoproteins), which is widespread among anaerobic and facultative prokaryotes in both the archaeal and bacterial domains (Saraiva *et al.*, 2004). More recently it was found that these enzymes have a significant nitric oxide reductase activity (Gomes *et al.*, 2002; Silaghi-Dumitrescu *et al.*, 2003, 2005), and there is increasing evidence that they are involved in microbial resistance to nitric oxide (Gardner *et al.*, 2002; Justino *et al.*, 2005)

A major question regarding these enzymes is whether the initially proposed oxygen reductase activity in sulphate reducers is physiologically relevant. The affinity for oxygen is lower than that for nitric oxide (Gomes *et al.*, 2002; Silaghi-Dumitrescu *et al.*, 2003; 2005) suggesting that NO is the preferred substrate. Of course it may be that these enzymes are bifunctional and act under both oxygen-derived and nitrosative stress conditions. In fact, in the absence of their native substrate, binuclear iron centres in other enzymes are also known to reduce oxygen to water and to bind nitric oxide, which is often used as a probe for diiron centres. Also, whereas the expression of flavorubredoxin is induced by nitrosative stress in *E. coli*, the corresponding enzymes in *D. gigas* and *D. vulgaris* may be constitutive, because they are present in the absence of added nitric oxide under the usual growth conditions for these organisms (Chen *et al.*, 1993a; Gomes *et al.*, 2002; Silaghi-Dumitrescu *et al.*, 2005). The *D. vulgaris* genome encodes two Roo homologues (Table 6.1: DVU3185 and DVU2014), one of which

(DVU3185) is downstream of the *sor-rub* operon (formerly referred to as the *rbo-rub* operon; Fu and Voordouw, 1997), encoding Sor and rubredoxin. There is also increasing evidence that Roo-like enzymes are responsive to oxidative stress in several anaerobes (Kawasaki *et al.*, 2005). Hence, the exact physiological function of Roo in sulphate reducers is still under active investigation.

6.4 PERIPLASMIC OXYGEN REDUCTION IN *DESULFOVIBRIO* SPP.

Evidence that *Desulfovibrio* spp. possess a highly efficient periplasmic oxygen reduction activity was first obtained by Baumgarten *et al.* (2001), who observed that washed cells of *D. vulgaris* Marburg reduced oxygen to water with hydrogen as electron donor and that 90% of this oxygen reduction activity was in the periplasmic fraction. The authors proposed a mechanism in which hydrogenases reduce cytochromes *c*, which in turn reduce oxygen to water although other unknown components could also be involved. The ROS production associated with periplasmic oxygen reduction has not been characterized, but non-specific reduction of oxygen by cytochromes (i.e. as opposed to the specific reduction of oxygen by Cox and Cbd) will very likely occur. As cytochromes and hydrogenases are distributed ubiquitously in the periplasm of *Desulfovibrio* spp., the periplasmic oxygen-reducing system might be widespread and crucial to the cell's defence against oxygen. Cytochromes and hydrogenases have long been associated only with the energy metabolism of these bacteria. The genome sequence of *D. vulgaris* has indicated the presence of four periplasmic hydrogenases: [NiFe] hydrogenase isoenzymes 1 and 2, [NiFeSe] hydrogenase and [Fe] hydrogenase, as well as two cytoplasm-facing membrane-bound hydrogenases. It has also indicated the presence of seventeen putative *c*-type cytochromes, which are either periplasmic or are periplasmically oriented and membrane-associated. Of all these components, only specific involvement of [Fe] hydrogenase in oxygen defence has been shown. When *D. vulgaris* was exposed to oxygen for up to one hour, [Fe] hydrogenase production appeared to be specifically induced in response to exposure to oxygen or to chromate, a strong chemical oxidant (Fournier *et al.*, 2004; 2006). Exposure to oxygen also increased the content of periplasmic cytochromes *c* in *D. vulgaris* (Fournier *et al.*, 2004) and in *D. desulfuricans*, where a continuous culture kept at a constant low partial oxygen pressure was found to have a 20% increase in cytochrome content (Abdollahi and Wimpenny, 1990).

6.5 REMOVAL OF REACTIVE OXYGEN SPECIES IN *DESULFOVIBRIO* SPP.

Except towards some radical forming enzymes, molecular oxygen (at least in its triplet ground state) is relatively harmless, but, being uncharged, is freely and rapidly diffusible across bacterial membranes. Cytoplasmic, membrane-bound and periplasmic oxygen reduction discussed in the previous sections, as well as non-specific reaction of oxygen with reduced targets, give rise to partially reduced reactive oxygen species (ROS), such as superoxide and hydrogen peroxide. Although ROS may arise in both the cytoplasm and the periplasm it appears that all of the ROS-removing defence proteins are cytoplasmic, with the exception of superoxide dismutase (Sod). The function of these ROS-removing enzymes in *Desulfovibrio* spp. and other anaerobes is considered in detail below.

6.5.1 Oxygen defence proteins and pathways in aerobic bacteria

In aerobic bacteria, endogenous oxidative stress is primarily due to short-circuiting of respiratory electron flow in the cytoplasmic membrane. This short-circuiting results in adventitious reduction of molecular oxygen producing superoxide and hydrogen peroxide on the inner side of the cytoplasmic membrane (Imlay, 2002a; b). While this adventitious dioxygen reduction typically consumes <1% of the flux of reducing equivalents passing through the *E. coli* aerobic respiratory chain, the resulting intracellular flux of superoxide and hydrogen peroxide is sufficient to produce oxidative stress under hyperoxic conditions or if scavenging and/or repair enzymes are disabled (Storz and Imlay, 1999; Park *et al.*, 2005).

Aerobic microorganisms were classically thought to follow what might be called a "disproportionation paradigm" for lowering intracellular superoxide and hydrogen peroxide to non-lethal levels (McCord *et al.*, 1971; Imlay, 2002a). Superoxide dismutases (Sods) catalyze the disproportionation of superoxide to hydrogen peroxide and dioxygen: $2O_2^- + 2H^+ \rightarrow O_2 + H_2O_2$. Bacterial Sods typically contain either non-haem iron or manganese active sites, although bacterial copper/zinc- and nickel-Sods are also known (Imlay and Imlay, 1996; Choudhury *et al.*, 1999; Miller, 2004). Catalases catalyze disproportionation of hydrogen peroxide to water and molecular oxygen: $2H_2O_2 \rightarrow O_2 + 2H_2O$. Classical catalases contain haem at their active sites (Zámocky and Koller, 1999; Loewen *et al.*, 2000). A non-haem, manganese-containing catalase has also been described in a few bacteria (Yoder *et al.*, 2000; Barynin *et al.*, 2001). Due to their relatively high Michaelis constants K_ms, haem catalases are efficient at scavenging millimolar levels of hydrogen peroxide. Scavenging of lower (micromolar) levels of hydrogen peroxide in most if not all known aerobic bacteria occurs by its two-electron reduction to water catalyzed by hydrogen peroxide reductases (peroxidases): $2e^- + 2H^+ + H_2O_2 \rightarrow 2H_2O$. The electron donor is usually either one of the reduced pyridine nucleotides, NAD(P)H (NAD(P)H peroxidase) or thiols (thiol peroxidases). Active sites in various aerobic bacterial peroxidases contain haem or cysteine residues (Alves *et al.*, 1999; Carmel-Harel and Storz, 2000; Loewen *et al.*, 2000; Parsonage *et al.*, 2005). Bacterial peroxiredoxins use cysteine residues at their active sites, and can be either thiol or NADH peroxidases (Wood *et al.*, 2003). Sods, catalases and peroxidases are ubiquitous among aerobic bacteria. Together they lower the 5–10 μM/sec fluxes of superoxide and hydrogen peroxide in aerobically growing *E. coli* to steady-state levels on the order of 0.1 nM and 20 nM, respectively (Imlay and Fridovich, 1991; Seaver and Imlay, 2001; Imlay, 2003). Anaerobic bacteria contain several of these enzymes, as well as entirely novel ones, which will now be considered in detail.

6.5.2 Removal of cytoplasmic reactive oxygen species in *Desulfovibrio* spp.

There is as yet no quantitative prototype of oxidative stress in strictly anaerobic or microaerophilic bacteria exposed to air. However, given the reducing nature of anaerobic growth environments, one expects, if anything, higher initial levels of reduced dioxygen species upon sudden exposure to air (Imlay, 2003). The transiently aerobic or microaerobic growth habitats to which SRB are exposed can presumably cause oxidative stress from the same damaging cascade of processes as for aerobic microbes. Thus, the oxygen reduction observed in some SRB, while perhaps being a means of lowering local environmental dioxygen levels (Cypionka, 2000; Fournier *et al.*, 2003), would also generate an intracellular flux of ROS. At least one crucial enzyme in the sulphate reduction pathway, adenylyl sulphate (adenosine 5′-phosphosulphate, APS) reductase, which contains two iron-sulphur clusters (Fritz *et al.*, 2002) and an active-site FAD, has actually been reported to generate superoxide *in vitro* during catalysis of the reverse reaction, i.e. aerobic oxidation of sulfite and adenosine monophosphate to APS (Bramlett and Peck, 1975).

6.5.2.1 Catalases and peroxiredoxins

Haem catalases have been isolated from *D. vulgaris* and *D. gigas* (Hatchikian *et al.*, 1977; Dos Santos *et al.*, 2000). However, the *D. vulgaris* catalase is encoded on a plasmid that also encodes nitrogen fixation enzymes (Heidelberg *et al.*, 2004), and this plasmid is reported to be lost when the organism is grown on ammonium-containing medium. No bifunctional haem catalases-peroxidases homologous to *E. coli* hydroperoxidase I or *M. tuberculosis* KatG (Zámocky and Koller, 1999; Loewen *et al.*, 2000), have been reported from SRB. No peroxiredoxins have as yet been isolated from SRB. However, at least two putative thiol peroxidase genes were found to be upregulated in response to oxygenation of *D. vulgaris* cultures (Fournier *et al.*, 2006), making it likely that thiol peroxidases participate in oxidative stress protection in at least this sulphate-reducer.

6.5.2.2 Non-haem iron oxidative stress protection enzymes: disproportionation versus reduction

As is typical for anaerobes, only iron-containing Sods (Fe-Sods) have been reported in SRB (Kirschvink *et al.*, 2000; Brioukhanov and Netrusov, 2004). Fe-Sods have been isolated from *D. desulfuricans* (Hatchikian and Henry, 1977), *D. gigas* (Dos Santos *et al.*, 2000), and two *D. vulgaris* strains, Hildenborough and Miyazaki (Lumppio *et al.*, 2001; Fournier *et al.*, 2003; Nakanishi *et al.*, 2003). The Fe-Sod gene from both *D. vulgaris* strains encodes a putative amino-terminal signal peptide indicative of a periplasmic localization; no recognizable cytoplasmic Sod is encoded in the *D. vulgaris* genome. An *sod*-knockout strain of *D. vulgaris* was more sensitive than the wild-type to prolonged air and air plus superoxide exposure, indicating an oxidative stress protective role for this Sod (Fournier *et al.*, 2003).

More recently a "non-haem iron reductive paradigm" for scavenging superoxide and hydrogen peroxide and possibly dioxygen in air-sensitive bacteria has become apparent (Kurtz, 2004). These enzymes include superoxide reductase (Sor), rubrerythrin (Rbr), functioning as a hydrogen peroxide reductase (peroxidase), and a flavo-diiron enzyme (Fpra or Roo), functioning as a four-electron dioxygen reductase or in reductive scavenging of intracellular nitric oxide, as discussed above.

6.5.2.3 Superoxide reductases

Superoxide reductases (Sors) have been found in many air-sensitive bacteria and archaea, including microaerophiles such as *Treponema pallidum* (Abreu *et al.*, 2002; Lombard *et al.*, 2000; Kurtz, 2004, Silva *et al.*, 2001).

Proteins which turned out to be Sors were first isolated from *D. vulgaris*, *D. desulfuricans* and *D. gigas* (Moura *et al.*, 1990; Chen *et al.*, 1994). A single Sor homologue is encoded in the *D. vulgaris* genome, and this Sor has been variously referred to as rubredoxin oxidoreductase (because its gene is co-transcribed with that for rubredoxin (Brumlik and Voordouw, 1989; Pianzzola *et al.*, 1996) and desulfoferrodoxin (Moura *et al.*, 1990). A protein given the trivial name, neelaredoxin from *D. gigas* (Chen *et al.*, 1994) has been shown to have an active site homologous to that of *D. vulgaris* Sor (Silva *et al.*, 2001). No recognizable signal peptide is encoded for any of these Sors, implying a cytoplasmic localization. The Sors from *D. vulgaris* and *D. gigas* (grown on lactate-sulphate), and *D. desulfuricans* (grown on lactate-nitrate) are constitutively expressed under anaerobic growth conditions (Moura *et al.*, 1990; Chen *et al.*, 1994; Silva *et al.*, 2001). The first clue to the function of Sors was provided by restoration of anaerobic growth to an *sod*-knockout strain of *E. coli* by insertion of a gene from the sulphate reducer, *Desulfoarculus (Da.) baarsii*, encoding what turned out be a Sor (Pianzzola *et al.*, 1996). Evidence for a superoxide reductase rather than dismutase function for this complementing *Da. baarsii* protein was subsequently provided (Liochev and Fridovich, 1997; Lombard *et al.*, 2000). The *D. gigas* Sor has also been shown to complement an *E. coli sod*-knockout strain (Silva *et al.*, 2001). A specific protective role for Sor against intracellularly generated superoxide in the native organism was first demonstrated by Lumppio *et al.* in *D. vulgaris* (Lumppio *et al.*, 2001), and later confirmed by Fournier *et al.* (2003). This latter work also demonstrated that the *D. vulgaris sor*-knockout strain was far less viable upon prolonged air exposure than was the wild-type (on both plates and in liquid culture), and that the intracellular level of Sor steadily decreased upon prolonged exposure of liquid cultures of the wild-type strain to air. These observations were reasonably interpreted as indicating a protective effect against oxidative stress by Sor's lowering of the steady state level of cytoplasmic superoxide, but they also indicate that the intracellular Sor is destroyed by prolonged exposure to hyperoxic conditions.

6.5.2.4 Rubrerythrins

The vernacular name, rubrerythrin (Rbr), was given to a protein originally isolated from *D. vulgaris* (LeGall *et al.*, 1988), and is a contraction of rubredoxin and hemerythrin, reflecting its two types of iron sites, namely, rubredoxin-type [Fe(Cys)$_4$] and non-haem, His,Glu-ligated diiron, respectively (deMaré *et al.*, 1996; Jin *et al.*, 2002). Peroxidase activity (hydrogen peroxide reductase) has been reproducibly observed in Rbrs from various sources (Coulter *et al.*, 1999; Coulter and Kurtz, 2001; Weinberg *et al.*, 2004). *In vitro* at least, Rbrs show efficient NAD(P)H peroxidase (when supplied with an intermediary reductase) and rubredoxin peroxidase activities, but no catalase activity (Pierik *et al.*, 1993; Coulter *et al.*, 1999; Coulter and Kurtz, 2001). Consistent with the peroxidase activity, a significant body of genetic and microbiological evidence supports a role for Rbrs in oxidative stress protection in air-sensitive bacteria and archaea (Alban and Krieg, 1998; Das *et al.*, 2001; Sztukowska *et al.*, 2002; Kawasaki *et al.*, 2004; May *et al.*, 2004), and at least two air-sensitive eukaryotic protozoans (Pütz *et al.*, 2005). With the exception of some cyanobacteria, genes encoding Rbr homologues (defined as containing the fused diiron and rubredoxin-like domains) have so far been found only in strict anaerobes or microaerophiles, where they are widespread (albeit not universal). Many of these genomes encode multiple Rbr homologues, some of which are tandemly encoded with Sor genes (Lumppio *et al.*, 2001). Among the sulphate-reducing bacteria, Rbrs have been isolated from *D. desulfuricans* (Moura *et al.*, 1994) and *D. vulgaris*. The latter genome encodes three Rbr homologues. One of these is the originally isolated Rbr (LeGall *et al.*, 1988), which has been studied the most extensively and is referred to as rbr1; a second homologue, nigerythrin (ngr), has also been isolated from *D. vulgaris* and characterized (Pierik *et al.*, 1993; Iyer *et al.*, 2005). Both rbr1 and ngr show the peroxidase activities described above *in vitro* (Coulter *et al.*, 1999). The third homologue, Rbr2, has not been isolated, but its expression has been detected on 2D gels (Fournier *et al.*, 2006). All three of these *D. vulgaris* Rbr homologues are constitutively expressed upon anaerobic growth on lactate and sulphate. The *ngr* gene is monocistronic, whereas the *Rbr1* gene is co-transcribed with a gene encoding a Fur (ferric uptake regulator) homologue (Lumppio *et al.*, 1997). However, this Fur homologue has at least as high sequence homology as PerR (Zou *et al.*, 1999), which regulates the peroxide stress response in some bacteria (van Vliet *et al.*, 1999). The *D. vulgaris* Fur/PerR homologue has not been isolated or further characterized nor has any Fur/PerR regulon been experimentally characterized in sulphate-reducing bacteria.

6.5.2.5 Roles of the various oxidative stress protection enzymes

A model indicating the functions of many of the components discussed so far is presented in Figure 6.1. A notable feature of the active sites of both Sor and Rbr is their high reduction potentials (>200 mV vs. NHE at pH 7). Assuming that their active sites are in equilibrium with the intracellular redox state, these high reduction potentials would keep these proteins largely reduced under all but the most oxidizing conditions. *Sor* and Rbr could, therefore, continue to function efficiently during oxidative stress so long as a supply of cytoplasmic reducing equivalents remains available. The ultimate source of intracellular reducing equivalents for these latter two enzymes remains mysterious, but *D. vulgaris* Sor and Rbr1 have been shown to function as the terminal components of an efficient NAD(P)H superoxide oxidoreductase and NAD(P)H hydrogen peroxide oxidoreductase, respectively, using the small electron transfer protein, rubredoxin, as proximal electron donor (Coulter *et al.*, 1999; Coulter and Kurtz, 2001; Emerson *et al.*, 2003). At least in *D. vulgaris* this *in vitro* Sor/rubredoxin redox partner pairing is likely to reflect that occurring *in vivo*, since, as noted above, the genes encoding Sor and rubredoxin are co-transcribed in *D. vulgaris*. Similarly, *A. fulgidus* Sor is efficiently reduced *in vitro* by rubredoxins from the same organism, which in turn are reduced by NADH oxidoreductases (Rodrigues *et al.*, 2005). On the other hand, the *D. gigas* Sor is not co-transcribed with a redox partner protein (Silva *et al.*, 2001) and the Sor complementations of the *sod*-knockout *E. coli* strain imply that this strain can supply reducing equivalents to support functional heterologous Sor activity *in vivo* without rubredoxin. These observations indicate that Sors can non-specifically scavenge reducing equivalents and funnel them into superoxide (Lombard *et al.*, 2000; Kurtz, 2004). The analogous conclusion may be true for the peroxidase activities of Rbrs. Although Fe-Sods typically also have active site reduction potentials exceeding +200 mV (Vance and Miller, 1998; Miller, 2004), there is no evidence that Fe-Sods function as Sors (Liochev and Fridovich, 2000).

Whether Sods are universally distributed in SRB and are always localized in the periplasm is currently unclear. However, the periplasmic localization of Sod can be rationalized as being required for scavenging superoxide generated within this compartment, because the negatively charged superoxide would have difficulty crossing the cytoplasmic membrane (Salvador *et al.*, 2001). In *D. vulgaris*, the Fe-Sod is thus likely to function as a "housekeeping" oxidative stress protection enzyme in the periplasm, as do Sor and Rbrs in the cytoplasm. The fact that dismutation requires two superoxides while the reduction requires only one may mean that dismutation is more efficient at higher superoxide concentrations and reductases are more efficient at lower superoxide concentrations (Imlay, 2002a). This observation provides a rationale for the periplasmic localization of Fe-Sod and cytoplasmic localization of Sor. At least under microaerobic conditions, the periplasm is likely to be exposed to a higher steady-state concentration of dioxygen, and, therefore, of superoxide than is the cytoplasm. Alternatively, the periplasm may have higher fluxes of reducing equivalents through enzymes that adventitiously produce superoxide. Hydrogen peroxide is more freely diffusible across bacterial membranes (Seaver and Imlay, 2001). The apparent absence of catalases and peroxidases from the periplasmic space of sulphate reducers could be due either to lower Fenton-reactive iron levels (discussed in Section 6.6) in that compartment or simply because the lethality of hydrogen peroxide is due primarily to DNA damage.

Along with the apparent destruction of Sor upon prolonged air exposure in *D. vulgaris* (Fournier *et al.*, 2003), Rbr1 and Rbr2 concentrations are also apparently lowered upon prolonged exposure of *D. vulgaris* to hyperoxic conditions (Fournier *et al.*, 2006). Two radically different explanations for this decrease in concentration have already been given in section 6.2. It is important to note that Fe-Sod, Sor and the Rbrs are constitutively expressed under anaerobic conditions. This is likely to protect the cell, at least initially, against the burst of cytoplasmic superoxide and hydrogen peroxide occurring upon exposure to air, or under prolonged microaerobic conditions. In view of the destruction of these enzymes upon continued aeration, other proteins, e.g. thiol peroxidases and iron-sulphur cluster repair enzymes, would assume this protective role upon prolonged air exposure.

6.6 PREVENTION OF FORMATION OF REACTIVE OXYGEN SPECIES BY IRON STORAGE ENZYMES

Oxygen and iron metabolism are intimately linked, mainly because ferrous ions catalyze the formation of ROS via Fenton chemistry (Imlay, 2003). Therefore, it is essential to any oxygen-exposed organism to avoid the presence of "free" ferrous iron in the cytoplasm. This is achieved by ferritins, iron storage proteins present in all domains of life. The genomes of *D. vulgaris* and *D. desulfuricans* G20 each contain genes encoding a ferritin and a bacterioferritin (Table 6.1: Bfr). Like other (bacterio)ferritins, the Bfr from *D. desulfuricans* ATCC 27774 was shown to be a 24-mer with one haem per dimer and one catalytic diiron site per monomer (Romão *et al.*, 2000a, Macedo *et al.*, 2003). Instead of the expected haem *b*, the dimers contained iron-coproporphyrin III, which had not previously been found in any biological system, indicating how little is known of haem biosynthesis in the SRB (Romão *et al.*, 2000a). In addition to the catalytic diiron site, *D. desulfuricans* Bfr was also shown to be able to incorporate an iron core (Romão *et al.*, 2000b). The *bfr* gene, which is highly expressed under anaerobic conditions, forms a di-cistronic transcriptional unit with a gene for a type II rubredoxin, which was found to act as the electron donor to Bfr in solution (da Costa *et al.*, 2001). The ferritin proteins are important for oxidative stress relief: in the presence of oxygen, iron ions are rapidly oxidized at the catalytic diiron site, and stored at the protein interior as harmless ferric oxyhydroxide minerals. This oxidation and sequestering of iron ions prevents formation of ROS via the Fenton reaction, while contributing to oxygen uptake. The high concentrations of Bfr under anaerobic conditions indicates it to be another housekeeping agent that plays an important role in the oxidative stress response.

6.7 DO *DESULFOVIBRIO* SPP. EVER LIVE AS GENUINE AEROBES?

As indicated in the preceding sections oxygen defence in *Desulfovibrio* spp. involves both cytoplasmic, membrane-bound and periplasmic oxygen defence proteins, as summarized in Figure 6.1. It should be noted that not all SRB have so many periplasmic redox proteins. For example, the periplasm of *Desulfotalea psychrophila* (Rabus *et al.*, 2004) appears almost devoid of *c*-type cytochromes, indicating that these are not strictly required for the anaerobic metabolism of SRB. This leads one to conclude that although the primary role of these redox proteins remains in anaerobic metabolism, they may also contribute to the remarkable aerotolerance of SRB which possess them, such as the *Desulfovibrio* spp.

The existence of a periplasmic oxygen reduction system that does not disappear upon oxygen exposure distinguishes *Desulfovibrio* spp. from facultative microorganisms like *E. coli*. When *E. coli* is grown anaerobically it probably also harbours redox proteins in its periplasm that can react with oxygen and thus constitute a periplasmic oxygen defence system. However, the outcome of exposing *E. coli* to oxygen is very different from exposing *D. vulgaris* to oxygen. In the case of *E. coli*, oxygen exposure will lead to repression of transcription of genes for oxygen-sensitive, anaerobic reductases and induction of genes required for aerobic electron transport components that have relatively low ROS production. In the case of *D. vulgaris*, other *Desulfovibrio* spp., and presumably many other SRB, a large fraction of reducing equivalents used for cytoplasmic sulphate reduction cycles routinely through periplasmic carriers as part of the anaerobic energy metabolism of these organisms (Voordouw, 2002). In the presence of increasing amounts of oxygen, an increasing flux of these reducing equivalents is directed towards periplasmic oxygen reduction. Although this redirection may protect the cytoplasmic sulphate reduction pathway from the harmful effects of oxygen, it does not contribute to growth. Genetic reprogramming to a growth-contributing aerobic energy metabolism, as in *E. coli*, has never been observed in sulphate reducers, even though the *D. vulgaris* genome harbours genes encoding homologues of well-known transcriptional regulators associated with the anaerobic to aerobic metabolic transition: FNR (DVU1083), ArcAB (DVU3045) and OxyR (DVU3313, DVU2111).

As indicated in section 6.1 many *Desulfovibrio* spp. live close to the edge in environments which experience periodic oxygenation. Availability of organic carbon and/or sulphate, formed by re-oxidation of sulphide, must serve as the attractants to such environments. Despite the high oxygen concentrations that can occur, we surmise that the total electron flux to oxygen remains a minor fraction of the total electron flux to sulphate in these aerated environments. In contrast to *E. coli*, which would reprogramme genetically under these conditions, SRB will move away from air once that condition is no longer satisfied.

ACKNOWLEDGEMENTS

GV acknowledges support of a Discovery Grant from the Natural Sciences and Engineering Research Council of Canada (NSERC). MT acknowledges support from Fundação para a Ciência e Tecnologia, Portugal. We would like to thank all our co-workers, whose names appear on the references.

REFERENCES

Abdollahi, H. and Wimpenny, J. W. T. (1990). Effects of the oxygen on the growth of *Desulfovibrio desulfuricans*. *J Gen Microbiol*, **136**, 1025–30.

Abreu, I. A., Xavier, A. V., LeGall, J., Cabelli, D. E. and Teixeira, M. (2002). Superoxide scavenging by neelaredoxin: dismutation and reduction activities in anaerobes. *J Biol Inorg Chem*, **7**, 668–74.

Alban, P. S. and Krieg, N. R. (1998). A hydrogen peroxide resistant mutant of *Spirillum volutans* has NADH peroxidase activity but no increased oxygen tolerance. *Can J Microbiol*, **44**, 87–91.

Alves, T., Besson, S., Duarte, L. C. *et al.* (1999). A cytochrome c peroxidase from *Pseudomonas nautica* 617 active at high ionic strength: expression, purification and characterization. *Biochim Biophys Acta*, **1434**, 248–59.

Barynin, V. V., Whittaker, M. M., Antonyuk, S. V. *et al.* (2001). Crystal structure of manganese catalase from *Lactobacillus plantarum. Structure (Camb.)*, **9**, 725–38.

Baumgarten, A., Redenius, I., Kranczoch, J. and Cypionka, H. (2001). Periplasmic reduction by *Desulfovibrio* species. *Arch Microbiol*, **176**, 306–9.

Bramlett, R. N. and Peck, H. D., Jr. (1975). Some physical and kinetic properties of adenylyl sulphate reductase from *Desulfovibrio vulgaris. J Biol Chem*, **250**, 2979–86.

Brioukhanov, A. L. and Netrusov, A. I. (2004). Catalase and superoxide dismutase: distribution, properties, and physiological role in cells of strict anaerobes. *Biochemistry (Mosc)*, **69**, 949–62.

Brumlik, M. J. and Voordouw, G. (1989). Analysis of the transcriptional unit encoding the genes for rubredoxin (*rub*) and a putative rubredoxin oxidoreductase (*rbo*) in *Desulfovibrio vulgaris* (Hildenborough). *J Bacteriol*, **171**, 4996–5004.

Canfield, D. E. and Des Marais, D. J. (1991). Aerobic sulphate reduction in microbial mats. *Science*, **251**, 1471–3.

Carmel-Harel, O. and Storz, G. (2000). Roles of the glutathione- and thioredoxin-dependent reduction systems in the *Escherichia coli* and *Saccharomyces cerevisiae* responses to oxidative stress. *Annu Rev Microbiol*, **54**, 439–61.

Caumette, P., Matheron, R., Raymond, N. and Relexans, J.-C. (1994). Microbial mats in the hypersaline ponds of Mediterranean salterns (Salins de Giraud, France). *FEMS Microbiol Ecol*, **13**, 273–86.

Chen, L., Liu, M.-Y., LeGall, J. *et al.* (1993a). Purification and characterization of an NADH-rubredoxin oxidoreductase involved in the utilization of oxygen by *Desulfovibrio gigas. Eur J Biochem*, **216**, 443–8.

Chen, L., Liu, M.-Y., LeGall, J. *et al.* (1993b). Rubredoxin oxidase, a new flavo-hemo-protein, is the site of oxygen reduction to water by the "strict anaerobe" *Desulfovibrio gigas. Biochem Biophys Res Commun*, **193**, 100–5.

Chen, L., Sharma, P., Le Gall, J. *et al.* (1994). A blue non-heme iron protein from *Desulfovibrio gigas. Eur J Biochem*, **226**, 613–18.

Choudhury, S. B., Lee, J. W., Davidson, G. *et al.* (1999). Examination of the nickel site structure and reaction mechanism in *Streptomyces seoulensis* superoxide dismutase. *Biochemistry*, **38**, 3744–52.

Coulter, E. D. and Kurtz, D. M., Jr. (2001). A role for rubredoxin in oxidative stress protection in *Desulfovibrio vulgaris*: catalytic electron transfer to rubrerythrin and two-iron superoxide reductase. *Arch Biochem Biophys*, **394**, 76–86.

Coulter, E. D., Shenvi, N. V. and Kurtz, D. M., Jr. (1999). NADH peroxidase activity of rubrerythrin. *Biochem Biophys Res Commun*, **255**, 317–23.

Cypionka, H. (2000). Oxygen respiration by *Desulfovibrio* species. *Annu Rev Microbiol*, **54**, 827–48.

da Costa, P. N., Romão, C. V., LeGall, J. *et al.* (2001). The genetic organization of *Desulfovibrio desulfuricans* ATCC 27774 bacterioferritin and rubredoxin-2 genes. Involvement of rubredoxin in the iron metabolism. *Mol Microbiol*, **41**, 217–29.

Das, A., Coulter, E. D., Kurtz, D. M., Jr. and Ljungdahl, L. G. (2001). Five-gene cluster in *Clostridium thermoaceticum* consisting of two divergent operons encoding rubredoxin oxidoreductase-rubredoxin and rubrerythrin-type A flavoprotein-high-molecular-weight rubredoxin. *J Bacteriol*, **183**, 1560–7.

deMaré, F., Kurtz, D. M., Jr. and Nordlund, P. (1996). The structure of *Desulfovibrio vulgaris* rubrerythrin reveals a unique combination of rubredoxin-like FeS_4 and ferritin-like diiron domains. *Nature Struct Biol*, **3**, 539–46.

Dos Santos, W. G., Pacheco, I., Liu, M. Y. *et al.* (2000). Purification and characterization of an iron superoxide dismutase and a catalase from the sulphate-reducing bacterium *Desulfovibrio gigas*. *J Bacteriol*, **182**, 796–804.

Emerson, J. P., Coulter, E. D., Phillips, R. S. and Kurtz, D. M., Jr. (2003). Kinetics of the superoxide reductase catalytic cycle. *J Biol Chem*, **278**, 39662–8.

Eschemann, A., Kèuhl, M. and Cypionka, H. (1999). Aerotaxis in *Desulfovibrio*. *Environ Microbiol*, **1**, 489–94.

Frazão, C., Silva, G., Gomes, C. M. *et al.* (2000). Structure of a dioxygen reduction enzyme from *Desulfovibrio gigas*. *Nature Struct Biol*, **7**, 1041–5.

Fournier, M., Aubert, C., Dermoun, Z. *et al.* (2006). Response of the anaerobe *Desulfovibrio vulgaris* Hildenborough to oxidative conditions: proteome and transcript analysis. *Biochimie*, **88**, 85–94.

Fournier, M., Dermoun, Z., Durand, M. C. and Dolla, A. (2004). A new function of the *Desulfovibrio vulgaris* Hildenborough Fe hydrogenase in the protection against oxidative stress. *J Biol Chem*, **279**, 1787–93.

Fournier, M., Zhang, Y., Wildschut, J. D. *et al.* (2003). Function of oxygen resistance proteins in the anaerobic, sulphate-reducing bacterium, *Desulfovibrio vulgaris* Hildenborough. *J Bacteriol*, **185**, 71–9.

Fritz, G., Bèuchert, T. and Kroneck, P. M. (2002). The function of the 4Fe-4S clusters and FAD in bacterial and archaeal adenylylsulphate reductases. Evidence for flavin-catalyzed reduction of adenosine 5′-phosphosulphate. *J Biol Chem*, **277**, 26066–73.

Fu, R. and Voordouw, G. (1997). Targeted gene-replacement mutagenesis of *dcrA*, encoding an oxygen sensor of the sulphate-reducing bacterium *Desulfovibrio vulgaris* Hildenborough. *Microbiology*, **143**, 1815–26.

Gardner, A. M., Helmick, R. A. and Gardner, P. R. (2002). Flavorubredoxin, an inducible catalyst for nitric oxide reduction and detoxification in *Escherichia coli*. *J Biol Chem*, **277**, 8172–7.

Gomes, C. M., Giuffrè, A., Forte, E. *et al.* (2002). A novel type of nitric oxide reductase: *Escherichia coli* flavorubredoxin, *J Biol Chem*, **277**, 25273–6.

Hatchikian, E. C. and Henry, Y. A. (1977). An iron-containing superoxide dismutase from the strict anaerobe *Desulfovibrio desulfuricans* (Norway 4). *Biochimie*, **59**, 153–61.

Hatchikian, C. E., LeGall, J. and Bell, G. R. (1977). Significance of superoxide dismutase and catalase activities in the strict anaerobes, sulphate-reducing bacteria. In A. M. Michael, J. M. McCord and I. Fridovich (eds.), *Superoxide and superoxide dismutase*. New York: Academic Press. pp. 159–72.

Heidelberg, J. F., Seshadri, R., Haveman, S. A. *et al.* (2004). The genome sequence of the anaerobic, sulphate-reducing bacterium *Desulfovibrio vulgaris Hildenborough*. *Nat Biotechnol*, **22**, 554–9.

Imlay, J. A. (2002a). How oxygen damages microbes: oxygen tolerance and obligate anaerobiosis. *Adv Microb Physiol*, **46**, 111–53.

Imlay, J. A. (2002b). What biological purpose is served by superoxide reductases? *J Biol Inorg Chem*, **7**, 659–63.

Imlay, J. A. (2003). Pathways of oxidative damage. *Annu Rev Microbiol*, **57**, 395–418.

Imlay, J. A. and Fridovich, I. (1991). Assay of metabolic superoxide production in *Escherichia coli*. *J Biol Chem*, **266**, 6957–65.

Imlay, K. R. C. and Imlay, J. A. (1996). Cloning and analysis of *sodC*, encoding the copper-zinc superoxide dismutase of *Escherichia coli*. *J Bacteriol*, **178**, 2564–71.

Iyer, R. B., Silaghi-Dumitrescu, R., Kurtz, D. M., Jr. and Lanzilotta, W. N. (2005). High-resolution crystal structures of *Desulfovibrio vulgaris* (Hildenborough) nigerythrin: facile, redox-dependent iron movement, domain interface variability, and peroxidase activity in the rubrerythrins. *J Biol Inorg Chem*, **10**, 407–16.

Jin, S., Kurtz, D. M., Jr., Liu, Z.-J., Rose, J. and Wang, B.-C. (2002). X-ray crystal structures of reduced rubrerythrin and its azide adduct: a structure-based mechanism for a non-heme diiron peroxidase. *J Am Chem Soc*, **124**, 9845–55.

Justino, M. C., Vicente, J. B., Teixeira, M. and Saraiva, L. M. (2005). New genes implicated in the protection of anaerobically grown *Escherichia coli* against nitric oxide. *J Biol Chem*, **280**, 2636–43.

Kawasaki, S., Ishikura, J., Watamura, Y. and Niimura, Y. (2004). Identification of O_2-induced peptides in an obligatory anaerobe, *Clostridium acetobutylicum*. *FEBS Lett*, **571**, 21–5.

Kawasaki, S., Watamura, Y., Ono, M. *et al.* (2005). Adaptive responses to oxygen stress in obligatory anaerobes *Clostridium acetobutylicum* and *Clostridium aminovalericum*. *Appl Environ Microbiol*, **71**, 8442–50.

Kirschvink, J. L., Gaidos, E. J., Bertani, L. E. *et al.* (2000). Paleoproterozoic snowball earth: extreme climatic and geochemical global change and its biological consequences. *Proc Natl Acad Sci USA*, **97**, 1400–5.

Kitamura, M., Mizugai, K., Taniguchi, M. *et al.* (1995). A gene encoding a cytochrome *c* oxidase-like protein is located closely to the cytochrome *c*-553 gene in the anaerobic bacterium, *Desulfovibrio vulgaris* (Miyazaki F). *Microbiol Immunol*, **39**, 75–80.

Kjeldsen, K. U., Joulian, C. and Ingvorsen, K. (2005). Effects of oxygen exposure on respiratory activities of *Desulfovibrio desulfuricans* strain DvO1 isolated from activated sludge. *FEMS Microbiol Ecol*, **53**, 275–84.

Krekeler, D., Sigalevich, P., Teske, A., Cypionka, H. and Cohen, Y. (1997). Sulphate-reducing bacterium form the oxic layer of a microbial mat from Solar Lake (Sinai), *Desulfovibrio oxyclinae* sp. nov. *Arch Microbiol*, **167**, 369–75.

Kumagai, H., Fujiwara, T., Matsubara, H. and Saeki, K. (1997). Membrane localization, topology, and mutual stabilization of the *rnfABC* gene products in *Rhodobacter capsulatus* and implications for a new family of energy-coupling NADH oxidoreductases. *Biochemistry*, **36**, 5509–21.

Kurtz, D. M., Jr. (2004). Microbial detoxification of superoxide: The non-heme iron reductive paradigm for combating oxidative stress. *Acc Chem Res*, **37**, 902–8.

LeGall, J., Prickril, B. C., Moura, I. *et al.* (1988). Isolation and characterization of rubrerythrin, a non-heme iron protein from *Desulfovibrio vulgaris* that contains rubredoxin centers and a hemerythrin-like binuclear iron cluster. *Biochemistry*, **27**, 1636–42.

Lemos, R. S., Gomes, C. M., LeGall, J., Xavier, A. V., Teixeira, M. (2002). The quinol:fumarate oxidoreductase from the sulphate reducing bacterium *Desulfovibrio gigas*: spectroscopic and redox studies. *J Bioenerg Biomemb*, **34**, 21–30.

Lemos, R. S., Gomes, C. M., Santana, M. *et al.* (2001). The 'strict' anaerobe *Desulfovibrio gigas* contains a membrane-bound oxygen-reducing respiratory chain. *FEBS Lett*, **496**, 40–3.

Liochev, S. I. and Fridovich, I. (1997). A mechanism for complementation of the *sodA sodB* defect in *Escherichia coli* by overproduction of the *rbo* gene product (desulfoferrodoxin) from *Desulfoarculus baarsii*. *J Biol Chem*, **272**, 25573–5.

Liochev, S. I. and Fridovich, I. (2000). Copper- and zinc-containing superoxide dismutase can act as a superoxide reductase and a superoxide oxidase. *J Biol Chem*, **275**, 38482–5.

Loewen, P. C., Klotz, M. G. and Hassett, D. J. (2000). Catalase – an "old" enzyme that continues to surprise us. *ASM News*, **66**, 76–82.

Lombard, M., Fontecave, M., Touati, D. and Nivière, V. (2000). Reaction of the desulfoferrodoxin from *Desulfoarculus baarsii* with superoxide anion. Evidence for a superoxide reductase activity. *J Biol Chem*, **275**, 115–21.

Lumppio, H. L., Shenvi, N. V., Garg, R. P., Summers, A. O. and Kurtz, D. M., Jr. (1997). A rubrerythrin operon and nigerythrin gene in *Desulfovibrio vulgaris* (Hildenborough). *J Bacteriol*, **179**, 4607–15.

Lumppio, H. L., Shenvi, N. V., Summers, A. O., Voordouw, G. and Kurtz, D. M., Jr. (2001). Rubrerythrin and rubredoxin oxidoreductase in *Desulfovibrio vulgaris*. A novel oxidative stress protection system. *J Bacteriol*, **183**, 101–8, and correction 2970.

Macedo, S., Romão, C. V., Mitchell, E. *et al.* (2003). The nature of the diiron site in the bacterioferritin from *Desulfovibrio desulfuricans*. *Nature Struct Biol*, **10**, 285–90.

Marschall, C., Frenzel, P. and Cypionka, H. (1993). Influence of oxygen on sulphate reduction and growth of sulphate-reducing bacteria. *Arch Microbiol*, **159**, 168–73.

May, A., Hillmann, F., Riebe, O., Fischer, R. J. and Bahl, H. (2004). A rubrerythrin-like oxidative stress protein of *Clostridium acetobutylicum* is encoded by a duplicated gene and identical to the heat shock protein Hsp21. *FEMS Microbiol Lett*, **238**, 249–54.

McCord, J. M., Keele, B. B., Jr. and Fridovich, I. (1971). An enzyme-based theory of obligate anaerobiosis: the physiological function of superoxide dismutase. *Proc Natl Acad Sci USA*, **68**, 1024–7.

Miller, A.-F. (2004). Superoxide Processing. In L. Que, Jr. and W. B. Tolman (eds.), *Comprehensive coordination chemistry II – from biology to nanotechnology*, Oxford, UK: Elsevier. pp. 479–506.

Moura, I., Tavares, P., Moura, J. J. *et al.* (1990). Purification and characterization of desulfoferrodoxin. A novel protein from *Desulfovibrio desulfuricans* (ATCC 27774) and from *Desulfovibrio vulgaris* (strain Hildenborough) that contains a distorted rubredoxin center and a mononuclear ferrous center. *J Biol Chem*, **265**, 21596–602.

Moura, I., Tavares, P. and Ravi, N. (1994). Characterization of 3 proteins containing multiple iron sites – rubrerythrin, desulfoferrodoxin, and a protein containing a six-iron cluster. *Methods Enzymol*, **243**, 216–40.

Nakanishi, T., Inoue, H. and Kitamura, M. (2003). Cloning and expression of the superoxide dismutase gene from the obligate anaerobic bacterium *Desulfovibrio vulgaris* (Miyazaki F). *J Biochem (Tokyo)*, **133**, 387–93.

Park, S., You, X. and Imlay, J. A. (2005). Substantial DNA damage from submicromolar intracellular hydrogen peroxide detected in Hpx-mutants of *Escherichia coli*. *Proc Natl Acad Sci USA*, **102**, 9317–22.

Parsonage, D., Youngblood, D. S., Sarma, G. N. *et al.* (2005). Analysis of the link between enzymatic activity and oligomeric state in AhpC, a bacterial peroxiredoxin. *Biochemistry*, **44**, 10583–92.

Pereira, M. M. and Teixeira, M. (2004). Proton pathways, ligand binding and dynamics of the catalytic site haem-copper superfamily of oxygen reductases. *Biochim Biophys Acta*, **1655**, 340–6.

Pereira, M. M., Bandeiras, T. M., Fernandes, A. S. *et al.* (2004). Respiratory chains from aerobic thermophilic prokaryotes. *J Bioenerg Biomemb*, **36**, 93–105.

Pereira, M. M., Carita, J. N. and Teixeira, M. (1999). Membrane-bound electron transfer chain of the thermohalophilic bacterium *Rhodothermus marinus*: a novel multihemic cytochrome *bc*, a new complex III. *Biochemistry*, **38**, 1268–75.

Pereira, M. M., Santana, M. and Teixeira, M. (2001). A novel scenario for the evolution of haem-copper oxidases. *Biochim Biophys Acta*, **1505**, 185–208.

Pianzzola, M. J., Soubes, M. and Touati, D. (1996). Overproduction of the *rbo* gene product from *Desulfovibrio* species suppresses all deleterious effects of lack of superoxide dismutase in *Escherichia coli*. *J Bacteriol*, **178**, 6736–42.

Pierik, A. J., Wolbert, R. B. G., Portier, G. L., Verhagen, M. F. J. M. and Hagen, W. R. (1993). Nigerythrin and rubrerythrin from *Desulfovibrio vulgaris* each contain two mononuclear iron centers and two dinuclear iron clusters. *Eur J Biochem*, **212**, 237–45.

Pütz, S., Gelius-Dietrich, G., Piotrowski, M. and Henze, K. (2005). Rubrerythrin and peroxiredoxin: two novel putative peroxidases in the hydrogenosomes of the microaerophilic protozoon *Trichomonas vaginalis*. *Mol Biochem Parasitol*, **142**, 212–23.

Rabus, R., Ruepp, A., Frickey, T. *et al.* (2004). The genome of *Desulfotalea psychrophila*, a sulphate-reducing bacterium from permanently cold Arctic sediments. *Environ Microbiol*, **6**, 887–902.

Ravenschlag, K., Sahm, K., Knoblauch, C., Jorgensen, B. B. and Amann, R. (2000). Community structure, cellular rRNA content, and activity of sulphate-reducing bacteria in marine arctic sediments. *Appl Environ Microbiol*, **66**, 3592–602.

Risatti, J. B., Capman, W. C. and Stahl, D. A. (1994). Community structure of a microbial mat: the phylogenetic dimension. *Proc Natl Acad Sci USA*, **91**, 10173–7.

Rodrigues, J. V., Abreu, I. A., Saraiva, L. M. and Teixeira, M. (2005). Rubredoxin acts as an electron donor for neelaredoxin in *Archaeoglobus fulgidus*. *Biochem Biophys Res Commun*, **329**, 1300–5.

Romão, C. V., Louro, R., Timkovich, R. *et al.* (2000a). Iron-coproporphyrin III is a natural cofactor in bacterioferritin from the anaerobic bacterium *Desulfovibrio desulfuricans*. *FEBS Lett*, **480**, 213–16.

Romão, C. V., Regalla, M., Xavier, A. V. *et al.* (2000b). A bacterioferritin from the strict anaerobe *Desulfovibrio desulfuricans* ATCC 27774. *Biochemistry*, **39**, 6841–9.

Salvador, A., Sousa, J. and Pinto, R. E. (2001). Hydroperoxyl, superoxide and pH gradients in the mitochondrial matrix: a theoretical assessment. *Free Radic Biol Med*, **31**, 1208–15.

Saraiva, L. M., Vicente, J. B. and Teixeira, M. (2004). The role of the flavodiiron proteins in microbial nitric oxide detoxification. *Advances in Microbial Physiology*, **49**, 77–129.

Saraste, M. (1999). Oxidative phosphorylation at the fin de siècle. *Science*, **283**, 1488–93.

Sass, H., Berchtold, M., Branke, J. *et al.* (1998). Psychrotolerant sulphate-reducing bacteria from an oxic freshwater sediment, description of *Desulfovibrio cuneatus* sp. nov. and *Desulfovibrio litoralis* sp. nov. *Syst Appl Microbiol*, **21**, 212–19.

Sass, H., Cypionka, H. and Babenzien, H.-D. (1997). Vertical distribution of sulphate-reducing bacteria at the oxic–anoxic interface in sediments of the oligotrophic Lake Stechlin. *FEMS Microbiol Ecol*, **22**, 245–55.

Seaver, L. C. and Imlay, J. A. (2001). Hydrogen peroxide fluxes and compartmentalization inside growing *Escherichia coli*. *J Bacteriol*, **183**, 7182–19.

Sigalevich, P. and Cohen, Y. (2000). Oxygen dependent growth of the sulphate-reducing bacterium *Desulfovibrio oxyclinae* in coculture with *Marinobacter* sp. strain MB in a aerated sulphate-depleted chemostat. *Appl Environ Microbiol*, **66**, 5019–23.

Silaghi-Dumitrescu, R., Coulter, E. D., Das, A. *et al.* (2003). A flavodiiron protein and high molecular weight rubredoxin from *Moorella thermoacetica* with nitric oxide reductase activity. *Biochemistry*, **42**, 2806–15.

Silaghi-Dumitrescu, R., Ng, K. Y., Viswanathan, R. and Kurtz, D. M., Jr. (2005). A flavodiiron protein from *Desulfovibrio vulgaris* with oxidase and nitric oxide reductase activities. Evidence for an *in vivo* nitric oxide scavenging function. *Biochemistry*, **44**, 3572–9.

Silva, G., LeGall, J., Xavier, A. V., Teixeira, M. and Rodrigues-Pousada, C. (2001). Molecular characterization of *Desulfovibrio gigas* neelaredoxin, a protein involved in oxygen detoxification in anaerobes. *J Bacteriol*, **183**, 4413–20.

Srinivasan, V., Rajendran, C., Sousa, F. L. *et al.* (2005). Structure at 1.3A resolution of *Rhodothermus marinus* caa_3 cytochrome *c* domain. *J Mol Biol*, **345**, 1047–57.

Storz, G. and Imlay, J. A. (1999). Oxidative stress. *Curr Opin Microbiol*, **2**, 188–94.

Sztukowska, M., Bugno, M., Potempa, J., Travis, J. and Kurtz, D. M., Jr. (2002). Role of rubrerythrin in the oxidative stress response of *Porphyromonas gingivalis*. *Mol Microbiol*, **44**, 479–88.

Teske, A., Ramsing, N. B., Habicht, K. *et al.* (1998). Sulphate-reducing bacteria and their activities in cyanobacterial mats of Solar Lake (Sinai, Egypt). *Appl Env Microbiol*, **64**, 2943–51.

van Vliet, A. H., Baillon, M. L., Penn, C. W. and Ketley, J. M. (1999). *Campylobacter jejuni* contains two Fur homologs: characterization of iron-responsive regulation of peroxide stress defense genes by the PerR repressor. *J Bacteriol*, **181**, 6371–6.

Vance, C. K. and Miller, A. F. (1998). A simple proposal that can explain the inactivity of metal-substituted superoxide dismutases. *J Am Chem Soc*, **120**, 461–7.

Voordouw, G. (2002). Carbon monoxide cycling by *Desulfovibrio vulgaris* Hildenborough. *J Bacteriol*, **184**, 5903–11.

Weinberg, M. V., Jenney, F. E., Cui, X. Y. and Adams, M. W. W. (2004). Rubrerythrin from the hyperthermophilic archaeon *Pyrococcus furiosus* is a rubredoxin-dependent, iron-containing peroxidase. *J Bacteriol*, **186**, 7888–95.

Wood, Z. A., Schröder, E., Robin Harris, J. and Poole, L. B. (2003). Structure, mechanism and regulation of peroxiredoxins. *Trends Biochem Sci*, **28**, 32–40.

Yoder, D. W., Hwang, J., and Penner-Hahn, J. E. (2000). Manganese catalases. In A. Sigel and H. Sigel. (eds.), *Metal Ions in Biological Systems*, Vol. 37. New York: Marcel Dekker, Inc. pp. 527–57.

Zámocky, M. and Koller, F. (1999). Understanding the structure and function of catalases: clues from molecular evolution and *in vitro* mutagenesis. *Prog Biophys Mol Biol*, **72**, 19–66.

Zou, P.-J., Borovok, I., Ortiz de Orué Lucana, D., Müller, D. and Schrempf, H. (1999). The mycelium-associated *Streptomyces reticuli* catalase-peroxidase, its gene and regulation by FurS. *Microbiology*, **145**, 549–59.

CHAPTER 7

Biochemical, genetic and genomic characterization of anaerobic electron transport pathways in sulphate-reducing *Delta proteobacteria*

Inês A. C. Pereira, Shelley A. Haveman and Gerrit Voordouw

7.1 INTRODUCTION

Sulphate-reducing bacteria (SRB) derive energy for growth by coupling the oxidation of hydrogen or organic compounds to the reduction of sulphate to sulphide. The bioenergetics and the global topology of energy-conserving reactions have already been discussed in Chapter 1. Understanding the bioenergetics of the coupling of hydrogen oxidation and sulphate reduction is simple, in principle. Four H_2 are oxidized by periplasmic hydrogenases and the eight protons and electrons are transferred to the cytoplasm through ATP synthase and transmembrane-electron-transfer complexes for sulphate reduction. This produces approximately three adenosine triphosphates (ATPs), of which two are needed to activate sulphate. Hence a net yield of one ATP is produced per sulphate reduced. Energy conservation by coupling the reduction of sulphate to the incomplete oxidation of lactate is more complex because the primary oxidation reactions are now also cytoplasmic. Because these yield two ATPs by substrate level phosphorylation, the same number as required for the activation of sulphate, a net energetic benefit can only be obtained by hydrogen cycling as proposed by Odom and Peck (Odom and Peck, 1981), cycling of formate or CO (Heidelberg *et al.*, 2004; Voordouw, 2002) or by electrogenic proton translocation associated with the electron transport chain for reduction of sulphate. The components that participate in these anaerobic electron transport pathways will be considered in detail here. Harry Peck and Jean LeGall, the pioneers of the biochemistry of SRB, contributed greatly by purifying and characterizing many of the redox proteins present in these organisms. The recently completed genome sequences of *Desulfovibrio vulgaris* Hildenborough (Heidelberg *et al.*, 2004), *D. desulfuricans* G20 (www.jgi.doe.gov) and *Desulfotalea psychrophila*

(*Dt. psychrophila*; Rabus *et al.*, 2004), all belonging to the *Deltaproteobacteria*, provide a comprehensive framework for appreciating and understanding these early biochemical studies. These completed genome sequences drive much of the current biochemical and genetic studies on these organisms. *D. vulgaris* and *D. desulfuricans* are closely related phylogenetically, whereas *Dt. psychrophila* is quite distant. Note that phylogenetic analysis of the 16S rRNA gene and the lack of genes for nitrate reduction (a defining characteristic of *D. desulfuricans* spp.) indicate that *D. desulfuricans* G20 needs to be reclassified. *D. alaskensis* G20 has been proposed, but not yet officially adopted. We will continue to refer to the organism as *D. desulfuricans* G20 here. Although we will focus on electron transport components of the *Desulfovibrio* spp. comparisons with *Dt. psychrophila* are of interest because the electron transport pathways in this organism are so different. These differences (Table 7.1) have been derived by genome comparison using MicrobesonLine (http://www.microbesonline.org/) or the Integrated Microbial Genome website of the DOE Joint Genome Institute (http://img.jgi.doe.gov/cgi-bin/pub/main.cgi).

7.2 PERIPLASMIC HYDROGENASES

The central role of hydrogen in *Desulfovibrio* metabolism is reflected in the high cellular content of hydrogenases in these organisms. The *Desulfovibrio* periplasmic hydrogenases of the [Fe] and [NiFe] classes, as defined by Vignais and Colbeau (2004), have been investigated in great structural and mechanistic detail (de Lacey *et al.*, 2005; Matias *et al.*, 2005; Volbeda and Fontecilla-Camps, 2005), and the first crystal structure to be obtained for a [NiFe] hydrogenase was for the enzyme from *D. gigas* (Volbeda *et al.*, 1995).

Early studies indicated that *Desulfovibrio* spp. can have periplasmic [NiFe], [NiFeSe] and [Fe] hydrogenases, with [NiFe] hydrogenase being always present in a screening of different species (Voordouw *et al.*, 1990). In *D. fructosovorans* a periplasmic [NiFe], a periplasmic [Fe] and a cytoplasmic NADP-reducing hydrogenase have been described (Casalot *et al.*, 1998; De Luca *et al.*, 1998; Hatchikian *et al.*, 1990; Rousset *et al.*, 1990). Mutant strains lacking genes for one, two or all three of these hydrogenases have been constructed (Casalot *et al.*, 2002a; 2002b; Malki *et al.*, 1997). The triple mutant still grew with hydrogen as sole electron donor with a specific growth rate comparable to that of the wild-type, but had a reduced molar growth yield with fructose as the electron donor, indicating the mutant cells to be less efficient in deriving energy from hydrogen cycling (Casalot *et al.*, 2002a).

Table 7.1. *Survey of redox proteins occurring in* D. vulgaris *Hildenborough (DV),* D. desulfuricans *G20 (DD) and* Dt. psychrophila *(DP). The survey is focused on redox proteins discussed in this chapter*

Class[a]/Name	DV	DD	DP	Locus tag	Location[b]
Periplasmic formate dehydrogenases					
FdhAB	X	XX		DVU0587–0588, Dde0717–0718, Dde3513–3516	peri
FdhABC			XX	DP1769–1767, DP2986–2988	peri, mem
CfdCDEAB	X			DVU2485–2481	peri, mem
FdhABC3	X			DVU2812–2809	peri
Periplasmic hydrogenases					
HydAB	X	XX		DVU1769–1770, Dde0081–0082, Dde2281–2280	peri
HysBA	X	X	X	DVU1917–1918, Dde2134–2135, DP0160–0159	peri, mem
HynBA	X	X		DVU1921–1922, Dde2137–2138	peri, mem
HynBAC			X	DP0574–0576	peri, mem
HynBAC3	X	X		DVU2524–2526, Dde3754–3756	peri
Periplasmic cytochromes					
Occ (8 haems)	X	XX		DVU3107, Dde0561, Dde0291	peri
TpI-c3 (4 haems)	X	X		DVU3171, Dde3182	peri
c553 (1 haem)	XX	X		DVU1817, DVU3041, Dde1821	peri
c554 (4 haems)	XX	X		DVU0702, DVU0922, Dde2858	peri
Cytoplasmic hydrogenases and formate dehydrogenases					
FhcABCD		X	X	Dde0473–0476, DP0481–0478	cyto
EchABCDEF	X			DVU0429–0434	mem, cyto
CooMKLXUH	X			DVU2286–2291	mem, cyto
NfdABC			X	DP0684–0682	mem, cyto
METC					
HmcABCDEF	X	X		DVU0536–0531, Dde0653–0648	peri, mem, cyto
DhcARnfCDGEAB	X	X		DVU2791–2797, Dde0580–0587	peri, mem, cyto
TmcCBA	X	X		DVU0265–0263, Dde3708–3710	peri, mem, cyto
DsrMKJOP	X	X	X	DVU1290–1286, Dde2271–2275, DP3075–3070	peri, mem, cyto

Table 7.1 (*cont.*)

Class[a]/Name	DV	DD	DP	Locus tag	Location[b]
QmoABC	X	X	X	DVU0848–850, Dde1111–1113. DP1106–1108	mem, cyto
OhcBAC	X			DVU3143–3145	peri, mem
CytBA			X	DP1985–1986	peri, mem
Sulphate reduction enzymes					
Sat	X	X	X	DVU1295, Dde2265, DP1472	cyto
PpaC	X	X	X	DVU1636, Dde1778, DP2180	cyto
ApsBA	X	X	X	DVU0846–0847, Dde1109–1110, DP1104–1105	cyto
DsrAB	X	X	X	DVU0402–0403, Dde0526–0527, DP0797–0798	cyto
DsrD	X	X	X	DVU0404, Dde0528, DP0799	cyto
DsrC	X	X	X	DVU2776, Dde0762, DP0997	cyto
Carbon metabolism enzymes					
LdhCAB	X	X	X	DVU3026–3028, Dde3238–3240, DP1023–1021	mem, cyto
Por	X	X	X	DVU3025, Dde3237, DP2886	cyto
Pta	X	X	X	DVU3029, Dde3241, DP0558	cyto
Ack	X	X	X	DVU3030, Dde3242, DP0559	cyto
Pfl-I	X	X	X	DVU2824, Dde1273, DP3027	cyto
Nitrite reductase					
NrfHA	X		X	DVU0624–0625, DP0343–0344	peri, mem

[a] METC is membrane-bound electron transfer complexes.
[b] Location: peri is periplasmic; mem is integrally or peripherally bound to the membrane; cyto is cytoplasmic.

The results led the authors to conclude that *D. fructosovorans* contains a fourth hydrogenase, a conclusion which could have been easily verified had the genome sequence been known. For instance, in *D. vulgaris* (Figure 7.1) a soluble periplasmic [Fe] hydrogenase and two periplasmic, membrane-associated enzymes of the [NiFe] and [NiFeSe] families were isolated and characterized (Huynh *et al.*, 1984; Romao *et al.*, 1997; Valente *et al.*, 2005;

Van der Westen *et al.*, 1978), whereas sequencing of the genome (Heidelberg *et al.*, 2004) revealed the presence of a second periplasmic [NiFe] isoenzyme (Table 7.1: HynBAC3) and two multi-subunit membrane-bound hydrogenases facing the cytoplasm (Table 7.1: EchABCDEF and CooMKLXUH). Expression of the [FeFe], [NiFe] and [NiFeSe] hydrogenases is affected by the presence of Ni and Se during growth (Valente *et al.*, 2006). So far only *hydAB* and *hynBA* mutant strains, lacking [Fe] hydrogenase or [NiFe] hydrogenase, respectively, have been constructed (Goenka *et al.*, 2005; Pohorelic *et al.*, 2002). Initial analysis indicated minor deficiencies in growth with either hydrogen or lactate as the electron donor. Hence, in the oxidation of externally provided hydrogen the lack of one (or more) periplasmic hydrogenases may be compensated by the presence of others revealing considerable plasticity of function. However, more detailed analysis of the *hydAB* mutant indicated a complex phenotype that includes transient production of CO and reduced expression of the cytoplasmic alcohol dehydrogenase DVU2405, when cells are grown on a defined lactate-sulphate medium (Haveman *et al.*, 2003; Voordouw, 2002). DVU2405 is among the most highly expressed genes in *D. vulgaris* and is present in three adjacent copies in *Dt. psychrophila* (DP0952, DP0951 and DP0950). Although the nature of the metabolic interaction between periplasmic HydAB and cytoplasmic DVU2405 Adh is not known, the observed coregulation exemplifies linkage between peri- and cytoplasmic redox reactions which is crucial to the bioenergetics of these organisms.

7.3 PERIPLASMIC FORMATE DEHYDROGENASES

Although *Desulfovibrio* periplasmic hydrogenases have been well studied, our knowledge of their periplasmic formate dehydrogenases is much more limited. Many formate dehydrogenases in other organisms are $\alpha\beta\gamma$ heterotrimers. The molybdopterin-containing α subunit (FdhA) catalyzes the oxidation of formate ($HCOO^- \rightarrow H^+ + CO_2 + 2e$), with the electrons being transported to the iron-sulphur cluster-containing β subunit (FdhB) from where they are transferred to the membrane-bound, haem *b*-containing γ subunit (FdhC). Two typical FdhABC-type formate dehydrogenases are present in *Dt. psychrophila*, but not in *D. desulfuricans* G20 or *D. vulgaris* (Table 7.1). In contrast, *D. vulgaris* has one and *D. desulfuricans* G20 has two periplasmic, soluble FdhABs, which presumably transfer electrons to the abundant periplasmic cytochromes *c*: further proof for this notion is provided by the fact that *D. vulgaris* has an

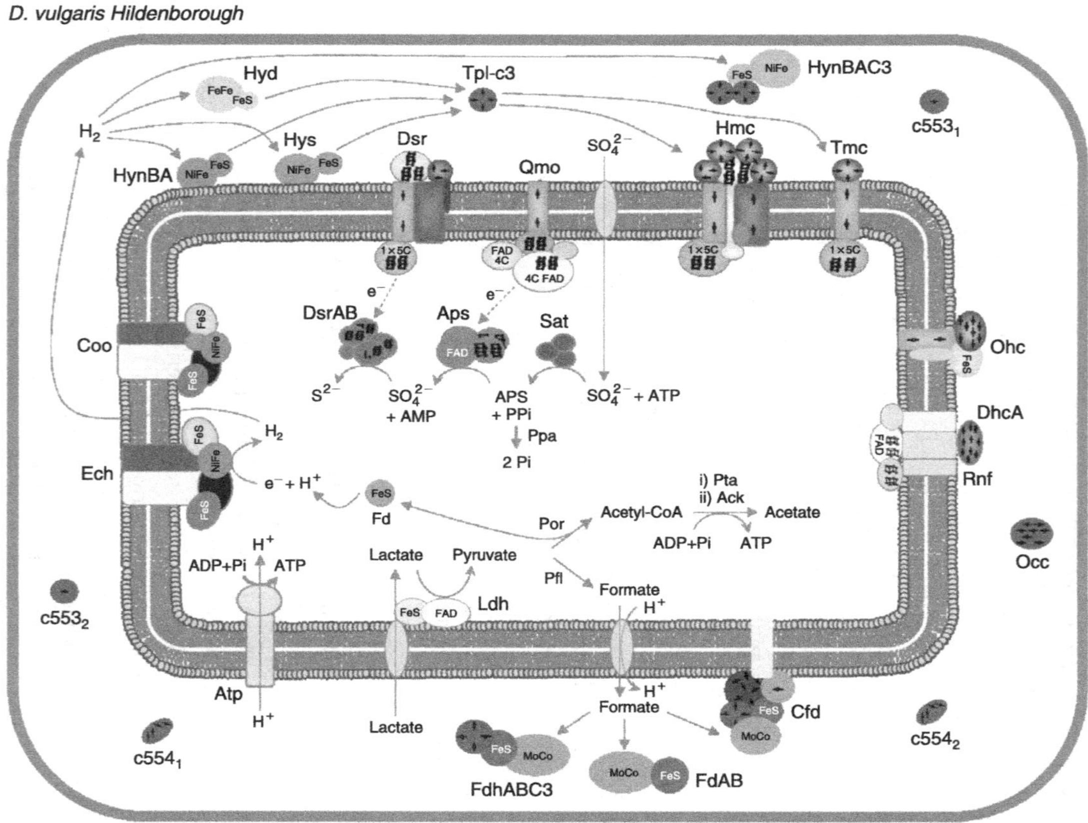

D. vulgaris Hildenborough
Hyd
Tpl-c3
HynBAC3
H_2
Hys
Dsr
Qmo
SO_4^{2-}
Hmc
Tmc
$c553_1$
HynBA
e^-
DsrAB
Aps
Sat
Coo
Ohc
S^{2-}
SO_4^{2-} + AMP
APS + PPi
SO_4^{2-} + ATP
Ppa
2 Pi
DhcA
H_2
Ech
$e^- + H^+$
Fd
Rnf
i) Pta
ii) Ack
Acetyl-CoA
Acetate
Por
ADP+Pi
ATP
Occ
H^+
ADP+Pi
ATP
Lactate
Pyruvate
Pfl
Formate
H^+
Ldh
$c553_2$
Atp
H^+
Formate
Cfd
Lactate
$c554_1$
$c554_2$
FdhABC3
FdAB

Figure 7.1. Bioenergetic model explaining how *D. vulgaris* Hildenborough derives energy for growth by coupling the oxidation of lactate to acetate and CO_2 with the reduction of sulphate to sulphide. Because ATP, formed by substrate level phosphorylation during conversion of acetyl-CoA to acetate, is used for activation of sulphate a net bioenergetic benefit is only realized by H_2 or formate cycling. H_2 cycling involves: (i) generation of reduced ferredoxin (Fd_{red}) during Por-catalyzed conversion of pyruvate; (ii) cytoplasmic production of H_2 from Fd_{red} by Ech or Coo Hase; (iii) periplasmic oxidation of H_2 by Hyd, HynBA, HynBAC3 or Hys Hases; (iv) transport of electrons to the cytoplasm through 6 possible membrane-bound complexes, Hmc, Tmc, Ohc, Rnf, Qmo and Dsr, of which the last two feed electrons into the sulphate-reduction pathway; the other four complexes may transfer electrons to Qmo and Dsr through the quinone pool (not shown) or may function in other cytoplasmic reduction reactions; (v) proton import through ATP synthase with associated energy conservation. Formate cycling involves: (i) generation of cytoplasmic formate by Pfl-catalyzed conversion of pyruvate; (ii) transport of formate to the periplasm by symport with a proton; (iii) periplasmic oxidation of formate by FdhAB, FdhABC3 or Cfd; steps (iv) and (v) are then as for hydrogen cycling.

FdhABC3-type formate dehydrogenase, which like HynBAC3, is likely to donate electrons to its dedicated cytochrome *c* subunit (FdhC3, Table 7.1: DVU2809). FdhC3 has been shown to belong to the cytochrome c_3 family (Elantak *et al.*, 2005; Sebban *et al.*, 1995) and to reduce monohemic cytochrome c_{553}, but not TpI-c_3 (Sebban *et al.*, 1995). *D. vulgaris* also has a periplasmic, membrane-bound cytochrome-formate dehydrogenase (Table 7.1: CfdCDEAB) in which FdhA- and FdhB-like subunits (CfdA and CfdB) are linked to CfdC, a membrane-bound mono-haem cytochrome, and CfdD, a periplasmic, 11-haem-containing *c*-type cytochrome. Hence the diversity of periplasmic hydrogenases in deltaproteobacterial SRB is matched by that of the periplasmic formate dehydrogenases indicating diverse functions in oxidation of hydrogen and formate, which may be externally supplied or may originate from the cytoplasm during hydrogen and formate cycling.

7.4 PERIPLASMIC CYTOCHROMES *C*

A striking difference between periplasmic hydrogenases and formate dehydrogenases of *Desulfovibrio* spp. and those of most other organisms is the lack of a haem *b*-containing, integral membrane subunit, which transfers electrons to the membrane quinone pool. *Desulfovibrio* spp. are also characterized by a high content of soluble and membrane-associated periplasmic *c*-type cytochromes, a feature which is not shared by all sulphate-reducing microorganisms. For example, *Archaeoglobus fulgidus* (Klenk *et al.*, 1997), and *Dt. psychrophila* have few external *c*-type cytochromes and, accordingly, have periplasmic [NiFe] hydrogenases and formate dehydrogenases with a membrane-anchored haem *b* subunit (Table 7.1: DP0576, subunit C of HynBAC and DP1767 and DP2988 of FdhABC). The *Desulfovibrio* spp. have heterodimeric hydrogenases (Table 7.1: HydAB, HysBA or HynBA) which donate electrons to the cytochrome c_3 network (Heidelberg *et al.*, 2004) of which type I cytochrome c_3 (Table 7.1: TpI-c_3, DVU3171, Dde3182) is the most prominent member. TpI-c_3 has a compact tetrahaem motif in which neighbouring haems are perpendicular to each other (reviewed in Matias *et al.*, 2005; Pereira and Xavier, 2005). It couples electron and proton transfer (the redox-Bohr effect), which may be relevant to energy transduction in *Desulfovibrio* spp. (Xavier, 2004). The c_3 network includes cytochromes c_3 that, judged by gene organization, are subunits of [NiFe] hydrogenase (Table 7.1: HynBAC3) or formate dehydrogenase (Table 7.1: FdhABC3). It also includes multi-haem *c*-type cytochromes that are part of transmembrane electron transport complexes (Table 7.1: METC). Of these the 16-haem, high molecular mass cytochrome HmcA and the 9-haem cytochrome *c* Nhc of *D. desulfuricans* ATCC 27774 (Czjzek *et al.*, 2002; Matias *et al.*, 1999a; 2002) have multiple c_3 domains. TpI-c_3 efficiently exchanges electrons with periplasmic hydrogenases (reviewed in Matias *et al.*, 2005). This process involves electrostatic interactions between the positive surface area surrounding haem 4 and a negative surface charge near the distal, most solvent exposed [4Fe-4S] centre of the periplasmic hydrogenases. It also serves as an intermediate carrier in electron transport from periplasmic hydrogenases to membrane-bound HmcA, Nhc and Type-II cytochrome c_3, (TpII-c_3; Table 7.1: TmcA). These cytochrome *c*-containing membrane-bound electron transfer complexes (METC) are present only in *Desulfovibrio*, and form conduits for electron transfer across the membrane, connecting the c_3 network to cytoplasmic redox reactions. HmcA has four c_3 domains (c_3-I to c_3-IV) with c_3-I containing only three haems. The c_3-III and c_3-IV domains have inserted regions which bind an isolated haem group. The N-terminal (c_3-I and c_3-II) and C-terminal (c_3-III and c_3-IV) are symetrically arranged such that haem-3 and haem-16, which are structurally equivalent to TpI-c_3 haem-4, are present at each extremity of the molecule. HmcA haem-4 and haem-8, equivalent to TpI-c_3 haem-1, are at the top of the junction between the N- and C-terminal regions, an arrangement that is likely to be relevant to the physiological role of HmcA (Matias *et al.*, 2002). The C-terminal part of HmcA strongly resembles the structure of the Nhc from *D. desulfuricans* ATCC 27774 (Matias *et al.*, 1999a; Umhau *et al.*, 2001). Nhc is also part of a membrane-bound redox protein complex (Saraiva *et al.*, 2001), which may replace the HmcABCDEF complex in this organism.

TpII-c_3, the smaller of the membrane-bound cytochromes, was isolated from *D. africanus*, *D. vulgaris* and *D. gigas* (Di Paolo *et al.*, 2005). It has a haem arrangement and overall protein fold similar to TpI-c_3 (Norager *et al.*, 1999). However, TpII-c_3 lacks the characteristic positive surface region around haem-4 of TpI-c_3 and has a more exposed haem-1 which is surrounded by a negative surface region, indicating that this may be the haem involved in electron transfer with the periplasmic c_3 network (Valente *et al.*, 2001). These features are shared with corresponding c_3 domains of HmcA and Nhc and are probably required for electron flow to and from TpI-c_3 (Matias *et al.*, 2005). Although TpII-c_3 and Nhc can be reduced directly by hydrogenases, the rate of reduction is considerably increased in the presence of catalytic amounts of TpI-c_3, suggesting an electron transfer sequence Hase $\rightarrow$ TpI-c_3 $\rightarrow$ TpII-c_3, HmcA or Nhc in vivo. Complex formation of

TpI-c_3 and TpII-c_3 involves interaction between the positive surface region of TpI-c_3 haem-4 and the negative surface region of TpII-c_3 haem-1 (Czjzek *et al.*, 2002; Matias *et al.*, 1999b; Pieulle *et al.*, 2005; Teixeira *et al.*, 2004). The simulated redox potential of the former increases by +80 mV upon complex formation, facilitating electron transfer to TpII-c_3 (Teixeira *et al.*, 2004). Interaction between TpI-c_3 and Nhc is also predicted to involve the negative surface regions close to haems-1 and -2 of the larger cytochrome (Matias *et al.*, 1999b).

In *D. vulgaris* and *D. desulfuricans* G20 the gene for cytochrome c_3 (M_r 26 000) or cytochrome cc_3 (Bruschi, 1994) is in an operon with genes for [NiFe] hydrogenase (Table 7.1: HynBAC3), suggesting it serves as a specific electron acceptor for this hydrogenase. HynC3 is a TpI-c_3-like tetrahaem cytochrome, which dimerizes by interactions of surface regions around haem-1 of each monomer, when purified free from its cognate hydrogenase (Czjzek *et al.*, 1996; Frazao *et al.*, 1999). As in TpI-c_3, the exposed haem-4 of each monomer is surrounded by a positive surface region, causing this cytochrome to be efficiently reduced by hydrogenases. However, reduction of HynC3 by [NiFeSe] hydrogenase is slower than reduction of TpI-c_3 in *Desulfomicrobium norvegicum*, and the rate of reduction of HynC3 is increased in the presence of small amounts of the latter (Aubert *et al.*, 2000). These results indicate that [NiFeSe] hydrogenase is not the cognate hydrogenase of HynC3. Purification of HynBAC3 and studying the kinetics of electron transfer from HynBA → HynC3 → TpI-c_3 may further reveal the specificity of electron exchanges in the *Desulfovibrio* c_3 network. A similarly dedicated c_3 subunit appears associated with one of the formate dehydrogenases in *D. vulgaris* and *D. desulfuricans* G20 (Table 7.1: FdhABC3).

Additional cytochromes uncovered by genome sequencing include tetrahaem cytochromes c_{554} (Table 7.1: DVU0702, DVU0922 and Dde2858), the electron acceptor for hydroxylamine oxidoreductase (Iverson *et al.*, 2001), octahaem cytochromes (Table 7.1: Occ DVU3107, Dde0561 and Dde0291), a decahaem cytochrome (DhcA) of the cytochrome c_3 family which is a subunit a membrane-bound redox protein complex DhcARnfCDGEAB (Table 7.1), and another octahaem cytochrome (DVU3144) also associated with a membrane redox complex present only in *D. vulgaris* (Table 7.1: OhcBAC). All these cytochromes are absent from *Dt. psychrophila*.

D. vulgaris has also a nitrite reductase complex (NrfHA, Table 7.1: DVU0624 and DVU0625) (Pereira *et al.*, 2000) composed of the pentahaem c NrfA catalytic subunit and the membrane-anchored tetrahaem c NrfH subunit which belongs to the NapC/NirT family. NrfHA enables this organism to prevent nitrite inhibition by nitrate-reducing, sulphide-oxidizing bacteria (Greene *et al.*, 2003). This complex is not found in *D. desulfuricans* G20, but is present in *Dt. psychrophila* (Table 7.1: DP0343 and DP0344). *D. desulfuricans* G20 has a similar complex with a NapC/NirT cytochrome (Dde02990-0300), but the larger 7-haem cytochrome (Dde0300) shows no similarity to NrfA and a low similarity to hydroxylamine oxidoreductase, so its function is uncertain.

7.5 MEMBRANE-BOUND, ELECTRON TRANSPORT COMPLEXES

Membrane-bound electron transport complexes (METC) in *Desulfovibrio* spp. include complexes that are partly in the cytoplasm (e.g. QmoABC), partly in the periplasm (e.g. Nhc) as well as complexes that are truly transmembrane (e.g. HmcABCDEF; Rossi *et al.*, 1993). Structural features of the 16-haem cytochrome HmcA of the Hmc-complex have already been discussed. HmcB is predicted to be a transmembrane protein with a periplasmic ferredoxin-like N-terminal domain with four $[4Fe\text{-}4S]^{2+/1+}$ centres and a cytoplasmic C-terminal tail (Keon and Voordouw, 1996). HmcC, HmcD and HmcE are all predicted to be integral membrane proteins with HmcE probably binding haems *b*, and HmcF is predicted to be a cytoplasmic FeS protein homologous to the HdrD subunit of heterodisulphide reductases (Hdr) present in methanogens (Kunkel *et al.*, 1997). Like HdrD, HmcF contains two $[4Fe4S]^{2+/1+}$ binding sites, but it has only one of the two five-cysteine-containing motifs, CXnCCXnCX2C, present in HdrD, where these are involved in binding of a special [4Fe4S] cluster in the catalytic site for heterodisulphide reduction (Duin *et al.*, 2002; Madadi-Kahkesh *et al.*, 2001; Shokes *et al.*, 2005). Expression of the *hmc* operon increased when hydrogen was the electron donor for sulphate reduction compared to when either lactate or pyruvate were used (Keon *et al.*, 1997). A *D. vulgaris* mutant with a deletion of the *rrf*1,2 genes, downstream of the *hmc* operon, encoding gene expression-regulating proteins, overexpressed the *hmc* operon threefold, and grew faster than the wild-type on hydrogen but slower on lactate (Keon *et al.*, 1997). An *hmc* operon deletion grew slower than the wild-type with hydrogen, but grew normally with lactate or pyruvate as the electron donor (Dolla *et al.*, 2000). Fermentative metabolism of lactate or pyruvate was changed in the *hmc* mutant, which was also found to form colonies on an agar surface more slowly.

These results implicate the Hmc complex in electron transfer from the peri- to the cytoplasm and *vice-versa*. *D. desulfuricans* ATCC 27774 harbours NhcABCD, a simpler METC related to the Hmc complex. NhcA is homologous to the C-terminus of HmcA, as already discussed, whereas NhcB and NhcC are homologous to HmcB and HmcC.

In addition to HmcABCDEF, *D. vulgaris* and *D. desulfuricans* G20 have TmcABCD and DhcARnfCDGEAB; only *D. vulgaris* has OhcABC (Table 7.1). The recently isolated $TmcA_2BCD$ complex includes the periplasmic tetrahaemic TmcA (TpII-c_3), the membrane-bound haem *b*-containing TmcC (an HmcE homologue), the cytoplasmic, HdrD-like FeS protein TmcB (an HmcF homologue) of the same family as DsrK and HdrD and the cytoplasmic TmcD (Pereira *et al.*, 2006). The sequence similarities between TmcC and HmcE, as well as between TmcB and HmcF are very high, indicating similar functions for the Hmc and Tmc complexes. A similar function in hydrogen oxidation is supported by the fact that both TmcA and HmcA are efficiently reduced by TpI-c_3. The *D. vulgaris* OhcBAC complex connects OhcA, a periplasmic, octahaemic *c*-type cytochrome with OhcB, an iron-sulphur integral membrane protein and OhcC, a haem *b*-containing integral membrane protein.

In *D. vulgaris* and *D. desulfuricans* G20 the decahaemic *c*–type cytochrome DhcA is associated with the RnfCDGEAB complex, which was first identified in *Rhodobacter capsulatus* as being involved in nitrogen fixation (Curatti *et al.*, 2005; Jeong and Jouanneau, 2000; Schmehl *et al.*, 1993). This complex, lacking the cytochrome, is present in *Dt. psychrophila* and many other microorganisms. In *E. coli*, where it is named Rsx, it is involved in keeping the redox-sensitive transcriptional factor SoxR in its inactive reduced state during aerobic growth (Koo *et al.*, 2003). The three integral membrane subunits RnfADE, and the cytoplasmic RnfG, show similarity to subunits of the Nqr complex in *Vibrio* spp., a Na^+-translocating NADH:quinone oxidoreductase (Kumagai *et al.*, 1997). Although the *Desulfovibrio* complexes are more closely related to the *Rh. capsulatus* Rnf than to the *Vibrio* spp. Nqr, they are unlikely to function in nitrogen fixation. In view of their association with DhcA they may be involved in electron transfer with the c_3 network, possibly coupled to extrusion of sodium ions.

Dt. psychrophila has a single *c*-type haem-containing METC (Table 7.1: CytBA) composed of DP1985 CytB, which has 7 haems in its periplasmic N-terminal domain and possibly haem *b* in its membrane-anchored C-terminal domain, and of DP1986 CytA, which is a periplasmic cytochrome subunit with 5 *c*-type haems. This complex may accept electrons from

HysBA (DP0160-0159), which lacks the membrane-bound, *b*-type haem containing gamma subunit present in all other periplasmic hydrogenases and formate dehydrogenases of *Dt. psychrophila* (Table 7.1).

Only two METC are strictly conserved among the three sequenced SRB (Table 7.1), and the sulphate-reducing archaeon *Archaeoglobus fulgidus:* the QmoABC complex (Pires *et al.*, 2003), and the DsrMKJOP complex in *Desulfovibrio* (Haveman *et al.*, 2004; Pires *et al.*, 2006), referred to as the Hme complex in *A. fulgidus* (Mander *et al.*, 2002). QmoABC may transfer electrons from the quinone pool to adenosine phosphosulphate (APS) reductase (ApsBA), whereas DsrMKJOP may be involved in electron transfer to the dissimilatory sulphite reductase (DsrAB). The strict conservation of these two complexes in all sulphate-reducing prokaryotes suggests an essential role in sulphate respiration.

QmoABC was isolated from *D. desulfuricans* ATCC 27774 (Pires *et al.*, 2003). Strikingly, all three subunits have homology to the HdrACE subunits of heterodisulphide reductase from methanogens. QmoA and QmoB are cytoplasmic, FAD-containing proteins related to HdrA. QmoB also has binding sites for two $[4Fe4S]^{2+/1+}$ centres. QmoC has an integral membrane domain that binds two haems *b* and a hydrophilic domain that binds two $[4Fe4S]^{2+/1+}$ centres. QmoABC has no periplasmic subunits and its haems are reduced by a menaquinol analogue, suggesting that it transfers electrons from the quinone pool to the cytoplasm. The fact that *qmoABC* genes are adjacent to the genes encoding APS reductase in the three sequenced deltaproteobacterial genomes (Table 7.1), and also in some sulphur-oxidizing prokaryotes, suggests that QmoABC may donate electrons to ApsBA.

The DsrMKJOP complex has been isolated from *A. fulgidus* (Hme; Mander *et al.*, 2002) and from *D. desulfuricans* ATCC 27774 (Pires *et al.*, 2006). In some microorganisms these genes are co-localized with the *dsrAB* genes encoding dissimilatory sulphite reductase (Dahl *et al.*, 2005). The DsrMKJOP complex is composed of functionally similar subunits as present in the Hmc complex, including a periplasmic cytochrome *c* (DsrJ), a periplasmic ferredoxin-like protein (DsrO), two integral membrane proteins (DsrM and DsrP), and a cytoplasmic FeS protein (DsrK) related to HdrD. However, the degree of sequence identity between corresponding subunits is low, indicating that the two complexes have distinct electron transfer functions. DsrJ, a small 3-haem cytochrome (15 kDa) is the most dissimilar. Spectroscopic characterization of DsrJ revealed that each of the three haems is differently coordinated, i.e. bis-His, His/Met and His/Cys (Pires *et al.*, 2006). Most multi-haem cytochromes *c* have bis-His coordination, and a His/Cys coordinated haem *c* is particularly unusual. This haem is either redox-inactive or has an extremely low redox potential since it is not completely reduced with dithionite. DsrO is a periplasmic ferredoxin-like protein, which by spectroscopic and sequence analysis contains three $[4Fe4S]^{2+/1+}$ centres, and not four as found in other proteins of this family. It is related to HmcB and NhcB but lacks the transmembrane helix and cytoplasmic domain found in these. DsrM binds two typical haems *b* (which are modified in *A. fulgidus* Hme) and belongs to the family of integral cytochrome *b* subunits of respiratory oxidoreductases which include the prototypical NarI (Berks *et al.*, 1995), as well as HmcE and HdrE. DsrP is a large integral membrane protein with ten transmembrane helices related to HmcC, NhcC and HybB, the membrane subunit of *E. coli* hydrogenase-2 which acts as a menaquinone reductase (Menon *et al.*, 1994). The cytoplasmic DsrK is, like HmcF and TmcB, homologous to HdrD, the catalytic subunit of heterodisulphide reductase of methanogens. DsrK, HmcF and TmcB all have binding sites for two $[4Fe4S]^{2+/1+}$ centres and one copy of the conserved five-cysteine motif, which is present in two copies in HdrD. EPR analysis of DsrMKJOP from *D. desulfuricans* ATCC 27774 and of the Hme complex of *A. fulgidus* has revealed the presence of a rhombic signal in the oxidized state. This signal is similar to that observed upon binding of HS-CoM to oxidized Hdr (Madadi-Kahkesh *et al.*, 2001). The paramagnetic species in CoM-Hdr is believed to be an intermediate in the reaction cycle, and has been shown to be due to a $[4Fe4S]^{3+}$ cluster in which one Fe site is pentacoordinated and has two thiolate ligands, one of which may be HS-CoM which binds directly to the [4Fe4S] cluster in the active site of the enzyme (Shokes *et al.*, 2005). Given their similarity to HdrD, it seems likely that DsrK, HmcF and TmcB may be catalytic subunits involved in thiol/disulphide redox chemistry. There is considerable evidence to suggest that DsrMKJOP donates electrons to dissimilatory sulphite reductase (DsrAB). Genomes from sulphate-reducing, as well as from sulphur-oxidizing prokaryotes have the *dsrAB* and *dsrMKJOP* genes, often in a single gene cluster that includes other conserved *dsr* genes (Dahl *et al.*, 2005; Mander *et al.*, 2002; Pires *et al.*, 2006). In *Allochromatium vinosum* the DsrKJO proteins were found associated with the DsrABC proteins, and the genes for the sulphite reductase and Dsr complex were found to be coordinately regulated by sulphide (Dahl *et al.*, 2005). However, the precise mechanism of electron transfer from DsrMKJOP to DsrAB, e.g. directly or through a sulphide/disulphide intermediate, remains to be established.

7.6 ENZYMES OF THE SULPHATE REDUCTION AND LACTATE OXIDATION PATHWAYS

As discussed in Chapter 1, sulphate reduction proceeds through reactions (5) to (8a). We will use the same numbering system here. Hence sulphate reduction requires (5) activation of sulphate to adenosine phosphosulphate (APS), (6) hydrolysis of pyrophosphate, (7) two electron reduction of APS to sulphite and AMP and (8a) six electron reduction of sulphite to sulphide, as catalyzed by the following cytoplasmic enzymes: (5) ATP sulphurylase or sulphate adenylyltransferase (Table 7.1: Sat), (6) pyrophosphatase (Table 7.1: PpaC), (7) Aps reductase (Table 7.1: ApsBA) and (8a) dissimilatory sulphite reductase (Table 7.1: DsrAB) (Figure 7.1). The sulphate-reducing *Deltaproteobacteria* have single copies of all genes for enzymes of the sulphate reduction pathway, in contrast to the redundancy of genes for enzymes involved in hydrogen and formate oxidation and associated electron transfer. The *dsrD* gene encoding a 9 kDa protein is present immediately downstream from the *dsrAB* genes. Because the structure of DsrD represents a winged helix motif it has been suggested that it may have a DNA or RNA-binding regulatory function (Mizuno *et al.*, 2003). DsrD is not found associated with dissimilatory sulphite reductase, which is an $\alpha_2\beta_2$ tetramer, following purification. In contrast, DsrC (γ) from *D. vulgaris* was found to associate with dissimilatory sulphite reductase, which was isolated as an $\alpha_2\beta_2\gamma_2$ complex (Pierik *et al.*, 1992). The *dsrC* gene is distant from the *dsrABD* genes in the genomes of the *Deltaproteobacterial* sulphate-reducers (Table 7.1). The structure of DsrC (12 kDa) indicates a compact protein with a flexible C-terminal tail containing a conserved cysteine as the penultimate residue (Cort *et al.*, 2001). Interestingly, the DsrC structure also contains a helix-turn-helix motif suggestive of DNA binding. It seems unlikely that both DsrD and DsrC are DNA-binding proteins. Perhaps the $\alpha_2\beta_2$ tetramer, of which the structure is unknown, has a surface feature that complements the helix-turn-helix motif of DsrC and DsrD, although the nature of the involvement of DsrD and DsrC in the catalytic cycle is currently unknown.

Membrane-bound lactate dehydrogenase has not been well studied biochemically (Stams and Hansen, 1982), and hence it was not easy to identify its genes in sequenced genomes of *Desulfovibrio* spp. DVU0600, identified by Heidelberg *et al.* (2004), is an unlikely candidate because it is not present in either *D. desulfuricans* G20 or *Dt. psychrophila*. Both *D. vulgaris* and *D. desulfuricans* G20 have an "organic acid oxidation region" in the genome

(Table 7.1: DVU3025–DVU3030 and Dde3237–3242), containing genes for pyruvate-ferredoxin-oxidoreductase (Por; DVU3025, Dde3237), putative lactate permease (DVU3026, Dde3238), putative glycolate oxidase subunit GlcD (DVU3027, Dde3239), putative iron-sulphur protein (DVU3028, Dde3240), phosphate acetyl transferase (Pta; DVU3029, Dde3241) and acetate kinase (Ack; DVU3030, Dde3242). DVU3027–3028 were proposed as likely candidate genes for lactate dehydrogenase (LdhAB) during discussions at the SRB Genome Annotation Jamboree at the Joint Genome Institute (Walnut Creek, CA) in April 2004. These may be bound to the membrane through interaction with DVU3026, the putative lactate permease LdhC, which is an integral membrane protein. Hence lactate may be imported and oxidized to pyruvate by membrane-bound lactate permease/ dehydrogenase LdhCAB (DVU3026–3028; Dde3238–3240), which is also conserved in *Dt. psychrophila* (DP1023–1021). *D. vulgaris* and *Dt. psychrophila* have another strongly conserved LdhC homologue (DVU2451, DP2077). The reactions catalyzed by these enzymes are (see also Chapters 1, 10 and 12–15):

LdhCAB lactate → pyruvate + 2 [H]; the 2 [H] are potentially transferred to the periplasmic c_3 pool.

Por pyruvate + CoA + $2Fd_{ox}$ → acetyl-CoA + CO_2 +$2Fd_{red}$ (Fd is ferredoxin)

Pta acetyl-CoA + P_i → acetylphosphate + CoA

Ack acetylphosphate + ADP → acetate + ATP

Overall lactate + ADP + P_i + $2Fd_{ox}$ +$2c_{3,ox}$ → acetate + CO_2 + ATP + $2Fd_{red}$ + $2c_{3,red}$

Hence, these reactions lead to ATP synthesis by substrate-level phosphorylation. Instead of the Por-catalyzed reaction, pyruvate can be converted into formate and acetate by pyruvate:formate lyase (Table 7.1: Pfl-I). *D. vulgaris* has one other isozyme (DVU2272), whereas *D. desulfuricans* G20 has three (Dde3039, Dde3055, Dde3282) and *Dt. psychrophila* has two (DP0616 and DP1823) other isozymes. It is unclear if these isozymes all catalyze the reaction indicated below or if some may, for instance, be specific for catalyzing the reverse reaction.

Pfl pyruvate + CoA→ acetyl-CoA +formate

Overall lactate + ADP + P_i +$2c_{3,ox}$ → acetate + formate + ATP + $2c_{3,red}$

Symport of cytoplasmically produced formate and a proton contributes to the $\Delta\mu H^+$ and allows further energy conservation by subsequent oxidation of formate by the periplasmic formate dehydrogenases (Table 7.1). The overall process, referred to as "formate cycling", is depicted in Figure 7.1.

7.7 CYTOPLASMIC HYDROGENASES AND FORMATE DEHYDROGENASES

The genome sequence has indicated that, in addition to four periplasmic hydrogenases, *D. vulgaris* has two membrane-bound, cytoplasmically oriented hydrogenases EchABCDEF (also identified in *D. gigas* (Rodrigues *et al.*, 2003)) and CooMKLXUH (Table 7.1). As discussed in Chapter 1, these could reoxidize $2Fd_{red}$ to form hydrogen while contributing to the $\Delta\mu H^+$:

Ech, Coo $2Fd_{red} + 2H^+ \rightarrow 2Fd_{ox} + H_2 + \Delta\mu H^+$

The cytoplasmically generated H_2 diffuses to the periplasm, where its oxidation by any of the four periplasmic hydrogenases contributes to further energy conservation by H_2 cycling (Figure 7.1). This mechanism of energy conservation is limited to *D. vulgaris*, as neither *D. desulfuricans* G20, nor *Dt. psychrophila* has homologues for these energy-conserving cytoplasmic hydrogenases. The latter two have a cytoplasmic formate dehydrogenase/hydrogenase (Table 7.1: FhcABCD) which could catalyze:

FhcABCD $H_2 + CO_2 \leftrightarrow \text{formate} + H^+$

This enzyme has yet to be isolated and its proposed function is thus presently hypothetical. If it also accepts electrons from Fd_{red} then it could also catalyze:

FhcABCD $2Fd_{red} + 2H^+ \rightarrow 2Fd_{ox} + H_2$

We thus propose that FhcABCD functionally substitutes for the energy-conserving cytoplasmic hydrogenases Ech and Coo found in *D. vulgaris*. Because FhcABCD is predicted to be soluble cytoplasmic, its reaction does not contribute directly to the $\Delta\mu H^+$. *Dt. psychrophila* also contains a cytoplasmic, putative NAD^+-reducing formate dehydrogenase (Table 7.1: NfdABC), which is lacking from the two sequenced *Desulfovibrio* spp. The putative function of this enzyme could be:

NfdABC $NAD^+ + \text{formate} \rightarrow NADH + CO_2$

The NADH formed can then be processed by a membrane-bound NADH dehydrogenase. The presence of NfdABC in *Dt. psychrophila* and its absence in *D. vulgaris* and *D. desulfuricans* G20 may indicate that the NADH/NAD^+ couple functions more prominently in the bioenergetics of the former than in that of the latter two organisms.

7.8 CONCLUSION

The sulphate-reducing *Deltaproteobacteria* employ different strategies for deriving energy for growth from coupling the oxidation of hydrogen and organic compounds to the reduction of sulphate. This is reflected in a different redox protein content between the *Desulfovibrio* spp., e.g. *D. vulgaris,* on the one hand and *Dt. psychrophila* on the other hand (Table 7.1). *D. vulgaris* has at least 17 periplasmic or membrane-bound *c*-type cytochromes, some of which are of the cytochrome c_3 family (the c_3 network), which are absent from *Dt. psychrophila*. *D. vulgaris* may feed electrons derived from periplasmic formate and hydrogen oxidation into this c_3 network, prior to transport across the membrane, whereas *Dt. psychrophila* feeds these electrons for the most part directly into the quinone pool through membrane-bound haem *b*-containing subunits, associated with its [NiFe] hydrogenase and formate dehydrogenases. Four METC with a periplasmic cytochrome *c* subunit, used by *D. vulgaris* for electron transport from the c_3 network to the quinone pool or to the cytoplasm, are also absent from *Dt. psychrophila.*

All sulphate reducers with a sequenced genome share the METC QmoABC and DsrMKJOP, which are considered essential for sulphate reduction. These do not depend on the c_3 network, as that is not conserved. Because the Qmo complex does not have a periplasmic subunit, it is likely to accept its electrons from the quinone pool. The Dsr complex may also draw electrons from this pool, as well as from the periplasm through its periplasmic DsrJ subunit, but TpI-c_3 does not serve as an electron donor to DsrJ (Pires *et al.*, 2006). It is possible that electron transfer through these complexes is coupled to proton translocation, contributing to the proton motive force. The Dsr, Hmc and Tmc complexes all share a cytoplasmic subunit (DsrK, HmcF and TmcB) related to HdrD heterodisulphide reductase. The similarity of pairs of DsrK sequences from different delta-proteobacterial SRB is an order of magnitude higher than the similarity of DsrK and HmcF or DsrK and TmcB sequences in either *D. vulgaris* or *D. desulfuricans* G20. Hence, although these complexes share mechanistic details of cytoplasmic electron transfer, the Hmc and Tmc complexes may target different electron acceptors, as discussed below.

On the cytoplasmic side *D. vulgaris* and *Dt. psychrophila* share all enzymes for sulphate reduction and lactate oxidation and associated carbon metabolism. *Dt. psychrophila* has putative cytoplasmic enzymes that transfer the reducing power of formate into either hydrogen or NADH, both of which are absent from *D. vulgaris*. However, *D. vulgaris* has two cytoplasmic energy-transducing hydrogenases, which are not found in *Dt. psychrophila*. Both organisms share the presence of a periplasmic, membrane-bound nitrite reductase (Table 7.1: NrfHA), which prevents inhibition of dissimilatory sulphite reductase DsrAB by nitrite (Haveman *et al.*, 2004). The striking differences between these two *Deltaproteobacterial* SRB can be further summarized by noting that *Dt. psychrophila* appears to have only five proteins with covalently-bound *c*-type haem (Table 7.1: DsrJ, CytB, CytA, NrfH and NrfA), whereas *D. vulgaris* has at least 17 of these.

Additional evidence that only QmoABC and DsrMKJOP are electron donors to the sulphate reduction pathway is provided by the effect of nitrite on gene expression in *D. vulgaris*. When DsrAB was inhibited by the presence of nitrite, transcription of genes for enzymes from the sulphate reduction pathway, for QmoABC and DsrMKJOP and for ATP synthase was downregulated, whereas transcription of genes for NrfHA nitrite reductase was upregulated (Haveman *et al.*, 2004; He *et al.*, 2006). A similar gene expression response was observed when the nitrite was generated by oxidation of sulphide with nitrate by nitrate-reducing, sulphide-oxidizing bacteria (Haveman *et al.*, 2005). Expression of genes for the four other METCs was not affected, supporting the idea that these have a different role. The Hmc complex lowers H_2 production in the absence of sulphate, i.e. during fermentative metabolism. The *hmc* mutant strain produced much larger amounts of hydrogen from lactate and pyruvate than wild-type *D. vulgaris* (Voordouw, 2002) under these conditions, indicating that the Hmc complex may contribute to reduction reactions of organic electron acceptors.

In conclusion it appears that the diverse strategies of sulphate respiration by deltaproteobacterial SRB continue to be as fascinating objects of study today as they have been for many years. Recently, we have seen much progress in our understanding of the anaerobic electron transport pathways of especially the *Desulfovibrio* spp. As always these advances give rise to many new questions of which understanding how a proton motive force is generated, or the mechanism by which QmoABC and DsrMJKOP interact with ApsBA and DsrAB, are among the most important.

ACKNOWLEDGEMENTS

IACP would like to thank António V. Xavier for his support, and Fundação para a Ciência e Tecnologia, Portugal, for funding. SAH and GV have benefitted from discussions during the SRB Genome Annotation Jamboree at the JGI in Walnut Creek (CA) in April 2004. GV acknowledges support of a Discovery Grant from the Natural Sciences and Engineering Research Council of Canada (NSERC). We would like to thank all our co-workers, whose names appear on the references.

REFERENCES

Aubert, C., Brugna, M., Dolla, A., Bruschi, M. and Giudici-Orticoni, M. T. (2000). A sequential electron transfer from hydrogenases to cytochromes in sulphate-reducing bacteria. *Biochim Biophys Acta*, **1476**, 85–92.

Berks, B. C., Page, M. D., Richardson, D. J. *et al.* (1995). Sequence analysis of subunits of the membrane-bound nitrate reductase from a denitrifying bacterium: the integral membrane subunit provides a prototype for the dihaem electron-carrying arm of a redox loop. *Mol Microbiol*, **15**, 319–31.

Bruschi, M. (1994). Cytochrome c_3 (*Mr* 26,000) isolated from sulphate-reducing bacteria and its relationship to other polyhemic cytochromes from *Desulfovibrio*. *Methods Enzym*, **243**, 140–55.

Casalot, L., Hatchikian, C. E., Forget, N. *et al.* (1998). Molecular study and partial characterization of iron-only hydrogenase in *Desulfovibrio fructosovorans*. *Anaerobe*, **4**, 45–55.

Casalot, L., De Luca, G., Dermoun, Z., Rousset, M. and De Philip, P. (2002a). Evidence for a fourth hydrogenase in *Desulfovibrio fructosovorans*. *J Bacteriol*, **184**, 853–6.

Casalot, L., Valette, O., De Luca, G. *et al.* (2002b). Construction and physiological studies of hydrogenase depleted mutants of *Desulfovibrio fructosovorans*. *FEMS Microbiol Lett*, **214**, 107–12.

Cort, J. R., Mariappan, S. V. S., Kim, C.-Y. *et al.* (2001). Solution structure of *Pyrobaculum aerophilum* DsrC, an archaeal homologue of the gamma subunit of dissimilatory sulfite reductase. *Eur J Biochem*, **268**, 55842–50.

Curatti, L., Brown, C. S., Ludden, P. W. and Rubio, L. M. (2005). Genes required for rapid expression of nitrogenase activity in *Azotobacter vinelandii*. *Proc Natl Acad Sci USA*, **102**, 6291–6.

Czjzek, M., Guerlesquin, F., Bruschi, M. and Haser, R. (1996). Crystal structure of a dimeric octaheme cytochrome c_3 (M(r) 26,000) from *Desulfovibrio desulfuricans* Norway. *Structure*, **4**, 395–404.

Czjzek, M., Elantak, L., Zamboni, V. *et al.* (2002). The crystal structure of the hexadeca-heme cytochrome Hmc and a structural model of its complex with cytochrome c_3. *Structure*, **10**, 1677–86.

Dahl, C., Engels, S., Pott-Sperling, A. S. *et al.* (2005). Novel genes of the *dsr* gene cluster and evidence for close interaction of Dsr proteins during sulfur oxidation in the phototrophic sulfur bacterium *Allochromatium vinosum. J Bacteriol*, **187**, 1392–404.

De Lacey, A. L., Fernandez, V. M. and Rousset, M. (2005). Native and mutant nickel-iron hydrogenases: unravelling structure and function. *Coordin Chem Rev*, **249**, 1596–608.

De Luca, G., Asso, M., Belaich, J. P. and Dermoun, Z. (1998). Purification and characterization of the HndA subunit of NADP-reducing hydrogenase from *Desulfovibrio fructosovorans* overproduced in *Escherichia coli. Biochemistry*, **37**, 2660–5.

Di Paolo, R. E., Pereira, P. M., Gomes, I. *et al.* (2006). Resonance Raman fingerprinting of multiheme cytochromes from the cytochrome c_3 family. *J Biol Inorg Chem*, **11**, 217–24

Dolla, A., Pohorelic, B. K. J., Voordouw, J. K. and Voordouw, G. (2000). Deletion of the *hmc* operon of *Desulfovibrio vulgaris* subsp. *vulgaris* Hildenborough hampers hydrogen metabolism and low-redox-potential niche establishment. *Arch Microbiol*, **174**, 143–51.

Duin, E. C., Madadi-Kahkesh, S., Hedderich, R., Clay, M. D. and Johnson, M. K. (2002). Heterodisulfide reductase from *Methanothermobacter marburgensis* contains an active-site 4Fe-4S cluster that is directly involved in mediating heterodisulfide reduction. *FEBS Lett*, **512**, 263–8.

Elantak, L., Dolla, A., Durand, M. C., Bianco, P. and Guerlesquin, F. (2005). Role of the tetrahemic subunit in *Desulfovibrio vulgaris* Hildenborough formate dehydrogenase. *Biochemistry*, **44**, 14828–34.

Frazao, C., Sieker, L., Sheldrick, G. *et al.* (1999). *Ab initio* structure solution of a dimeric cytochrome c_3 from *Desulfovibrio gigas* containing disulfide bridges. *J Biol Inorg Chem*, **4**, 162–5.

Goenka, A., Voordouw, J. K., Lubitz, W., Gartner, W. and Voordouw, G. (2005). Construction of a NiFe-hydrogenase deletion mutant of *Desulfovibrio vulgaris* Hildenborough. *Biochem Soc Trans*, **33**, 59–60.

Greene, E. A., Hubert, C., Nemati, M., Jenneman, G. E. and Voordouw, G. (2003). Nitrite reductase activity of sulphate-reducing bacteria prevents their inhibition by nitrate-reducing, sulfide-oxidising bacteria. *Environ Microbiol*, **5**, 607–17.

Hatchikian, C. E., Traore, A. S., Fernandez, V. M. and Cammack, R. (1990). Characterization of the nickel-iron periplasmic hydrogenase from *Desulfovibrio fructosovorans. Eur J Biochem*, **187**, 635–43.

Haveman, S. A., Brunelle, V., Voordouw, J. K. *et al.* (2003). Gene expression analysis of energy metabolism mutants of *Desulfovibrio vulgaris*

Hildenborough indicates an important role for alcohol dehydrogenase. *J Bacteriol*, **195**, 4345–53.

Haveman, S. A., Greene, E. A., Stilwell, C. P., Voordouw, J. K. and Voordouw, G. (2004). Physiological and gene expression analysis of inhibition of *Desulfovibrio vulgaris* Hildenborough by nitrite. *J Bacteriol*, **186**, 7944–50.

Haveman, S. A., Greene, E. A. and Voordouw, G. (2005). Gene expression analysis of the mechanism of inhibition of *Desulfovibrio vulgaris* Hildenborough by nitrate-reducing, sulfide-oxidizing bacteria. *Environ Microbiol*, **7**, 1461–5.

He, Q., Huang, K. H., He, Z. *et al.*, (2006). Energetic Consequences of Nitrite Stress in *Desulfovibrio vulgaris* Hildenborough, Inferred from Global Transcriptional Analysis. *Appl Environ Microbiol*, **72**, 4370–81.

Heidelberg, J. F., Seshadri, R., Haveman, S. A. *et al.* (2004). The genome sequence of the anaerobic, sulphate-reducing bacterium *Desulfovibrio vulgaris* Hildenborough. *Nat Biotechnol*, **22**, 554–9.

Huynh, B. H., Czechowski, M. H., Kruger, H. J. *et al.* (1984). *Desulfovibrio vulgaris* hydrogenase – a nonheme iron enzyme lacking nickel that exhibits anomalous electron-paramagnetic-res and Mossbauer-spectra. *Proc Natl Acad Sci-Biol*, **81**, 3728–32.

Iverson, T. M., Hendrich, M. P., Arciero, D. M., Hooper, A. B. and Rees, D. C. (2001). Cytochrome c_{554}. In A. Messerschmidt, R. Huber, T. Poulos and K. Wieghardt (eds.), New York: Wiley. pp. 136–46.

Jeong, H. S. and Jouanneau, Y. (2000). Enhanced nitrogenase activity in strains of *Rhodobacter capsulatus* that overexpress the *rnf* genes. *J Bacteriol*, **182**, 1208–14.

Keon, R. G. and Voordouw, G. (1996). Identification of the HmcF and topology of the HmcB subunit of the Hmc complex of *Desulfovibrio vulgaris. Anaerobe*, **2**, 231.

Keon, R. G., Fu, R. and Voordouw, G. (1997). Deletion of two downstream genes alters expression of the *hmc* operon of *Desulfovibrio vulgaris* subsp. *vulgaris* Hildenborough. *Arch Microbiol*, **167**, 376–83.

Klenk, H. P., Clayton, R. A., Tomb, J. F. *et al.* (1997). The complete genome sequence of the hyperthermophilic, sulphate-reducing archaeon *Archaeoglobus fulgidus. Nature*, **390**, 364–70.

Koo, M. S., Lee, J. H., Rah, S. Y. *et al.* (2003). A reducing system of the superoxide sensor SoxR in *Escherichia coli. EMBO J*, **22**, 2614–22.

Kumagai, H., Fujiwara, T., Matsubara, H. and Saeki, K. (1997). Membrane localization, topology, and mutual stabilization of the rnfABC gene products in *Rhodobacter capsulatus* and implications for a new family of energy-coupling NADH oxidoreductases. *Biochemistry*, **36**, 5509–21.

Kunkel, A., Vaupel, M., Heim, S., Thauer, R. K. and Hedderich, R. (1997). Heterodisulfide reductase from methanol-grown cells of *Methanosarcina barkeri* is not a flavoenzyme. *Eur J Biochem,* **244**, 226–34.

Madadi-Kahkesh, S., Duin, E. C., Heim, S. *et al.* (2001). A paramagnetic species with unique EPR characteristics in the active site of heterodisulfide reductase from methanogenic archaea. *Eur J Biochem,* **268**, 2566–77.

Malki, S., Deluca, G., Fardeau, M. L. *et al.* (1997). Physiological characteristics and growth behavior of single and double hydrogenase mutants of *Desulfovibrio fructosovorans. Arch Microbiol,* **167**, 38–45.

Mander, G. J., Duin, E. C., Linder, D., Stetter, K. O. and Hedderich, R. (2002). Purification and characterization of a membrane-bound enzyme complex from the sulphate-reducing archaeon *Archaeoglobus fulgidus* related to heterodisulfide reductase from methanogenic archaea. *Eur J Biochem,* **269**, 1895–904.

Matias, P. M., Coelho, R., Pereira, I. A. *et al.* (1999a). The primary and three-dimensional structures of a nine-haem cytochrome *c* from *Desulfovibrio desulfuricans* ATCC 27774 reveal a new member of the Hmc family. *Structure,* **7**, 119–30.

Matias, P. M., Saraiva, L. M., Soares, C. M. *et al.* (1999b). Nine-haem cytochrome *c* from *Desulfovibrio desulfuricans* ATCC 27774: primary sequence determination, crystallographic refinement at 1.8 and modelling studies of its interaction with the tetrahaem cytochrome c_3. *J Biol Inorg Chem,* **4**, 478–94.

Matias, P. M., Coelho, A. V., Valente, F. M. A. *et al.* (2002). Sulphate respiration in *Desulfovibrio vulgaris* Hildenborough: structure of the 16-heme cytochrome *c HmcA* at 2.5A resolution and a view of its role in transmembrane electron transfer. *J Biol Chem,* **277**, 47907–16.

Matias, P. M., Pereira, I. A., Soares, C. M. and Carrondo, M. A. (2005). Sulphate respiration from hydrogen in *Desulfovibrio* bacteria: a structural biology overview. *Prog Biophys Mol Biol,* **89**, 292–329.

Menon, N. K., Chatelus, C. Y., Dervartanian, M. *et al.* (1994). Cloning, sequencing, and mutational analysis of the *hyb* operon encoding *Escherichia coli* hydrogenase 2. *J Bacteriol,* **176**, 4416–23.

Mizuno, N., Voordouw, G., Miki, K., Sarai, A. and Higuchi, Y. (2003). Crystal structure of dissimilatory sulfite reductase D (DsrD) protein – possible interaction with B- and Z-DNA by its winged helix motif. *Structure,* **11**, 1133–40.

Norager, S., Legrand, P., Pieulle, L., Hatchikian, C. and Roth, M. (1999). Crystal structure of the oxidised and reduced acidic cytochrome c_3 from *Desulfovibrio africanus. J Mol Biol,* **290**, 881–902.

Odom, J. M. and Peck Jr., H. D. (1981). Hydrogen cycling as a general mechanism for energy coupling in the sulphate-reducing bacteria, *Desulfovibrio* sp. *FEMS Microbiol Lett*, **12**, 47–50.

Pereira, I. A. C., Romão, C. V., Xavier, A. V., Legall, J. and Teixeira, M. (2000). Characterization of a heme *c* nitrite reductase from a non-ammonifying microorganism, *Desulfovibrio vulgaris* Hildenborough. *Biochim Biophys Acta*, **1481**, 119–30.

Pereira, I. A. C. and Xavier, A. V. (2005). Multi-Heme *c* cytochromes and enzymes. In R. B. King (ed.), *Encyclopedia of inorganic chemistry*, 2nd edn. John Wiley & Sons.

Pereira, P. M., Teixeira, M., Xavier, A. V. *et al.*, (2006). The Tmc complex from *Desulfovibrio vulgaris* Hildenborough is involved in transmembrane electron transfer from periplasmic hydrogen oxidation. *Biochemistry*, **45**, 10359–67.

Pierik, A. J., Duyvis, M. G., van Helvoort, J. M. L. M., Wolbert, R. B. G. and Hagen, W. R. (1992). The third subunit of desulfoviridin-type dissimilatory sulfite reductases. *Eur J Biochem*, **205**, 111–15.

Pieulle, L., Morelli, X., Gallice, P. *et al.* (2005). The type I/type II cytochrome c_3 complex: an electron transfer link in the hydrogen-sulphate reduction pathway. *J Mol Biol*, **354**, 73–90.

Pires, R. H., Lourenco, A. I., Morais, F. *et al.* (2003). A novel membrane-bound respiratory complex from *Desulfovibrio desulfuricans* ATCC 27774. *Biochim Biophys Acta*, **1605**, 67–82.

Pires, R. H., Venceslau, S., Morais, F. *et al.* (2006). Characterization of the *Desulfovibrio desulfuricans* ATCC 27774 DsrMKJOP complex – a membrane-bound redox complex involved in the sulphate respiratory pathway. *Biochemistry*, **45**, 249–62.

Pohorelic, B. K., Voordouw, J. K., Lojou, E. *et al.* (2002). Effects of deletion of genes encoding Fe-only hydrogenase of *Desulfovibrio vulgaris* Hildenborough on hydrogen and lactate metabolism. *J Bacteriol*, **184**, 679–686.

Rabus, R., Ruepp, A., Frickey, T. *et al.* (2004). The genome of *Desulfotalea psychrophila*, a sulphate-reducing bacterium from permanently cold Arctic sediments. *Environ Microbiol*, **6**, 887–902.

Rodrigues, R., Valente, F. M., Pereira, I. A. C., Oliveira, S. and Rodrigues-Pousada, C. (2003). A novel membrane-bound Ech NiFe hydrogenase in *Desulfovibrio gigas*. *Biochem Biophys Res Commun*, **306**, 366–75.

Romao, C. V., Pereira, I. A., Xavier, A. V., Legall, J. and Teixeira, M. (1997). Characterization of the NiFe hydrogenase from the sulphate reducer *Desulfovibrio vulgaris* Hildenborough. *Biochem Biophys Res Commun*, **240**, 75–9.

Rossi, M., Pollock, W. B., Reij, M. W. *et al.* (1993). The *hmc* operon of *Desulfovibrio vulgaris* subsp. *vulgaris* Hildenborough encodes a potential transmembrane redox protein complex. *J Bacteriol*, **175**, 4699–711.

Rousset, M., Dermoun, Z., Hatchikian, C. E. and Belaich, J. P. (1990). Cloning and sequencing of the locus encoding the large and small subunit genes of the periplasmic Nife hydrogenase from *Desulfovibrio fructosovorans*. *Gene*, **94**, 95–101.

Saraiva, L. M., Da Costa, P. N., Conte, C., Xavier, A. V. and Legall, J. (2001). In the facultative sulphate/nitrate reducer *Desulfovibrio desulfuricans* ATCC 27774, the nine-haem cytochrome *c* is part of a membrane-bound redox complex mainly expressed in sulphate-grown cells. *Biochim Biophys Acta*, **1520**, 63–70.

Schmehl, M., Jahn, A., Vilsendorf, A. M. Z. *et al.* (1993). Identification of a new class of nitrogen-fixation genes in *Rhodobacter capsulatus* – a putative membrane complex involved in electron-transport to nitrogenase. *Mol Gen Genet*, **241**, 602–15.

Sebban, C., Blanchard, L., Bruschi, M. and Guerlesquin, F. (1995). Purification and characterization of the formate dehydrogenase from *Desulfovibrio vulgaris* Hildenborough. *FEMS Microbiol Lett*, **133**, 143–9.

Shokes, J. E., Duin, E. C., Bauer, C. *et al.* (2005). Direct interaction of coenzyme M with the active-site Fe-S cluster of heterodisulfide reductase. *FEBS Lett*, **579**, 1741–4.

Stams, A. J. M. and Hansen, T. A. (1982). Oxygen-labile lactate dehydrogenase activity in *Desulfovibrio desulfuricans*. *FEMS Microbiol Lett*, **13**, 389–94.

Teixeira, V. H., Baptista, A. M. and Soares, C. M. (2004). Modeling electron transfer thermodynamics in protein complexes: interaction between two cytochromes *c*(3). *Biophys J*, **86**, 2773–85.

Umhau, S., Fritz, G., Diederichs, K. *et al.* (2001). Three-dimensional structure of the nonaheme cytochrome *c* from *Desulfovibrio desulfuricans* Essex in the Fe(III) state at 1.89 A resolution. *Biochemistry*, **40**, 1308–16.

Valente, F. M. A., Saraiva, L. M., Legall, J. *et al.* (2001). A membrane-bound cytochrome c_3: a type II cytochrome c_3 from *Desulfovibrio vulgaris* Hildenborough. *Chembiochem*, **2**, 895–905.

Valente, F. M. A., Oliveira, A. S. F., Gnadt, N. *et al.* (2005). Hydrogenases in *Desulfovibrio vulgaris* Hildenborough: structural and physiologic characterisation of the membrane-bound NiFeSe hydrogenase. *J Biol Inorg Chem*, **10**, 667–82.

Valente, F. M. A., Almeida, C. C., Pacheco, I. *et al.*, (2006). Selenium is involved in regulation of periplasmic hydrogenase gene expression in *Desulfovibrio vulgaris* Hildenborough. *J Bacteriol*, **188**, 3228–35.

Van der Westen, H. M., Mayhew, S. G. and Veeger, C. (1978). Separation of hydrogenase from intact cells of *Desulfovibrio vulgaris* – purification and properties. *FEBS Lett*, **86**, 122–6.

Vignais, P. M. and Colbeau, A. (2004). Molecular biology of microbial hydrogenases. *Curr Issues Mol Biol*, **6**, 159–88.

Volbeda, A., Charon, M. H., Piras, C. *et al.* (1995). Crystal structure of the nickel-iron hydrogenase from *Desulfovibrio gigas*. *Nature*, **373**, 580–7.

Volbeda, A. and Fontecilla-Camps, J. C. (2005). Structure-function relationships of nickel-iron sites in hydrogenase and a comparison with the active sites of other nickel-iron enzymes. *Coordin Chem Rev*, **249**, 1609–19.

Voordouw, G., Niviere, V., Ferris, F. G., Fedorak, P. M. and Westlake, D. W. S. (1990). Distribution of hydrogenase genes in *Desulfovibrio* spp. and their use in identification of species from the oil field environment. *Appl Environ Microbiol*, **56**, 3748–54.

Voordouw, G. (2002). Carbon monoxide cycling by *Desulfovibrio vulgaris* Hildenborough. *J Bacteriol*, **184**, 5903–11.

Xavier, A. V. (2004). Thermodynamic and choreographic constraints for energy transduction by cytochrome *c* oxidase. *Biochim Biophys Acta*, **1658**, 23–30.

CHAPTER 8

Dissimilatory nitrate and nitrite ammonification by sulphate-reducing eubacteria

José J. G. Moura, Pablo Gonzalez, Isabel Moura and Guy Fauque

8.1 INTRODUCTION

Anaerobic respiration with sulphate as terminal electron acceptor is a central component of the global sulphur cycle and is exhibited exclusively by prokaryotes (Fauque, 1995; Widdel, 1988). Sulphate-reducing bacteria (SRB) are thus of major functional and numerical importance in many ecosystems including cyanobacterial microbial mats, marine sediments, oil fields environments, deep-sea hydrothermal vents and even in human diseases (Loubinoux *et al.*, 2002; Ollivier *et al.*, Chapter 9 of this book; Widdel, 1988). For a long time, SRB were assumed to be a much specialized group of microorganisms using only a limited spectrum of organic substrates with sulphate as terminal electron acceptor. Today, SRB appear to be the microorganisms that reduce the greatest number of different terminal electron acceptors, including inorganic sulphur compounds and various other organic and inorganic compounds (Fauque *et al.*, 1991; Fauque and Ollivier, 2004; LeGall and Fauque, 1988).

In this chapter, we focus on the dissimilatory reduction of nitrate and nitrite by sulphate-reducing eubacteria because, to our knowledge, no sulphate-reducing archaebacteria (belonging to the *Archaeoglobaceae* family) has been reported so far to utilize this type of nitrogen metabolism.

ABBREVIATIONS

sulphate-reducing bacteria	SRB
periplasmic nitrate reductase	Nap
multihaem nitrite reductase	*cc*Ni
nitrate-reducing, sulphide-oxidizing bacteria	NR-SOB

Desulfovibrio	*D.*
Escherichia	*E.*
Wolinella	*W.*
Sulfurospirillum	*S.*
Paracoccus	*P.*
molybdopterin guanine dinucleotide	MGD
electron paramagnetic resonance	EPR
formate dehydrogenase	Fdh
dimethyl sulphoxide	DMSO
sulphite oxidase	SO.

8.2 PHYSIOLOGICAL AND BIOENERGETICAL CONSIDERATIONS OF NITRATE AND NITRITE AMMONIFICATION BY SULPHATE-REDUCING EUBACTERIA

The dissimilatory reduction of nitrate and nitrite (also called ammonification) can function as the sole energy-conserving process in some sulphate-reducing eubacteria (Fauque and Ollivier, 2004; Moura *et al.*, 1997). Nitrate or nitrite reduction has been first demonstrated in washed cells or cell extracts of different *Desulfovibrio* species when employing cells from a lactate-sulphate culture (Barton *et al.*, 1983; Liu and Peck, 1981; Senez and Pichinoty, 1958). Later, nitrate has been reported to be reduced to ammonia (with nitrite as intermediate) by a few *Desulfovibrio* species belonging mainly to *D. desulfuricans* but also by *D. furfuralis, D. profundus, D. oxamicus, D. simplex,"D. multispirans"* and *D. termitidis* (Dalsgaard and Bak, 1994; Keith and Herbert, 1983; Lopez-Cortès *et al.*, 2006; McCready *et al.*, 1983; Mitchell *et al.*, 1986; Seitz and Cypionka, 1986; Trinkerl *et al.*, 1990). A dissimilatory nitrate reduction has also been reported more recently with *Desulfotomaculum thermobenzoicum* (Plugge *et al.*, 2002), *Desulfobacterium catecholicum* (Widdel, 1988), *Desulforhopalus singaporenssi* (Lie *et al.*, 1999), *Thermodesulfovibrio islandicus* (Sonne-Hansen and Ahring, 1999), *Thermodesulfobium narugense* (Mori *et al.*, 2003) and *Desulfobulbus propionicus* (Widdel, 1988). *Desulfovibrio desulfuricans* strains ATCC 27774 and Essex 6, and *"D. multispirans"* (NCIB 12078) are also able to use cathodic hydrogen from mild steel as sole source of energy for growth and nitrate reduction (Rajagopal and LeGall, 1989). Dissimilatory nitrite reduction by SRB is widespread, but strains capable of nitrate ammonification are far less common (Mitchell *et al.*, 1986; Moura *et al.*, 1997). Depending on the microorganism, sulphate or nitrate may be the preferred terminal electron acceptor, or both electron acceptors can be reduced concomitantly (Cypionka, 1995; Krekeler and Cypionka, 1995; McCready *et al.*, 1983; Seitz and Cypionka, 1986). In *D. desulfuricans* strain Essex 6, nitrite reductase is synthesized constitutively, whereas nitrate reductase is inducible by nitrite or nitrate (Seitz and Cypionka, 1986). Adenosine triphosphate (ATP) synthesis, coupled to the dissimilatory reduction of nitrite to ammonia, was obtained with a membrane fraction of lactate-sulphate grown cells of *D. gigas* and $P/2e^-$ values of 0.16–0.4 were observed in the hydrogen-nitrite system (Barton *et al.*, 1983). A vectorial proton translocation during nitrate or nitrite reduction has been demonstrated with whole cells of different *Desulfovibrio* species (Cypionka, 1995; Steenkamp and Peck, 1981). Proton translocation during dissimilatory nitrate reduction has not yet been observed.

Nitrite ammonification at the expense of sulphide oxidation to sulphate was described for *Desulfobulbus propionicus* and *D. desulfuricans* (Dannenberg *et al.*, 1992; Greene *et al.*, 2003). A novel type of metabolism connecting the nitrogen and sulphur cycles has been discovered in *D. desulfuricans* CSN which is able to oxidize sulphite and thiosulphate with nitrate and nitrite as electron acceptors (Krekeler and Cypionka, 1995). An oxidation of sulphide coupled to nitrate or nitrite reduction was shown in *Desulfobulbus propionicus* and *D. desulfuricans* CSN (Fuseler *et al.*, 1996). Nitrate-reducing, sulphide-oxidizing bacteria (NR-SOB) are able to inhibit SRB in the presence of nitrate (Haveman *et al.*, 2005). This inhibition could be due to the production of nitrite by the NR-SOB or to an increase in redox potential (Greene *et al.*, 2003). Nitrite (but not nitrate) is known to inhibit the last step in the dissimilatory sulphate reduction pathway (reduction of sulphite to sulphide by the enzyme dissimilatory sulphite reductase) (Havemen *et al.*, 2004). The production of hydrogen sulphide by SRB in oil reservoirs (souring) and the microbially induced corrosion can be controlled through nitrate or nitrite addition (Hubert *et al.*, 2005; Jenneman *et al.*, 1986).

Desulfovibrio desulfuricans subsp. *desulfuricans* DSM 6949 (ATCC 27774) is the ammonifying strain of sulphate reducer that has been the most characterized from a physiological and a biochemical point of view. The growth yield of *D. desulfuricans* ATCC 27774 grown in lactate nitrate medium is higher than when sulphate is the terminal electron acceptor (Fauque *et al.*, 1991). The free energy change per hydrogen oxidized is about four times higher with nitrate than with sulphate. *Desulfovibrio desulfuricans* ATCC 27774 is also active in the bidirectional transformation of aromatic aldehydes (such as benzaldehyde, 3-hydroxybenzaldehyde) under nitrate-respiring conditions and the direction of transformation (i.e. oxidation or reduction) is mainly regulated by reductant availability (Parekh *et al.*, 1996).

8.3 CHARACTERIZATION OF THE DISSIMILATORY NITRATE REDUCTASE ISOLATED FROM *DESULFOVIBRIO DESULFURICANS* ATCC 27774

All the reductive routes of the N-cycle involve the conversion of nitrate to nitrite. This step is carried out by nitrate reductases (NR), which catalyze the unique reaction:

$$NO_3^- + 2H^+ + 2e^- \rightarrow NO_2^- + H_2O \qquad E^\circ = +420\,\text{mV}$$

NR are mononuclear Mo-containing enzymes that, according to the Hille's classification (Hille, 1996), belong to the DMSO reductase family with the only exception being eukaryotic nitrate reductases, which belong to the SO family (Figure 8.1). Several nitrate reductases have been obtained from both prokaryotic and eukaryotic organisms. The prokaryotic enzymes (Figure 8.1) can also be sub-grouped as respiratory nitrate reductase (Nar), Periplasmic Nitrate Reductases (Nap, EC 1.7.99.4) and Assimilatory Nitrate Reductases (Nas) (Stolz and Basu, 2002). Their characterization gave important information on the molecular basis of nitrate reduction involved in all the branches of the N-cycle.

Nap from the *D. desulfuricans* ATCC 27774 constitutes currently the only characterized NR from SRB (Bursakov *et al.*, 1995; 1997; Dias *et al.*, 1999). Well-studied enzymes of this class are the ones from *Paracoccus (P.) pantotrophus* (Butler *et al.*, 1999; 2002) and *Rhodobacter sphaeroides* (Arnoux *et al.*, 2003; Frangioni *et al.*, 2004).

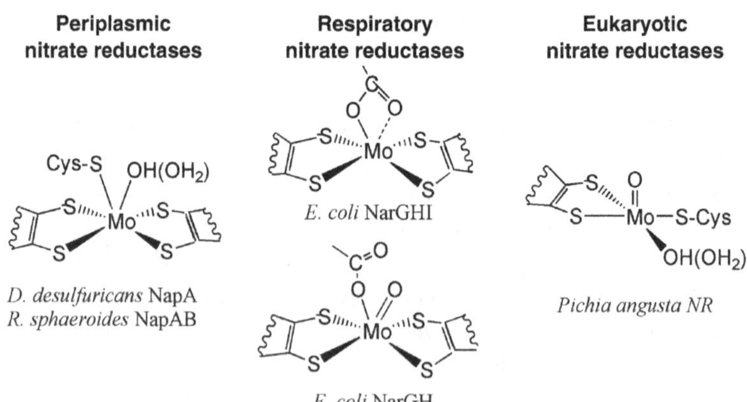

Figure 8.1. Active site structures of Mo-pterin co-factors in periplasmic, respiratory and eukaryotic nitrate reductases.

We review here the biochemical and molecular properties of the dissimilatory periplasmic nitrate reductase isolated from *D. desulfuricans* ATCC 27774, which is able to grow on nitrate, inducing the enzymes responsible for the conversion of nitrate to nitrite and for the reduction of nitrite to ammonia.

8.3.1 Structure of *D. desulfuricans* ATCC 27774 nitrate reductase

NapA from *D. desulfuricans* ATCC 27774 was the first reported structure for a periplasmic nitrate reductase (Figure 8.2) (Dias *et al.*, 1999). *Desulfovibrio desulfuricans* NapA is a monomeric protein organized in four domains, all involved in cofactor binding. The structure of NapA reveals the details of the catalytic molybdenum site, which is coordinated to two molybdopterin guanine dinucleotide (MGD) cofactors, a sulphur atom from Cysteine 140, and a hydroxo/(water) ligand (Figure 8.1). In addition, the protein has an iron-sulphur cluster of the type [4Fe-4S], involved in electron transfer. The [4Fe-4S] centre is located near the periphery of the molecule, whereas the MGD cofactor extends across the interior of the molecule interacting with residues from all four domains. The molybdenum atom is located at the bottom of a 15 Å deep crevice (domains II and III form a funnel-like cavity that extends from the surface to the catalytic site), and is positioned 12 Å from the [4Fe-4S] cluster. The hydroxo/(water) ligand of the molybdenum atom, which is supposed to be the position

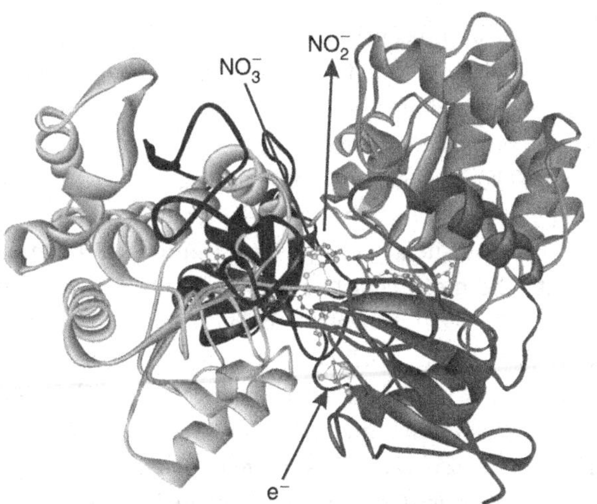

Figure 8.2. 3D structure of NapA isolated from *D. desulfuricans* ATCC 27774.

where the substrate binds, points into this channel, suggesting that nitrate entrance and nitrite exit would be via this channel. A facile electron-transfer pathway through bonds connects the molybdenum and the [4Fe-4S] cluster.

The polypeptide fold of NapA and the arrangement of the cofactors are related to that of formate dehydrogenase H (Fdh-H) from *Escherichia coli* K12 (Moura *et al.*, 2004), and distantly resembles dimethylsulphoxide reductase. The close structural homology of NapA and Fdh-H shows how small changes in the vicinity of the molybdenum catalytic site are enough for the substrate specificity.

Naps isolated from *Rhodobacter (R.) sphaeroides* (Reyes *et al.*, 1996), *Wautersia eutropha* (formerly *Alcaligenes eutrophus*) (Siddiqui *et al.*, 1993) and *Paracoccus (P.) pantotrophus* (formerly *Tiosphaera pantotropha*) (Berks *et al.*, 1994) are heterodimeric proteins with a large (80–90 kDa) and small (~17 kDa) subunits. The only crystal structure available for a heterodimeric Nap belongs to *R. sphaeroides* NapAB which was determined at a resolution of 3.2 Å (Arnoux *et al.*, 2003). The arrangement of the catalytic subunits of *D. desufuricans* NapA and *R. sphaeroides* NapA are very similar in terms of metal cofactor content, global folding and domain organization. A structural comparison between all these proteins was recently reported by Moura *et al.* (Moura *et al.*, 2004).

8.3.2 Gene organization and electron pathway for nitrate reduction

The *nap* genes have been identified in several prokaryotic organisms (Gonzalez *et al.*, 2006a). As found in other bacteria, in *D. desulfuricans* ATCC 27774 the genes that code the Nap system for the reduction of nitrate are grouped in an operon. This cluster comprises the six genes *napCMADGH* (Marietou *et al.*, 2005). The respective proteins organize in the cell as depicted in Figure 8.3 (left). The *napA* gene codes the catalytic subunit that contains the Mo-*bis*MGD active site and the [4Fe-4S] centre. The assembling of the apo-protein with the metal cofactors take place in the cytoplasm assisted by the chaperone NapD and then the folded holoprotein is transported to the periplasm through the TAT (Twin Arginine Translocator) system by recognizing the signal peptide present in the N-terminal region of NapA (Thomas *et al.*, 1999). The first gene in the operon codes for *c*-type tetrahaemic NapC (~25 kDa). This protein belongs to the NapC/NirT family, which are periplasmic proteins anchored to the membrane and responsible for transferring electrons from the menaquinone pool to

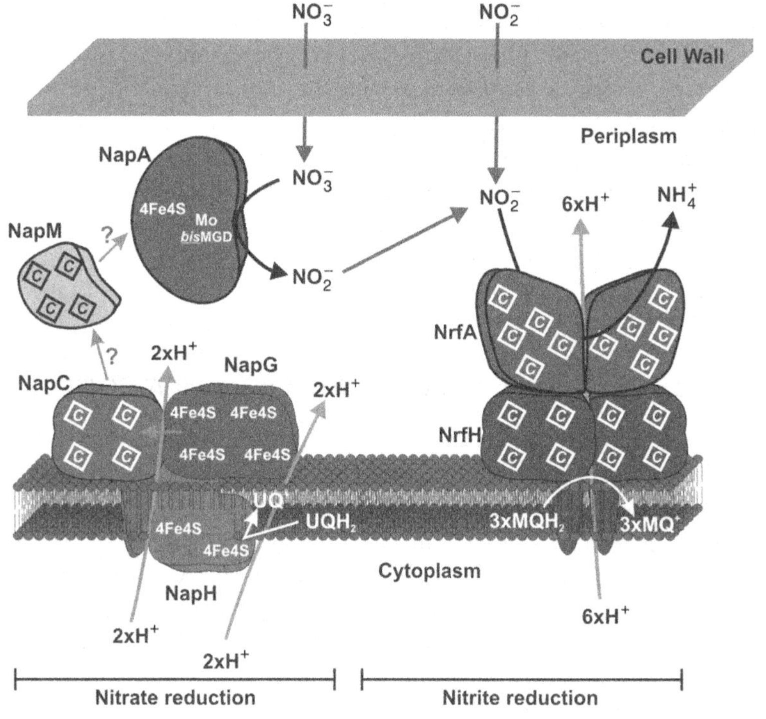

Figure 8.3. Dissimilatory nitrate reduction (Nap enzymes) coupled to nitrite utilization.

periplasmic reductases. The last two genes codes for the ferredoxins NapG and NapH. NapG is a periplasmic ferredoxin having 4×[4Fe-4S] clusters and its proposed function is transferring electrons to either NapA or NapC. NapH is an integral membrane protein with 2×[4Fe-4S] clusters exposed to the cytoplasm, and its role could be taking electrons from the ubiquinone pool and transferring them to NapG (Potter *et al.*, 1999; Gonzalez *et al.*, 2006a). The *napM* gene codes for a putative *c*-type tetrahaemic cytochrome with periplasmic predicted localization. Its probable task is to deliver electrons from either NapC or an electron carrier to the monomeric NapA (Marietou *et al.*, 2005).

All these proteins are expressed only when *D. desulfuricans* ATCC 27774 is cultured under anaerobic conditions using nitrate as final electron acceptor, since, when nitrate is replaced for sulphate, no nitrate reductase activity can be detected in cells or soluble extract. Then, the regulation of the transcription must be under the control of a nitrate:nitrite sensing system.

The currently accepted model (Figure 8.3, left) for electrons delivery to nitrate is the one proposed for *E. coli* K12 when cultured in nitrate-limiting conditions (Gonzalez *et al.*, 2006a; Potter *et al.*, 1999). Here, NapH oxidizes the ubiquinone pool (E° = +90 mV) in the cytoplasmic face of the cell membrane, which is accompanied with the translocation of protons to the periplasm. Next, NapH transfer the electrons to NapG that yields the electrons to NapC, transporting more protons to the periplasm. Having into consideration that reduction of nitrate to nitrite consumes $2H^+$; a net gradient of $2H^+$ is obtained for each nitrate molecule reduced. This means that in energetic terms, Nap system is as efficient as the NarGHI system.

8.3.3 EPR Studies in NapA from *D. desulfuricans* ATCC 27774

As seen in other periplasmic nitrate reductases, *D. desulfuricans* NapA yield several Mo(V) species designated as: *resting, low potential, high potential, nitrate, cyanide* and *turn-over* signals (Table 8.1) (Gonzalez *et al.*, 2006b). The resting signal can be seen only in as-isolated preparations of the enzyme (<0.05 spin/molecule) and cannot be restored after redox cycling the enzyme.

The low potential Mo(V) signal is rhombic and highly anisotropic, and has the same g-*values* of the rhombic II signal from *D. desulfuricans* Fdh. Low potential signal is only seen when the enzyme is poised at redox potentials below −400 mV (vs. NHE). At potentials as low as −500 mV, all the FeS clusters are found in the $[4Fe\text{-}4S]^{1+}$ state (S = 1/2), but only a 10–15% of the total Mo is in the paramagnetic +5 oxidation state, indicating that the Mo(IV) rather occur in vivo, and so rule out this species for having a relevant role in catalysis (Table 8.1).

The nitrate signal is obtained when a dithionite-reduced sample of *D. desulfuricans* NapA is reacted in anaerobic conditions with nitrate or

Table 8.1. *EPR characterized Mo(V) species in* D. desulfuricans *ATCC 27774 periplasmic nitrate reductase*

Enzyme	EPR signal	g_1	g_2	g_3	A_1	A_2	A_3
Dd NaA	Low potential	2.016 (5.5)	1.987 (5.5)	1.964 (5.5)	–	–	–
	High potential	2.019 (5.5)	1.988 (5.5)	1.960 (5.5)	–	–	–
	Nitrate	2.000 (4.4)	1.990 (3.6)	1.981 (3.4)	4.6	5.0	4.6
	Turn-over	1.999 (5.0)	1.990	1.982	5.0	6.0	5.0
14N/15N			(5.0/4.0)	(5.0/4.0)			
Cyanide	2.024 (4.5)	2.001 (4.0)	1.995 (4.0)	8.0	7.5	6.0	

oxidized with air. EPR-monitored potentiometric titration of this Mo(V) signal showed that it is redox inactive in the potential range of +200 to −500 mV (vs. NHE). Moreover, the enzyme cannot reduce nitrate when dithionite is used as sole electron donor, indicating that this Mo(V) state is not involved in catalysis. Furthermore, this signal resembles the High g species found in *Pp* NapAB, which suggest that those species could not be involved in catalysis also.

EPR studies on *D. desulfuricans* NapA using methyl viologen (MV) as electron donor instead of sodium dithionite showed that FeS cluster is not paramagnetic upon MV + enzyme incubation. However, nitrate addition oxidizes the electron donor through the Nap-catalyzed reaction, and yields the turn-over signal. This novel Mo(V) species has never been reported before in any other mononuclear Mo-containing enzyme, and can be simulated assuming the interaction with a I=1 nuclear spin, supporting the existence of a Mo(V)-nitrate catalytic intermediary in the mechanism of nitrate reduction.

The cyanide Mo(V) species found in *D. desulfuricans* NapA is highly similar to the ones found in cyanide-treated *P. pantotrophus* NapAB, air-oxidized NarB from *Synechococcus sp.* and dithionite-reduced Fdh from *Methanobacterium formicicum*. These, in addition to the recent studies made in *D. desulfuricans* NapA using both ^{13}C and ^{15}N labelled cyanide (no changes in hyperfine structure are visible), indicates that the cyanide Mo(V) species does not have the CN molecule bound.

8.3.4 Mechanism of nitrate reduction

Based on the molecular structure of the active site in *D. desulfuricans* ATCC 27774 NapA, the proposed reaction mechanism for nitrate reduction implies the replacement of a OH/H_2O Mo ligand by a nitrate molecule (Figure 8.4), which is then reduced to nitrite, with the abstraction of an O atom from the substrate. In this process, the electrons necessary for nitrate reduction are obtained from the Mo site which receives them from an external electron donor by means of an electron transfer reaction mediated by metal cofactors such as FeS clusters and haems.

The reaction scheme proposed in Figure 8.4 is a simplified version of the actual mechanism. The evaluation of the spectroscopic properties of these enzymes under turnover conditions (substrate binding) is underway (Gonzalez *et al.*, 2006b) and catalytic electrochemical data has been published on the interaction of nitrate with the different redox states of the enzyme (Frangioni *et al.*, 2004). These results will help in the clarification of unveiled

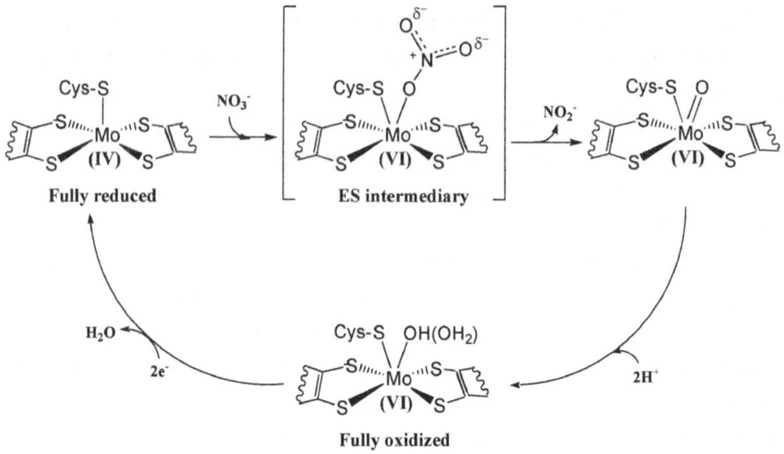

Figure 8.4. Simplified mechanism proposed for nitrate utilization by NapA.

aspects regarding the mechanism of action of nitrate reductases and mononuclear Mo-containing enzymes in general.

8.4 CHARACTERIZATION OF THE CYTOCHROME *C* NITRITE REDUCTASE PURIFIED FROM *D. DESULFURICANS* ATCC 27774

The multihaem nitrite reductases (*cc*Nir) [(EC 1.7.2.2 nitrite reductase) (cytochrome; ammonia-forming)] act on the dissimilative ammonification process, where they catalyze the reduction of nitrite to ammonia in a unique six-electron step:

$$NO_2^- + 8H^+ + 6e^- \rightarrow NH_4^+ + 2H_2O \quad E^\circ = +330\,mV$$

They occur mainly in proteobacteria of subdivision δ, γ and ε, and can be isolated from the soluble or the membrane extracts, although in most of the cases the enzyme is found in the periplasmic space.

The first biochemical characterization of a *cc*Nir from SRB of the genus *Desulfovibrio* was carried out in the laboratory of H. D. Peck Jr. (Liu and Peck, 1981). After this discovery, several *cc*Nirs were isolated from different microorganisms. Over more than a decade *cc*Nirs were erroneously considered to be monomers of ~60 kDa containing six *c*-type haems (Kajie and Anraku, 1986; Rehr and Klemme, 1986; Liu *et al.*, 1988; Liu *et al.*, 1983; Schumacher and Kroneck, 1991). But twelve years later (1993) the DNA sequence of the gene coding for the protein from *E. coli* was obtained.

From the primary sequence four *c*-type haem attachment motifs (CXXCH) were identified (Darwin *et al.*, 1993). Following this interpretation, *cc*Nir from *Sulforospirillum (S.) deleyianum* and *Wollinella (W.) succinogenes* (Schumacher *et al.*, 1994) were considered to contain four haems also. The analysis of the Mossbauer data allowed us to conclude later that the spectra consist of six components (Costa *et al.*, 1990b; 1996), which was confirmed later with the re-interpretation of the gene sequence of *E. coli* Nir. In this sense, a fifth haem motif ligation was identified, where a lysine is the axial ligand instead of a histidine (Darwin *et al.*, 1993). In 1999 the first X-ray structure was obtained for the *S. deleyianum* *cc*Nir, which clarified the situation showing the presence of five haems per subunit (Einsle *et al.*, 1999).

8.4.1 Characterization of the subunit composition of *D. desulfuricans* ATCC 27774 *cc*Nir

Desulfovibrio desulfuricans ATCC 27774 *cc*Nir is obtained by membrane solubilization. A careful analysis of the SDS-PAGE profile showed that, besides the 60 kDa subunit, another band with ~20 kDa is present, and this was not reported in the first preparations of the enzyme (Almeida *et al.*, 2003). The as-isolated *cc*Nir is a high molecular mass oligomer having 890 kDa and higher mass (Liu and Peck, 1981) which was concluded to be a mixture of $\alpha_2\beta_2$ and α_2 complexes (Almeida *et al.*, 2003). The two subunits which compose this oligomer are difficult to separate and SDS (sodium dodecyl sulphate) treatment is the only way to achieve it. The primary sequence of both subunits revealed that NrfA (gene encoding cytochrome C nitrate reductase) bind five haems, four of each are *bis*-histidine coordinated, and one just by a lysine. NrfH contains 4 × bis-hystidinyl coordinated haems. The N-terminal amino sequence of NrfH forms a hydrophobic helix which attaches it to the membrane. The UV-Vis and EPR spectra of the subunits separated by SDS treatment, are not very informative because they represent denatured forms. Nevertheless, the subunit NrfA shows a high spin contribution that should correspond to the catalytic site (Almeida *et al.*, 2003).

8.4.2 Structure of *D. desulfuricans* ATCC 27774 *cc*Nir

Currently, the X-ray structures of *cc*Nir from *E. coli* K12 (Bamford *et al.*, 2002), *S. deleyianum* (Einsle *et al.*, 1999), *W. succinogenes* (Einsle *et al.*, 2000) and *D. desulfuricans* ATCC 27774 are reported (Cunha *et al.*, 2003).

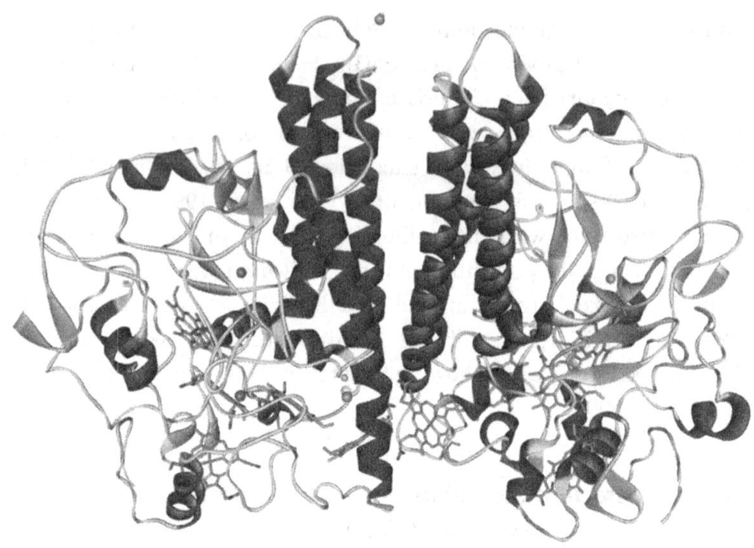

Figure 8.5. 3D structure of *cc*Nir isolated from *D. desulfuricans* ATCC 27774.

The structure of NrfA from *D. desulfuricans* ATCC 27774 was solved at 2.3 Å (Figure 8.5). The protein crystallizes as a homodimer, showing the ten haems in a very dense packing, as indicated by the distances between the iron atoms that vary from 9 to 12.5 Å. With the exception of the pentacoordinated haem 1 (lysine in axial position), the remaining haems are coordinated by two axial histidines with different relative orientation. There are one anionic and one cationic channels of access to the active centre, which were proposed to be the entrance of the substrate and the exit of the product, respectively. The electrostatic potential at the surface of the protein is dominated by the positively charged region around the channel leading to the active site and by the negative electrostatic potential around the putative product exit. These electrostatic features conserved in this family of enzymes, are considered to be relevant to attract the negatively charged nitrite ions to the active site and to drive the positively charged ammonium ions to the exterior of the enzyme. Two calcium ions are detected, one very close to the active site.

8.4.3 Spectroscopic properties of *D. desulfuricans* ATCC 27774 *cc*Nir

From all the multihaem *cc*Nir, the one from D. *desulfuricans* ATCC 27774 represents one of the best characterized enzymes from the spectroscopic point of view. The UV-Vis spectrum of the native enzyme exhibits absorption bands characteristics of typical haem *c* containing proteins. Typical bands are observed at 532,409 (Soret) and 278 nm in the oxidized form. In addition, a band due to high-spin haem *c* contribution can be also observed at 610 nm. After reduction, the Soret band is right-shifted to 420 nm with increased intensity, and the α and β peaks can be observed at 552 and 523 nm, respectively (Liu and Peck, 1981).

The first EPR and Mössbauer studies (Costa *et al.*, 1990b; 1996) on this enzyme revealed a complex system, involving multiple haem species with detectable magnetic interactions between them (Table 8.2, Figure 8.6). From these studies, it was concluded that: (i) only one of the haems is in a high-spin state, and is involved in substrate binding; (ii) a complex EPR signal was detected with *g*-values $g = 9.36$ and 3.85 due to a magnetic interaction between a pair of haems, one of them is the high-spin; (iii) a signal is observed at $g = 4.8$; (iv) only one of the low-spin haems is observed by EPR as magnetically isolated having $g_{max} = 2.96$, $g_{med} = 2.28$ and $g_{min} = 1.50$, (v) in half-reduced samples it is possible to identify another low spin-haem with $g_{max} = 3.00$ and $g_{med} = 2.14$, meaning that this haem is coupled to another one in the native state and; (vi) Mössbauer data indicate that two of the low spin haems have abnormally high *g*-values (3.6 and 3.5) indicating a *bis*-histidinyl coordination, with the histindinyl planes almost perpendicular to each one.

The EPR of the native complex exhibits a derivative-type signal with zero-crossing at $g = 4.8$ (Costa *et al.*, 1990b) which is absent in the spectra of *cc*NiR preparations from the soluble fraction of other microorganisms such as *E. coli* K-12 (Bamford *et al.*, 2002), *W. succinogenes* (Liu *et al.*, 1987) and *S. deleyianum* (Schumacher *et al.*, 1994), exclusively constituted by the

Table 8.2. *EPR assignments and redox properties of the haem system in* D. desulfuricans *ATCC 27774 nitrite reductase (NrfA and NrfH)*

Subunit	Haem	Angles(°)	Solvent(Å^2)	g_{max}	E'_m(mV)
NrfA	1		34.0	6.12	−80
	2	23.3	96.0	2.96	−50
	3	54.6	2.5	3.20	c. −480
	4	74.8	3.0	3.60	c. −400
	5	78.8	70.0	3.50	150
NrfH	1			3.55	>0
	2,3,4			3.00	c. −300

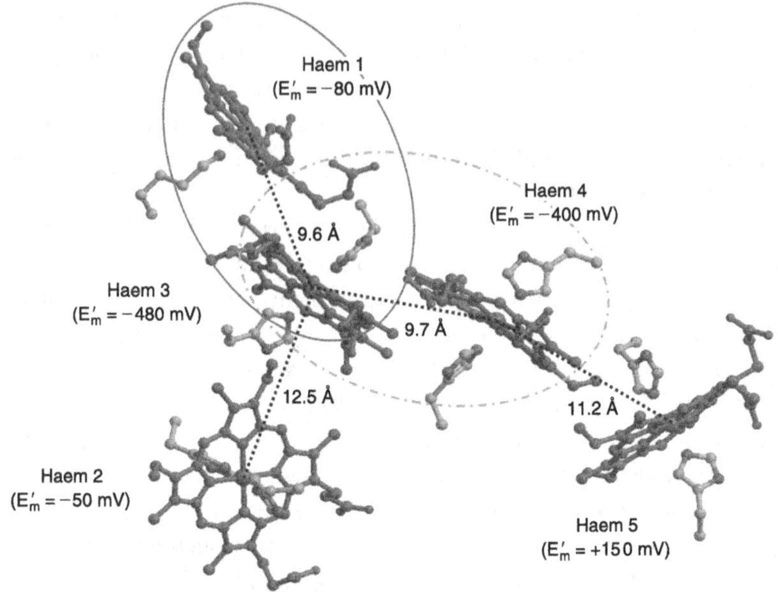

Figure 8.6. Haem orientation in NrfA from *D. desulfuricans* ATCC 277774 (including redox potential assignments).

periplasmic NrfA subunit. Then, this resonance should be originated by internal magnetic coupling from NrfH haems or, if in close proximity to the NrfA, could arise from spin coupling between haems from both NrfA subunits.

The *D. desulfuricans cc*Nir Mössbauer spectra obtained in the presence of a strong magnetic field were originally interpreted as a superposition of six spectral components of equal intensity (16.6%) and distinct hyperfine parameters (at the time the enzyme was considered as a monomer containing six *c*-type haems) (Costa *et al.*, 1990b). Following our present analysis, the enzyme is a complex of two different subunits – the pentahaemic NrfA and the tetrahaemic NrfH – implying the existence of nine different haems. Thus, a re-evaluation of the Mössbauer data in the native state and in samples poised at different reduction potentials was undertaken. According to the real stoichiometry, each NrfA haem corresponds to 14% of the total iron absorption, and each NrfH haem corresponds to 7%. From previous EPR studies (Costa *et al.*, 1990b), two sets of low-spin ferric g-values (g_{max} at 2.96 and 3.20) and a high-spin haem were observed, here assigned to the NrfA subunit and therefore contributing with 14% each.

The original Mössbauer studies on NrfHA complex identified two low-spin haems with g_{max} values at 3.60 and 3.50. The work of Walker *et al.* on low-spin ferric haem model compounds with axial imidazole ligands correlated the g_{max} values larger than 3.3 with perpendicularly aligned axial imidazole planes (Walker *et al.*, 1986). As the NrfA crystal structure shows two *bis*-His ligated haems with orthogonal imidazole geometry (Cunha *et al.*, 2003), we assigned the $g_{max} = 3.60$ and 3.50 signals to the NrfA subunit.

The proposed Mössbauer spectra simulation, based on the current biochemical characterization of *D. desulfuricans* ATCC 27774 *cc*NiR complex (NfrHA), plus several structural considerations on NrfA, fit as well to the former simulation as they do to the experimental data (Almeida *et al.*, 2003). The large subunit NrfA contains the high-spin haem and four low-spin *c*-type haems with $g_{max} = 3.6$, 3.50, 3.2 and 2.96; while the NrfH subunit encloses four haem groups in a low-spin configuration, one with $g_{max} = 3.55$ and a positive midpoint reduction potential (>0 mV), and three with $g_{max} = 3.00$, with a midpoint reduction potential of approximately -300 mV (Figure 8.6).

8.4.4 Spectroscopy and structural correlations

The determination of the 3D structure of *D. desulfuricans* ATCC 27774 NrfA (Cunha *et al.*, 2003) enabled the discovery of the spatial characterization of the five haems, namely their proximity and the axial histidine plane angles (Figure 8.5). A correlation between individual haems obtained by spectroscopy (EPR and Mössbauer), with known reduction potentials, was then undertaken.

Haem 1 (according to *D. desulfuricans* ATCC 27774 NrfA amino-acid numbering) has the sixth axial position vacant. Thus, is the site of substrate binding (high-spin haem, -80 mV). The EPR results revealed that this haem is pairwise coupled with a $g_{max} = 3.20$ low-spin haem (Costa *et al.*, 1990b). Regarding the distance and the relative orientation to haem 1 (Figure 8.6), the main candidate to be magnetic coupled is haem 3 (approximately -480 mV). Haem 2 (-50 mV) is the one with $g_{max} = 2.96$ that is EPR detectable and is magnetically isolated. Structurally, it is distant from the remaining haems and the dihedral angles of the axially coordinated His support this assignment (Figure 8.6). The Mössbauer data reveals the presence of two low-spin ferric haems with large g_{max} (3.50 and 3.60). The haems which satisfy such requirements are haems 4 and 5 (Figure 8.6). One of these haems is magnetically isolated (g_{max} 3.50) (Costa *et al.*, 1996).

As haems 1, 3 and 4 are almost coplanar and haem 5 is slightly apart; this is, probably, the magnetically isolated one. As seen by UV-Vis and Mössbauer spectroscopy, this haem is reduced at a positive reduction potential (+150 mV) which is unusual for a haem with *bis*-His axial ligation. Haem 4 should have g_{max} at 3.60 and reduction potential of approximately −400 mV. Due to high reduction potential and haem solvent exposure, it was proposed that haem 5 is the site of electron entrance from the redox partner NrfH (Figure 8.6).

The solvent exposure calculations did not consider the presence of the NrfH subunit; which decrease the solvent accessibility of haems 2 and 5, which are located near the putative surface contact (Liu *et al.*, 1987). The redox potential of *c*-type cytochromes can be tuned by approximately 500 mV through variations in the haem exposure to solvent (Kennedy and Gibney, 2001; Tezcan *et al.*, 1998). The encapsulation of the haem group in a hydrophobic environment causes a positive shift in the reduction potential, up to approximately 240 mV in cytochrome *c* (Tezcan *et al.*, 1998). This may explain the atypical positive reduction potential of haem 5, if in close proximity with the hydrophobic transmembrane NrfH subunit. However, these suggestions are purely speculative and haem 2 should not be excluded as a candidate for the electron entrance, as postulated (Bamford *et al.*, 2002; Einsle *et al.*, 2000).

In the NrfH subunit, it was not possible to perform the structural assignment of the haem spectroscopic and reduction potentials, as there are no structures available for NrfH like proteins or any member of the NapC/NirT family. Nevertheless, crystals of *W. succinogenes* NrfHA complex have been reported recently (Einsle *et al.*, 2002b). Figure 8.6 describes the relationship between the haem core description and the spectroscopic and redox properties of each identified haem from the NrfHA complex.

8.4.5 Reactivity and mechanism

Besides nitrite, *cc*Nirs are also able to reduce other substrates (Costa *et al.*, 1990a; Liu *et al.*, 1987). Nitric oxide and hydroxylamine are probably intermediates of the reaction of reduction of nitrite to ammonia, but they were never isolated. The enzyme can also reduce sulphite to sulphide in a six-electron step (Einsle *et al.*, 2002a; Pereira *et al.*, 1996).

Based on the structural characterization of the complexes of *cc*Nir with nitrite and hydroxylamine together with DFT calculations, a mechanistic proposal was put forward (Pereira *et al.*, 1996). In this hypothesis, nitrite

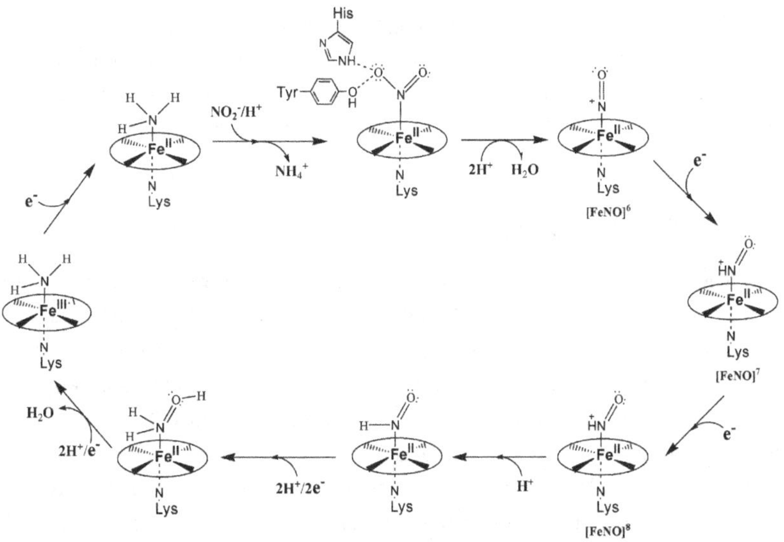

Figure 8.7. Mechanism proposed for nitrite utilization by *cc*Nir (multihaemic).

binds to the reduced haem 1 (the high spin) through the nitrogen atom. Subsequently, there is a cleavage with formation of a $[FeNO]^6$ species and a water molecule. Next, two electrons enter forming $[FeNO]^8$, followed by a protonation step leading to an Fe(II)-HNO adduct. Next, a step consuming $2e^-$, $2H^+$ will form hydroxylamine which is dehydrated in the final two electron-reduction step to give ammonia and an additional water molecule. The catalytic cycle is closed by an additional reduction step by one electron (Figure 8.7).

8.5 CHARACTERIZATION OF A CYTOCHROME C NITRITE REDUCTASE FROM *D. VULGARIS* HILDENBOROUGH, A NON-AMMONIFYING SULPHATE REDUCER

A membrane-bound *cc*Nir has been purified and characterized from *D. vulgaris* Hildenborough, a sulphate reducer not able to grow with nitrate respiration (Pereira *et al.*, 2000). This membrane-bound complex of 760 kDa contains 2 cytochrome *c* subunits of 18 and 56 kDa and has both nitrite and sulphite reductases activities (Pereira *et al.*, 2000). Multiple haem–haem magnetic interactions are observed in the EPR spectra of oxidized and partially reduced forms of this *cc*Nir. The physiological role of a constitutive periplasmic *cc*Nir expressed in *D. vulgaris* Hildenborough grown on lactate-sulphate medium is still in question. Even if this *cc*Nir is bifunctional, the fact that it is more active in reducing nitrite than sulphite favours its physiological role as a detoxifying enzyme (Pereira *et al.*, 2000).

8.6 CONCLUSION

Microorganisms displaying dissimilatory sulphate reduction constitute a highly diverse group with broad metabolic capabilities. SRB are not restricted to energy conservation by sulphate reduction and the dissimilatory nitrate or nitrite reduction to ammonia can function as the sole energy-conservation process in some sulphate reducers. The growth yield of SRB with nitrate as alternative electron acceptor is even higher than for those with sulphate. The ecological significance of the ability of some SRB to reduce nitrate into ammonium is not well known, but we may hypothesize that most probably such SRB display this reductive process at lower redox potentials in sediments than that of sulphate reduction. In this way, SRB in the global nitrogen cycle within the ecosystems they inhabit might have been underestimated.

In this review, we have also described the biochemical and structural aspects of nitrate and nitrite reductases, the two key enzymes involved in the ammonification process by SRB of the genus *Desulfovibrio*.

REFERENCES

Almeida, M. G., Macieira, S., Gonçalves, L. L. *et al.* (2003). The isolation and characterization of cytochrome *c* nitrite reductase subunits (NrfA and NrfH) from *Desulfovibrio desulfuricans* ATCC 27774. Re-evaluation of the spectroscopic data and redox properties. *European Journal of Biochemistry*, **270**, 3904–15.

Arnoux, P., Sabaty, M., Alric, J. *et al.* (2003). Structural and redox plasticity in the heterodimeric periplasmic nitrate reductase. *Nature Structural Biology*, **10**, 928–34.

Bamford, V. A., Angove, H. C., Seward, H. E. *et al.* (2002). Structure and spectroscopy of the periplasmic cytochrome *c* nitrite reductase from *Escherichia coli*. *Biochemistry*, **41**, 2921–31.

Barton, L. L., LeGall, J., Odom, J. M. and Peck, H. D. Jr. (1983). Energy coupling to nitrite respiration in the sulphate-reducing bacterium *Desulfovibrio gigas*. *Journal of Bacteriology*, **153**, 867–71.

Berks, B. C., Richardson, D. J., Robinson, C. *et al.* (1994). Purification and characterization of the periplasmic nitrate reductase from *Thiosphaera pantotropha*. *European Journal of Biochemistry*, **220**, 117–24.

Bursakov, S. A., Carneiro, C., Almendra, M. J. *et al.* (1997). Enzymatic properties and effect of ionic strength on periplasmic nitrate reductase (NAP) from *Desulfovibrio desulfuricans* ATCC 27774. *Biochemical and Biophysical Research Communications*, **239**, 816–22.

Bursakov, S. A., Liu, M., Payne, W. J. *et al.* (1995). Isolation and preliminary characterization of a soluble nitrate reductase from the sulphate reducing organism *Desulfovibrio desulfuricans* ATCC 27774. *Anaerobe*, **1**, 55–60.

Butler, C. S., Charnock, J. M., Bennett, B. *et al.* (1999). Models for molybdenum coordination during the catalytic cycle of periplasmic nitrate reductase from *Paracoccus denitrificans* derived from EPR and EXAFS spectroscopy. *Biochemistry*, **38**, 9000–12.

Butler, C. S., Fairhurst, S. A., Ferguson, S. J. *et al.* (2002). Mo(V) co-ordination in the periplasmic nitrate reductase from *Paracoccus pantotrophus* probed by electron nuclear double resonance (ENDOR) spectroscopy. *Biochemical Journal*, **363**, 817–23.

Costa, C., Macedo, A., Moura, I. *et al.* (1990a). Regulation of the hexahaem nitrite/nitric oxide reductase of *Desulfovibrio desulfuricans, Wolinella succinogenes* and *Escherichia coli*. A mass spectrometry study. *FEBS Letters*, **276**, 67–70.

Costa, C., Moura, J. J. G., Moura, I. *et al.* (1990b). Hexahaem nitrite reductase from *Desulfovibrio desulfuricans*. Mössbauer and EPR characterization of the haem groups. *Journal of Biological Chemistry*, **254**, 14382–7.

Costa, C., Moura, J. J. G., Moura, I. *et al.* (1996). Redox properties of cytochrome *c* nitrite reductase from *Desulfovibrio desulfuricans* ATCC 27774. *Journal of Biological Chemistry*, **271**, 23191–6.

Cunha, C. A., Macieira, S., Dias, J. M. *et al.* (2003). Cytochrome *c* nitrite reductase from *Desulfovibrio desulfuricans* ATCC 27774. The relevance of the two calcium sites in the structure of the catalytic subunit (NrfA). *Journal of Biological Chemistry*, **278**, 17455–65.

Cypionka, H. (1995). Solute transport and cell energetics. In L. L. Barton (ed.), *Biotechnology handbooks, volume 8, Sulphate-reducing bacteria*, New York: Plenum Press. pp. 151–84.

Dalsgaard, T. and Bak, F. (1994). Nitrate reduction in a sulphate-reducing bacterium, *Desulfovibrio desulfuricans*, isolated from rice paddy soil: sulfide inhibition, kinetics, and regulation. *Applied and Environmental Microbiology*, **60**, 291–7.

Dannenberg, S., Kroder, M., Dilling, W. and Cypionka, H. (1992). Oxidation of H_2, organic compounds and inorganic sulfur compounds coupled to reduction of O_2 or nitrate by sulphate-reducing bacteria. *Archives of Microbiology*, **158**, 93–9.

Darwin, A., Hussain, H., Griffiths, L. *et al.* (1993). Regulation and sequence of the structural gene for cytochrome c_{552} from *Escherichia coli*: not a hexahaem but a 50 kDa tetrahaem nitrite reductase. *Molecular Microbiology*, **9**, 1255–65.

Dias, J. M., Than, M. E., Humm, A. *et al.* (1999). Crystal structure of the first dissimilatory nitrate reductase at 1.9 A solved by MAD methods. *Structure*, **7**, 65–79.

Einsle, O., Messerschmidt, A., Huber, R., Kroneck, P. M. H. and Neese, F. (2002a). Mechanism of the six-electron reduction of nitrite to ammonia by cytochrome *c* nitrite reductase. *Journal of American Chemical Society*, **124**, 11737–45.

Einsle, O., Messerschmidt, A., Stach, P. *et al.* (1999). Structure of cytochrome *c* nitrite reductase. *Nature*, **400**, 476–80.

Einsle, O., Stach, P., Messerschmidt, A. *et al.* (2002b). Crystallization and preliminary X-ray analysis of the membrane-bound cytochrome *c* nitrite reductase complex (NrfHA) from *Wolinella succinogenes*. *Acta Crystallographica Section D*, **58**, 341–2.

Einsle, O., Stach, P., Messerschmidt, A. *et al.* (2000). Cytochrome *c* nitrite reductase from *Wolinella succinogenes*. Structure at 1.6 A resolution, inhibitor binding, and haem-packing motifs. *Journal of Biological Chemistry*, **275**, 39608–16.

Fauque, G. D. (1995). Ecology of sulphate-reducing bacteria. In L. L. Barton (ed.), *Biotechnology handbooks, volume 8, Sulphate-reducing bacteria*, New York: Plenum Press. pp. 217–41.

Fauque, G., LeGall, J. and Barton, L. L. (1991). Sulphate-reducing and sulfur-reducing bacteria. In J. M. Shively and L. L. Barton (eds.), *Variations in autotrophic life*. London: Academic Press Limited. pp. 271–337.

Fauque, G. and Ollivier, B. (2004). Anaerobes: the sulphate-reducing bacteria as an example of metabolic diversity. In A. T. Bull (ed.), *Microbial diversity and bioprospecting*. Washington, DC: ASM Press. pp. 169–76.

Frangioni, B., Arnoux, P., Sabaty, M. *et al.* (2004). In *Rhodobacter sphaeroides* respiratory nitrate reductase, the kinetics of substrate binding favors intramolecular electron transfer. *Journal of the American Chemical Society*, **126**, 1328–9.

Fuseler, K., Krekeler, D., Sydow, U. and Cypionka, H. (1996). A common pathway of sulfide oxidation by sulphate-reducing bacteria. *FEMS Microbiology Letters*, **144**, 129–34.

González, P. J., Correia, C., Moura, I., Brondino, C. D. and Moura, J. J. G. (2006a). Bacterial nitrate reductases: molecular and biological aspects of nitrate reduction. *Journal of Inorganic Biochemistry*, **100**, 1015–23.

Gonzalez, P. J., Rivas, M. G., Bursakov, S. A. *et al.* (2006b). EPR and redox properties of periplasmic nitrate reductase from *Desulfovibrio desulfuricans* ATCC 27774. *Journal of Biological Inorganic Chemistry*, **11**, 609–16.

Greene E. A., Hubert, C., Nemati, M., Jenneman, G. E. and Voordouw, G. (2003). Nitrite reductase activity of sulphate-reducing bacteria prevents their inhibition by nitrate-reducing, sulphide-oxidizing bacteria. *Environmental Microbiology*, **5**, 607–17.

Haveman, S. A., Greene, E. A., Stilwell, C. P., Voordouw, J. K. and Voordouw, G. (2004). Physiological and gene expression analysis of inhibition of *Desulfovibrio vulgaris* Hildenborough by nitrite. *Journal of Bacteriology*, **186**, 7944–50.

Haveman, S. A., Greene, E. A. and Voordouw, G. (2005). Gene expression analysis of the mechanism of inhibition of *Desulfovibrio vulgaris* Hildenborough by nitrate-reducing, sulfide-oxidizing bacteria. *Environmental Microbiology*, **7**, 1461–5.

Hille, R. (1996). The mononuclear molybdenum enzymes. *Chemical Reviews*, **96**, 2757–816.

Hubert, C., Nemati, M., Jenneman, G. and Voordouw, G. (2005). Corrosion risk associated with microbial souring control using nitrate or nitrite. *Applied Microbiology and Biotechnology*, **68**, 272–82.

Jenneman, G. E., McInerney, M. J. and Knapp, R. M. (1986). Effect of nitrate on biogenic sulfide production. *Applied and Environmental Microbiology*, **51**, 1205–11.

Kajie, D. and Anraku, Y. (1986). Purification of a hexahaem cytochrome c_{552} from *Escherichia coli* K12 and its properties as a nitrite reductase. *European Journal of Biochemistry*, **154**, 457–63.

Keith, S. M. and Herbert, R. A. (1983). Dissimilatory nitrate reduction by a strain of *Desulfovibrio desulfuricans*. *FEMS Microbiology Letters*, **18**, 55–9.

Kennedy, M. L. and Gibney, B. R. (2001). Metalloprotein and redox protein design. *Current Opinions in Structural Biology*, **11**, 485–90.

Krekeler, D. and Cypionka, H. (1995). The preferred electron acceptor of *Desulfovibrio desulfuricans* CSN. *FEMS Microbiology Ecology*, **17**, 271–8.

LeGall, J. and Fauque, G. (1988). Dissimilatory reduction of sulfur compounds. In A. J. B. Zehnder (ed.), *Biology of anaerobic microorganims*. New York: John Wiley and Sons, Inc. pp. 587–639.

Lie, T. J., Clawson, M. L., Godchaux, W. and Leadbetter, E. R. (1999). Sulfdidogenesis from 2-aminoethanesulfonate (taurine) fermentation by a morphologically unusual sulphate-reducing bacterium, *Desulforhopalus singaporensis* sp. nov. *Applied and Environmental Microbiology*, **65**, 3328–34.

Liu, M.-C., Bakel, B. W., Liu, M.-Y. and Dao, T. N. (1988). Purification of *Vibrio fischeri* nitrite reductase and its characterization as a hexahaem *c*-type cytochrome. *Archives of Biochemistry and Biophysics*, **262**, 259–65.

Liu, M.-C., Liu, M.-Y., Payne, W. J., Peck, H. D. Jr. and LeGall, J. (1983). *Wolinella succinogenes* nitrite reductase: purification and properties. *FEMS Microbiology Letters*, **19**, 201–6.

Liu, M.-C., Liu, M.-Y., Payne, W. J. *et al.* (1987). Comparative EPR studies on the nitrite reductases from *Escherichia coli* and *Wolinella succinogenes*. *FEBS Letters*, **218**, 227–30.

Liu, M. C. and Peck, H. D. Jr. (1981). The isolation of a hexahaem cytochrome from *Desulfovibrio desulfuricans* and its identification as a new type of nitrite reductase. *Journal of Biological Chemistry*, **256**, 13159–64.

Lopez-Cortès, A., Fardeau, M.-L., Fauque, G., Joulian, C. and Ollivier, B. (2006). Reclassification of the sulphate-, nitate-reducing bacterium *Desulfovibrio vulgaris* subsp. *oxamicus* as *Desulfovibrio oxamicus* sp. nov.comb. nov. *International Journal of Systematic and Evolutionary Microbiology*, **56**, 1495–9.

Loubinoux, J., Bronowicki, J.-P., Pereira, I. A. C., Mougenel, J.-L. and LeFaou, A. E. (2002). Sulphate-reducing bacteria in human feces and their association with inflammatory bowel diseases. *FEMS Microbiology Ecology*, **40**, 107–12.

Marietou, A., Richardson, D. J., Cole, J. and Mohan, S. (2005). Nitrate reduction by *Desulfovibrio desulfuricans*: a periplasmic nitrate reductase system that lacks NapB, but includes a unique tetrahaem *c*-type cytochrome, NapM. *FEMS Microbiology Letters*, **248**, 217–25.

McCready, R. G. L., Gould, W. D. and Cook, F. D. (1983). Respiratory nitrate reduction by *Desulfovibrio* sp. *Archives of Microbiology*, **135**, 182–5.

Mitchell, G. J., Jones, J. G. and Cole, J. A. (1986). Distribution and regulation of nitrate and nitrite reduction by *Desulfovibrio* and *Desulfotomaculum* species. *Archives of Microbiology*, **144**, 35–40.

Mori, K., Kim, H., Kakegawa, T. and Hanada, S. (2003). A novel lineage of sullate-reducing microorganisms: *Thermodesulfobiaceae* fam. nov., *Thermodesulfobium narugense*, gen. nov., sp. nov., a new thermophilic isolate from a hot spring. *Extremophiles*, **7**, 283–90.

Moura, J. J. G., Brondino, C. D., Trincao, J. and Romao, M. J. (2004). Mo and W *bis*-MGD enzymes: nitrate reductases and formate dehydrogenases. *Journal of Biological Inorganic Chemistry*, **9**, 791–9.

Moura, I., Bursakov, S., Costa, C. and Moura, J. J. G. (1997). Nitrate and nitrite utilization in sulphate-reducing bacteria. *Anaerobe*, **3**, 279–290.

Parekh, M., Drake, H. L. and Daniel, S. L. (1996). Bidirectional transformation of aromatic aldehydes by *Desulfovibrio desulfuricans* under nitrate-dissimilating conditions. *Letters in Applied Microbiology*, **22**, 115–20.

Pereira, I. C., Abreu, I. A., Xavier, A. V. M., LeGall, J. and Teixeira, M. (1996). Nitrite reductase from *Desulfovibrio desulfuricans* (ATCC 27774) – a heterooligomer haem protein with sulfite reductase activity. *Biochemical and Biophysical Research Communications*, **224**, 611–18.

Pereira, I. A. C., LeGall, J., Xavier, A. V. and Teixeira, M. (2000). Characterization of a haem *c* nitrite reductase from a non-ammonifiying microorganism, *Desulfovibrio vulgaris* Hildenborough. *BBA – Protein Structure and Molecular Enzymology*, **1481**, 119–30.

Plugge, C. M., Balk, M. and Stams, A. J. M. (2002). *Desulfotomaculum thermobenzoicum* subsp. thermosyntrophicum subsp. nov., a thermophilic, syntrophic, propionate-oxidizing, spore-forming bacterium. *International Journal of Systematic and Evolutionary Microbiology*, **52**, 391–9.

Potter, L. C., Millington, P., Griffiths, L., Thomas, G. H. and Cole, J. A. (1999). Competition between *Escherichia coli* strains expressing either a periplasmic or a membrane-bound nitrate reductase: does Nap confer a selective advantage during nitrate-limited growth? *Biochemical Journal*, **344**, 77–84.

Rajagopal, B. S. and LeGall, J. (1989). Utilization of cathodic hydrogen by hydrogen-oxidizing bacteria. *Applied Microbiology and Biotechnology*, **31**, 406–12.

Rehr, B. and Klemme, J.-H. (1986). Metabolic role and properties of nitrite reductase of nitrate-ammonifying marine *Vibrio* species. *FEMS Microbiology Letters*, **35**, 325–8.

Reyes, F., Roldan, M. D., Klipp, W., Castillo, F. and Moreno-Vivian, C. (1996). Isolation of periplasmic nitrate reductase genes from *Rhodobacter sphaeroides* DSM 158: structural and functional differences among prokaryotic nitrate reductases. *Molecular Microbiology*, **19**, 1307–18.

Schumacher, W., Hole, U. and Kroneck, P. M. H. (1994). Ammonia-forming cytochrome *c* nitrite reductase from *Sulfurospirillum deleyianum* is a tetrahaem protein: new aspects of the molecular composition and spectroscopic properties. *Biochemical and Biophysical Research Communications*, **205**, 911–16.

Schumacher, W. and Kroneck, P. M. H. (1991). Dissimilatory hexahaem *c* nitrite reductase of '*Spirillum*' strain 5175: purification and properties. *Archives of Microbiology*, **156**, 70–4.

Seitz, H.-J. and Cypionka, H. (1986). Chemolithotrophic growth of *Desulfovibrio desulfuricans* with hydrogen coupled to ammonification of nitrate or nitrite. *Archives of Microbiology*, **146**, 63–7.

Senez, J. C. and Pichinoty, F. (1958). Reduction of nitrite at the expense of molecular hydrogen by *Desulfovibrio desulfuricans* and other bacterial species. *Bulletin de la Société Chimique et Biologique de Paris*, **40**, 2099–17.

Siddiqui, R. A., Warnecke-Eberz, U., Hengsberger, A. *et al.* (1993). Structure and function of a periplasmic nitrate reductase in *Alcaligenes eutrophus* H16. *Journal of Bacteriology*, **175**, 5867–76.

Sonne-Hansen, J. and Ahring, B. K. (1999). *Thermodesulfobacterium hveragerdense* sp. nov., and *Thermodesulfovibrio islandicus* sp. nov., two thermophilic sulphate-reducing bacteria isolated from an Icelandic hot spring. *Systematic and Applied Microbiology*, **22**, 559–64.

Steenkamp, D. J. and Peck, H. D. Jr. (1981). Proton translocation associated with nitrite respiration in *Desulfovibrio desulfuricans*. *Journal of Biological Chemistry*, **256**, 5450–8.

Stolz, J. F. and Basu, P. (2002). Evolution of nitrate reductase: molecular and structural variations on a common function. *ChemBioChem*, **3**, 198–206.

Tezcan, F. A., Winkler, J. R. and Gray, H. B. (1998). Effects of ligation and folding on reduction potentials of haem proteins. *Journal of American Chemical Society*, **120**, 13383–8.

Thomas, G., Potter, L. and Cole, J. A. (1999). The periplasmic nitrate reductase from *Escherichia coli:* a heterodimeric molybdoprotein with a double-arginine signal sequence and an unusual leader peptide cleavage site. *FEMS Microbiology Letters*, **174**, 167–71.

Trinkerl, M., Breunig, A., Schauder, R. and Konig, H. (1990). *Desulfovibrio termitidis* sp. nov., a carbohydrate-degrading sulphate-reducing bacterium from the hindgut of a termite. *Systematic and Applied Microbiology*, **13**, 372–7.

Walker, F. A., Huynh, B. H., Scheidt, W. R. and Osvath, S. R. (1986). Models of the cytochromes *b*. Effect of axial ligand plane orientation on the EPR and Mössbauer spectra of low-spin ferrihaems. *Journal of American Chemical Society*, **108**, 5288–97.

Widdel, F. (1988). Microbiology and ecology of sulphate- and sulfur-reducing bacteria. In A. J. B. Zehnder (ed.), *Biology of anaerobic microorganims.* New York: John Wiley and Sons, Inc. pp. 469–585.

CHAPTER 9

Anaerobic degradation of hydrocarbons with sulphate as electron acceptor

Friedrich Widdel, Florin Musat, Katrin Knittel and Alexander Galushko

9.1 INTRODUCTION

Sulphate-reducing bacteria (SRB), or more generally speaking sulphate-reducing prokaryotes (SRP), are terminal oxidizers in the natural recycling of bio-organic compounds to CO_2 in anoxic environments, in particular in marine sediments. SRP play this geochemically important role because they make use of a globally abundant electron acceptor, sulphate (in seawater up to 28 mM), and possess numerous degradative (oxidative) capacities with respect to electron donors. The study of the degradative potentials of SRP via de novo enrichment (including direct counting) and isolation from natural samples has been of interest over some decades and formed the basis for our knowledge of the phylogenetic diversity of SRP. Common electron donors and carbon sources of SRP are the low-molecular mass products from the primary anaerobic (fermentative) breakdown of polysaccharides, proteins, lipids and other substances of dead biomass. Several of the involved degradative capacities, for instance complete oxidation or the channelling of branched-chain fatty acids or aromatic compounds into the central metabolism, require special enzymatic reactions (for overview see Rabus *et al.*, 2000) which are not encountered in fermentative bacteria. The study of such and other metabolic capacities in SRP has led to the recognition of principles of general importance or heuristic value in our understanding of the biochemistry and energetics of anaerobes.

A chemical class of organic substrates which have become of interest relatively recently in the study of SRP (and other anaerobes) are hydrocarbons, in particular those from crude oil (petroleum). Hydrocarbons are naturally widespread transformation products of biomass in sediments; they constitute 85% or more of crude oil and gas (Tissot and Welte, 1984) and as such represent a major "driving force" of our technological society. Hydrocarbons are also directly formed by living organisms (for references see Widdel and Rabus, 2001), mostly not at high concentrations as in oil, but more widespread in terrestrial and marine environments. Hydrocarbons are, therefore, ubiquitous, naturally, as well as arising from human activities.

Hydrocarbons being constituted exclusively of carbon and hydrogen are devoid of functional groups and largely apolar. As a result, hydrocarbons (excepting the synthetic alkynes) exhibit low chemical reactivity and are poorly soluble or essentially insoluble in water. Geochemical and biological reactions leading to hydrocarbons are defunctionalization reactions that eliminate CO_2, H_2O or other polar residues (Tissot and Welte, 1984) by mechanisms which, with a few exceptions (e.g. biological methanogenesis), are poorly understood. Vice versa, any biological utilization of a hydrocarbon as a growth substrate requires an initial functionalization (activation) reaction, the introduction of a polar group which is essential for further processing in the metabolism. Activation reactions of hydrocarbons are, like their formation reactions, by no means trivial but represent one of the biochemically most intriguing mechanisms, in particular in anaerobes. These mechanisms are not only of basic scientific interest; they may, in the longer run, also inspire biomimetic approaches for the design of novel chemical catalysts for the functionalization of hydrocarbons, a research issue in organic chemistry (Jones, 2000; Shilov and Shul'pin, 1997).

9.2 PRINCIPAL ENERGETIC AND MECHANISTIC ASPECTS OF HYDROCARBON ACTIVATION

Growth by oxidation of hydrocarbons coupled to sulphate reduction may, at a glance, appear almost a paradox from the bioenergetic point of view. The net free energy gain (per mol of a substrate), in particular with hydrocarbons, in SRP is low (Table 9.1) in comparison with that in aerobes, whereas the biochemical activation of chemically unreactive compounds often requires a significant investment of energy to overcome the activation barrier. Known examples are the nitrogenase reaction or the reductive dearomatization of the benzoate ring in denitrifying bacteria (Boll, 2005). Also the activation of hydrocarbons in aerobic bacteria is associated with an enormous "investment" of energy. Their activation reactions, which are long-known and biochemically well-understood mechanisms, all make use of enzymatically activated O_2 as co-substrate, resulting in the introduction of one or two HO-functions via mono- or di-oxygenases, respectively, into the apolar molecules (Groves, 2006). The insertion reactions of oxygen as one of

Table 9.1. *Net free energies of the oxidation of some hydrocarbons with sulphate*[a,b]

		Free energy (kJ mol^{-1})			
		per mol hydrocarbon		per mol sulphate	
Hydrocarbon	Stoichiometric equation of oxidation	$\Delta G^{o\prime}$	$\Delta G'$	$\Delta G^{o\prime}$	$\Delta G'$
Methane (g)	$CH_4 + SO_4^{2-} \rightarrow HCO_3^- + HS^- + H_2O$	−16.6	−31.5	−16.6	−31.5
Ethane (g)[c]	$4C_2H_6 + 7SO_4^{2-} \rightarrow 8HCO_3^- + 7HS^- + H^+ + 4H_2O$	−63.8	−93.1	−36.4	−53.2
Propane (g)	$2C_3H_8 + 5SO_4^{2-} \rightarrow 6HCO_3^- + 5HS^- + H^+ + 2H_2O$	−102.4	−146.1	−41.0	−58.4
n-Hexane (l)	$4C_6H_{14} + 19SO_4^{2-} \rightarrow 24HCO_3^- + 19HS^- + 5H^+ + 4H_2O$	−210.1	−297.1	−44.2	−62.5
Cyclohexane (l)	$2C_6H_{12} + 9SO_4^{2-} \rightarrow 12HCO_3^- + 9HS^- + 3H^+$	−202.6	−289.1	−45.0	−64.2
n-Hexadecane (l)	$4C_{16}H_{34} + 49SO_4^{2-} \rightarrow 64HCO_3^- + 49HS^- + 15H^+ + 4H_2O$	−559.2	−790.3	−45.7	−64.5
Ethene (g)[c]	$2C_2H_4 + 3SO_4^{2-} \rightarrow 4HCO_3^- + 3HS^- + H^+$	−126.7	−155.6	−84.5	−103.7
Benzene (l)	$4C_6H_6 + 15SO_4^{2-} + 12H_2O \rightarrow 24HCO_3^- + 15HS^- + 9H^+$	−186.2	−271.5	−49.7	−72.4
Toluene (l)	$2C_7H_8 + 9SO_4^{2-} + 6H_2O \rightarrow 14HCO_3^- + 9HS^- + 5H^+$	−204.8	−304.6	−45.5	−67.7
m-Xylene (l)	$4C_8H_{10} + 21SO_4^{2-} + 12H_2O \rightarrow 32HCO_3^- + 21HS^- + 11H^+$	−228.0	−342.2	−43.4	−65.2
Ethylbenzene (l)	$4C_8H_{10} + 21SO_4^{2-} + 12H_2O \rightarrow 32HCO_3^- + 21HS^- + 11H^+$	−240.3	−354.5	−45.8	−67.5
Naphthalene (c)	$C_{10}H_8 + 6SO_4^{2-} + 6H_2O \rightarrow 10HCO_3^- + 6HS^- + 4H^+$	−265.8	−407.5	−44.3	−67.9
1-Methylnaphthalene (l)	$4C_{11}H_{10} + 27SO_4^{2-} + 24H_2O \rightarrow 44HCO_3^- + 27HS^- + 17H^+$	−283.5	−439.7	−42.0	−65.1

Table 9.1. (*cont.*)

Hydrocarbon	Stoichiometric equation of oxidation	Free energy (kJ mol^{-1})			
		per mol hydrocarbon		per mol sulphate	
		$\Delta G^{\circ\prime}$	$\Delta G'$	$\Delta G^{\circ\prime}$	$\Delta G'$
2-Methylnaphthalene (c)	$4C_{11}H_{10} + 27SO_4^{2-} + 24H_2O \rightarrow 44HCO_3^- + 27HS^- + 17H^+$	−286.7	−442.9	−42.5	−65.6
Phenanthrene (c)	$4C_{14}H_{10} + 33SO_4^{2-} + 36H_2O \rightarrow 56HCO_3^- + 33HS^- + 23H^+$	−336.2	−534.4	−40.8	−64.8
Graphite (c)[c,d]	$2C + SO_4^{2-} + 2H_2O \rightarrow 2HCO_3^- + HS^- + H^+$	(undefined)	(undefined)	+17.5	−10.5
Hydrogen (g)[e]	$4H_2 + SO_4^{2-} + H^+ \rightarrow HS^- + 4H_2O$	–	–	−151.8	−153.5

[a]General stoichiometry: $4C_cH_h + (2c + \frac{1}{2}h)SO_4^{2-} \rightarrow 4cHCO_3^- + (2c + \frac{1}{2}h)HS^- + (2c - \frac{1}{2}h)H^+ + 2(h - 2c)H_2O$ (may be divided for convenience so as to achieve smallest whole numbers); if a calculated quantity is negative (as H_2O in the case of many aromatic compounds), this is transferred as positive to the opposite side.

[b]Values were calculated for standard conditions at pH = 7 ($\Delta G^{\circ\prime}$), and for exemplified "real" conditions which may be close to those in a marine sediment or enrichment culture ($\Delta G'$). Assumptions in the latter case: hydrocarbons as pure compounds in their standard states (25°C, 1 kPa; g, gaseous; l, liquid; c, crystalline) in seawater with 10 mM SO_4^{2-}, 10 mM HCO_3^-, and 1 mM HS^-, having activity coefficients of 0.1, 0.5, and 0.5, respectively. Calculation of $\Delta G^{\circ\prime}$ is from Gf^{o} values given by d'Ans and Lax (1983), Dean (1992), Thauer *et al.* (1977), and Zengler *et al.* (1999). See also Spormann and Widdel (2000). Activity coefficients are approximated from Stumm and Morgan (1981).

[c]Utilization not documented so far.

[d]The infinitely extended polycyclic aromatic structure.

[e]For comparison.

the strongest oxidants are highly exergonic and irreversible (Table 9.2, lower part). At our present level of knowledge, the energy is not conserved and is thus completely dissipated (wasted) from the net free energy. Nevertheless, such a waste of free energy is "affordable" in the aerobic metabolism since it still leaves enough for conservation, i.e. for growth, even with the smallest hydrocarbon, methane. Even though methane activation by methane monooxygenase has a free energy change as high as $\Delta G^{\circ\prime} = -344\,\text{kJ mol}^{-1}$ (Table 9.2), this is less than half of the net free energy of aerobic methane oxidation ($CH_4 + 2O_2 \rightarrow CO_2 + 2H_2O$) which is $\Delta G^{\circ} = -817\,\text{kJ mol}^{-1}$. In the anaerobic oxidation of hydrocarbons with sulphate as a "low-potential" electron acceptor, the affordable energy dissipation is much more restricted and necessarily decreases with the number of carbon atoms in the hydrocarbon substrate (Table 9.1).

A use by SRP of strongly oxidizing agents such as reactive oxygen species or the oxygen-coordinating high-valent heam-iron (+IV, +V; Groves, 2006) for hydrocarbon activation would be not only problematic with respect to the net energy balance; also the generation of such agents in a strictly anaerobic metabolism is essentially impossible. Our present understanding is that reactive oxygen species such as haem-bound O atoms or $HO^{\cdot}$ radicals can be generated only from O_2 or from incompletely reduced oxygen such as H_2O_2.

Such principal considerations show that the agents for anaerobic hydrocarbon activation in SRP must differ completely from those in aerobic activation. Indeed, all anaerobic mechanisms for hydrocarbon activation detected in SRP (and also in the frequently studied denitrifiers) or hypothesized (Table 9.2) are far less exergonic than the aerobic ones or even occur near the equilibrium. With one exception, these reactions are unprecedented in biochemistry; the exception is methane activation which is a reversal of the well-studied methane formation reaction (Shima and Thauer, 2005), but as such is again of particular mechanistic, energetic and kinetic interest.

9.3 ANAEROBIC OXIDATION OF METHANE

9.3.1 General and habitat-related aspects

Methane is chemically the most stable hydrocarbon and represents the energetically stable state of carbon under reducing conditions in the absence of oxidized forms of oxygen, nitrogen, iron and sulphur; just as CO_2 is the energetically stable form of carbon under oxidizing conditions. Methane is also the most abundant hydrocarbon on earth (in subsurface reservoirs) and presumably also in the cosmos.

Table 9.2. *Free energies of activation reactions of saturated, monounsaturated and aromatic hydrocarbons*[a]

Type of activation Compound	Equation[b]	ΔG° or $\Delta G^{\circ\prime}$ (kJ mol^{-1})	Reference for free energy value[c]
Methyl-coenzyme M reductase reaction (reversal)			
Methane	$\mathbf{CH_4}$ + CoM–S–S–CoB → CoM–S–$\mathbf{CH_3}$ + HS–CoB	+30	Shima and Thauer, 2005
Addition to fumarate			
Methane[d]	$\mathbf{CH_4}$ + $^{-}$OOC–CH=CH–COO^{-} → $^{-}$OOC–CH_2–[$\mathbf{CH_3}$]CH–COO^{-}	−27 to −31	Rabus *et al.*, 2001
n-Alkane	**R–$\mathbf{CH_2CH_3}$** + $^{-}$OOC–CH=CH–COO^{-} → $^{-}$OOC–CH_2–[**(R)($\mathbf{CH_3}$)CH**]CH–COO^{-}	−35 to −39	Rabus *et al.*, 2001
Toluene[e]	$\mathbf{C_6H_5CH_3}$ + $^{-}$OOC–CH=CH–COO^{-} → $^{-}$OOC–CH_2–[$\mathbf{C_6H_5CH_2}$]CH–COO^{-}	−31 to −35	Rabus *et al.*, 2001
Addition of water to isolated double bond			
Ethene[d]	$\mathbf{CH_2{=}CH_2}$ + H_2O → $\mathbf{CH_3{-}CH_2}$OH	−13	This chapter
Butene[d]	$\mathbf{CH_3 - CH_2{-}CH{=}CH_2}$ + H_2O → $\mathbf{CH_3 - CH_2{-}CH_2{-}CH_2}$OH	−7	This chapter
Carboxylation			
Benzene	$\mathbf{C_6H_6}$ + HCO_3^- → $\mathbf{C_6H_5}$COO^{-} + H_2O	−4	This chapter
Benzene[f]	$\mathbf{C_6H_6}$ + *Carrier*-COO^{-} → $\mathbf{C_6H_5}$COO^{-} + *Carrier*-H	−31	This chapter

Methylation			
Benzene[g]	$\mathbf{C_6H_6}$ + *Carrier*-CH_3 → $\mathbf{C_6H_5}CH_3$ + *Carrier*-H	−24 to −72	This chapter
Naphthalene[g]	$\mathbf{C_{10}H_8}$ + *Carrier*-CH_3 → $\mathbf{C_{10}H_7}CH_3$ + *Carrier*-H	−21 to −79	This chapter
Aerobic hydroxylation[h]			
Methane[i]	$\mathbf{CH_4}$ + O_2 + NADH + H^+ → $\mathbf{CH_3}OH$ + H_2O + NAD^+	−344	This chapter
Benzene	$\mathbf{C_6H_6}$ + O_2→ (Intermediates) → *o*–$\mathbf{C_6H_4}(OH)_2$	−335	This chapter

[a] Purely hypothetical reactions have also been included. Values are given for standard conditions (pH = 7, if protons are involved).

[b] Maintenance of the hydrocarbon skeleton is visualized in bold; for convenience, the fate of the removed hydrogen atom has been ignored.

[c] Calculations are based on Gf° -values (d'Ans and Lax, 1983; Dean, 1992; Mavrovounoitis, 1991; Thauer *et al.*, 1977). The value for the benzoate anion in its standard state (aq), −229.3 kJ mol^{-1}, was calculated from that of the crystalline acid (−245.6 kJ mol^{-1}) via its saturation solubility (3.4 g l^{-1}) and K_a-value (6.5×10^{-5}) at 25°C.

[d] Reaction is purely hypothetical.

[e] The analogous activations of xylenes, ethylbenzene and 2-methylnaphthalene are not included; their reactions are expected to have very similar ΔG°-values.

[f] Since energy values of a common carboxyl donor, carboxy-biotin, are not available, the calculation was formally done with oxaloacetate (yielding pyruvate) as an energetically presumably equivalent carboxyl donor.

[g] It is assumed that the CH_3-group in the carrier is bound to a heteroatom. To explore different possibilities, nitrogen, oxygen, or sulphur have been assumed as heteroatoms. Since the actual carrier is unknown and published energy values for methyl carriers are scarce, methylammonium, methanol and methionine were formally used as energetically nearly equivalent carriers yielding ammonium, water or homocysteine, respectively. For methionine, a Gf°-value of −314.8 kJ mol^{-1} was used (Mavrovounoitis, 1991); there are other incorrect values in the literature.

[h] For comparison.

[i] The reaction can be regarded as the sum of the two formal (mechanistically not occurring) reactions $CH_4 + \frac{1}{2}O_2 \rightarrow CH_3OH$ ($\Delta G^\circ = -124.6$ kJ mol) and $NADH + H^+ + \frac{1}{2}O_2 \rightarrow H_2O + NAD^+$ ($\Delta G^\circ = -219.6$ kJ mol^{-1}), such that also their free energies can be added; calculation is based on Gf°- and $E^{\circ\prime}$-data, respectively (Thauer *et al.*, 1977).

Methane is of abiotic (geochemic, thermogenic) and microbial origin. Thermogenic methane either results from chemical transformation reactions (catagenesis, metagenesis; Tissot and Welte, 1984) of buried organic carbon which also yield higher hydrocarbons, or from the interaction of carbon dioxide with water and iron(II) as reductant (yielding, for instance, magnetite, Fe_3O_4) at several hundred degrees Centigrade (Holm and Charlou, 2001). Microbial methane formation is the terminal mineralization process in the absence of electron acceptors other than CO_2, and involves a unique C_1-biochemistry that is one of the best-studied biochemical processes in prokaryotes (Thauer, 1998; Wolfe, 1991). Microbial origin of methane is usually evident from low $^{13}C/^{12}C$ ratios and the absence of ethane, propane and butane.

Large reservoirs of methane are contained in the sediments of the continental shelves where it most likely results from the long-term microbial degradation of the buried organic fraction that has not been mineralized with oxygen and sulphate. High methane concentrations may lead to hydrates. Their mass in marine sediments may exceed that of conventional fossil fuels in various reservoirs by at least a factor of two (Kvenvolden, 1999).

Despite permanent thermogenic or microbial production and upward migration in marine sediments, most of the methane disappears before any contact with O_2 from ocean water is possible. Evidence for such an anaerobic oxidation of methane (AOM) was furnished by geochemical studies (Barnes and Goldberg, 1976; Martens and Berner, 1974; Reeburgh, 1976) long before this process was considered as feasible in microbial physiology and biochemistry. The only electron acceptor that can account for the striking disappearance of methane in such sediments is sulphate. More recently, nitrate also was shown to serve as an electron acceptor for methane oxidation which occurred in an enriched freshwater community (Raghoebarsing *et al.*, 2006). In the marine sediments, however, nitrate does not play a role as an electron acceptor for methane oxidation. The zone of AOM with sulphate in depth profiles of sediments is usually evident from a concave-up curvature of the methane concentration coinciding with an increased sulphate reduction rate (Alperin and Reeburgh, 1985; Iversen and Jørgensen, 1985). In zones where biogenic methane with its naturally low $^{13}C/^{12}C$ ratio disappears, isotopically light dissolved inorganic carbon and precipitated carbonates have been detected (Paull *et al.*, 1992; Reeburgh, 1980; Ritger *et al.*, 1987). This finding, as well as the formation of radiolabelled CO_2 upon injection of ^{14}C-methane into samples from anoxic sediments, provided further evidence for AOM (Iversen and Jørgensen, 1985; Reeburgh, 1980). Later, AOM with the postulated stoichiometry (Table 9.1) was demonstrated in strictly anoxic laboratory incubations by following directly the consumption of methane and production of sulphide from sulphate using natural samples from methane seepages (Nauhaus *et al.*, 2002; 2005).

At submarine cold gas seepage areas, AOM may occur at high rates in "sharp" methane/sulphate gradients, even though AOM at the much lower rate in regular (non-seepage) sediments with extended gradients may contribute more to the global recycling of methane carbon. The most striking habitats with intense AOM so far demonstrated are surface sediments above methane hydrates at Hydrate Ridge (Cascadia Margin, Oregon) in the Northeast Pacific (Boetius *et al.*, 2000), mud volcanoes in the North Atlantic (K. Knittel, T. Lösekann, personal communication), and the gas seeps of the Northwestern Black Sea shelf (Michaelis *et al.*, 2002). In these, AOM is the basis of microbial life. If AOM occurs a few millimetres to centimetres from the oxic sediment surface, sulphide may nourish a secondary microbial community of chemolithotrophic sulphide oxidizers which may occur free-living or in obligate symbiosis with bivalves. In the Black Sea, AOM occurs several tens of metres below the oxic water. Since AOM leads to a significant increase of inorganic carbon and alkalinity, calcium ions from seawater (usually near 10 mM) tend to be precipitated according to the net equation

$$Ca^{2+} + CH_4 + SO_4^{2-} \rightarrow CaCO_3 + H_2S + H_2O \quad (9.1)$$

The absence of a re-oxidation of hydrogen sulphide to sulphuric acid in the direct vicinity of the methane-oxidizing communities may explain the massive deposition of carbonate plates and chimney-like structures in the anoxic part of the Black Sea. Acidification by a nearby aerobic complete re-oxidation of hydrogen sulphide would counteract such formation of precipitates.

Another unique yet largely unexplored habitat where AOM may be one of the processes that sustains archaeal and bacterial communities is the deep marine sediment biosphere (see Chapter 11, this volume).

Despite the long-known existence of AOM with sulphate in various anoxic sediments, the microorganisms responsible have not been isolated to date in axenic cultures. Nevertheless, many insights into the affiliation, physiology and biochemistry can be attained by the study of sediments with AOM in situ and in vitro, as described in the following.

9.3.2 Organisms involved and their physiology

Labelling studies with ^{14}C-methane in cultures of methanogenic archaea suggested that methanogenesis is reversible to a minor extent during net formation of methane (Harder, 1997; Moran *et al.* 2005; Zehnder and Brock, 1979). The occurrence of such "mini-reversibility"[1] originally suggested that methanogens themselves may catalyze a slow net oxidation of methane with an appropriate electron acceptor under special circumstances still to be elucidated. Hoehler *et al.* (1994) proposed that AOM is performed by archaea and SRB forming a consortium in which the former release an intermediate that is scavenged by the latter; a catabolic interaction comparable to the syntrophism based on interspecies hydrogen transfer. Further evidence for the existence of methanotrophic archaea was provided by the finding of strongly ^{13}C-depleted, apparently methane-derived lipids with isoprenoid structure as common in archaea, and of 16S rRNA gene sequences related to those of Methanosarcinales (Elvert and Suess, 1999; Hinrichs and Boetius, 2002; Hinrichs *et al.*, 1999; Pancost *et al.*, 2000). Microscopy of whole-cell hybridization assays with 16S rRNA-targeted fluorescent probes revealed consortia of archaeal and bacterial cells belonging to the Methanosarcinales and to the *Desulfosarcina*/*Desulfococcus* branch of the Deltaproteobacteria, respectively (Boetius *et al.*, 2000). Hence, there is little doubt that the observed archaea which are related to methanogens represent the anaerobic methanotrophs, often referred to as "ANME". The phylogenetic affiliation of the presently known ANME groups involved in AOM with sulphate is shown in Figure 9.1. Other phylogenetically related Archaea were detected in enrichment cultures oxidizing methane with nitrate (Raghoebarsing *et al.*, 2006). The associated apparent sulphate reducers often belong to the *Desulfosarcina*/*Desulfococcus* group (Knittel *et al.*, 2003); however, bacteria of the *Desulfobulbus* group have also been observed together with ANME archaea (K. Knittel, T. Lösekann, personal communication). Often, the archaeal and bacterial cells from one type of habitat associate in a particular manner (Knittel *et al.*, 2005). For instance, sphere-like consortia with an inner core of densely packed archaeal ANME-II cells surrounded by the bacteria are characteristic of the gas hydrate area of the Cascadia margin, whereas extended, biofilm-like associations (mats) of ANME-I cells and bacteria in a somewhat slimy matrix have been observed in the Black Sea.

The intermediate channelled from methane oxidation into sulphate reduction, and the location of each process is still a matter of discussion. In vitro "feeding" attempts with the conventional methanogenic substrates, H_2, formate, acetate, or methanol, in the absence of methane suggested that

[1] According to the principle of "mini-reversibility" (in particular in weakly exergonic processes), the microorganisms performing AOM are expected to form methane from carbon dioxide to a minor extent during net oxidation of methane.

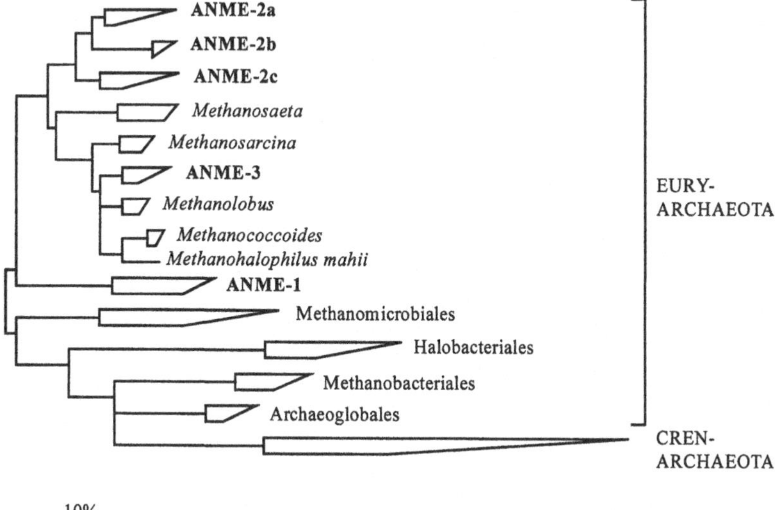

Figure 9.1. Phylogenetic 16S rRNA-based affiliations of archaeal clone sequences associated with an anaerobic oxidation of methane (ANME groups, bold) to selected reference sequences within the Archaea. The tree was calculated by maximum-likelihood analysis in combination with filters excluding highly variable positions on a subset of 82 nearly full-length sequences (> 1350 bp) from nucleotide sequence databases (DDBJ, EMBL, and GenBank). Bar, 10% sequence divergence.

none of these compounds is an intermediate during AOM (Nauhaus *et al.*, 2002). Also, on the basis of calculated diffusion limits and kinetic predictions, a role for these compounds as intermediates has been viewed critically (Boetius *et al.*, 2000; Sørensen *et al.*, 2001; Spormann and Widdel, 2000). Three principal possibilities for such a transfer can be envisaged (Figure 9.2). Firstly, a syntrophic interaction may occur by a transfer of reducing equivalents only via electron or hydrogen shuttles (but not molecular H_2) which are associated with the cell surfaces. Instead of mobile redox-active compounds for electron transfer, also fixed structures such as nanowires may be envisaged: these were recently suggested for electron transfer from organotrophic iron(III)-reducing bacteria to an external abiotic acceptor such as an anode (Reguera *et al.*, 2005). In the case of electron or hydrogen transfer, the archaeal partner would form the methane-derived CO_2. Secondly, a syntrophic interaction may occur by a transfer of a methane-derived organic carbon compound which, according to the experiments, is not acetate or methanol. In this case, the bacterial partner (the sulphate reducer) would form the methane-derived CO_2. Thirdly, methane oxidation

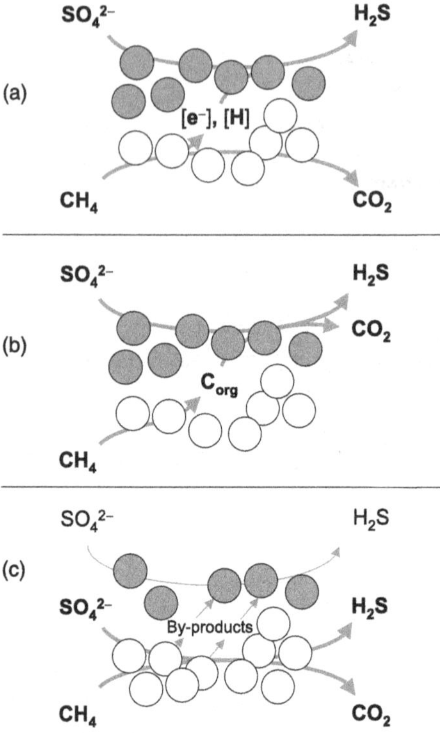

Figure 9.2. Possibilities for the coupling between methane oxidation and sulphate reduction. Archaea, white; Delta-proteobacteria, grey.

as well as sulphate reduction may both take place in the archaeal cells. Growth of the sulphate-reducing partner could be explained by scavenge and utilization of a certain amount of reduced, so far unknown metabolites, i.e. as a kind of metabolic parasitism or commensalism. This model is favoured by the finding that archaeal cells in some sediments with AOM are not closely associated with bacterial cells (Orphan *et al.*, 2002).

Some basic growth parameters in association with AOM have been estimated. A doubling time of some weeks was estimated from the early increase in the numbers of *mcr* genes (encoding the tentative methane-activating enzyme; see next section) during incubation of sediment with initially no detectable methane-oxidizing consortia (Girguis *et al.*, 2005). With sediment from the Hydrate Ridge area, growth of the microscopically countable consortia and a steady increase in the AOM rate was observed during more than 2 years of incubation with methane. The estimated doubling time, t_d, was 7 months. From this and the specific AOM rate,

V (relative to the volume-derived biomass of consortia), a molar growth yield, Y_{mol}, of approximately 0.6 g cell dry mass per mol sulphate reduced was estimated (Nauhaus *et al.*, 2007) using the relation $Y_{mol} = \ln 2/(t_d\ V)$. Hence, AOM is associated with the slowest growth and lowest growth yield established so far for the oxidation of an organic substrate with sulphate. For comparison, another slow organism, a naphthalene-oxidizing sulphate-reducing bacterium, grew with a doubling time of 7–8 days (Galushko *et al.*, 1999). Growth yields of SRB growing with various organic substrates mostly range between 4 and 18 g dry mass per mol sulphate (Rabus *et al.*, 2000; Widdel, 1988), depending on the "energy content" of the electron donor.

9.3.3 Biochemistry

The hypothesis that AOM is, in principle, a reversal of methanogenesis was supported by the analysis of genes and biochemical components in samples from habitats where methane-oxidizing microbial communities are abundant (Hallam *et al.*, 2003; 2004; Krüger *et al.*, 2003). The prerequisite for these analyses and their interpretation was the detailed knowledge of the biochemistry of methanogenesis (Thauer, 1998; Wolfe, 1991).

With AOM being regarded as a reversal of methanogensis, the activation of methane as the most intriguing step is expected to be a reversal of the terminal step in methanogenesis, the methyl-coenzyme M reductase reaction ($CoM{-}S{-}CH_3 + HS{-}CoB \rightarrow CoM{-}S{-}S{-}CoB + CH_4$; $\Delta G° = -30\,[\pm 10]\ kJ\ mol^{-1}$). From habitats with AOM, apparently ANME-affiliated genes were retrieved which were very similar to those encoding the three subunits of methyl-coenzyme M reductase (MCR; composition, $\alpha_2\beta_2\gamma_2$), the terminal enzyme in methanogenesis. The deduced proteins were phylogenetically related to MCR subunits of the Methanosarcinales, but clearly represented novel lines of descent. Furthermore, from a methane seep of the Black Sea, sufficient active methane-oxidizing biomass was obtained to enable protein purification. The dominant protein consisted of three subunits with N-terminal amino acid sequences matching those of the retrieved genes (Krüger *et al.*, 2003). This protein harboured a nickel factor which was apparently a heavier (951 Da) variant of factor F_{430} (905 Da), the unique nickel porphinoid in MCR (two molecules F_{430} per molecule of MCR; Hedderich and Whitman, 2006; Thauer, 1998). Furthermore, the gene-deduced large subunit contained a motif with four cysteines instead of one as in the case of conventional MCR. Hence, special archaea seem to catalyze the initial reaction of AOM by an enzyme that shares an evolutionary origin with

MCR in conventional methanogens, but which may have been "optimized" for reverse methanogenesis.

The mechanistic models of the terminal reaction in methanogenesis and AOM and their comparison have mutually stimulated and advanced their recent development so as to achieve a unifying hypothesis for the forward and back reactions. In methanogenesis, reaction models with methane formation either by protonation of the methylated F_{430} in its Ni(II) state in MCR ($F_{430}[Ni^{2+}]$–CH_3), or by hydrogen addition from a cofactor thiol to a free methyl radical ($^{\bullet}CH_3$) have been favoured during several years (Horng *et al.*, 2001; Pelmenschikov *et al.*, 2002; Thauer, 1998). For AOM, however, a reversal of the first mechanism was recently considered highly unlikely because the Ni(II) state is not a strong enough electrophile to attack methane (Shima and Thauer, 2005). Such an argument favours the second and other mechanisms in AOM and methanogenesis (provided that the electron and hydrogen transfers in methanogenesis are exactly a reversal of those in AOM). Even though there is not yet a final model, an accepted principle in the reverse MCR reaction is the reductive cleavage of the heterodisulphide (CoM–S–S–CoB) by F_{430} in its strongly reducing Ni(I) state, with involvement of organic sulphur and/or disulphide radicals. Also, according to the net reaction, the hydrogen atom or proton from methane has to appear in the HS-group of coenzyme B, even though intermediate acceptors may be involved. In one hypothesized concrete mechanism (Krüger *et al.*, 2003, supplement; Widdel *et al.*, 2004) methane is attacked by the coenzyme B thiyl radical (from heterodisulphide reduction) according to

$$CoB-S^{\bullet} + CH_4 \rightarrow CoB-SH + {}^{\bullet}CH_3 \tag{9.2}$$

The methyl radical then combines with Ni(II)-bound coenzyme M thiolate to yield CoM–S–CH_3 and the Ni (I) state. Most recently, however, such a mechanism has been viewed critically by Shima and Thauer (2005) because the C–H bond in methane is significantly stronger than the S–H bond (Table 9.3). Rather, it has been suggested that reduction of the disulphide by F_{430} leads to the highest oxidation state of nickel, Ni(III), which as a strong electrophile forms a metal–organic bond with the methyl group (as CH_3^- equivalent) according to

$$F_{430}[Ni^{3+}] + CH_4 \rightarrow F_{430}[Ni^{3+}]-CH_3 + H^+(\text{bound}) \tag{9.3}$$

followed by transfer of the methyl group (now a CH_3^+ equivalent) to the coenzyme M thiolate according to

$$F_{430}[Ni^{3+}]-CH_3 + CoM-S^- \rightarrow F_{430}[Ni^+] + CoM-S-CH_3 \tag{9.4}$$

Table 9.3. *Energies for homolytic C–H bond dissociation (or C–H bond energies, with negative sign) of some hydrocarbons*[a]

Formed radical	Energy (kJ mol^{-1})
Phenyl$^\bullet$	473[b], 464[c,d]
2-Naphthyl$^\bullet$	469[e]
$^\bullet CH{=}CH_2$	461[c], 444[d]
$^\bullet CH_3$	440[c], 438[d]
$^\bullet CH_2{-}CH_3$	411[c], 420[d]
$^\bullet CH_2{-}CH_2{-}CH_3$	410[c], 417[d]
$CH_3{-}^\bullet CH{-}CH_3$	398[c], 401[d]
Phenyl–$^\bullet CH_2$	368[c]
$CH_2{=}CH{-}^\bullet CH_2$	361[c,d]
Phenyl–$^\bullet CH{-}CH_3$	358[c]
H–S$^\bullet$	381[c]
CH_3–S$^\bullet$	370[d]
Phenyl–S$^\bullet$	349[d]
H–O$^\bullet$	498[d]
CH_3–O$^\bullet$	437[d]
Phenyl–O$^\bullet$	362[d]

[a] Only the radical after abstraction of H$^\bullet$ is indicated. Energies for **S–H** and **O–H** bond dissociation with the influence of a substituent are indicated for comparison. (For conceptual problems and controversies in bond energy determination see indicated references.)

[b] From Davico *et al.* (1995)

[c] From McGillen and Golden (1982)

[d] From other references compiled by Weast (1989)

[e] From Reed and Kass (2000)

Upon subsequent transfer of the methyl group from methyl-coenzyme M, presumably to tetrahydromethanopterin for further processing, the free coenzyme M and coenzyme B would be oxidized to the heterodisulphide (Hedderich and Whitman, 2006; Thauer *et al.*, 1998) that would enter a new round of methane activation. All reducing equivalents are finally channelled into sulphate reduction.

Methane activation via a reversal of the MCR reaction not only challenges mechanistic but also kinetic views. The negative standard free energy change of the exergonic MCR reaction necessarily reverses its sign for the opposite process, methane activation (CH_4 + CoM—S—S—CoB → CoM—S—CH_3 + H—S—CoB; $\Delta G° = +30\ [\pm 10]\ kJ\ mol^{-1}$). This sets severe limits to the rate of the formation of the initial intermediate, methyl-coenzyme M, as discussed elsewhere (Shima and Thauer, 2005). According to the Haldane equation, which connects the catalytic efficiencies (k_{cat}/K_M) of the forward and back reactions through an enzyme with the thermodynamic equilibrium constant, the first step in AOM may be slower by a factor between 10^{-3} and 10^{-7} than the final step in methanogenesis. Also, the low equilibrium concentrations of methyl-coenzyme M and coenzyme B may drastically limit the rate of subsequent enzymatic steps. Higher rates could be achieved by a coupling of methane activation to an energy yielding step in the further metabolism in such way that, upon substrate binding, the reverse MCR would undergo an externally effected change in favour of the release of the activation product. In the studied pathway of methanogenesis, however, energetic coupling has not been observed with MCR but rather with enzymes of preceding reactions (Hedderich and Whitman, 2006). From a thermodynamic point of view, postulation of an energetically coupled MCR would not be out of place because dissipation of the free energy of the rather exergonic final step in methanogenesis may be critical under low-energy conditions (e.g. with acetate, or H_2 of low pressure).

Also, enzymes potentially involved in the further metabolism of the activation product, methyl-coenzyme M, have been investigated at the level of genes retrieved from a habitat with AOM (Hallam *et al.*, 2004). Again, the underlying hypothesis was that enzymes known from methanogenesis (or structurally closely related ones) are involved. With one exception, all genes encoding potential enzymes for oxidation of the methyl group to CO_2 were detected on a chromosomal contig obviously belonging to ANME archaea. The only missing gene was that encoding an enzyme for interconversion of the methylene and methyl group (in methanogens, *mer*, encoding methyl tetrahydromethanopterin reductase).

9.4 ANAEROBIC OXIDATION OF NON-METHANE HYDROCARBONS

9.4.1 General and habitat-related aspects

Studies of the anaerobic degradation of non-methane hydrocarbons by sulphate-reducing microorganisms were significantly motivated by questions as to what extent such microbial processes can, on the one hand, occur in oil fields, and, on the other hand, play a role in the anaerobic bioremediation of fuel hydrocarbons in oxygen-free groundwater. Since hydrocarbons in oil (crude oil, petroleum) and the oil-derived fuels are either saturated (aliphatic and alicyclic) or aromatic (unsubstituted and alkyl-substituted), most studies have been dealing with these classes of hydrocarbons (9.4.3, 9.4.5), and little is known about sulphate-dependent oxidation of alkenes (9.4.4).

Sulphide formation in oil fields, often referred to as souring, is a frequent yet undesirable phenomenon (see Chapter 10, this volume), because sulphide is corrosive (Chapter 16, this volume), toxic and leads to sulphur dioxide upon burning of the produced fuels. Bastin *et al.* (1926) first attributed the sulphide formation in oil production to bacterial sulphate reduction, but the electron donors for the process remained unknown. In the 1940s and 1950s, the possibility that SRB oxidize hydrocarbons was discussed and investigated (for overview see ZoBell 1946; 1959). The microbial reduction of sulphate with hydrocarbons to sulphide, and its subsequent incomplete oxidation by limited oxygen has been regarded as an origin of sulphur in oil (Ruckmick *et al.*, 1979). Consumption of *n*-alkanes and alkylbenzenes directly from crude oil was demonstrated with both a pure culture and an enrichment culture of SRB (Rueter *et al.*, 1994). Evidence that such processes may have also occurred in petroleum reservoirs comes from the finding of selective hydrocarbon depletion without apparent oxygen penetration (Connan *et al.*, 1996; Head *et al.*, 2003), and the detection of metabolites from the anaerobic degradation of naphthalene or methylnaphthalene (Aitken *et al.*, 2004).

The capacity of SRB for the utilization of various hydrocarbons from crude oil has to be reconciled with the observation that such hydrocarbons are still present in most oils, and have not undergone a "bioremediation" as is desirable in fuel-polluted aquifers (see next paragraph). Explanations are: strong limitations in the diffusion of hydrocarbons trapped in rocks to aqueous phases; geothermal "sterilization" of oil reservoirs; unfavourable salinity and temperature for growth; and sulphate limitation (Widdel *et al.*, 2004). The latter explanation is illustrated by a calculation: for the anaerobic degradation of 10% of the organic carbon (average sum formula $\langle CH_2 \rangle$) in 1 litre crude oil (density 0.85 kg l^{-1}), the total sulphate from 163 litres seawater (with 28 mM SO_4^{2-}) would be needed. On the other hand, the latter explanation would leave the open question as to why various hydrocarbons were not depleted from oil by methanogenic communities, as observed in contaminated anaerobic soil (Townsend *et al.*, 2003).

Sulphate-dependent oxidation of hydrocarbons as a natural or stimulated bioremediation process has been of interest in respect of aromatic hydrocarbons because of their toxicity. In particular, benzene, toluene, ethylbenzene and xylenes (BTEX), as well as naphthalene and methylnaphthalenes, have been studied: these compounds exhibit a certain solubility in water and are therefore of concern as toxic compounds transported in aquifers. Alkanes are far less toxic (and less water-soluble), but they may act as "storage matrix" for aromatic hydrocarbons such that alkane consumption would lead to better accessibility of the aromatic fraction. The study of the anaerobic degradation of hydrocarbons from contaminating fuel in aquifers has significantly contributed to our knowledge of anaerobic metabolism of hydrocarbons. Furthermore, insights into stable isotope fractionation (^{12}C vs. ^{13}C) during anaerobic hydrocarbon degradation have been gained from in situ studies (Meckenstock *et al.*, 2004; Richnow *et al.*, 2003). In situ research of anaerobic hydrocarbon bioremediation has developed into a vast area of research that cannot be reviewed within the scope of this chapter. The reader is referred to other articles (Gieg and Suflita, 2002; Hunkeler *et al.*, 2002; Reusser *et al.*, 2002; Richnow *et al.*, 2003; Townsend *et al.*, 2003).

9.4.2 Organisms involved – overview and general remarks

The anaerobic utilization of non-methane hydrocarbons has been documented in the bacterial domain with microorganisms using nitrate, ferric iron, sulphate or protons (in interspecies hydrogen transfer) as electron acceptor, or growing phototrophically (for an overview see Widdel *et al.*, 2004). Among these, SRB so far exhibit the highest diversity, both in respect of the range of utilizable hydrocarbons (Table 9.4) and phylogenetically (Figure 9.3). Even though some strains, e.g. those utilizing *m*-xylene, ethylbenzene and naphthalene, form striking clusters, hydrocarbon-using SRB on the whole are irregularly distributed over the phylogenetic tree of Deltaproteobacteria, and even occur among Gram-positives. Physiologically, hydrocarbon-degrading SRB have in common the characteristic that they oxidize their organic substrates completely to CO_2, for which they use the carbon monoxide dehydrogenase pathway (see Chapter 1, this volume). With regard to the range of substrates and their oxidation, hydrocarbon-degrading SRB differ significantly from *Desulfovibrio* species, and early reports of alkane utilization by *Desulfovibrio* strains have been viewed critically, or could not be repeated (Aeckersberg *et al.*, 1991; 1998).

The majority of hydrocarbon-degrading SRB have been directly enriched and isolated with hydrocarbons; exceptions are *Desulfobacula phenolica* (formerly *Desulfobacterium phenolicum*; Bak and Widdel, 1986) and

Desulfosarcina cetonica (formerly *Desulfobacterium cetonicum*; Galushko and Rozanova, 1991) which, after their isolation with phenol or butyrate, respectively, were later shown to utilize toluene also (Harms *et al.*, 1999; Rabus *et al.*, 1993). Hydrocarbon-degrading SRB generally grow rather slowly, and if the hydrocarbon is substituted by a polar, more soluble substrate (for instance, a carboxylic acid), the growth rate is usually not significantly enhanced. Linear (i.e. non-exponential) growth, which is not uncommon among SRB (Postgate, 1984), is pronounced in hydrocarbon-degrading SRB, such that minimum doubling times or maximum growth rates are often only approximate values from an early growth interval. For instance, the estimated doubling times of *Desulfobacula toluolica* on toluene (Rabus *et al.*, 1993) and strain NaphS2 on naphthalene (Galushko *et al.*, 1999), were 35 and 180 h, respectively. Such slow growth and the relatively low cell densities render biochemical studies on hydrocarbons difficult. Nevertheless, several intermediates that gave decisive hints on the anaerobic degradation pathways have been first identified in sulphate-reducing microcosms or enrichment cultures (Annweiler *et al.*, 2000, 2002; Beller *et al.*, 1992; Meckenstock *et al.*, 2000; Zhang and Young, 1997), that is even without the isolation of a pure strain. On the other hand, our present knowledge and proposed hypotheses of hydrocarbon pathways in SRB are based on analogies to denitrifying bacteria in which the metabolism of selected hydrocarbons (in particular that of toluene) has been studied in much detail. Hydrocarbon-degrading denitrifiers grow much faster than their sulphate-reducing counterparts such that biochemical studies are more straightforward. However, detectable metabolites indicate that SRB make use of the same pathways as denitrifiers. A striking exception is the different use of ethylbenzene in denitrifiers and SRB (see 9.4.5.2).

9.4.3 Alkanes

Alkane-utilizing SRB apparently exhibit specificities for a certain, rather narrow range of chain lengths in the case of alkanes (Table 9.4). Besides alkanes, also 1-alkenes often serve as growth substrates. Rather commonly utilized substrates are fatty acids, the range of chain lengths being wider than in the case of alkanes (see references given in Table 9.1).

Knowledge of the metabolism of toluene in denitrifiers (Boll *et al.*, 2002; Selmer *et al.*, 2005), which involves a radical-catalyzed addition to fumarate, has been of great value for the understanding of the activation of *n*-alkanes in SRB. The formation of alkyl-substituted succinate in sulphidogenic enriched

Table 9.4. *Cultures of sulphate-reducing bacteria that degrade hydrocarbons*[a]

Class of hydrocarbon/ species or strain[b]	Hydrocarbon utilized	Reference
Alkanes		
Lake	*n*-Alkanes C_6–C_{10}	Davidova and Suflita, 2005
ALDC	*n*-Alkanes C_6–C_{12}	Davidova and Suflita, 2005
TD3	*n*-Alkanes C_6–C_{16}	Rueter *et al.*, 1994
Hxd3	*n*-Alkanes C_{12}–C_{20}	Aeckersberg *et al.*, 1991
AK-01	*n*-Alkanes C_{13}–C_{18}	So and Young, 1999
Desulfatibacillum aliphaticivorans	*n*-Alkanes C_{13}–C_{18}	Cravo-Laureau *et al.*, 2004b
Pnd3	*n*-Alkanes C_{14}–C_{17}	Aeckersberg *et al.*, 1998
Alkenes		
Desulfatibacillum aliphaticivorans	*n*-1-Alkenes C_7–C_{23}	Cravo-Laureau *et al.*, 2004b
Desulfatibacillum alkenivorans	*n*-1-Alkenes C_8–C_{23}	Cravo-Laureau *et al.*, 2004c
Hxd3	*n*-1-Alkenes C_{14}–C_{17}	Aeckersberg *et al.*, 1991; 1998
AK-01	*n*-1-Alkenes C_{15}, C_{16}	So and Young, 1999
Pnd3	*n*-1-Hexadecene	Aeckersberg *et al.*, 1998
Aromatic hydrocarbons		
Desulfobacula toluolica	Toluene	Rabus *et al.*, 1993
Desulfobacula phenolica	Toluene	Rabus *et al.*, 1993
PRTOL1	Toluene	Beller *et al.*, 1996
Desulfosarcina cetonica	Toluene	Harms *et al.*, 1999
TRM1[c]	Toluene	Meckenstock *et al.*, 1999
mXyS1	Toluene, *m*-xylene, *m*-ethyltoluene, *m*-isopropyltoluene	Harms *et al.*, 1999

Table 9.4. (*cont.*)

Class of hydrocarbon/ species or strain[b]	Hydrocarbon utilized	Reference
oXyS1	Toluene, *o*-xylene, *o*-ethyltoluene	Harms *et al.*, 1999
Desulfotomaculum sp. OX39	Toluene, *m*-xylene, *o*-xylene	Morasch *et al.*, 2004
Enrichment culture[c]	*p*-Xylene	Morasch and Meckenstock, 2005
EbS7	Ethylbenzene	Kniemeyer *et al.*, 2003
NaphS2	Naphthalene, tetralin, 2-methylnaphthalene	Galushko *et al.*, 1999; and unpublished data
N47 (highly enriched)[c,d]	Naphthalene, tetralin, 2-methylnaphthalene	Annweiler *et al.*, 2000; 2002; Meckenstock *et al.*, 2000
RS2MN (highly enriched)[c,d]	2-Methylnaphthalene	Galushko *et al.*, 2003

[a] For phylogenetic relationships see Figure 9.3.

[b] If a species and/or genus name has not been proposed yet, only the strain designation is given.

[c] Phylogeny of bacteria of the strains or enrichment cultures has not been studied yet.

[d] One morphological type of cells dominated the enrichment culture.

microcosms (Kropp *et al.*, 2000) and pure cultures (Cravo-Laureau *et al.*, 2005; Davidova and Suflita, 2005) with individual petroleum alkanes indicated that alkane activation in SRP also involves an addition to fumarate. A likely mechanism and route (Davidova *et al.*, 2005) is that suggested for an alkane-utilizing denitrifier where carbon-2 of the *n*-alkane is linked to fumarate yielding a methyl-alkyl-succinate (Rabus *et al.*, 2001; Wilkes *et al.*, 2002; Figure 9.4a). However, an additional route may exist in strain Hxd3. Upon growth with C-odd alkanes, lipid fatty acids were C-even, and upon growth

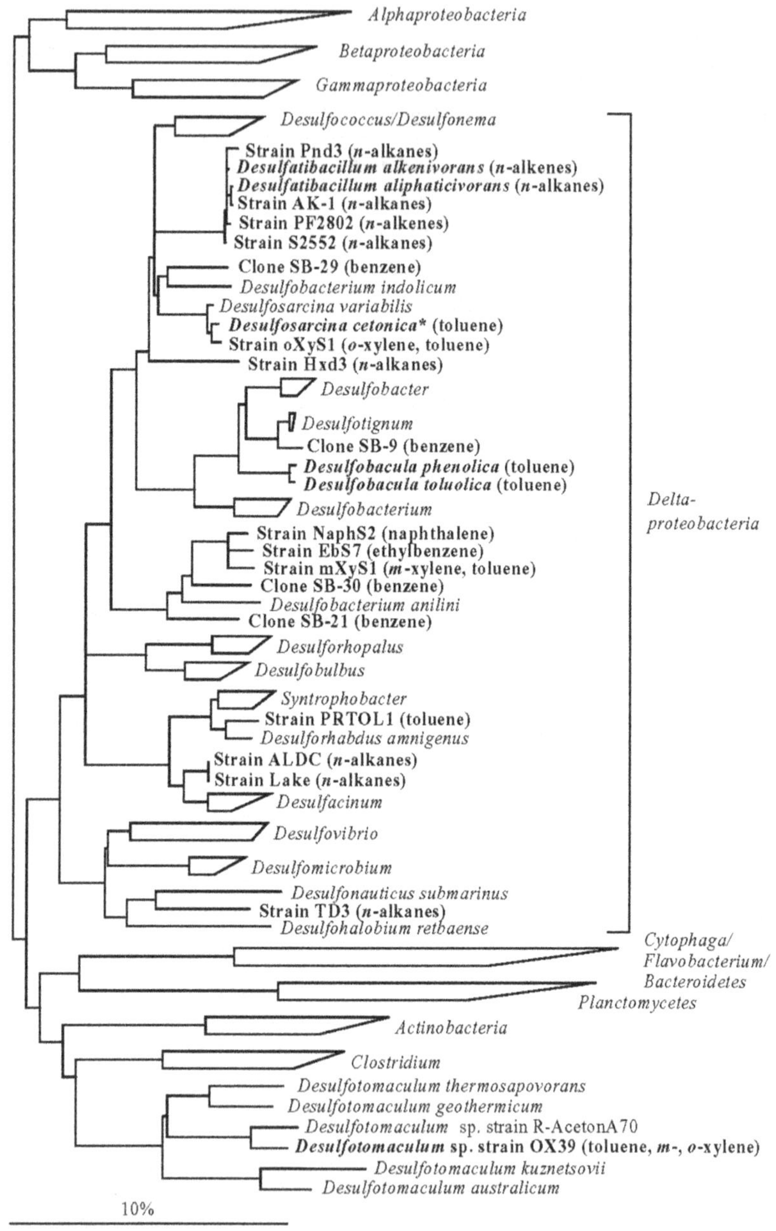

Figure 9.3. Phylogenetic 16S rRNA-based affiliations of SRB with the ability to utilize hydrocarbons (bold) to selected reference sequences within the Bacteria. Clones associated with sulphate-dependent benzene degradation in enrichment cultures have been included also. Species names and designation "Strain" indicate pure cultures; designation "Clone"

with C-even alkanes, lipid fatty acids were C-odd (Aeckersberg *et al.*, 1998). One explanation is that alkanes in this strain are activated at carbon-3 and then metabolized analogous to alkanes activated at carbon-2; after removal of the ethyl-branch by a β-oxidation, the formed *n*-acyl chain would differ from the parental *n*-alkane by three carbon atoms (Wilkes *et al.*, 2002; Figure 9.4b). Its utilization for lipid biosynthesis (with the common elongation by C_2-units) could thus explain the fatty acid pattern observed by Aeckersberg *et al.* (1998). Based on labelling studies and fatty acids analyses, So *et al.* (2003) suggested a different, still unexplored activation mechanism involving a CO_2-derived carboxyl group in strain Hxd3.

9.4.4 Alkenes

Alkenes are common products in plants. Petroleum usually does not contain alkenes because they were converted to saturated or aromatic compounds by geochemical processes (so-called diagenesis and catagenesis; Tissot and Welte, 1984). Utilization of monoterpenes, the naturally most abundant alkenes, has been studied in denitrifiers (Hylemon and Harder, 1998) but not in sulphate reducers. With sulphate as electron acceptor, thus far simple 1-alkenes were tested as model compounds. Some SRB isolated on *n*-alkanes were also able to grow with 1-alkenes. In principle, such sulphate reducers may activate 1-alkenes at the saturated end like an alkane, or employ a specific activation mechanism which makes use of the double bond. In *Desulfatibacillum aliphaticivorans* (isolated as an alkane-degrader), the capacity for 1-alkene degradation was apparently inducible, suggesting specific alkene activation (Cravo-Laureau *et al.*, 2004a). *Desulfatibacillum alkenivorans* was directly enriched on 1-tetradecene and does not use *n*-alkanes, which indicates an activation mechanism that makes use of the double bond. As one possibility for alkene activation in anaerobic bacteria, the addition of water yielding a primary or secondary alcohol has been suggested (Hylemon and Harder, 1998; Schink, 1985). For an activation by

Caption for Figure 9.3 (*cont.*)
indicates sequence retrieved from enriched samples without strain isolation. The tree was calculated by neighbour-joining analysis in combination with filters excluding highly variable positions on a subset of 109 nearly full-length sequences (>1400 bp) from nucleotide sequence databases (DDBJ, EMBL, and GenBank). The subtrees of *Deltaproteobacteria* and *Actinobacteria/Firmicutes* were calculated by maximum-likelihood analysis and are shown with identical tree topologies. Bar, 10% sequence divergence.
**D. cetonica* is the former *Desulfobacterium cetonicum.*

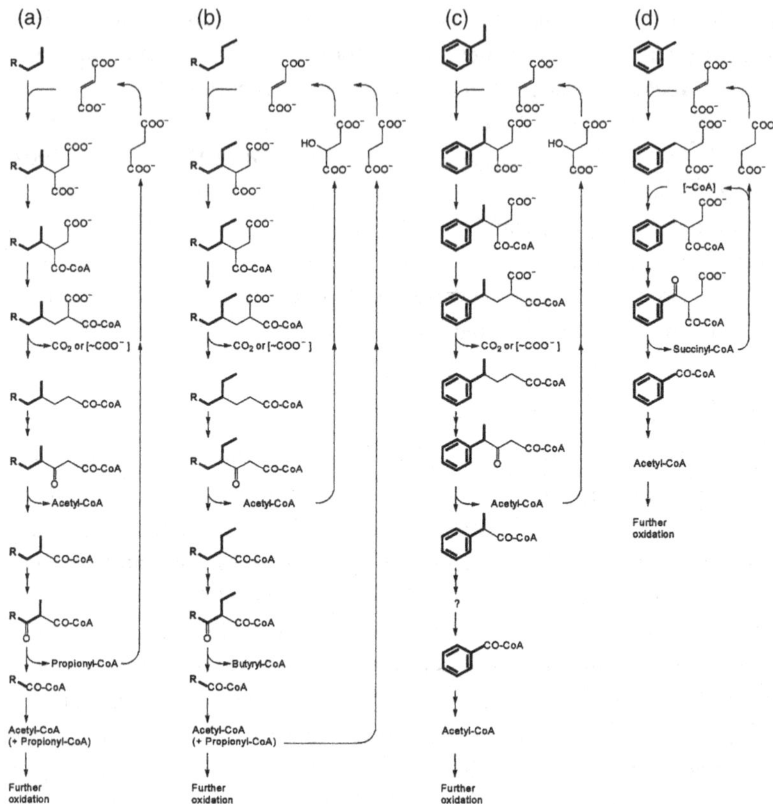

Figure 9.4. Suggested anaerobic pathways of *n*-alkanes (a, b) and ethylbenzene (c) that apparently use analogous steps in sulphate-reducing bacteria. Suggestions are based on metabolite analyses (Elshahed *et al.*, 2001; Kropp *et al.*, 2000; Kniemeyer *et al.*, 2003) and parallels to a denitrifying strain (Rabus *et al.*, 2001; Wilkes *et al.*, 2002). The proven pathway of toluene (d) with different steps after activation (Boll *et al.*, 2002; Selmer *et al.*, 2005), is included for comparison. *n*-Alkane activation at carbon-3 has been observed as a by-reaction in a denitrifyer and has been suggested as the principal activation in sulphate-reducing strain Hxd3 (Rabus *et al.*, 2001); however, another still unknown initial mechanism has been suggested for the same strain (So *et al.*, 2003). The co-substrate, fumarate, can be regenerated from propionyl-CoA (oxidatively, via methylmalonyl-CoA and succinyl-CoA), or, if this is not available, from acetyl-CoA (reductively, via pyruvate, phosphoenolpyruvate, oxaloacetate and malate).

initial H-atom abstraction, the carbon atom adjacent to the double bond would be the most favourable, similar to toluene activation (see below), whereas C–H bond cleavage at an sp^2-carbon atom of the double bond would require an enormous activation energy (Table 9.3).

9.4.5 Aromatic hydrocarbons

9.4.5.1 Benzene, naphthalene, phenanthrene

Anaerobic biodegradation of benzene, the most stable aromatic hydrocarbon, under conditions of sulphate-reduction has been repeatedly demonstrated in incubations with sediments and in enrichment cultures (Edwards and Grbic-Galic, 1992; Lovley *et al.*, 1995; Phelps *et al.*, 1996; for an overview see Coates *et al.*, 2002). Pure cultures of SRP that use benzene have not been isolated to date. Molecular characterization of an enrichment culture by 16S rRNA gene sequencing has not identified unambiguously the bacteria responsible for benzene degradation. The retrieved phylotype most likely to be responsible for benzene degradation (Phelps *et al.*, 1998) affiliates with purified strains of SRB which degrade ethylbenzene, *m*-xylene and naphthalene (Figure 9.3). The enrichment of naphthalene-degrading SRB was already attempted several decades ago (Tauson and Veselov, 1934). More recently, naphthalene oxidation with sulphate as electron acceptor has been demonstrated repeatedly in various sediments and enrichment cultures (Coates *et al.*, 1996; 1997; Hayes *et al.*, 1999; Zhang and Young 1997), in a highly enriched culture that was used for several metabolic studies (Annweiler *et al.*, 2000; 2002; Meckenstock *et al.*, 2000; Safinowski and Meckenstock, 2004; 2006), and in a pure culture of a marine SRB (Galushko *et al.*, 1999). Also, the anaerobic oxidation of other unsubstitued aromatic compounds, phenanthrene (Coates *et al.*, 1996; Hayes *et al.*, 1999; Zhang and Young, 1997), fluorene and fluoranthene (Coates *et al.*, 1997) in sediments has been reported.

Hints as to the activation of the unsubstituted hydrocarbons in SRB result from studies with enriched bacterial communities, and are supported by analogies to reactions in other organisms. Conversion to carboxylates is a frequently suggested mechanism for benzene (Caldwell and Suflita, 2000; Phelps *et al.*, 2001) as well as for naphthalene (Meckenstock *et al.*, 2000; Zhang and Young, 1997), the products being benzoate or 2-naphthoate, respectively. However, labelling experiments with a benzene-degrading enrichment culture showed that the carboxyl group was derived from products of benzene degradation and not from CO_2 which would exclude direct carboxylation (Caldwell and Suflita, 2000). Another attractive activation reaction of unsubstituted aromatic hydrocarbons is methylation. In the case of benzene, hints at methylation to form toluene came from studies in a denitrifier (Coates *et al.*, 2002), and in nitrate-reducing and methanogenic enrichment cultures (Ulrich *et al.*, 2005). Earlier, a methylation of benzene, possibly as a detoxification mechanism, was shown to occur in human bone marrow (Flesher and Myers, 1991). Results pointing at naphthalene methylation to yield 2-methylnaphthalene were obtained directly with a highly enrichend culture (Safinowski and Meckenstock 2006). Methylation may be explained as an electrophilic substitution, in analogy to a Friedel–Crafts type reaction, by a carrier-bound methyl group (CH_3^+ equivalent); possible carriers are methyl-coenzyme B_{12} or *S*-adenosylmethionine. It is true that the latter can, in principle, act upon a π-electron system as in the conversion of unsaturated fatty acids to cyclopropyl fatty acids, which subsequently may be reduced to methyl-branched saturated fatty acids (Grogan and Cronan, 1997). However, methylation of the stabilized aromatic π-electron system would be a biochemically more demanding reaction, despite the clearly exergonic character of the net reaction (Table 9.2). For the further metabolism of formed toluene and 2-methylnaphthalene, the only mechanism presently conceivable is their addition to fumarate (Figures 9.4 and 9.5; see also next section). Since these would lead to benzoyl-CoA or 2-naphthoyl-CoA, the finding of benzoate and 2-naphthoate may be explained by a hydrolysis. A direct addition of benzene or naphthalene to fumarate in a radical reaction is unlikely because of the high energy barrier for the needed C–H-bond cleavage leading to a phenyl or naphthyl radical, respectively (Table 9.3). On the other hand, there are also observations which are not in agreement with 2-methylnaphthalene as an intermediate in anaerobic naphthalene degradation: e.g. marine sulphate-reducing isolates including strain NaphS2 exhibited significant lag-phases when naphthalene-grown cells were exposed to 2-methylnaphthalene (A. Galushko, F. Musat, unpublished results).

9.4.5.2 Toluene, xylenes, ethylbenzene, 2-methylnaphthalene

Alkyl-substituted aromatic hydrocarbons, which combine properties of aromatic and aliphatic compounds, occur in a large variety of isomers and homologues in crude oil. Cultures degrading representatives of such compounds have been isolated repeatedly, and the study of their metabolism is more advanced than that of the unsubstituted aromatic hydrocarbons.

Toluene is by far the most frequently studied hydrocarbon in cultures of anaerobic microorganisms. The frequent usage of toluene by anaerobes may be explained by its wide occurrence (even though at low concentrations) as a metabolite of phenylacetate from the anaerobic degradation of phenylalanine (Fischer-Romero *et al.*, 1996; Mrowiec *et al.*, 2005; Selmer and Andrei, 2001).

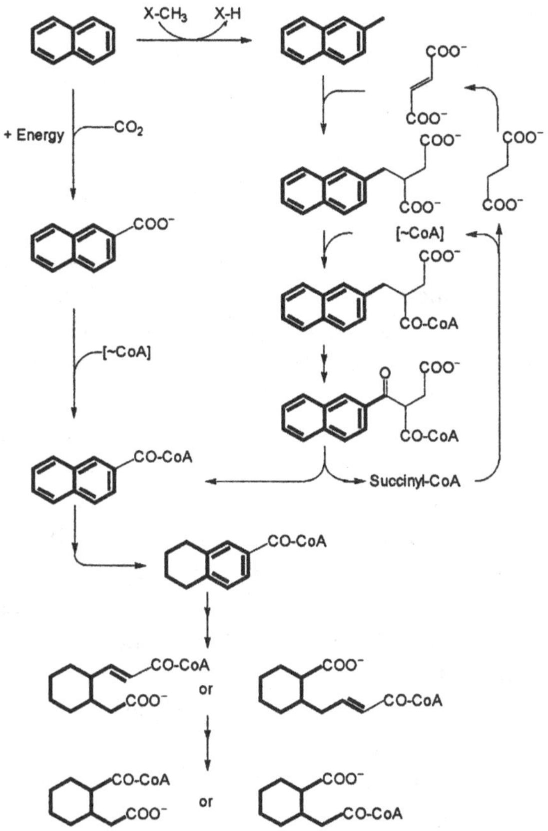

Figure 9.5. Suggested possibilities for the anaerobic activation of naphthalene via carboxylation or methylation, and the further metabolism.

The pathway of anaerobic toluene degradation from toluene activation yielding benzylsuccinate to the formation of benzoyl-CoA (Figure 9.4d) has been studied in depth in denitrifying bacteria (Boll *et al.*, 2002; Selmer *et al.*, 2005). So far, all metabolite analyses indicate that essentially the same pathway is operative in SRB (Beller and Spormann, 1997; Rabus and Heider, 1998); the initial activation product, benzylsuccinate, was even originally detected in a sulphate-reducing enrichment (Beller *et al.*, 1992). Also the further metabolism of benzoyl-CoA in SRB is likely to involve essentially a dearomatization and ring cleavage as in denitrifiers, but with modifications which allow dearomatization with a much lower energy demand than in dentrifiers (Peters *et al.*, 2004; Wischgoll *et al.*, 2005). Xylenes are expected to follow essentially the pathway of toluene. In the case of *o*-xylene and *p*-xylene, the methyl branch is expected to require additional mechanisms to circumvent blockage of the β-oxidation steps following dearomatization and ring cleavage.

The activation of ethylbenzene in SRB is apparently analogous to that of toluene, i.e. by the addition of the benzyl carbon atom (C-atom next to the benzene ring) to fumarate (Elshahed *et al.*, 2001; Kniemeyer *et al.*, 2003; Figure 9.4c). This mechanism is thus completely different from that in denitrifiers that dehydrogenate ethylbenzene anaerobically to yield *S*-1-phenylethanol (Johnson and Spormann, 1999; Kniemeyer and Heider, 2001). A plausible explanation is that the dehydrogenation has a relatively positive redox potential (phenylethanol/ethylbenzene, $E^{\circ\prime} = -0.01$ to $+0.03$ V); for effective conversion, the redox potential of the in vivo electron acceptor may have to be even more positive (estimate: $+0.3$ V), a value that is hardly achieved within the metabolism of SRB. In contrast, the assumed enzyme radical needed for ethylbenzene addition is formed reductively and thus is well compatible with a strongly reducing cell environment. Reactions for the further metabolism have been suggested (Kniemeyer *et al.*, 2003).

2-Methylnaphthalene degradation with sulphate has been documented in freshwater (Annweiler *et al.*, 2000; Galushko *et al.*, 2003) and marine (Sullivan *et al.*, 2001) enrichment cultures. Activation of methylnaphthalene by fumarate addition to the methyl group has been already suggested somewhat earlier (Schmitt *et al.*, 1998). The initial reaction sequence of 2-methylnaphthalene degradation by freshwater SRB follow essentially those of toluene. Metabolite analyses suggest that formed naphthoyl-CoA first undergoes reduction of the unsubstituted ring, as depicted in Figure 9.5. With the marine enrichment culture, hints at carboxylation were obtained (Sullivan *et al.*, 2001).

9.5 CONCLUDING REMARKS: THE EVOLUTIONARY PERSPECTIVE

Hydrocarbons shown to be degraded anaerobically by SRP are chemically rather diverse, as are the phylogenetic lineages where these capacities occur (Figures 9.1, 9.3). There may be further, hitherto unrecognized capacities and species with respect to the anaerobic degradation of hydrocarbons with sulphate. Such apparent diversity may be viewed as the result of a long period of metabolic evolution during which SRP were exposed to hydrocarbons in their environments. Dissimilatory sulphate reduction represents an "old" metabolic trait in prokaryotes. It may have been present already $\sim 3.5 \times 10^9$ years ago in the Achaean, presumably before the onset of cyanobacterial oxygen production and when oceanic levels of sulphate (e.g. from anoxygenic photosynthesis) were below 1 mM and may have reached higher concentrations only in rather isolated pools (Habicht *et al.*, 2002; Shen *et al.*, 2001). Hydrocarbons from geochemical and biological processes presumably also represent "old" substrates in the biosphere. In the case of non-methane hydrocarbons, one may, therefore, assume an origin of the hydrocarbon-activating enzymes, which are the most crucial ones for anaerobic hydrocarbon metabolism, in primordial Deltaproteobacteria. Such an assumption is in line with the finding of methanogenesis from non-methane hydrocarbons (for references see Widdel and Rabus, 2001; Widdel *et al.*, 2004) which depends on proton-reducing (hydrogen-producing, syntrophic) bacteria. These bacteria are even independent of sulphate, and many of them, including those detected in methanogenesis from an alkane (Zengler *et al.*, 1999), belong to the Deltaproteobacteria (Schink and Stams, 2002), just as SRB do. From primordial Deltaproteobacteria, the capacities for the anaerobic degradation of hydrocarbons may have spread to denitrifiers. A point that is presently beyond our perception of evolutionary events is the steps that have led from simpler (non-hydrocarbon-activating) precursor enzymes to the hydrocarbon-activating enzymes as mechanistically delicate biocatalysts. Only in the case of methane is the origin of the capacity for its anaerobic oxidation almost at hand. At present, we cannot perceive another origin than from the methanogenic pathway (see section 9.3.3), the latter as a sulphate-independent process being presumably the more ancient one.

REFERENCES

Aeckersberg, F., Bak, F. and Widdel, F. (1991). Anaerobic oxidation of saturated hydrocarbons to CO_2 by a new type of sulphate-reducing bacteria. *Arch Microbiol*, **156**, 5–14.

Aeckersberg, F., Rainey, F. A. and Widdel, F. (1998). Growth, natural relationships, cellular fatty acids and metabolic adaptation of sulphate-reducing bacteria that utilize long-chain alkanes under anoxic conditions. *Arch Microbiol*, **170**, 361–9.

Aitken, C. M., Jones, D. M. and Larter, S. R. (2004). Anaerobic hydrocarbon biodegradation in deep subsurface oil reservoirs. *Nature*, **431**, 291–4.

Alperin, M. J. and Reeburgh, W. S. (1985). Inhibition experiments on anaerobic methane oxidation. *Appl Environ Microbiol*, **50**, 940–5.

Annweiler, E., Materna, A., Safinowski, M. *et al.* (2000). Anaerobic degradation of 2-methylnaphthalene by a sulphate-reducing enrichment culture. *Appl Environ Microbiol*, **66**, 5329–33.

Annweiler, E., Michaelis, W. and Meckenstock, R. U. (2002). Identical ring cleavage products during anaerobic degradation of naphthalene, 2-methylnaphthalene, and tetralin indicate a new metabolic pathway. *Appl Environ Microbiol*, **68**, 852–8.

Bak, F. and Widdel, F. (1986). Anaerobic degradation of phenol and phenol derivates by *Desulfobacterium phenolicum* sp. nov. *Arch Microbiol*, **146**, 177–80.

Barnes, R. O. and Goldberg, E. D. (1976). Methane production and consumption in anoxic marine sediments. *Geology*, **4**, 297–300.

Bastin, E. S., Greer, F. E., Merritt, C. A. and Moulton, G. (1926). The presence of sulphate reducing bacteria in oil field waters. *Science*, **63**, 21–4.

Beller, H. R., Reinhard, M. and Grbic'-Galic', D. (1992). Metabolic by-products of anaerobic toluene degradation by sulphate-reducing enrichment cultures. *Appl Environ Microbiol*, **58**, 3192–5.

Beller, H. and Spormann, A. (1997). Benzylsuccinate formation as a means of anaerobic toluene activation by sulphate-reducing strain PRTOL1. *Appl Environ Microbiol*, **63**, 3729–31.

Beller, H., Spormann, A., Sharma, P., Cole, J. and Reinhard, M. (1996). Isolation and characterization of a novel toluene-degrading, sulphate-reducing bacterium. *Appl Environ Microbiol*, **62**, 1188–96.

Boetius, A., Ravenschlag, K., Schubert, C. J. *et al.* (2000). A marine microbial consortium apparently mediating anaerobic oxidation of methane. *Nature*, **407**, 623–6.

Boll, M. (2005). Key enzymes in the anaerobic aromatic metabolism catalysing Birch-like reductions. *Biochim Biophys Acta (BBA) – Bioenergetics*, **1707**, 34–50.

Boll, M., Fuchs, G. and Heider, J. (2002). Anaerobic oxidation of aromatic compounds and hydrocarbons. *Curr Opin Chem Biol*, **6**, 604–11.

Caldwell, M. E. and Suflita, J. M. (2000). Detection of phenol and benzoate as intermediates of anaerobic benzene biodegradation under different terminal electron-accepting conditions. *Environ Sci Technol*, **34**, 1216–20.

Coates, J., Anderson, R. and Lovley, D. (1996). Oxidation of polycyclic aromatic hydrocarbons under sulphate-reducing conditions. *Appl Environ Microbiol*, **62**, 1099–101.

Coates, J. D., Chakraborty, R. and McInerney, M. J. (2002). Anaerobic benzene degradation – a new era. *Res Microbiol*, **153**, 621–8.

Coates, J., Woodward, J., Allen, J., Philp, P. and Lovley, D. (1997). Anaerobic degradation of polycyclic aromatic hydrocarbons and alkanes in petroleum-contaminated marine harbor sediments. *Appl Environ Microbiol*, **63**, 3589–93.

Connan, J., Lacrampe-Coulome, G. and Magot, M. (1996). Origin of gases in reservoirs. In D. Dolenc (ed.), *Proceedings of the 1995 International Gas Research Conference* Vol. 1. Rockville: Government Institutes. pp. 21–62.

Cravo-Laureau, C., Grossi, V., Raphel, D., Matheron, R. and Hirschler-Rea, A. (2005). Anaerobic *n*-alkane metabolism by a sulphate-reducing bacterium, *Desulfatibacillum aliphaticivorans* strain CV2803^T. *Appl Environ Microbiol*, **71**, 3458–67.

Cravo-Laureau, C., Hirschler-Rea, A., Matheron, R. and Grossi, V. (2004a). Growth and cellular fatty-acid composition of a sulphate-reducing bacterium, *Desulfatibacillum aliphaticivorans* strain CV2803^T, grown on *n*-alkenes. *Comptes Rendus Biologies*, **327**, 687–94.

Cravo-Laureau, C., Matheron, R., Cayol, J.-L., Joulian, C. and Hirschler-Rea, A. (2004b). *Desulfatibacillum aliphaticivorans* gen. nov., sp. nov., an *n*-alkane- and *n*-alkene-degrading, sulphate-reducing bacterium. *Int J Syst Evol Microbiol*, **54**, 77–83.

Cravo-Laureau, C., Matheron, R., Joulian, C., Cayol, J.-L. and Hirschler-Rea, A. (2004c). *Desulfatibacillum alkenivorans* sp. nov., a novel *n*-alkene-degrading, sulphate-reducing bacterium, and emended description of the genus *Desulfatibacillum*. *Int J Syst Evol Microbiol*, **54**, 1639–42.

d'Ans, J. and Lax, E. (1983). *Taschenbuch für Chemiker und Physiker*, Bd. 2 (Ed, Synowietz, C.). Berlin: Springer.

Davico, G. E. B., Veronica, M., DePuy, C. H., Ellison, G. B. and Squires, R. R. (1995). The C–H bond energy of benzene. *J Am Chem Soc*, **117**, 2590–9.

Davidova, I. A., Gieg, L. M., Nanny, M. *et al.*, (2005). Stable isotope studies of *n*-alkane metabolism by a sulphate-reducing bacteria enrichment culture. *Appl Environ Microbiol*, **71**, 8174–82.

Davidova, I. A. and Suflita, J. M. (2005). In J. R. Leadbetter (ed.), *Methods in enzymology*, Vol. 397. Amsterdam and London: Elsevier Academic Press. pp. 17–34.

Dean, J. A. (1992). *Lange's handbook of chemistry*. New York: McGraw-Hill.

Edwards, E. A. and Grbić-Galić, D. (1992). Complete mineralization of benzene by aquifer microorganisms under strictly anaerobic conditions. *Appl Environ Microbiol*, **58**, 2663–6.

Elshahed, M. S., Gieg, L. M., McInerney, M. J. and Suflita, J. M. (2001). Signature metabolites attesting to the in situ attenuation of alkylbenzenes in anaerobic environments. *Environ Sci Technol*, **35**, 682–9.

Elvert, M., Suess, E. and Whiticar, M. J. (1999). Anaerobic methane oxidation associated with marine gas hydrates: superlight C-isotopes from saturated and unsaturated C_{20} and C_{25} irregular isoprenoids. *Naturwissenschaften*, **86**, 295–300.

Fischer-Romero, C., Tindall, B. and Jüttner, F. (1996). *Tolumonas auensis* gen. nov., sp. nov., a toluene-producing bacterium from anoxic sediments of a freshwater lake. *Int J Syst Bacteriol*, **46**, 183–8.

Flesher, J. W. and Myers, S. R. (1991). Methyl-substitution of benzene and toluene in preparations of human bone marrow. *Life Sci*, **48**, 843–50.

Galushko, A. S., Kiesele-Lang, U. and Kappler, A. (2003). Degradation of 2-methylnaphthalene by a sulphate-reducing enrichment culture of mesophilic freshwater bacteria. *Polyc Arom Comp*, **23**, 207–18.

Galushko, A., Minz, D., Schink, B. and Widdel, F. (1999). Anaerobic degradation of naphthalene by a pure culture of a novel type of marine sulphate-reducing bacterium. *Environ Microbiol*, **1**, 415–20.

Galushko, A. S. and Rozanova, E. P. (1991). *Desulfobacterium cetonicum* sp. nov. – a sulphate–reducing bacterium which oxidizes fatty acids and ketones. *Microbiol* (Engl.Transl. Mikrobiologiya (USSR)), **60**, 742–6.

Gieg, L. M. and Suflita, J. M. (2002). Detection of anaerobic metabolites of saturated and aromatic hydrocarbons in petroleum-contaminated aquifers. *Environ Sci Technol*, **36**, 3755–62.

Grogan, D. W. and Cronan, J. E., Jr. (1997). Cyclopropane ring formation in membrane lipds of bacteria. *Microbiol Mol Biol Rev*, **61**, 429–41.

Girguis, P. R., Cozen, A. E. and DeLong, E. F. (2005). Growth and population dynamics of anaerobic methane-oxidizing archaea and sulphate-reducing bacteria in a continuous-flow bioreactor. *Appl Environ Microbiol*, **71**, 3725–33.

Groves, J. T. (2006). High-valent iron in chemical and biological oxidations. *J Inorg Biochem*, **100**, 434–47.

Habicht, K. S., Gade, M., Thamdrup, B., Berg, P. and Canfield, D. E. (2002). Calibration of sulphate levels in the Archaean ocean. *Science*, **298**, 2372–4.

Hallam, S. J., Girguis, P. R., Preston, C. M., Richardson, P. M. and DeLong, E. F. (2003). Identification of methyl coenzyme M reductase A (*mcrA*) genes associated with methane-oxidizing Archaea. *Appl Environ Microbiol*, **69**, 5483–91.

Hallam, S. J., Putnam, N., Preston, C. M. *et al.* (2004). Reverse methanogenesis: testing the hypothesis with environmental genomics. *Science*, **305**, 1457–62.

Harder, J. (1997). Anaerobic methane oxidation by bacteria employing ^{14}C-methane uncontaminated with ^{14}C-carbon monoxide. *Marine Geol*, **137**, 13–23.

Harms, G., Zengler, K., Rabus, R. *et al.* (1999). Anaerobic oxidation of *o*-xylene, *m*-xylene, and homologous alkylbenzenes by new types of sulphate-reducing bacteria. *Appl Environ Microbiol*, **65**, 999–1004.

Hayes, L. A., Nevin, K. P. and Lovley, D. R. (1999). Role of prior exposure on anaerobic degradation of naphthalene and phenanthrene in marine harbor sediments. *Org Geochem*, **30**, 937–45.

Head, I. M., Jones, D. M. and Larter, S. R. (2003). Biological activity in the deep subsurface and the origin of heavy oil. *Nature*, **426**, 344–52.

Hedderich, R. and Whitman, W. B. (2006). Physiology and biochemistry of the methane-producing archaea. In M. Dworkin, E. Rosenberg, K.-H. Schleifer and E. Stackebrandt (eds.), *The prokaryotes*, electronic edition. New York: Springer. (URL: http://141.150.157.117:8080/prokPUB/index.htm).

Hinrichs, K. U. and Boetius, A. B. (2002). In G. Wefer, D. Hebbeln, B. B. Jørgensen, M. Schlüter and T. van Weering (eds.), *Ocean margin systems*, Heidelberg: Springer-Verlag. pp. 457–77.

Hinrichs, K. U., Hayes, J. M., Sylva, S. P., Brewer, P. G. and DeLong, E. F. (1999). Methane-consuming archaebacteria in marine sediments. *Nature*, **398**, 802–5.

Hoehler, T. M., Alperin, M. J., Albert, D. B. and Martens, C. S. (1994). Field and laboratory studies of methane oxidation in an anoxic marine sediment: evidence for a methanogen-sulphate reducer consortium. *Global Biogeochem Cycles*, **8**, 451–63.

Holm, N. G. and Charlou, J. L. (2001). Initial indications of abiotic formation of hydrocarbons in the Rainbow ultramafic hydrothermal system, Mid-Atlantic Ridge. *Earth Planet Sci Lett*, **191**, 1–8.

Horng, Y.-C., Becker, D. F. and Ragsdale, S. W. (2001). Mechanistic studies of methane biogenesis by methyl-coenzyme M reductase: evidence that coenzyme B participates in cleaving the C–S bond of methyl-coenzyme M. *Biochemistry*, **40**, 12875–85.

Hunkeler, D., Höhener, P. and Zeyer, J. (2002). Engineered and subsequent intrinsic in situ bioremediation of a diesel fuel contaminated aquifer. *J Contam Hydrol*, **59**, 231–45.

Hylemon, P. B. and Harder, J. (1998). Biotransformation of monoterpenes, bile acids, and other isoprenoids in anaerobic ecosystems. *FEMS Microbiol Rev*, **22**, 475–88.

Iversen, N. and Jørgensen, B. B. (1985). Anaerobic methane oxidation rates at the sulphate methane transition in marine-sediments from Kattegat and Skagerrak (Denmark). *Limnol Oceanogr*, **30**, 944–55.

Johnson, H. A. and Spormann, A. M. (1999). In vitro studies on the initial reactions of anaerobic ethylbenzene mineralization. *J Bacteriol*, **181**, 5662–8.

Jones, W. D. (2000). Conquering the carbon–hydrogen bond. *Science*, **287**, 1942–3.

Kniemeyer, O., Fischer, T., Wilkes, H., Glöckner, F. O. and Widdel, F. (2003). Anaerobic degradation of ethylbenzene by a new type of marine sulphate-reducing bacterium. *Appl Environ Microbiol*, **69**, 760–8.

Kniemeyer, O. and Heider, J. (2001). Ethylbenzene dehydrogenase, a novel hydrocarbon-oxidizing molybdenum/iron-sulfur/heme enzyme. *J Biol Chem*, **276**, 21381–6.

Knittel, K., Boetius, A., Lemke, A. *et al.* (2003). Activity, distribution, and diversity of sulphate reducers and other bacteria in sediments above gas hydrates (Cascadia Margin, Oregon). *Geomicrobiol J*, **20**, 269–94.

Knittel, K., Lösekann, T., Boetius, A., Kort, R. and Amann, R. (2005). Diversity and distribution of methanotrophic Archaea at cold seeps. *Appl Environ Microbiol*, **71**, 467–79.

Kropp, K. G., Davidova, I. A. and Suflita, J. M. (2000). Anaerobic oxidation of *n*-dodecane by an addition reaction in a sulphate-reducing bacterial enrichment culture. *Appl Environ Microbiol*, **66**, 5393–8.

Krüger, M., Meyerdierks, A., Glöckner, F. O. *et al.* (2003). A conspicuous nickel protein in microbial mats that oxidize methane anaerobically. *Nature*, **426**, 878–81.

Kvenvolden, K. A. (1999). Potential effects of gas hydrate on human welfare. *PNAS*, **96**, 3420–6.

Lovley, D. R., Coates, J. D., Woodward, J. C. and Phillips, E. J. P. (1995). Benzene oxidation coupled to sulphate reduction. *Appl Environ Microbiol*, **61**, 953–8.

Martens, C. S. and Berner, R. A. (1974). Methane production in the interstitial waters of sulphate-depleted marine sediment. *Science*, **185**, 1167–9.

Mavrovounoitis, M. L. (1991). Estimation of standard Gibbs energy changes of biotransformations. *J Biol Chem*, **266**, 14440–5.

McGillen, D. F. and Golden, D. M. (1982). Hydrocarbon bond dissociation energies. *Ann Rev Phys Chem*, **33**, 493–532.

Meckenstock, R. U., Annweiler, E., Michaelis, W., Richnow, H. H. and Schink, B. (2000). Anaerobic naphthalene degradation by a sulphate-reducing enrichment culture. *Appl Environ Microbiol*, **66**, 2743–7.

Meckenstock, R. U., Morasch, B., Griebler, C. and Richnow, H. H. (2004). Stable isotope analysis as a tool to monitor biodegradation in contaminated aquifers. *J Contam Hydrol*, **75**, 215–55.

Meckenstock, R. U., Morasch, B., Warthmann, R. *et al.* (1999). $^{13}C/^{12}C$ isotope fractionation of aromatic hydrocarbons during microbial degradation. *Environ Microbiol*, **1**, 409–14.

Michaelis, W., Seifert, R., Nauhaus, K. *et al.* (2002). Microbial reefs in the Black Sea fueled by anaerobic oxidation of methane. *Science*, **297**, 1013–15.

Moran, J. J., House, C., Freeman, K. H. and Ferry, J. J. (2005). Trace methane oxidation studied in several Euryarchaeota under diverse conditions. *Archaea*, **1**, 303–9.

Morasch, B. and Meckenstock, R. U. (2005). Anaerobic degradation of *p*-xylene by a sulphate-reducing enrichment culture. *Curr Microbiol*, **51**, 127–30.

Morasch, B., Schink, B., Tebbe, C. and Meckenstock, R. U. (2004). Degradation of *o*-xylene and *m*-xylene by a novel sulphate-reducer belonging to the genus *Desulfotomaculum*. *Arch Microbiol*, **181**, 407–17.

Mrowiec, B., Suschka, J. and Keener, T. C. (2005). Formation and biodegradation of toluene in the anaerobic sludge digestion process. *Water Environ Res*, **77**, 274–8.

Nauhas, K., Albrecht, M. and Elvert, M. *et al.* (2007). *In vitro* cell growth of marine archaeal-bacterial consortia during anaerobic oxidation of methane with sulfate. *Environ Microbiol*, **9**, 187–96.

Nauhaus, K., Boetius, A., Kruger, M. and Widdel, F. (2002). In vitro demonstration of anaerobic oxidation of methane coupled to sulphate reduction in sediment from a marine gas hydrate area. *Environ Microbiol*, **4**, 296–305.

Nauhaus, K., Treude, T., Boetius, A. and Krüger, M. (2005). Environmental regulation of the anaerobic oxidation of methane: a comparison of ANME-I and ANME-II communities. *Environ Microbiol*, **7**, 98–106.

Orphan, V. J., House, C. H., Hinrichs, K. U., McKeegan, K. D. and DeLong, E. F. (2002). Multiple archaeal groups mediate methane oxidation in anoxic cold seep sediments. *PNAS*, **99**, 7663–8.

Pancost, R. D., Sinninghe Damste, J. S., de Lint, S., van der Maarel, M. J. E. C., Gottschal, J. C. and The Medinaut Shipboard Scientific Party. (2000). Biomarker evidence for widespread anaerobic methane oxidation in Mediterranean sediments by a consortium of methanogenic Archaea and Bacteria. *Appl Environ Microbiol*, **66**, 1126–32.

Paull, C. K., Chanton, J., Neumann, A. C. *et al.* (1992). Indicators of methane-derived carbonates and chemosynthetic organic carbon deposits: examples from the Florida Escarpment. *Palaios*, **7**, 361–75.

Pelmenschikov, V., Blomberg, M. R. A., Siegbahn, P. E. M. and Crabtree, R. H. (2002). A mechanism from quantum chemical studies for methane formation in methanogenesis. *J Am Chem Soc*, **124**, 4039–49.

Peters, F., Rother, M. and Boll, M. (2004). Selenocysteine-containing proteins in anaerobic benzoate metabolism of *Desulfococcus multivorans*. *J Bacteriol*, **186**, 2156–63.

Phelps, C. D., Kazumi, J. and Young, L. Y. (1996). Anaerobic degradation of benzene in BTX mixtures dependent on sulphate reduction. *FEMS Microbiol Lett*, **145**, 433–7.

Phelps, C. D., Kerkhof, L. J. and Young, L. Y. (1998). Molecular characterization of a sulphate-reducing consortium which mineralizes benzene. *FEMS Microbiol Ecol*, **27**, 269–79.

Phelps, C. D., Zhang, X. and Young, L. Y. (2001). Use of stable isotopes to identify benzoate as a metabolite of benzene degradation in a sulphidogenic consortium. *Environ Microbiol*, **3**, 600–3.

Postgate, J. R. (1984). *The sulphate-reducing bacteria*. Cambridge, UK: Cambridge University Press.

Rabus, R., Hansen, T. and Widdel, F. (2000). Dissimilatory sulphate- and sulfur-reducing prokaryotes. In M. Dworkin, E. Rosenberg, K.-H. Schleifer and E. Stackebrandt (eds.), *The prokaryotes*, electronic edition. New York: Springer. (URL: http://141.150.157.117:8080/prokPUB/index.htm).

Rabus, R. and Heider, J. (1998). Initial reactions of anaerobic metabolism of alkylbenzenes in denitrifying and sulphate-reducing bacteria. *Arch Microbiol*, **170**, 377–84.

Rabus, R., Nordhaus, R., Ludwig, W. and Widdel, F. (1993). Complete oxidation of toluene under strictly anoxic conditions by a new sulphate-reducing bacterium. *Appl Environ Microbiol*, **59**, 1444–51.

Rabus, R., Wilkes, H., Behrends, A. *et al.* (2001). Anaerobic initial reaction of *n*-alkanes in a denitrifying bacterium: evidence for (1-methylpentyl) succinate as initial product and for involvement of an organic radical in *n*-hexane metabolism. *J Bacteriol*, **183**, 1707–15.

Raghoebarsing, A. A., Pol, A., van de Pas-Schoonen, K. T. *et al.* (2006). A microbial consortium couples anaerobic methane oxidation to denitrification. *Nature*, **440**, 918–21.

Reeburgh, W. S. (1976). Methane consumption in Cariaco Trench waters and sediments. *Earth Planet Sci Lett*, **28**, 337–44.

Reeburgh, W. S..(1980). Anaerobic methane oxidation: rate depth distributions in Skan Bay sediments. *Earth Planet Sci Lett*, **47**, 345–52.

Reed, D. R. and Kass, S. R. (2000). Experimental determination of the α and β C–H bond dissociation energies in naphthalene. *J Mass Spectr*, **35**, 534–9.

Reguera, G., McCarthy, K. D., Mehta, T. *et al.* (2005). Extracellular electron transfer via microbial nanowires. *Nature*, **435**, 1098–101.

Reusser, D. E., Istok, J. D., Beller, H. R. and Field, J. A. (2002). In situ transformation of deuterated toluene and xylene to benzylsuccinic acid analogues in BTEX-contaminated aquifers. *Environ Sci Technol*, **36**, 4127–34.

Richnow, H. H., Annweiler, E., Michaelis, W. and Meckenstock, R. U. (2003). Microbial in situ degradation of aromatic hydrocarbons in a contaminated aquifer monitored by carbon isotope fractionation. *J Contam Hydrol*, **65**, 101–20.

Ritger, S., Carson, B. and Suess, E. (1987). Methane-derived authigenic carbonates formed by subduction-induced pore-water expulsion along the Oregon/Washington margin. *Geol Soc Am Bull*, **98**, 147–56.

Ruckmick, J. C., Wimberly, B. H. and Edwards, A. F. (1979). Classification and genesis of biogenic sulfur deposits. *Econ Geol*, **74**, 469–74.

Rueter, P., Rabus, R., Wilkes, H. *et al.* (1994). Anaerobic oxidation of hydrocarbons in crude oil by new types of sulphate-reducing bacteria. *Nature*, **372**, 455–8.

Safinowski, M. and Meckenstock, R. U. (2004). Enzymatic reactions in anaerobic 2-methylnaphthalene degradation by the sulphate-reducing enrichment culture N 47. *FEMS Microbiol Lett*, **240**, 99–104.

Safinowski, M. and Meckenstock, R. U. (2006). Methylation is the initial reaction in anaerobic naphthalene degradation by a sulphate-reducing enrichment culture. *Environ Microbiol*, **8**, 347–52.

Schink, B. (1985). Degradation of unsaturated hydrocarbons by methanogenic enrichment cultures. *FEMS Microbiol Lett*, **31**, 69–77.

Schink, B. and Stams, A. F. (2002). Structure and growth dynamics of syntrophic associations. In M. Dworkin, E. Rosenberg, K.-H. Schleifer and E. Stackebrandt (eds.), *The prokaryotes*, electronic edition. New York: Springer. (URL: http://141.150.157.117:8080/prokPUB/index.htm).

Schmitt, R., Langguth, H. R. and Püttmann, W. (1998). Abbau aromatischer Kohlenwasserstoffe und Metabolitenbildung im Grundwasserleiter eines ehemaligen Gaswerkstandorts *Grundwasser*, **3**, 78–86.

Selmer, T. and Andrei, P. I. (2001). *p*-Hydroxyphenylacetate decarboxylase from *Clostridium difficile*: a novel glycyl radical enzyme catalysing the formation of *p*-cresol. *Eur J Biochem*, **268**, 1363–72.

Selmer, T., Pierik, A. and Heider, J. (2005). New glycyl radical enzymes catalysing key metabolic steps in anaerobic bacteria. *Biol Chem*, **386**, 981–8.

Shen, Y., Buick, R. and Canfield, D. E. (2001). Isotopic evidence for microbial sulphate reduction in the early Archaean era. *Nature*, **410**, 77–81.

Shilov, A. E. and Shul'pin, B. (1997). Activation of C–H bonds by metal complexes. *Chem Rev*, **97**, 2879–932.

Shima, S. and Thauer, R. K. (2005). Methyl-coenzyme M reductase and the anaerobic oxidation of methane in methanotrophic Archaea. *Curr Opin Microbiol*, **8**, 643–8.

So, C. M., Phelps, C. D. and Young, L. Y. (2003). Anaerobic transformation of alkanes to fatty acids by a sulphate-reducing bacterium, strain Hxd3. *Appl Environ Microbiol*, **69**, 3892–900.

So, C. M. and Young, L. Y. (1999). Isolation and characterization of a sulphate-reducing bacterium that anaerobically degrades alkanes. *Appl Environ Microbiol*, **65**, 2969–76.

Sørensen, K. B., Finster, K. and Ramsing, N. B. (2001). Thermodynamic and kinetic requirements in anaerobic methane oxidizing consortia exclude hydrogen, acetate and methanol as possible shuttles. *Microbial Ecol*, **42**, 1–10.

Spormann, A. M. and Widdel, F. (2000). Metabolism of alkylbenzenes, alkanes, and other hydrocarbons in anaerobic bacteria. *Biodegradation*, **11**, 85–105.

Stumm, W. and Morgan, J. J. (1981). *Aquatic Chemistry*, 2nd edn. New York: John Wiley & Sons.

Sullivan, E. R., Zhang, X., Phelps, C. and Young, L. Y. (2001). Anaerobic mineralization of stable-isotope-labeled 2-methylnaphthalene *Appl Environ Microbiol*, **67**, 4353–7.

Tauson, V. O., Veselov, I. Ya. (1934). O bakterialnom razlozhenii tsiklicheskikh soyedineniy pri vosstanovlenii sulfatov. (On the bacteriology of the decomposition of cyclical compounds at the reduction of sulphates.) *Mikrobiologiya* (in Russian), **3**, 360–9.

Tissot, B. P. and Welte, D. H. (1984). *Petroleum formation and occurrence*. Berlin: Springer.

Thauer, R. K., Jungermann, K. and Decker, K. (1977). Energy conservation in anaerobic bacteria. *Bacteriol Rev*, **41**, 100–80.

Thauer, R. K. (1998). Biochemistry of methanogenesis: a tribute to Marjory Stephenson. *Microbiology*, **144**, 2377–406.

Townsend, G. T., Prince, R. C. and Suflita, J. M. (2003). Anaerobic oxidation of crude oil hydrocarbons by the resident microorganisms of a contaminated anoxic aquifer. *Environ Sci Technol*, **37**, 5213–18.

Ulrich, A. C., Beller, H. R. and Edwards, E. A. (2005). Metabolites detected during biodegradation of $^{13}C_6$-benzene in nitrate-reducing and methanogenic enrichment cultures. *Environ Sci Technol*, **39**, 6681–91.

Weast, R. C. (1989). *Handbook of chemistry and physics*. Boca Raton, USA: CRC Press.

Widdel, F. (1988). Microbiology and ecology of sulphate- and sulfur-reducing bacteria. In A. J. B. Zehnder (ed.), *Biology of anaerobic microorganisms*. New York: John Wiley & Sons. pp 469–585.

Widdel, F., Boetius, A. and Rabus, R. (2004). Anaerobic biodegradation of hydrocarbons including methane. In M. Dworkin, E. Rosenberg, K.-H. Schleifer and E. Stackebrandt (eds.), *The prokaryotes*, electronic edition. New York: Springer. (URL: http://141.150.157.117:8080/prokPUB/index.htm).

Widdel, F. and Rabus, R. (2001). Anaerobic biodegradation of saturated and aromatic hydrocarbons. *Curr Opin Biotechnol*, **12**, 259–76.

Wilkes, H., Rabus, R., Fischer, T. *et al.* (2002). Anaerobic degradation of *n*-hexane in a denitrifying bacterium: Further degradation of the initial intermediate (1-methylpentyl) succinate via c-skeleton rearrangement. *Arch Microbiol*, **177**, 235–43.

Wischgoll, S., Heintz, D., Peters, F. *et al.* (2005). Gene clusters involved in anaerobic benzoate degradation of *Geobacter metallireducens*. *Mol Microbiol*, **58**, 1238–52.

Wolfe, R. S. (1991). My kind of biology. *Annu Rev Microbiol*, **45**, 1–35.

Zehnder, A. J. and Brock, T. D. (1979). Methane formation and methane oxidation by methanogenic bacteria. *J Bacteriol*, **137**, 420–32.

Zengler, K., Richnow, H. H., Rosselló-Mora, R., Michaelis, W. and Widdel, F. (1999). Methane formation from long-chain alkanes by anarobic microorganisms. *Nature*, **401**, 266–9.

Zhang, X. and Young, L. Y. (1997). Carboxylation as an initial reaction in the anaerobic metabolism of naphthalene and phenanthrene by sulfidogenic consortia. *Appl Environ Microbiol*, **63**, 4759–64.

ZoBell, C. E. (1946). Action of microörganisms on hydrocarbons. *Bacteriol Rev*, **10**, 1–49.

ZoBell, C. E. (1959). Ecology of sulphate reducing bacteria. *Prod Mon*, **22**: 12–29.

CHAPTER 10

Sulphate-reducing bacteria from oil field environments and deep-sea hydrothermal vents

Bernard Ollivier, Jean-Luc Cayol and Guy Fauque

10.1 INTRODUCTION

Early in 1886, it was demonstrated that the addition of gypsum ($CaSO_4 \cdot 2H_2O$) to anaerobic mud enrichments containing cellulose led to the production of the malodorous gas, hydrogen sulphide (Hoppe-Seyler, 1886). Soon after, Beijerinck first provided evidence of a microorganism reducing sulphate into sulphide, named as *Spirillum desulphuricans* (Beijerinck, 1895), which was the first sulphate-reducing bacterium (SRB) isolated in the world. As pointed out by Voordouw (1995), Beijerinck had already addressed, at the end of the nineteenth century, questions to the scientific community with regard to the metabolism and ecological distribution of the SRB, which are still nowadays themes of debate. SRB were first believed to use a limited range of substrates as energy sources (e.g. hydrogen, lactate, ethanol, etc. . . .), but recent biochemical and microbiological studies have greatly extended the range of electron donors and electron acceptors known to be used by SRB (Fauque *et al.*, 1991; Widdel, 1988). Indeed the latter may have an autotrophic, lithoautotrophic, heterotrophic, or respiration type of life under anaerobiosis and their possible microaerophilic nature has been discussed in the literature (Fauque and Ollivier, 2004). Besides their common ability to use sulphate as terminal electron acceptor, many of them were shown to utilize other mineral sulphur compounds, including elemental sulphur, thiosulphate, sulphite, polythionates and polysulphide (Le Faou *et al.*, 1990). In addition, SRB have been demonstrated to reduce a wide range of heavy metals and radionuclides including Fe(III), and U(VI). The dissimilatory reduction of nitrate or nitrite to ammonia was also reported as an energy-yielding reaction for some SRB (Fauque and Ollivier, 2004; Moura *et al.*, 1997; and Chapter 8 of this book). Of interest was the recent demonstration that few SRB, first considered as strict anaerobes, were able to perform a microaerobic respiration coupled to energy conservation. However no growth of SRB has been obtained so far under microaerobic conditions (Fauque and Ollivier, 2004). There is a wide range of electron acceptors used by SRB as almost one hundred compounds, including volatile fatty acids (e.g. acetate, propionate, butyrate), alcohols (e.g. methanol, ethanol, etc. . . .), aromatic compounds (phenol, etc. . . .), and sugars (e.g. fructose, glucose, etc. . . .), are potential electron donors for SRB. This suggests that their metabolical and ecological role in nature is of great importance. SRB exhibit various metabolical features and are widely distributed in marine, terrestrial, and subterrestrial ecosystems. Their contribution to the total carbon mineralization process in marine sediments, where sulphate is not limiting, was estimated to be up to 50% (Fauque, 1995; Widdel, 1988). They may grow in different physicochemical conditions, thus inhabiting the most extreme environments of our planet such as the cold, hot, saline and/or alkaline ecosystems. In addition, recent work on the geological and biogeochemical data from the 3.7-Ga North Pole barite deposit indicate that sulphate reduction would have been performed early in the evolution of life, thus suggesting that SRB should be considered as ancestral microorganisms, which have participated actively in the primordial biogeochemical cycle for sulphur as soon as life emerged on the planet (Shen and Buick, 2004). Today around 40 genera of SRB have been reported. They belong to four phylogenetically distinct groups of the domain *Bacteria* with most members being represented within the *delta* subdivision of the *Proteobacteria* and including the most studied genus within the SRB, *Desulfovibrio*. The three other groups comprise (i) the low-G + C Gram-positive *Bacillus*/*Clostridium* group containing several spore formers (e.g. *Desulfotomaculum* or *Desulfosporosinus*); (ii) the deepest-branching bacterial group consisting of the genera *Thermodesulfobacterium* and *Thermodesulfatator*; and (iii) members of the genus *Thermodesulfovibrio* belonging to the *Nitrospira* phylum or the recently described genus *Thermodesulfobium* peripherically related to this latter phylum (Castro *et al.*, 2000). Besides the *Bacteria* domain, there are also SRB pertaining to the *Archaea* domain. They include *Archaeoglobus* spp. together with a *Caldivirga* species (Birkeland, 2005).

SRB are metabolically and phylogenetically versatile and may represent the first respiring microorganisms with subsequent role to be played in the biogeochemistry of the various environments they inhabit. This is particularly true for oil field environments, and deep-sea hydrothermal vents where SRB have to face drastic physicochemical conditions combining high temperatures and high pressure. In such extreme ecosystems, sulphide production resulting from their metabolism may have beneficial or pernicious effects (e.g metal precipitation in deep-sea hydrothermal vents or oil souring in petroleum reservoirs). This chapter focuses on SRB living in deep and hot environments such as these.

10.2 THE SULPHATE-REDUCING MICROORGANISMS FROM OIL RESERVOIRS

Oil reservoirs constitute extreme subterrestrial ecosystems which are considered as essentially anaerobic, with temperatures ranging from 30 to 180 °C (temperature increases with depth at a mean rate of 3 °C per 100 m). Beside high temperatures, high salinities up to saturation (around 300 g/l NaCl) have been reported in oil waters. Therefore, it is not surprising that both thermophilic and/or halophilic anaerobic microorganisms are frequently recovered from these ecosystems (Magot *et al.*, 2000; Ollivier and Cayol, 2005). Because of depth, pressure up to several hundred bars should also affect the physiology of subterranean microbes. Despite the extreme physicochemical conditions occuring in oil reservoirs, the existence of microbial life has been long established in these ecosystems. Studies conducted for more than 50 years now indicated that a wide range of anaerobic microorganisms belonging to the *Bacteria* and *Archaea* domains are commonly distributed in oil reservoirs (Magot *et al.*, 2000). However, some data suggest that the presence of truly indigenous microorganisms in oil fields could be limited to a threshold temperature between 80 and 90 °C. This is in agreement in particular with Philippi (1977) who noted that in situ oil biodegradation was never observed in reservoirs where the temperature exceeds 82 °C. Microbes originating from oilfield facilities include fermentative, methanogenic, and sulphate-reducing microorganisms. Among them, great attention has been paid to SRB, which have been rapidly recognized as pernicious agents in the oil industry because of the production of H_2S within reservoirs or top facilities, leading to (i) the reduction of oil quality; (ii) the corrosion of steel material; and (iii) the generation of health problems for workers in contact with this highly toxic compound during oil production (Magot *et al.*, 2000). These health problems are particularly important when the concentration of sulphide is increased in production fluids after a long period of water injection, leading to what is commonly called "the oilfield reservoir souring". To decrease this souring phenomenon, oil companies have recently successfully used nitrate (Vance and Trasher, 2005). Nitrate not only inhibited the growth of SRB,

but also activated nitrate-reducing microbial populations capable of sulphide oxidation. Simultaneously, nitrate addition might help in microbial enhanced oil recovery.

Bastin first reported on the widespread presence of SRB within the oil reservoirs (Bastin, 1926). He questioned whether these microorganisms were indigenous, or introduced by waters descending from the surface, or by drilling operations during oil production, but indicated that it might never be possible to answer this question. Recently (Basso *et al.*, 2005) tried to solve the problems of contamination when sampling the deep subsurface water. They demonstrated that cleaning and disinfecting the sampling well was very helpful for diagnosing which are the indigenous microbial communities inhabiting the deep water studied. However the technique requires so many precautions when sampling that it seems unlikely to be usable by most scientists or limited to some oil reservoirs. The operations requested to prevent any contamination are too costly and depends on too many actors to be performed in the oil industry.

The survival or growth of SRB resulted from the availability of electron donors and acceptors within the oil reservoirs (Birkeland, 2005). Therefore sulphate contained in stratal waters or in the injection water during oil production may sustain growth and sulphide production by SRB, making sulphate reduction a major metabolic process within the oil reservoirs. When accidental introduction of oxygen into oilfield facilities occurs, thiosulphate is possibly produced through chemical oxidation of sulphide, thus giving to SRB or to thiosulphate-, non sulphate-reducing microorganisms the opportunity to reduce another sulphur compound other than sulphate into sulphide. It is noteworthy that serious biocorrosive problems observed in the oil industry were recently attributed to microorganisms able to use thiosulphate as terminal electron acceptor, but in this case SRB were not incriminated (Magot *et al.*, 2000). The potential electron donors for SRB include CO_2, and H_2 of geochemical or bacterial origin, and numerous organic molecules. In particular volatile fatty acids are known to be oxidized by the sulphate-reducing microflora. They include acetate as the major compound detected with concentration exceeding 20 mM in some reservoirs, (Barth 1991; Barth and Riis, 1992). Formate, propionate, butyrate, and benzoate were also found at lower concentrations. Among organic molecules found in crude oil, *n*-alkanes were shown to be possibly oxidized by SRB (Rueter *et al.*, 1994). However their involvement in anaerobic hydrocarbon biodegradation in situ is still unclear, as no organism that degrades these compounds under the conditions existing in deep petroleum reservoirs has yet been isolated (Aitken *et al.*, 2004).

The SRB isolated so far from the oilfield facilities belong to the *delta-proteobacteria*, the low G+C DNA containing Gram-positive group, the genus *Thermodesulfobacterium* representing the deepest phylogenetical branch within the domain *Bacteria*, and representatives of the domain *Archaea* consisting of the genus *Archaeoglobus*. SRB belonging to the *Nitrospira* phylum (e.g. *Thermodesulfovibrio*) or peripherically related to this phylum (e.g. *Thermodesulfobium*) have been recovered only from hot terrestrial ecosystems (e.g. thermal springs) (Figure 10.1).

10.2.1 The *delta-proteobacteria*

The most frequently isolated SRB from oilfield production waters belong to the *delta-proteobacteria*. They comprise mesophilic members of the genera *Desulfovibrio, Desulfobulbus, Desulfobacterium, Desulfobacter* and thermophilic members of the genera *Thermodesulforhabdus* and *Desulfacinum*. *Desulfovibrio* spp. are the most frequently isolated microorganisms. For example, in an extensive study of SRB, 21 strains isolated from 15 production waters by Tardy-Jacquenod *et al.* (1996a), were identified as *Desulfovibrio* species based on 16S rRNA gene sequence analysis (Magot *et al.*, 2000). Representatives of this genus within the oilfield environment consist of 7 species (Table 10.1). All are incomplete oxidizers and use a limited range of

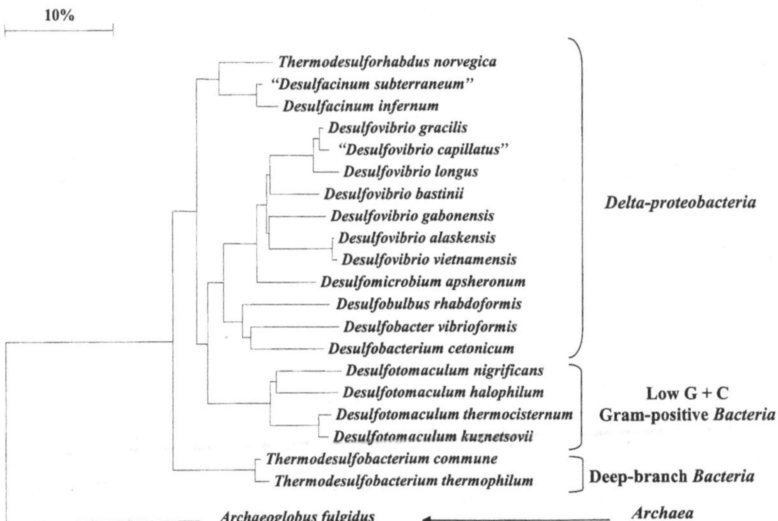

Figure 10.1. Phylogenetic dendrogram showing the position of the sulphate-reducing microorganisms isolated from petroleum reservoirs.

Table 10.1. *Sulphate-reducing prokaryotes recovered from oilfield environments*

Species	Salinity (%)		Temperature (°C)		Complete Oxidizer	References
	Range	Optimum	Range	Optimum		
Delta-proteobacteria						
Thermodesulforhabdus norvegica	0–5.6	1,6	44–74	60	+	Beeder *et al.*, 1995
"*Desulfacinum subterraneum*"	0–5	0.5	45–65	60	+	Rozanova *et al.*, 2001
Desulfacinum infernum	0–5	1	40–65	60	+	Rees *et al.*, 1995
Desulfovibrio gracilis	2–12	5–6	20–40	37–40	–	Magot *et al.*, 2004
"*Desulfovibrio capillatus*"	0.5–10	3	20–50	40	–	Miranda-Tello *et al.*, 2003
Desulfovibrio longus	0–8	1–2	10–40	35	–	Magot *et al.*, 1992
Desulfovibrio bastinii	1–12	4	20–50	35–40	–	Magot *et al.*, 2004
Desulfovibrio gabonensis	1–17	5–6	15–40	30	–	Tardy-Jacquenod *et al.*, 1996b
Desulfovibrio alaskensis	0–10	2, 5	10–45	37	–	Feio *et al.*, 2004

Desulfovibrio vietnamensis	0–10	5	12–45	37	–	Nga *et al.*, 1996
Desulfomicrobium apsheronum	0–8	1	4–40	25–30	–	Rozanova *et al.*, 1988
Desulfobulbus rhabdoformis	ND	1.5–2	10–40	31	–	Lien *et al.*, 1998
Desulfobacter vibrioformis	1–5	ND	5–38	33	+	Lien and Beeder, 1997
Desulfobacterium cetonicum	0–5	1	4–37	30	+	Galushlo and Rozanova, 1991
Low G+C Gram-positive Bacteria						
Desulfotomaculum nigrificans	0.5–4	1	40–70	60	–	Nazina and Rozanova, 1978
Desulfotomaculum halophilum	1–14	4–6	30–40	35	–	Tardy-Jacquenod *et al.*, 1998
Desulfotomaculum thermocisternum	0–5	0.3–1.2	41–75	62	–	Nilsen *et al.*, 1996b
Desulfotomaculum kuznetsovii	0–3	0	50–85	60–65	+	Nazina *et al.*, 1988

Table 10.1 (*cont.*)

Species	Salinity (%)		Temperature (°C)		Complete Oxidizer	References
	Range	Optimum	Range	Optimum		
Deep-branch Bacteria						
Thermodesulfobacterium commune	0–2	0	60–82	70	–	L'Haridon *et al.*, 1995
Thermodesulfobacterium thermophilum	0–2	0.1	44–85	65	–	Rozanova and Khudyakova, 1974
						Rozanova and Pivovarova, 1988
Archaea						
Archaeoglobus fulgidus	0.02–3	2	60–85	76	+	Beeder *et al.*, 1994

ND: not determined.

substrates, which include lactate, pyruvate, and hydrogen, which is most probably their primary source of energy in situ. Most of them are halotolerant or moderately halophilic. *Desulfovibrio vietnamensis* (Nga *et al.*, 1996), *D. capillatus* (Miranda-Tello *et al.*, 2003), and *D. longus* (Magot *et al.*, 1992) are considered halotolerant, whereas *D. gabonensis* (Tardy-Jacquenod *et al.*, 1996b), *D. bastinii* (Magot *et al.*, 2004), *D. gracilis* (Magot *et al.*, 2004), and *D. alaskensis* (Feio *et al.*, 2004) are considered as moderate halophiles requiring 2 to 6% NaCl for optimal growth. Other strains were identified as representatives of species isolated from ecosystems not related to petroleum reservoirs and included *D. desulfuricans, D. oxyclinae* or *D. longreachii* (Magot *et al.*, 2000).

Desulfomicrobium apsheronum (Rozanova *et al.*, 1988), a halotolerant SRB isolated from stratal waters in the Apsheron peninsula, tolerates up to 8% NaCl for growth and oxidizes incompletely lactate to acetate. It grows autotrophically and hence can be easily distinguished from the *Desulfovibrio* species described above. Microorganisms pertaining to the genus *Desulfomicrobium* were also recovered from four samples from two different North Sea oil fields, thus suggesting that they could play an important role in the generation of sulphide (Leu *et al.*, 1999).

Members of the genus *Desulfobacter* have been identified by oligonucleotide probes in oilfield environments (Telang *et al.*, 1997), and *Desulfobacter vibrioformis* has been isolated from a water/oil separation system (Lien and Beeder, 1997). It is a moderate halophile with optimal growth occurring between 1 and 5% NaCl. Acetate is the only carbon and energy source used in dissimilatory sulphate reduction, thus suggesting that this microorganism may be of ecological significance in oil reservoirs where acetate may accumulate at high concentrations (see discussion above). *Desulfobacterium cetonicum*, which was isolated from a flooded oil stratum, displays the unusual ability to oxidize ketones (Galushko and Rozanova, 1991). Finally, *Desulfobulbus rhabdoformis* isolated from a water/oil separation system was recognized as a propionate-oxidizing sulphate reducer which incompletely oxidizes propionate to acetate (Lien *et al.*, 1998). This latter species also uses fumarate and malate as carbon and energy sources.

There are few thermophiles recovered from oilfield environments which also belong to the *delta* subdivision of the *Proteobacteria*. They include *Desulfacinum infernum*, "*Desulfacinum subterraneum*", and *Thermodesulforhabdus norvegica* (previously named *Thermodesulforhabdus norvegicum*). *Desulfacinum infernum* (Rees *et al.*, 1995) and *T. norvegica* (Beeder *et al.*, 1995) were isolated from North Sea hot oil reservoirs whereas "*D. subterraneum*" (Rozanova *et al.*, 2001) was isolated from a high temperature oilfield in

Vietnam. *Desulfacinum infernum,* together with "*D. subterraneum*", and *T. norvegica* are complete oxidizers. *Desulfacinum* species are autotrophic microorganisms which use acetate, butyrate, and alcohols and grow optimally at 60 °C. In contrast to the *Desulfacinum* species, *T. norvegica* does not oxidize hydrogen and is unable to use thiosulphate as terminal electron acceptors (Table 10.1).

Beside classical microbiological studies, molecular studies also indicated that members of the *delta-proteobacteria* are common inhabitants of oilfield environments. Microbial 16S rRNA gene studies from several different oilfields by Voordouw *et al.* (1996) indicated that all the Gram-negative SRB detected belonged to the family *Desulfovibrionaceae* or *Desulfobacteriaceae*. Using a similar approach in high-temperature oil-bearing formations, Orphan *et al.* (2000), and Watanabe *et al.* (2002) detected the presence of *Desulfomicrobium* and *Desulfovibrio* species respectively. Genus-specific antibodies against the genus *Thermodesulforhabdus* have been used for analyzing the distribution of SRB in oilfield waters sampled from 10 different wells in the Gullfaks field in the North Sea (Nilsen *et al.*, 1996a). *Thermodesulforhabdus* strains were detected in 4 of 16 samples analyzed. Finally, when analyzing the fatty acids of bacteria, Osipov *et al.* (1995) demonstrated that SRB belonging to the genera *Desulfovibrio* and *Desulfobacter* inhabited the near bottom zone of the injection wells of the Talinsk oilfield in West Siberia.

10.2.2 The low G+C DNA containing Gram-positive group

Three genera of Gram-positive endospore-forming SRB closely related to the *Bacillus-Clostridium* group have been described so far. They include *Desulfosporosinus, Thermoacetogenium,* and *Desulfotomaculum*. However, up to now microbiological and molecular studies performed in oilfield environments only revealed the presence of *Desulfotomaculum* species (Christensen *et al.*, 1992; Nilsen *et al.*, 1996b; Rosnes *et al.*, 1991; Rozanova and Nazina, 1979; Watanabe *et al.*, 2002). Among the species recovered, most of them are thermophilic (*Desulfotomaculum kuznetsovii, D. thermocisternum, D. nigrificans* subsp. *salinus*), only one being mesophilic and moderately halophilic (*D. halophilum*) growing optimally in the presence of 6% NaCl (Table 10.1). *Desulfotomaculum kuznetsovii,* which was initially isolated from an underground thermal mineral water ecosystem (Nazina *et al.*, 1988), subsequently has been isolated from two wellhead samples from non-waterflooded oil fields in the Paris Basin (Magot *et al.*, 2000). It grows optimally at 60–65 °C and completely oxidizes its substrates. In contrast the thermophiles, *D. thermocisternum* (Nilsen *et al.*, 1996b) and *D. nigrificans* subsp. *salinus* (Nazina and Rozanova, 1978) are incomplete oxidizers growing optimally at 62 °C and 60 °C respectively. The former was isolated below the sea floor of the Norwegian sector of the North Sea, whereas the latter was recovered from oil sample originating from Western Siberia. *Desulfotomaculum thermocisternum* uses a wide range of substrates including lactate, ethanol, butanol, and carboxylic acids (C3 to C10 and C14 to C17) (Nilsen *et al.*, 1996b), while *D. nigrificans* subsp. *salinus* was reported as halotolerant and using lactate, ethanol, and hydrogen. Finally, the moderately halophilic *D. halophilum* (Tardy-Jacquenod *et al.*, 1998) was isolated from oilfield brines with total salinity of 70 g l^{-1}. It oxidizes hydrogen, formate, lactate and ethanol and ferments pyruvate. Surprisingly, the use of enrichment cultures and genus-specific fluorescent antibodies produced against these microorganisms by Nilsen *et al.* (1996a) did not lead to the detection of any members of the genus *Desulfotomaculum*. In contrast, by analyzing the 16S rRNA fragments of DNA extracted from an underground oil-storage cavity, Watanabe *et al.* (2002) obtained *Desulfotomaculum* spp. as the major sequences. *Desulfotomaculum* species were also recovered from a high-temperature North Sea oil well, and from a continental high-temperature oil reservoir in Western Siberia by using (i) genus-specific antibodies or (ii) an oligonucleotide microchip method respectively (Bonch-Osmolovskaya *et al.*, 2003; Nielsen *et al.*, 1996a).

10.2.3 The *Thermodesulfobacterium* division

The genus *Thermodesulfobacterium* consists of four species. Among them, two were isolated from oilfield environments (*Thermodesulfobacterium commune* and *T. thermophilum*) whereas *T. hydrogenophilum* was isolated from a deep-sea hydrothermal vent and *T. hveragerdense* from a terrestrial hot spring. *Thermodesulfobacterium* species are generally considered as the most thermophilic sulphate-reducing microorganisms of the domain *Bacteria*, with an upper limit temperature for growth around 80 °C. *Thermodesulfobacterium thermophilum* (previously named as *Desulfovibrio thermophilus*. Rozanova and Khudyakova, 1974; Rozanova and Pivovarova, 1988) was isolated from a North Sea oil reservoir (Christensen *et al.*, 1992) and *T. commune* from a continental oil reservoir located in the East Paris Basin (L'Haridon *et al.*, 1995). Both species are rod-shaped and utilize a limited range of substrates, including H_2, formate, lactate and pyruvate (Table 10.1). Strains of both species have also been isolated from other non-waterflooded reservoirs in the Paris Basin (Magot *et al.*, 2000). The analysis of the diversity of the uncultivated microorganisms inhabiting a continental oil reservoir in Russia by an oligonucleotide microchip method revealed the presence of *Thermodesulfobacterium* species, considered as indigenous to this deep ecosystem (Bonch-Osmolovskaya *et al.*, 2003).

10.2.4 The genus *Archaeoglobus*

There are only three validated species pertaining to the genus *Archaeoglobus*. Among them, *A. fulgidus* and *A. profundus*, but not *A. veneficus*, reduce sulphate into sulphide. *Archaeoglobus fulgidus* was isolated from a shallow hydrothermal system in the Mediterranean Sea, whereas both *A. veneficus* and *A. profundus* were isolated from deep-sea hydrothermal vents. Among these three species only *A. fulgidus* (strain 7324) has been recovered from hot oilfield. Geographically, this species has been located in different sectors of the North Sea (Beeder *et al.*, 1994; Stetter *et al.*, 1993). *Archaeoglobus profundus* and "*A. lithotrophicus*" were also found in North Sea wells (Stetter *et al.*, 1993), but their occurrence was mostly attributed to anthropogenic contaminations due to genomic similarities with inhabitants of deep-sea hydrothermal vents (e.g. injection of seawater, drilling operations, etc. . . .) (Stetter *et al.*, 1993). In contrast to *A. fulgidus* type species, *A. fulgidus* strain 7324 has a lower growth temperature optimum (76 °C) and cannot grow autotrophically on H_2+CO_2. It grows on lactate and pyruvate as carbon and energy sources. Growth on valerate has only been observed when hydrogen is present (Beeder *et al.*, 1994). *Archaeoglobus* strains have been detected by genus-specific antibodies in North Sea oil reservoirs (Nilsen *et al.*, 1996a).

10.3 THE SULPHATE-REDUCING MICROORGANISMS FROM DEEP-SEA HYDROTHERMAL VENTS

It was only at the end of the 1970s that deep-sea hydrothermal vents were discovered along the Galapagos Rift. It was an astonishing experience for scientists participating in these unique expeditions to observe a luxurious biomass of invertebrates living in the surroundings of what are commonly called "black smokers" which emit very hot fluids into the dark deep ocean (Jeanthon, 2000). Numerous questions were instantly addressed to the scientific community with regard (i) to the source of energy that may allow life to emerge far from the sunlight and (ii) to the adaptation of both Eukaryotes and Prokaryotes to these extreme conditions where temperatures over 300 °C have been measured frequently within hydrothermal fluids at the top of the chimneys. In addition, these fluids delivered to this deep ecosystem significant amounts of a mixture of sulphur-reduced compounds and heavy metals, in particular ones which are known to be quite toxic for microorganisms (Jannasch and Mottl, 1985). Overall, these physicochemical constraints and also the high hydrostatic pressure existing in situ appeared as important physical phenomena to which deep-sea macro- and microfauna should adapt. Due to high temperatures and immediate contact with seawater containing oxygen, the deep-sea hydrothermal ecosystem exhibits complex gradients of temperature, oxygen, and chemical composition (Jannasch and Mottl, 1985) making the emerging of aerobic or anaerobic life under mesophilic or thermophilic conditions possible. It has been clearly established today that the primary source of energy in this deep ecosystem is related to the use of sulphur-reduced compounds, which are oxidized by specialized microorganisms living symbiotically with invertebrates in the colder parts of this environment (Jannasch and Mottl, 1985; Jeanthon, 2000). Drawing nearer to the top of chimneys, there is an increase in temperature and a subsequent decrease in oxygen concentration, giving the opportunity for free-living microrganisms to grow microaerobically or anaerobically at high temperatures. It is of interest that microbial studies which have been carried out in the deep-sea hydrothermal vents have demonstrated for the first time that life could occur at a temperature over 100 °C (Jeanthon, 2000). As discussed above, the delivery of huge amounts of sulphur-reduced compounds from hot fluids in seawater may result in the abiotic oxidation of these compounds (e.g. thiosulphate, sulphite, and elemental sulphur) which become suitable electron donors for the anaerobic microbial community. Taking into account that sulphate is also a common constituent of seawater, all these oxidized forms of sulphur most probably participate, through the anaerobic microbial activity, in the overall biogeochemistry of the deep-sea hydrothermal environments. Among the microorganisms reducing sulphur compounds in this hot environment, much attention has been paid to those reducing elemental sulphur, belonging mainly to the *Archaea* domain (e.g. *Thermococcus* and *Pyrococcus*) and considered as chemoorganoheterotrophs (Jeanthon, 2000). As well as sulphur compounds used as possible terminal electron acceptors by anaerobic microorganisms, it is noteworthy that hydrogen (part of the gas emitted from smokers) together with volatile fatty acids (e.g. acetate) resulting from biotic and thermal abiotic reactions may act as potential electron donors for this microbial community. Here we will focus on the sulphate-reducing microorganisms belonging to the *Bacteria* or the *Archaea* domains which have been cultivated or retrieved by molecular studies from the deep hot ocean. The SRB isolated from deep-sea hydrothermal vents belong to the *delta-proteobacteria*, the genera *Thermodesulfobacterium* and *Thermodesulfatator* representing the deepest phylogenetical branches within the domain *Bacteria*, and representatives of the domain *Archaea* consisting of the genus *Archaeoglobus* (Figure 10.2).

10.3.1 The *delta-proteobacteria*

The SRB isolated from deep-sea hydrothermal vents belonging to the *delta-proteobacteria* comprise mesophilic members of the genus *Desulfovibrio*, and thermophilic members of the genera *Desulfonauticus* and *Desulfacinum* (Table 10.2). It is only recently that the first member of the genus *Desulfovibrio*, *D. hydrothermalis* has been isolated from a deep-sea hydrothermal chimney sample (latitude 13 °N along the East Pacific Rise at 2600 m) (Alazard *et al.*, 2003). It is a hydrogenotrophic microorganism oxidizing a limited range of substrates including lactate, glycerol, and choline. *D. zoesterae* is its closest phylogenic relative. Its barophilic nature was ascertained by a higher growth rate at a hydrostatic pressure of 260 atm (pressure from its extracted sediments) than at atmospheric pressure. The authors also reported on the isolation of other *Desulfovibrio* strains, phylogenetically and genomically similar to *D. profundus* which was first isolated from deep sediment layers in the Japan Sea (Bale *et al.*, 1997), thus suggesting that this species might have been of ecological significance

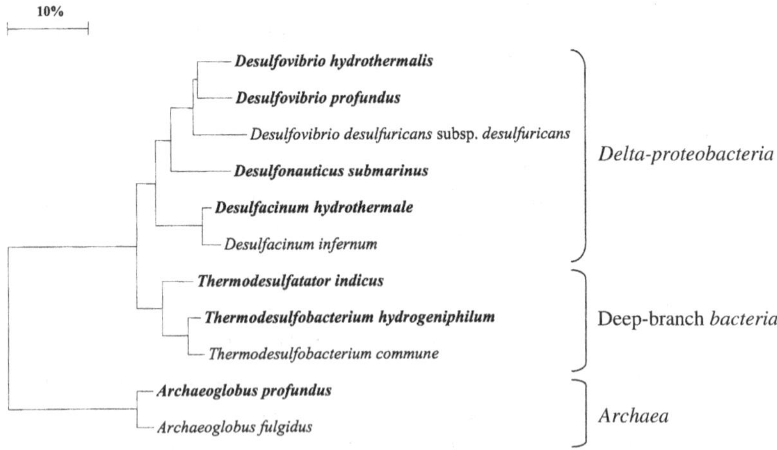

Figure 10.2. Phylogenetic dendrogram showing the position of the sulphate-reducing microorganisms (boldface) isolated from the deep-sea hydrothermal vents.

Table 10.2. *Sulphate-reducing prokaryotes recovered from deep-sea hydrothermal vents*

Species	Salinity (%)		Temperature (°C)		Complete	References
	Range	Optimum	Range	Optimum	Oxidizer	
Delta-proteobacteria						
Desulfovibrio hydrothermalis	0–4	2.5	20–40	35	–	Alazard *et al.*, 2003
Desulfovibrio profundus	0.2–10	0.6–8	15–65	25	–	Bale *et al.*, 1997
Desulfonauticus submarinus	0–5	2	30–60	45	–	Audiffrin *et al.*, 2003
Desulfacinum hydrothermale	1.5–7.8	3.2–3.6	37–64	60	+	Sievert and Kuever, 2000
Deep-branch Bacteria						
Thermodesulfatator indicus	1–3.5	2.5	55–80	70	–	Moussard *et al.*, 2004
Thermodesulfobacterium hydrogeniphilum	0.5–5.5	3	50–80	75	–	Jeanthon *et al.*, 2002
Archaea						
Archaeoglobus profundus	0.9–3.6	1.8	65–90	82	–	Burggraf *et al.*, 1990

in participating in the oxidation of organic matter in the deep marine environments.

Amongst the thermophilic SRB, *Desulfacinum hydrothermale* was recovered from an active, marine, shallow-water hydrothermal vent (Sievert and Kuever, 2000). It is a complete oxidizer with the ability to use various substrates including volatile fatty acids (acetate, isobutyrate, isovalerate), alcohols (ethanol, propanol, butanol) and to grow autotrophically on hydrogen. This species was closely phylogenetically related to *Desulfacinum infernum*, a thermophilic isolate from a petroleum reservoir. In contrast, *Desulfonauticus submarinus*, isolated from matrixes of *Alvinella* and *Riftia* which originated from deep-sea hydrothermal vent samples collected along the East Pacific Rise, was found to oxidize only hydrogen (formate poorly used) in the presence of acetate as carbon source, thus suggesting that its metabolic role in hydrothermal vents is probably limited to the oxidation of hydrogen (Audiffrin *et al.*, 2003). This bacterium is moderately thermophilic and is located phylogenetically within the family *Desulfohalobiaceae*, the alkalophilic *Desulfonatronovibrio hydrogenovorans* being its closest phylogenetic relative. Finally, a thermophilic SRB, designated strain TD3, growing optimally between 55 and 65 °C, was recovered from Guaymas Basin sediments (Rueter *et al.*, 1994). This strain, which belongs to the *delta-proteobacteria*, was unable to use hydrogen, lactate, and ethanol, but oxidized *n*-alkanes from C_6 to C_{16} and fatty acids from C_4 to C_{18}. Strain TD3 has not been assigned to any novel genus or novel species within the SRB so far.

In recent years, several authors have provided evidence on the occurrence of SRB originating from deep-sea hydrothermal vents pertaining to the delta-proteobacteria using molecular techniques. Analyses of both 16S rRNA and dissimilatory sulphite reductase (*dsrAB*) clone libraries from sediments of the Guaymas Basin, a hydrocarbon-enriched deep-sea hydrothermal environment, resulted in the identification of members of *Desulfobacter* and *Desulfobacterium* genera, order *Desulfobacteriales*, known as acetate-oxidizing SRB (Dhillon *et al.*, 2003). It is only the *dsr* clones analyzed which revealed in this same ecosystem the presence of the delta-proteobacterial species *Desulforhabdus amnigena* and *Thermodesulforhabdus norvegica*, the latter being an isolate from oilfield reservoirs (Dhillon *et al.*, 2003). In addition, when studying the community of SRB associated with the deep-sea hydrothermal vent polychaete annelid *Alvinella pompejana* by constructing a clone library of bisulfite reductase gene PCR (Polymerase chain reaction) products, members of the genera *Desulfovibrio* (e.g. *D. vulgaris* and *D. gigas*), *Desulfobacter* (e.g. *D. latus*), and *Desulfobacterium* (e.g. *D. autotrophicum*) were retrieved (Cottrell and Cary, 1999). This diversity of SRB in the bacterial community on the back of annelids suggest them as playing a prominent role in the ecology of *Alvinella pompejana* (e.g. detoxification process).

10.3.2 The *Thermodesulfobacterium/Thermodesulfatator* division

Most of the *Thermodesulfobacterium* species have been isolated only from terrestrial hot springs and oil reservoirs and it is only recently that a novel species of this genus, *T. hydrogenophilum* has been recovered from Guaymas Basin hydrothermal vent sites, where it may contribute to the primary production (Jeanthon *et al.*, 2002). *Thermodesulfobacterium hydrogenophilum* is a slightly halophilic thermophilic bacterium, which grows optimally at 75 °C, using hydrogen as the only energy source and sulphate as the only electron acceptor (thiosulphate, sulphite, and elemental sulphur are not used). Another thermophilic SRB has been isolated from the Central Indian Ridge having metabolical features similar to those of *T. hydrogenophilum* (Table 10.2). However, based on phylogenetic characteristics, this isolate was recognized as belonging to a novel genus within the family *Thermodesulfobacteriaceae*, and named *Thermodesulfatator indicus* (Moussard *et al.*, 2004).

Finally, with regard to the domain Bacteria, there are few reports based only on molecular characterization, indicating that SRB belonging to the the low G+C DNA containing Gram-positive group (e.g. *Desulfotomaculum*) or the *Nitrospira* phylum (e.g. *Thermodesulfovibrio*) may inhabit the deep-sea hydrothermal vents (Cottrell and Cary, 1999; Dhillon *et al.*, 2003).

10.3.3 The *Archaeoglobus* genus

The only sulphate-reducing archaeon isolated so far at great depths from hydrothermal vents is *Archaeoglobus profundus* (Burggraf *et al.*, 1990). This species was recovered from the walls of active smokers and from sediments from a deep-sea hydrothermal system at Guaymas Basin. *Archaeoglobus profundus* is an hyperthermophile, which grows optimally at 82 °C using hydrogen as energy source only in the presence of an organic compound (e.g., acetate or lactate) as carbon source (Table 10.2). Both thiosulphate and sulphite were also used as terminal electron acceptors. *Archaeoglobus veneficus* was isolated from the deep-sea hot ocean (Mid-Atlantic Ridge), but was described as a non sulphate-reducing microorganism with the ability to reduce only thiosulphate and sulphite (Huber *et al.*, 1997). Based on molecular analyses, it has also been clearly established that members of the

Archaeoglobales were inhabitants of the deep hot marine biosphere (Cottrell and Cary, 1999; Jeanthon, 2000; Reysenbach *et al.*, 2000).

10.4 CONCLUSION

In addition to their ability to use sulphate as terminal electron acceptor and to oxidize both mineral and organic compounds, SRB are recognized mainly as anaerobic metabolically versatile microorganisms ranging from autotrophs to chemoorganoheterotrophs, some of them being capable of microaerobic respiration (Fauque and Ollivier, 2004). It is most probably because of the wide range of substrates and electron acceptors used that SRB have adapted to almost all the ecosystems of the planet, including the deep extreme niches such as oilfield environments and deep-sea hydrothermal vents. In both these environments, SRB have to cope with drastic physicochemical conditions (e.g. temperature, pressure). However, in contrast to oil reservoirs, which are considered as essentially anaerobic with salinity up to saturation measured in waters, oxygen, together with highly reduced compounds and slight saline conditions, are commonly found in the deep, hot ocean. It is most probably because of the physicochemical differences observed between these two ecosystems that a specific community of SRB belonging to the domain *Bacteria* and *Archaea* has been recovered from these hot environments, thus suggesting that some of them, and particularly the thermophilic SRB, should be considered as indigenous to these deep habitats. It is noteworthy that whereas hydrocarbon-oxidizing SRB have been isolated from various marine sediments, including deep-sea hydrothermal systems (e.g. Guaymas Basin sediments), such microorganisms have not been recovered so far from oil reservoirs (Head *et al.*, 2003). It is now almost certain that oil biodegradation in deep subsurface petroleum reservoirs procceeds through anaerobic microbial metabolism rather than through aerobic mechanisms, thus suggesting that SRB might have been good candidates together with other anaerobes for degrading hydrocarbons (Head *et al.*, 2003). Metabolical features of SRB isolated from oilfield environments and deep-sea hydrothermal vents indicate that they are most probably involved in hydrogen and/or acetate oxidation in these ecosystems as both these compounds may be delivered biotically or abiotically in situ. SRB contribute therefore to the complete oxidation of organic matter and participate through sulphide production and/or metal reduction, in particular in deep-sea hydrothermal vents, to the overall biogeochemistry of these extreme environments. Despite the several microbiological and molecular studies of SRB inhabiting the deep subterrestrial ecosystems which have been carried out, it is clear that the complete diversity and the ecological significance of these microorganisms is still only very partially understood. The deep hot biosphere awaits additional work by microbiologists, biochemists, and geochemists to fully explain the ecological role of SRB in the subterrestrial environment.

ACKNOWLEDGMENTS

Many thanks are extended to Dr. Pierre Roger for carefully revising the manuscript.

REFERENCES

Aitken, C. M., Jones, D. M. and Larter, S. R. (2004). Anaerobic hydrocarbon biodegradation in deep subsurface oil reservoirs. *Nature*, **431**, 291–4.

Alazard, D., Dukan, S., Urios, A. *et al.* (2003). *Desulfovibrio hydrothermalis* sp. nov., a novel sulphate-reducing bacterium isolated from hydrothermal vents. *International Journal of Systematic and Evolutionary Microbiology*, **53**, 173–8.

Audiffrin, C., Cayol, J.-L., Joulian, C. *et al.* (2003). *Desulfonauticus submarinus* gen. nov., sp. nov., a novel sulphate-reducing bacterium isolated from a deep-sea hydrothermal vent. *International Journal of Systematic and Evolutionary Microbiology*, **53**, 1585–90.

Bale, S. J., Goodman, K., Rochelle, P. A. *et al.* (1997). *Desulfovibrio profundus* sp. nov., a novel barophilic sulphate-reducing bacterium from deep sediment layers in the Japan Sea. *International Journal of Systematic Bacteriology*, **47**, 515–21.

Barth, T. (1991). Organic acids and inorganic ions in waters from petroleum reservoirs, Norwegian continental shelf: a multivariate statistical analysis and comparison with American reservoir formation waters. *Applied Geochemistry*, **6**, 1–15.

Barth, T. and Riis, M. (1992). Interactions between organic acid anions in formation waters and reservoir mineral phases. *Organic Geochemistry*, **19**, 455–82.

Basso, O., Lascourrèges, J.-F., Jarry, M. and Magot M. (2005). The effect of cleaning and disinfecting the sampling well on the microbial communities of deep subsurface water samples. *Environmental Microbiology*, **7**, 13–21.

Bastin, E. S. (1926). The problem of the natural reduction of sulphates. *Bulletin of the American Association of Petroleum Geologists*, **10**, 1270–99.

Beeder, J., Nilsen, R. K., Rosnes, J. T., Torsvik, T. and Lien, T. (1994). *Archaeglobus fulgidus* isolated from hot North Sea oil field water. *Applied and Environmental Microbiology*, **60**, 1227–31.

Beeder, J., Torsvik, T. and Lien, T. (1995). *Thermodesulforhabdus norvegicus* gen. nov., sp. nov., a novel thermophilic sulphate-reducing bacterium from oil field water. *Archives of Microbiology*, **164**, 331–6.

Beijerinck, W. M. (1895). Ueber *Spirillum desulphuricans* als ursache von sulfat-reduction. *Zentralblatt für Bakteriologie und Parasitenkunde*, **1**, 1–9, 49–59 and 104–14.

Birkeland, N.-K. (2005). Sulphate-reducing *Bacteria* and *Archaea*. In *Petroleum Microbiology*, B. Ollivier and M. Magot (eds.). Washington, D.C.: ASM Press, pp. 35–54.

Bonch-Osmolvskaya, E. A., Miroshnichenko, M. L., Lebedinsky, A. V. *et al.* (2003). Radioisotopic, culture-based, and oligonucleotide microchip analyses of thermophilic microbial communities in a continental high-temperature petroleum reservoir. *Applied and Environmental Microbiology*, **69**, 6143–51.

Burggraf, S., Jannasch, H. W., Nicolaus, B. and Stetter, K. O. (1990). *Archaeoglobus profundus* sp. nov., represents a new species within the sulphate-reducing Archaebacteria. *Systematic and Applied Microbiology*, **13**, 24–8.

Castro, H. F., Williams, N. H. and Ogram, A. (2000). Phylogeny of sulphate-reducing bacteria. *FEMS Microbiology Ecology*, **31**, 1–9.

Christensen, B., Torsvik, T. and Lien, T. (1992). Immunomagnetically captured thermophilic sulphate-reducing bacteria from North Sea oil field waters. *Applied and Environmental Microbiology*, **58**, 1244–8.

Cottrell, M. T. and Cary, S. G. (1999). Diversity of dissimilatory bisulfite reductase genes of bacteria associated with the deep-sea hydrothermal vent polychaete annelid *Alvinella pompejana*. *Applied and Environmental Microbiology*, **65**, 1127–32.

Dhillon, A., Teske, A., Dillon, J., Stahl, D. A. and Sogin, M. L. (2003). Molecular characterization of sulphate-reducing bacteria in the Guaymas Basin. *Applied and Environmental Microbiology*, **69**, 2765–72.

Fauque, G. (1995). Ecology of sulphate-reducing bacteria. In L. L. Barton (ed.), *Sulphate-reducing bacteria*. New York and London: Plenum Press. pp. 217–41.

Fauque, G., LeGall, J. and Barton, L. L. (1991). Sulphate-reducing and sulfur-reducing bacteria. In J. M. Shively and L. L. Barton (eds.), *Variations in autotrophic life*. London: Academic Press. pp. 271–337.

Fauque, G. and Ollivier, B. (2004). Anaerobes: the sulphate-reducing bacteria as an example of metabolic diversity. In A. Bull (ed.), *Microbial diversity and bioprospecting*. Washington, DC: ASM Press, pp. 169–76.

Feio, M. J., Zinkevich, V., Beech, I. W. *et al.* (2004). *Desulfovibrio alaskensis* sp. nov., a sulphate-reducing bacterium from a soured oil reservoir. *International Journal of Systematic and Evolutionary Microbiology*, **54**, 1747–52.

Galushko, A. S. and Rozanova, E. P. (1991). *Desulfobacterium cetonicum.* sp. nov: a sulphate-reducing bacterium which oxidizes fatty acids and ketones. *Microbiology*, **60**, 102–7.

L'Haridon, S., Reysenbach, A. L., Glénat, P., Prieur, D. and Jeanthon, P. (1995). Hot subterranean biosphere in a continental oil reservoir. *Nature*, **377**, 223–4.

Head, I. M., Jones, D. M. and Larter, S. R. (2003). Biological activity in the deep subsurface and the origin of heavy oil. *Nature*, **426**, 344–52.

Hoppe-Seyler, F. (1886). Ueber die gährung der Cellulose mit Bildung von methan und Kohlensaüre: II. Der Zerfall der Cellulose durch Gährung unter Bildung von Methan und Kohlensaüre und die Erscheinungen, welche dieser Process veranlasst. *Zeitschrift für Physiologische Chemie*, **10**, 401–40.

Huber, H., Jannasch, H., Rachel, R., Fuchs, T. and Stetter, K. O. (1997). *Archaeoglobus veneficus* sp. nov., a novel facultative chemolithoautotrophic hyperthermophilic sulfite reducer, isolated from abyssal black smokers. *Systematic and Applied Microbiology*, **20**, 374–80.

Jannasch, H. W. and Mottl, M. J. (1985). Geomicrobiology of deep-sea hydrothermal vents. *Science*, **229**, 717–25.

Jeanthon, C. (2000). Molecular ecology of hydrothermal vent microbial communities. *Antonie van Leeuwenhoek*, **77**, 117–33.

Jeanthon, C., L'Haridon, S., Cueff, V. *et al.* (2002). *Thermodesulfobacterium hydrogeniphilum* sp. nov., a thermophilic, chemolithoautotrophic, sulphate-reducing bacterium isolated from a deep-sea hydrothermal vent at Guaymas Basin, and emendation of the genus *Thermodesulfobacterium. International Journal of Systematic and Evolutionary Microbiology*, **52**, 765–72.

Le Faou, A., Rajagopal, B. S., Daniels, L. and Fauque, G. (1990). Thiosulphate, polythionates and elemental sulfur assimilation and reduction in the bacterial world. *FEMS Microbiology Reviews*, **75**, 351–82.

Leu, J.-Y, McGovern-Traa, C. P., Porter, A. J. R. and Hamilton, W. A. (1999). The same species of sulphate-reducing *Desulfomicrobium* occur in different oil field environments in the North Sea. *Letters in Applied Microbiology*, **29**, 246–52.

Lien, T. and Beeder, J. (1997). *Desulfobacter vibrioformis* sp. nov., a sulphate-reducer from a water–oil separation system. *International Journal of Systematic Bacteriology*, **47**, 1124–8.

Lien, T., Madsen, M., Steen, I. H. and Gjerdevik, K. (1998). *Desulfobulbus rhabdoformis* sp. nov., a sulphate reducer from a water–oil separation system. *International Journal of Systematic Bacteriology*, **48**, 469–74.

Magot, M., Basso, O., Tardy-Jacquenod, C. and Caumette, P. (2004). *Desulfovibrio bastinii* sp. nov. and *Desulfovibrio gracilis* sp. nov., moderately halophilic, sulphate-reducing bacteria isolated from deep subsurface oilfield water. *International Journal of Systematic and Evolutionary Microbiology*, **54**, 1693–7.

Magot, M., Caumette, P., Desperrier, J. M. *et al.* (1992). *Desulfovibrio longus* sp. nov., a sulphate-reducing bacteria isolated from oil-producing well. *International Journal of Systematic Bacteriology*, **42**, 398–403.

Magot, M., Ollivier, B. and Patel B. K. C. (2000). Microbiology of petroleum reservoirs. *Antonie van Leeuwenhoek*, **77**, 103–16.

Miranda-Tello, E., Fardeau, M.-L., Fernandez, L. *et al.* (2003). *Desulfovibrio capillatus* sp. nov., a novel sulphate-reducing bacterium isolated from an oil field separator located in the Gulf of Mexico. *Anaerobe*, **9**, 97–103.

Moura, I., Bursakov, S., Costa, C. and Moura, J. J. G. (1997). Nitrate and nitrite utilization in sulphate-reducing bacteria. *Anaerobe*, **3**, 279–90.

Moussard, H., L'Haridon, S., Tindall, B. J. *et al.* (2004). *Thermodesulfatator indicus* gen. nov., sp. nov., a novel thermophilic chemolithoautotrophic sulphate-reducing bacterium isolated from the Central Indian Ridge. *International Journal of Systematic and Evolutionary Microbiology*, **54**, 227–33.

Nazina, T. N., Ivanova, A. E., Kanchaveli, L. P. and Rozanova, E. P. (1988). A new sporeforming thermophilic methylotrophic sulphate-reducing bacterium, *Desulfotomaculum kuznetsovii* sp. nov. *Microbiology*, **57**, 823–7.

Nazina, T. N. and Rozanova, E. P. (1978). Thermophilic sulphate-reducing bacteria from oil strata. *Microbiology*, **47**, 142–8.

Nga, D. P., Cam Ha, D. T., Hien, L. T. and Stan-Lotter, H. (1996). *Desulfovibrio vietnamensis* sp. nov., a halophilic sulphate-reducing bacterium from Vietnamese oil fields. *Anaerobe*, **2**, 385–92.

Nilsen, R. K., Beeder, J., Thostenson, T. and Torsvik, T. (1996a). Distribution of thermophilic marine sulphate reducers in North Sea oil field waters and oil reservoirs. *Applied and Environmental Microbiology*, **62**, 1793–8.

Nilsen, R. K., Torsvik, T. and Lien, T. (1996b). *Desulfotomaculum thermocisternum* sp. nov., a sulphate reducer isolated from a hot North Sea oil reservoir. *International Journal of Systematic Bacteriology*, **46**, 397–402.

Ollivier, B. and Cayol, J.-L. (2005). The fermentative, iron-reducing, and nitrate-reducing microorganisms. In B. Ollivier and M. Magot (eds.), *Petroleum microbiology*. Washington, DC: ASM Press. pp. 71–88.

Orphan, V. J., Taylor, L. T., Hafenbradl, D. and Delong, E. F. (2000). Culture-dependent and culture-independent characterization of microbial assemblages associated with high-temperature petroleum reservoirs. *Applied and Environmental Microbiology*, **66**, 700–11.

Osipov, G. A., Nazina, T. N. and Ivanova, A. E. (1995). Study of species composition of microbial community of water-flooded oil field by chromato-mass spectrometry. *Microbiology*, **63**, 490–3.

Philippi, G. T. (1977). On the depth, time, and mechanism of origin of the heavy to medium gravity naphtenic crude oil. *Geochimica Cosmochimica Acta*, **41**, 33–52.

Rees, G. N., Grassia, G. S., Sheehy, A. J., Dwivedi, P. P. and Patel, B. K. C. (1995). *Desulfacinum infernum* gen. nov., sp. nov., a thermophilic sulphate-reducing bacterium from a petroleum reservoir. *International Journal of Systematic Bacteriology*, **45**, 85–9.

Reysenbach, A.-L., Longnecker, K. and Kirshtein, J. (2000). Novel bacterial and archaeal lineages from an in situ growth chamber deployed at a Mid-Atlantic Ridge hydrothermal vent. *Applied and Environmental Microbiology*, **66**, 3798–806.

Rosnes, J. T., Torsvik, T. and Lien, T. (1991). Spore-forming thermophilic sulphate-reducing bacteria isolated from North Sea oil field waters. *Applied and Environmental Microbiology*, **57**, 2302–7.

Rozanova, E. P. and Khudyakova, A. I. (1974). A new nonspore-forming thermophilic sulphate-reducing organism, *Desulfovibrio thermophilus* nov. sp. *Microbiology*, **43**, 1069–75.

Rozanova, E. P. and Nazina, T. N. (1979). Occurrence of thermophilic sulphate-reducing bacteria in oil-bearing strata. *Microbiology*, **48**, 907–11.

Rozanova, E. P., Nazina, T. N. and Galushko, A. S. (1988). Isolation of a new genus of sulphate-reducing bacteria and description of a new species of this genus, *Desulfomicrobium apsheronum* gen. nov. sp. nov. *Microbiology*, **57**, 634–41.

Rozanova, E. P. and Pivovarova, T. A. (1988). Reclassification of *Desulfovibrio thermophilus* (Rozanova, Khudyakova, 1974). *Microbiology*, **57**, 102–6.

Rozanova, E. P., Tourova, T. V. *et al.* (2001). *Desulfacinum subterraneum* sp. nov., a new thermophilic sulphate-reducing bacterium isolated from a high-temperature oil field. *Microbiology*, **70**, 466–71.

Rueter, P., Rabus, R., Wilkes, H. *et al.* (1994). Anaerobic oxidation of hydrocarbons in crude oil by new types of sulphate-reducing bacteria. *Nature,* **372**, 455–8.

Shen, Y. and Buick, R. (2004). The antiquity of microbial sulphate-reduction. *Earth Science Reviews,* **64**, 243–72.

Sievert, S. M. and Kuever, J. (2000). *Desulfacinum hydrothermale* sp. nov., a thermophilic, sulphate-reducing bacterium from geothermally heated sediments near Milos Island (Greece). *International Journal of Systematic and Evolutionary Microbiology,* **50**, 1239–46.

Stetter, K. O., Huber, R., Blöchl, E. *et al.* (1993). Hyperthermophilic Archaea are thriving in deep North Sea and Alaskan oil reservoirs. *Nature,* **365**, 743–5.

Tardy-Jacquenod, C., Caumette, P., Matheron, R. *et al.* (1996a). Characterization of sulphate-reducing bacteria isolated from oil-field waters. *Canadian Journal of Microbiology,* **42**, 259–66.

Tardy-Jacquenod, C., Magot, M., Laigret, F. *et al.* (1996b). *Desulfovibrio gabonensis* sp. nov., a new moderately halophilic sulphate-reducing bacterium isolated from an oil pipeline. *International Journal of Systematic Bacteriology,* **46**, 710–15.

Tardy-Jacquenod, C., Magot, M., Patel, B. K. C., Matheron, R. and Caumette, P. (1998). *Desulfotomaculum halophilum* sp. nov., a halophilic sulphate-reducing bacterium isolated from oil production facilities. *International Journal of Systematic Bacteriology,* **48**, 333–8.

Telang, A. J., Ebert, S., Foght, J. M. *et al.* (1997). Effect of nitrate injection on the microbial community in an oil field as monitored by reverse sample genome probing. *Applied and Environmental Microbiology,* **63**, 1785–93.

Vance, I. and Trasher, D. R. (2005). Reservoir souring: mechanisms and prevention. In B. Ollivier and M. Magot (eds.), *Petroleum Microbiology,* Washington, DC: ASM Press. pp. 123–42.

Voordouw, G. (1995). The genus *Desulfovibrio*: the centennial. *Applied and Environmental Microbiology,* **61**, 2813–19.

Voordouw, G., Armstrong, S. M., Reimer, M. F. *et al.* (1996). Characterization of 16S rRNA genes from oil field microbial communities indicates the presence of a variety of sulphate-reducing, fermentative, and sulfide-oxidizing bacteria. *Applied and Environmental Microbiology,* **62**, 1623–9.

Watanabe, K., Kodama, Y. and Kaku, N. (2002). Diversity and abundance of bacteria in an underground oil-storage cavity. *BMC Microbiology,* **2**, 23–32.

Widdel, F. (1988). Microbiology and ecology of sulphate- and sulfur-reducing bacteria. In A. J. B. Zehnder (ed.), *Biology of Anaerobic Microorganisms.* New York: John Wiley and Sons, Inc. pp. 469–585.

CHAPTER 11

The sub-seafloor biosphere and sulphate-reducing prokaryotes: their presence and significance

R. John Parkes and Henrik Sass

11.1 GENERAL INTRODUCTION

Approximately 70% of the Earth's environment is marine, which includes substantial sediment deposits, some of which can be greater than 10 km in depth (Fowler, 1990). Although these sediments contain the largest global organic carbon reservoir (~15 000 × 10^{18} g C, Hedges and Keil, 1995), apart from shallow margin sediments (to 200 m water depth), they have been considered to be relatively biogeochemically inactive. For example, Jørgensen (1983) calculated that margin sediments accounted for 83% of global marine sediment oxygen uptake whilst only representing 8.6% of global sediment area. In contrast, deeper sediments (200 to >4000 m water depths), despite being ~91% of marine sediment area, accounted for only 17% of global oxygen uptake. The situation was considered even more extreme for rates of sulphate reduction, with this being responsible for, respectively, 50% and 0% of all organic matter being degraded in margin and deep water sediments (>4000 m water depths) (Jørgensen, 1983). This low activity was consistent with results demonstrating the limited depth distribution of prokaryotic populations in deep sediments. Morita and ZoBell (1955) concluded that the marine biosphere ended at 7.47 m deep, based on their inability to culture bacteria at this or greater depths. Reports of prokaryotes being isolated from deeper sediments were considered to be contaminants introduced during sampling, or dormant organisms being re-activated (ZoBell, 1938).

Although this situation fitted with the conventional separation of biosphere and geosphere, it resulted in a puzzling anomaly; apparently little was happening between the cold (~4 °C), surface biosphere processes in the top few metres, and thermogenic geosphere processes at several kilometres depth and temperatures around 100 °C and above (Quigley and Mackenzie, 1988). Extreme conditions, such as high pressure and initial low temperatures, combined with removal of limited energy supplies by efficient near-surface microbial communities, was accepted as an explanation for this paradox, despite indirect geochemical evidence suggesting the presence of microbial processes at considerable depths within sediments (e.g. chemical changes in pore water, gas production, modification of organic biomarkers, stable isotopic evidence). It was not until direct measurement of microbial populations and processes on deep sediment samples from the Ocean Drilling Program (ODP) was conducted, using a comprehensive range of microbial ecological techniques, that an extensive deep prokaryotic biosphere was shown to exist globally in marine sediments (Coolen *et al.*, 2002; Cragg *et al.*, 1997; Cragg *et al.*, 1998; D'Hondt *et al.*, 2004; Parkes *et al.*, 1994; 2000; 2005; Schippers *et al.*, 2005; Wellsbury *et al.*, 1997; 2002). Including subsurface prokaryotes in global estimates of the magnitude and distribution of prokaryotes astonishingly suggested that 90% of all prokaryotes reside in subsurface sediments and rocks (Whitman *et al.*, 1998), with the majority being in the marine subsurface (60%). Hence, this recent research demonstrated that deep sub-seafloor sediments were not a prokaryotic desert but probably the largest prokaryotic habitat on earth, with significant populations of sulphate-reducing bacteria (SRB, e.g. Bale *et al.*, 1997).

11.2 GEOCHEMICAL EVIDENCE FOR DEEP SULPHATE REDUCTION

Removal of pore water sulphate with increasing sediment depth can indicate the intensity of sulphate reduction in deep sediments (e.g. Figure 11.1). Analysis of these sulphate gradients on a global distribution of ODP sites suggest rates of subsurface sulphate reduction are at least 2–3 orders of magnitude higher in ocean margin sites than in sulphate-rich open ocean sites (Canfield, 1991; D'Hondt *et al.*, 2002). However, using this approach also suggests that subsurface sulphate reduction is restricted in both locations: (1) although high organic matter input to ocean margin sites fuels intense sulphate reduction, this rapidly removes sulphate (within a few tens of metres), thus restricting activity to near surface sediments; (2) conversely, in open ocean sites low organic matter input severely restricts microbial activity, and sulphate reduction rates can be almost indistinguishable from zero. Beneath 1.5 metres below the seafloor (mbsf) most sulphate is thought to be used in the anaerobic oxidation of CH_4 (AOM) formed in the

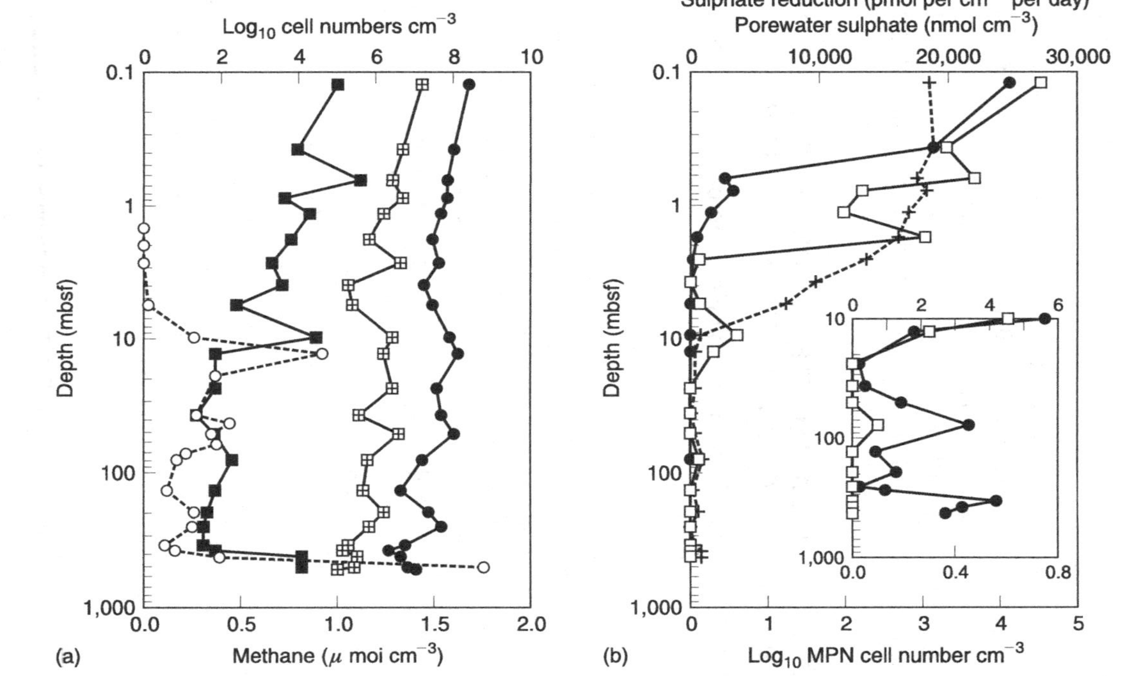

Figure 11.1. Depth profiles and subsurface stimulation of bacterial activity, including sulphate reduction in Japan Sea sediments (Parkes *et al.*, 1994). (a) Total prokaryotic population (●), numbers of dividing and divided cells (⊞), viable anaerobic heterotrophic bacteria (■), methane concentrations (○). (b) Sulphate reduction rates (●), viable SRB (□), porewater sulphate concentrations (+).

sediments below, presumably by a consortium of prokaryotes including SRB (Boetius *et al.*, 2000). Overall modelled sulphate reduction rates are so low as to suggest that most cells detected in the subsurface must be inactive or adapted for extraordinarily low metabolic activity (D'Hondt *et al.*, 2002). However, this may also reflect limitations of the modelling approach due to, for example, non-steady state conditions in the sediment; anaerobic oxidation of the sulphides formed back to sulphate (Bottrell *et al.*, 2000); flow of sulphate-containing fluids at depth (Parkes *et al.*, 2000; 2005; see below), or other anaerobic processes, in addition to sulphate reduction, being important in sub-seafloor sediments, for example, metal reduction, nitrification and nitrate respiration (Cragg *et al.*, 2003; D'Hondt *et al.*, 2004).

11.3 PRESENCE AND DISTRIBUTION OF PROKARYOTIC SULPHATE REDUCTION IN DEEP MARINE SEDIMENTS

11.3.1 Sensitivity of geomicrobiological approaches to detect sulphate reduction

Measurement of sulphate reduction rates by radiotracer techniques, although having their own limitations (e.g. laboratory incubation conditions are not those in situ, particularly with regard to pressure; disturbance during coring; long incubations (days to months) which may change sedimentary conditions) which preclude measurement of true in situ activities, they are extremely valuable, especially on a comparative basis (changes with depth, site to site etc). They are also much more sensitive than modelling based on sulphate removal. For example, in the Japan Sea (ODP Leg 128, site 798B in 900 m water depth Figure 11.1, Parkes et al., 1994), sulphate was rapidly removed with depth and reached minimum concentrations by 9.62 mbsf (0.78 mM). Hence, it could be considered that there was no sulphate-reduction or SRB in deeper layers. However, although radiotracer-measured rates of sulphate reduction correlated with the sulphate profile ($P<0.05$) and had highest rates in the top 5–6 mbsf, very low rates continued all the way down the core (bottom, 503.45 mbsf, average rate ~2 pmol cm^{-3} d^{-1}) in the continuous presence of sulphate, albeit at low concentrations (max 0.92 mM). In addition, there was a small stimulation of sulphate reduction rates towards the bottom of the core (Figure 11.1) and this was consistent with increases in iron sulphides in the same layer. These data also correlated with viable populations of SRB, although in the very deepest layers SRB were below detection limits. Despite this, SRB were enriched from both 80 and 500 mbsf and these bacteria had characteristics which matched their deep sedimentary habitat (Bale *et al.*, 1997). Together with deep stimulation of sulphate reduction, there was a significant increase in viable bacterial populations below ~360 mbsf, particularly of anaerobic heterotrophic bacteria ($P<0.01$), which coincided with a large increase in methane (Figure 11.1) and other thermogenic hydrocarbons (Ingle *et al.*, 1990). This suggested that these thermogenic products were stimulating deep subsurface SRB-containing bacterial communities in sediments ~ 4.3 million years old (Parkes *et al.*, 1994).

11.3.2 Stimulation of sulphate reduction in deep sub-seafloor gas hydrate formations

If CH_4 is an important energy source for sub-seafloor prokaryotes, then prokaryotic processes might be stimulated in deep sediments containing gas hydrates, which globally have been estimated to contain twice the amount of organic carbon in conventional fossil fuels (Kennicutt *et al.*, 1993). Scientists of the ODP conducted the first microbiological investigation on subsurface gas hydrates in 1992 (Cragg *et al.*, 1995; 1996). The study site on the Cascadia Margin accretionary system (Pacific Ocean, site 889/890, water depth 1320 m) contained a discrete hydrate layer between 215 and 225 mbsf (Figure 11.2). Prokaryotic populations and activities were significantly stimulated within this zone, including rates of anaerobic oxidation of methane (~9 times the average rate at other depths). Presumably, the elevated rates of AOM were due to the high CH_4 concentrations within the hydrate zone and this was responsible for a stimulation in deep prokaryotic populations (tenfold increase). Despite SRB populations also being stimulated within this zone, it was unclear whether their activity was directly associated with AOM, as in CH_4 seep sites and near surface gas hydrates (Boetius *et al.*, 2000; Knittel *et al.*, 2005), since rates of sulphate reduction were stimulated above rather than within the hydrate layer (Figure 11.2). Possibly fluid flow into accretionary wedge sediments is important in providing electron acceptors for continued deep AOM and thus making the deep hydrate zone so biogeochemically active (Cragg *et al.*, 1996). This was reinforced by rates of deep AOM increasing as gas and fluid venting increased at sites in this location (Table 11.1).

Deep stimulation of biogeochemical activity in gas hydrate containing sediments was confirmed at Blake Ridge in the Atlantic Ocean (ODP Leg 164, site 995, water depth 2778 m, Figure 11.3, Wellsbury *et al.*, 2000). Although sulphate reduction rates were maximal near the surface (400 nmol cm^{-3} d^{-1}) and decreased with increasing depth, at and below the base of the hydrate

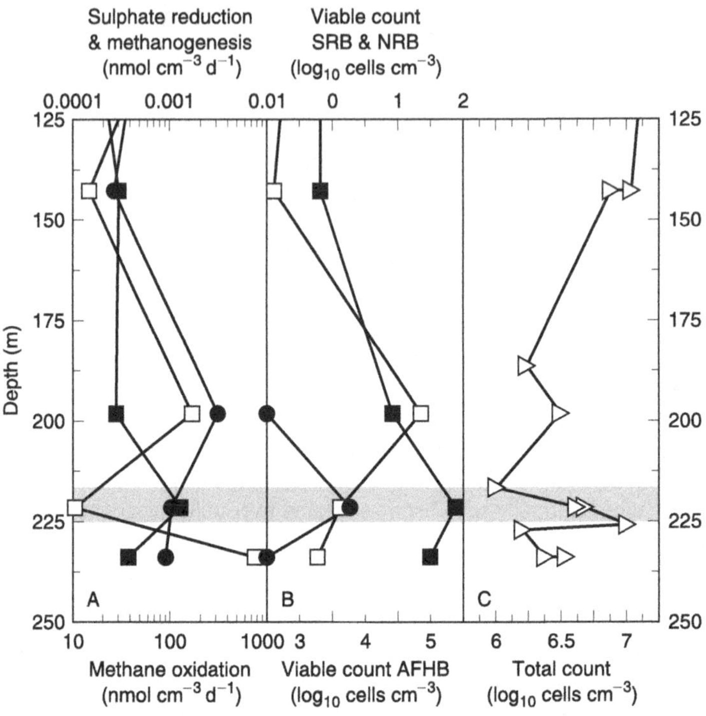

Figure 11.2. Depth profiles and subsurface stimulation of bacterial activity, including sulphate reduction in Cascadia Margin hydrate sediments (Cragg *et al.*, 1996). (A) Rates of sulphate reduction (●), methanogenesis (□), methane oxidation (●). (B) Viable SRB (●), nitrate-reducing (■) and fermentative heterotrophs (□). (C) Total prokaryotic population (▷). Shaded area denotes a zone of gas hydrates.

zone (between 195 and 450 mbsf), there was stimulation in sulphate reduction with maximum subsurface rates in the deepest sample (691 mbsf, 11 nmol cm^{-3} d^{-1}). This was associated with increases in a range of other activities (Figure 11.3), including AOM and increases in the total bacterial population (Wellsbury *et al.*, 2000). These results are similar to those for the Cascadia Margin, however, with a more extensive hydrate layer, stimulation was clearly shown to be below the base of the hydrate layer, suggesting that it is the high free CH_4 gas concentration (Dickens *et al.*, 1997) which stimulates prokaryotic activity rather than the CH_4 locked within the hydrate ice structure. Rates of sulphate reduction correlated with both numbers of SRB (deepest 439 mbsf) and sulphate concentrations ($P<0.0002$). Although sulphate concentrations decreased rapidly with depth, low concentrations persisted in some deeper layers (0–0.2 mM).

Table 11.1. *The impact of increasing subsurface gas hydrates and fluid venting on deep bacterial activity at three sites on the Cascadia Margin subduction zone sediments (ODP Leg 146, (Cragg et al., 1996)*

			Rates of activity nmol cm^{-3} d^{-1}		
Site	Mean organic carbon (wt%)	Depth (mbsf)	Sulphate reduction	Methano-genesis	Methane oxidation
888	0.56	Near-surface	84.50	0.0071	0.033
		>200	0.00	0.0029	0.71
891	0.61	Near-surface	34.96	0.0048	0.006
		>200	0.00004	0.00087	76.91
889/890	0.64	Near-surface	6.85	0.027	0.124
		>200	0.0009	0.0077	134.54

Gas and fluid venting increases from Site 888 to 891 to 889/890.

This sulphate, and other electron acceptors, might be provided by deep fluid flow coupled with deep anaerobic oxidation of sulphides (Bottrell *et al.*, 2000). Increasing temperature with depth can also increase the reactivity of buried organic matter, and Wellsbury *et al.* (1997) demonstrated that significant acetate concentrations can be produced during even gentle heating and that this processes probably directly involves prokaryotic activity. In Blake Ridge large concentrations of acetate and other volatile fatty acids were produced at depth (Figure 11.4), probably because there was a dramatic increase in the mass accumulation rate for organic matter at depth (Paull *et al.*, 1996) which would greatly facilitate organic acid generation. This would be further enhanced by deep fluid flow (Egeberg and Barth, 1998) and together these factors would facilitate the deep stimulation of prokaryotic activity and populations, including deep CH_4 production which may contribute to biogenic gas hydrate formation in this region (Wellsbury and Parkes, 2000).

11.3.3 Deep supply of sulphate

Sulphate re-supply at depth and deep stimulation of sulphate reduction occur in a number of different locations, especially if sulphate diffuses into CH_4-rich sediments. For example, deep ancient brines occur on the Peru Margin (~150 m water depth) and result in two sulphate:CH_4 interfaces where there was significant deep stimulation of prokaryotic activity

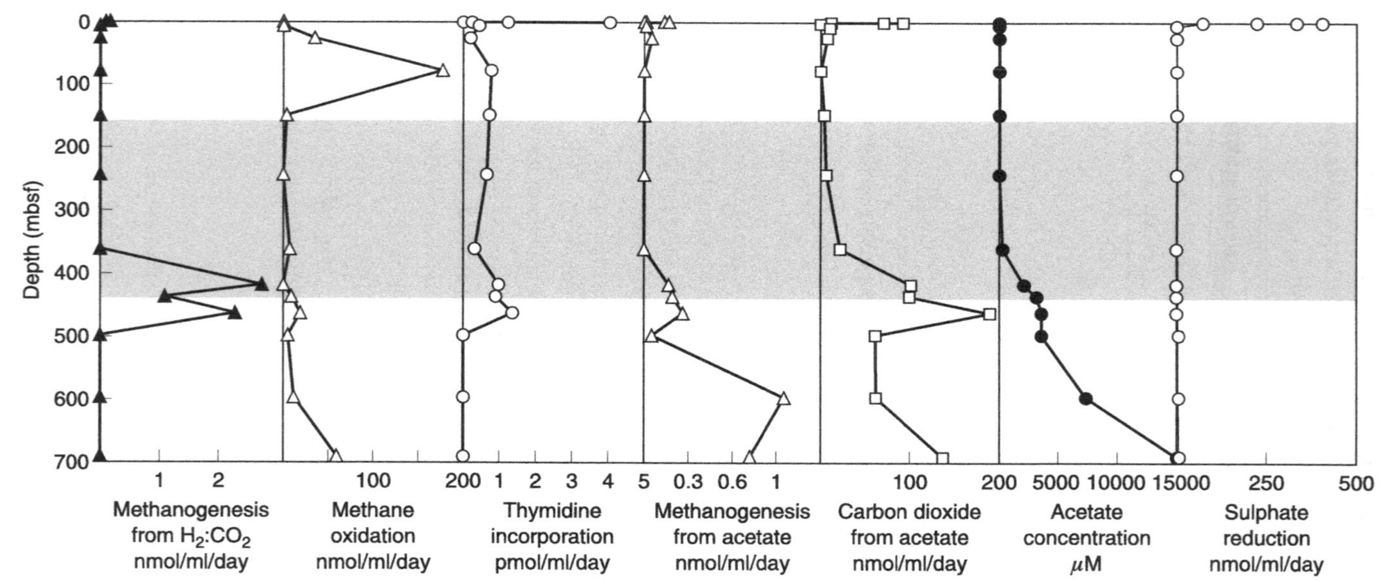

Figure 11.3. Depth profiles of rates of bacterial activity, including sulphate reduction, and concentrations of porewater acetate in gas-hydrate sediments from Blake Ridge, ODP Leg 164. The shaded area denotes the boundaries of the gas-hydrate stability field. (Modified from Wellsbury *et al.*, 1997; Wellsbury *et al.*, 2000.)

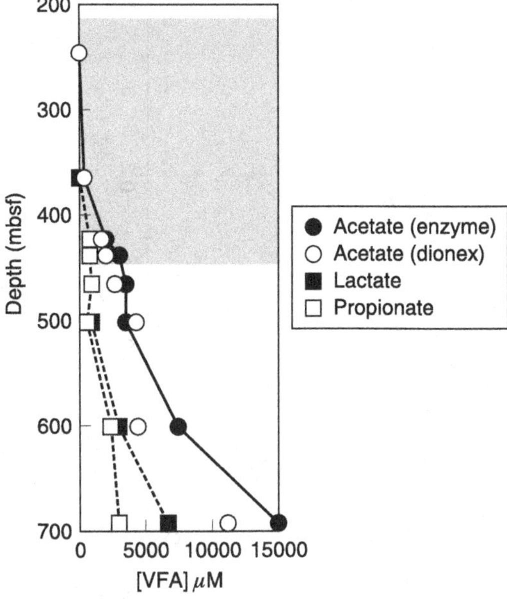

Figure 11.4. Depth profiles of volatile fatty acids in gas-hydrate sediments from Blake Ridge, ODP Leg 164. The shaded area denotes the boundaries of the gas-hydrate stability field.

and biomass, presumably fuelled by AOM (Figure 11.5, Parkes *et al.*, 2005). The population increase was such that at the lower sulphate:CH_4 interface at ~90 mbsf, the total prokaryotic populations was larger than those near the sediment surface (~13-fold). Reinforcing the dynamic nature of these subsurface interfaces, there are considerable changes in bacterial biodiversity despite the lower interface being deposited about 0.8 Mya ago (Figure 11.5). Just below this interface, in the deep sulphate zone, there was a peak in sulphate reduction. As previously described for gas hydrate sites anaerobic sulphide oxidation can also provide sulphate in deep sediment layers (Bottrell *et al.*, 2000).

It has been estimated that more than one-third of the entire seafloor is underlain by active convective circulation, including ridge-crest and ridge-flank systems (Anderson *et al.*, 1979). On the eastern flank of the Juan de Fuca Ridge (ODP Leg 168, site 1027, water depth 2658 m) this results in sulphate diffusing from the basement below into CH_4-rich sediments (Figure 11.6, Mather and Parkes, 2000). Total prokaryotic populations are stimulated ($P<0.05$) in this zone below ~527 mbsf, presumably due to AOM-associated sulphate reduction, even under the prevailing thermophilic

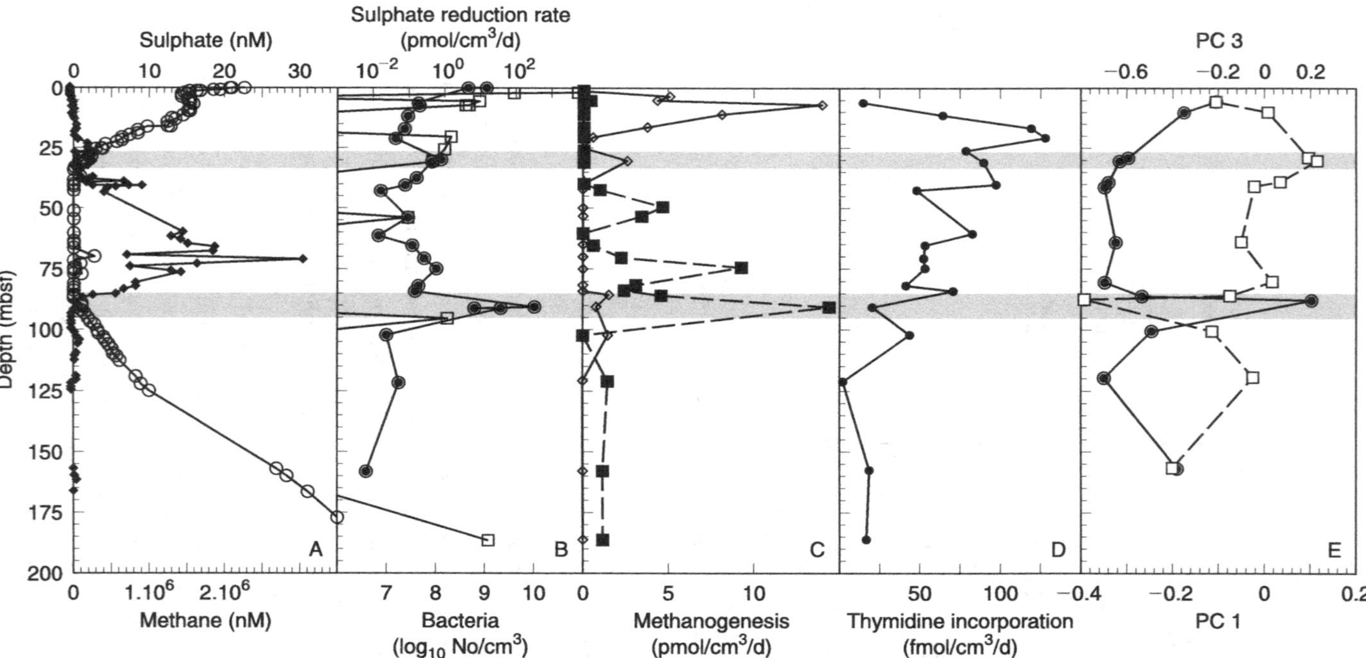

Figure 11.5. Stimulation of sulphate reduction and other deep biogeochemical process coupled to changes in bacterial biodiversity at geochemical interfaces in Peru Margin Site (ODP 1229). (A) geochemistry: (○) pore water sulphate (mM), (◆) CH_4 (nM). (B) (□) sulphate reduction rates (pmol/cm^3/d), (●) total population (Log$_{10}$ Nos./cm^3). (C) methanogenic rates (pmol/cm^3/d) (◇) H_2/CO_2, (■) acetate. (D) growth rates (●) thymidine incorporation (fmol/cm^3/d). (E) Principal components profile of diversity of Bacteria from DGGE analysis of 16S rRNA gene sequences: (●) Component 1 (56% of variation), (□) Component 3 (9% of variation); Component 2 (24% of variation) has a similar profile to Component 1. Shaded boxes highlight elevated prokaryotic processes and sulphate:methane interfaces (Parkes *et al.*, 2005).

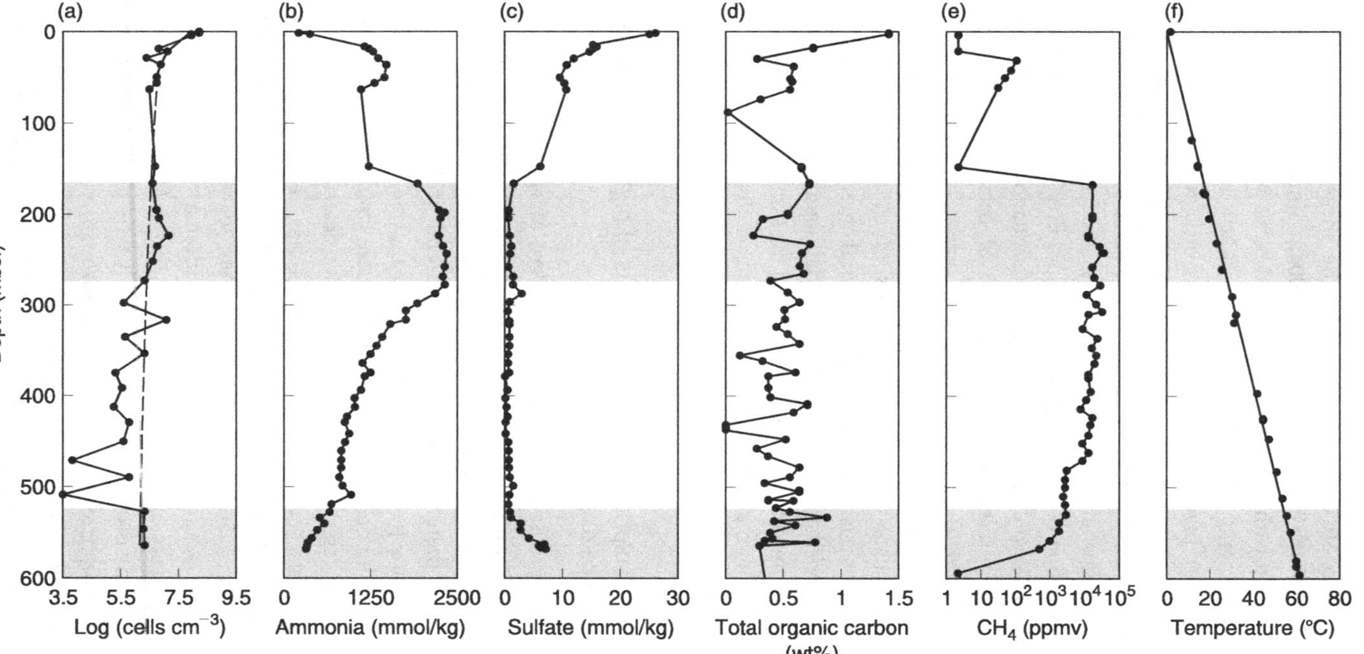

Figure 11.6. The stimulation of deep prokaryotic populations due to sulphate diffusing into CH_4 rich sediments from underlying bedrock at ODP Site 1027 (Leg 168, Juan de Fuca Ridge, Pacific Ocean). Shaded areas highlight zones of elevated prokaryotic populations (Mather and Parkes, 2000). In addition to the stimulation due to the deep sulphate source there was an increase in populations between ~166 & ~273 m depth. The reason for this was unclear as organic carbon concentrations were similar both above and below this zone. However, the increased concentrations in ammonia in this zone also suggested a more active bacterial population in this zone.

conditions (~55 to 60 °C, which is within the measured temperature maximum for AOM (Kallmeyer and Boetius, 2004)). Similar deep sulphate supply accompanied with CH_4 removal occurred below ~700 mbsf in deep Pacific Ocean sediments of the Woodlark Basin (ODP Leg 180, site 1118, 2303 m water depth), together with the presence of significant prokaryotic populations (3.2×10^5 cm^{-3}) in the deepest samples so far studied, 842 mbsf (Wellsbury *et al.*, 2002). As such a large area of ocean sediment is thought to have similar fluid flow along the basement, this could result in a globally significant stimulation of deep sulphate reduction. Even in the absence of significant CH_4 in sediments above the basement, oxidation of the sediments due to fluid flow results in deep stimulation of bacterial activity (e.g. nitrification, D'Hondt *et al.*, 2004). Profiles of activity upwards from the basement are interestingly a mirror image of the depth profile of bacterially driven redox reactions in near surface sediments, which is also as a result of oxygen diffusion from seawater. Hence, in these and other deepwater, Pacific Ocean sediments (Wellsbury *et al.*, 2002), bacterial populations and activity are present in the complete sediment column and the same situation is likely to occur in most other marine sediments, including sediments much deeper than those analyzed to date.

11.3.4 Deep sulphate reduction in layers of high organic matter and persistence over geological time

In deepwater Pacific Ocean sites off the Peru Margin (3297 m water depth) organic matter input is much less than in margin sites, and there is only limited sulphate removal (Figure 11.7). However, in three discrete zones down to 400 mbsf there is a diatom-rich layer and prokaryotic activities (e.g. thymidine incorporation growth and manganese reduction) and populations are consistently stimulated in these zones. This includes sulphate reduction rates which increase to above detection limits. It may be that the diatomaceous organic material is less reactive than other sedimentary organic matter and thus survives to fuel low, but continuing prokaryotic activity, including sulphate reduction, for at least 7–11 Mya (Parkes *et al.*, 2005). A similar situation occurs with marine sapropels which are also discrete layers of high organic matter. In the Mediterranean Sea, sapropels can contain over 10 to a maximum of 30 wt% total organic carbon and, in addition, sulphate concentrations increase rather than decrease due to upwards diffusion from ancient buried evaporite deposits. However, the organic matter in the sapropels must be resistant to degradation under sulphate-reducing conditions in order to have survived the long burial times

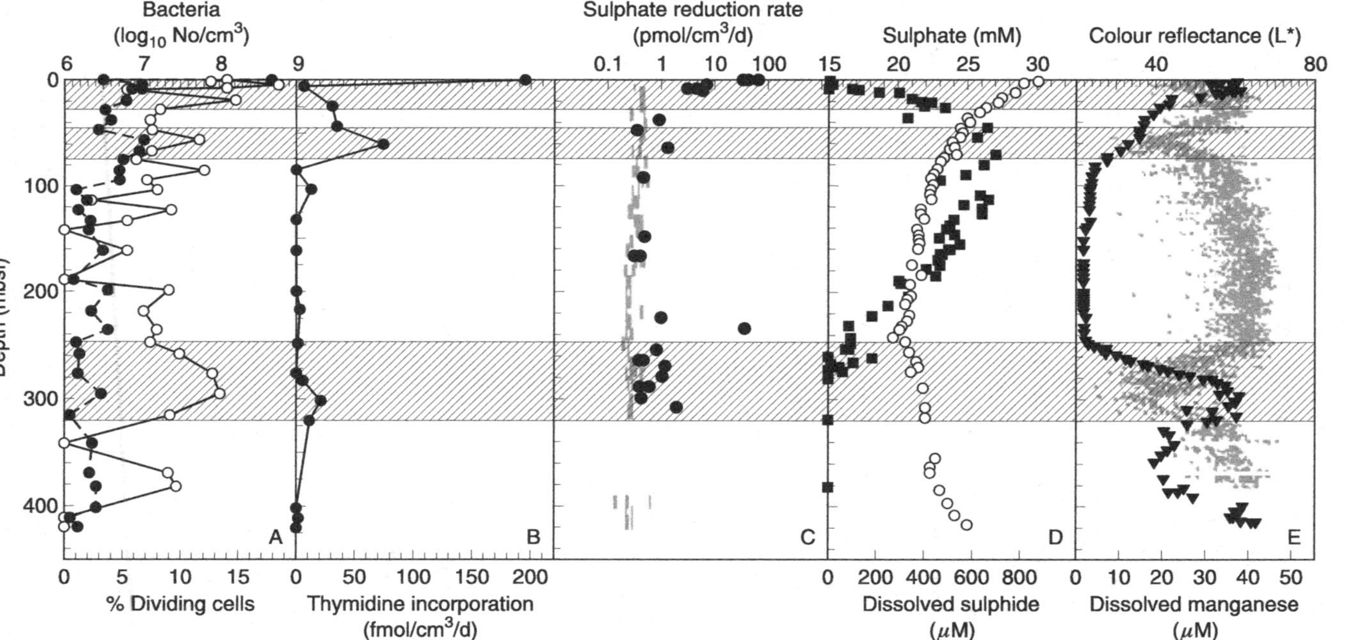

Figure 11.7. Stimulation of sulphate reduction and other deep biogeochemical process in diatom rich lithologies at a Pacific Open Ocean Site (ODP 1226). (A) Prokaryotic profiles: (●) total prokaryotic population (Log_{10} Nos./cm^3), (○) percentage dividing and divided cells. (B) Rates of prokaryotic growth (●) thymidine incorporation (fmol/cm^3/d). (C) (●) sulphate reduction rates, (▮) minimum detection limits (pmol/cm^3/d). (D) geochemistry: (○) porewater sulphate (mM), and (■) hydrogen sulphide (μM). (E) Colour reflectance as a measure of diatom abundance, low reflectance equals high diatom abundance, (▼) porewater manganese (μM). Shaded boxes highlight elevated prokaryotic processes and high diatom layers (Parkes *et al.*, 2005).

(millions of years), and sulphate reduction proceeds only slowly, as reflected in depth increases in total sulphur. Despite this recalcitrance, prokaryotic populations were significantly elevated in sapropel compared to non-sapropel layers (Coolen *et al.*, 2002; Cragg *et al.*, 1998). In addition, prokaryotic numbers (compared to adjacent non-sapropel layers) increased with sapropel depth/age, enabling a growth rate to be calculated. Growth rates equated to an average division time for active cells of ~105 000 years (Parkes *et al.*, 2000). As the sapropel environment probably provides both substrate and electron acceptor in excess, this growth rate should reflect the optimal rate for sulphate reducing prokaryotic communities growing on recalcitrant sapropel organic matter. Although present at much lower concentrations, normal sedimentary organic matter is probably equally recalcitrant as sapropel organic matter, and hence SRB in normal sub-seafloor sediments will probably grow, on average, at rates even slower than those in sapropels. The survival of very slowly degradable marine organic matter over geological time frames is consistent with the continuing microbiological activity of Atlantic Coastal Plain sediments, originally deposited in the marine environment during the Cretaceous period (65–144 Mya, Chapelle and Bradley, 1996), but now on land. Even in consolidated marine Cretaceous shale rocks (170–300 m depth), sulphate reduction can be stimulated at interfaces with sandstone layers which supply sulphate to oxidize fermentation products slowly diffusing from the low porosity shales (Krumholz *et al.*, 1997).

11.3.5 Sulphate reduction in the oceanic crust

It has been suggested that prokaryotes also inhabit the basement rock beneath marine sediments (Fisk *et al.*, 1998), that biotic alteration of glass dominates the upper 250 m of the crust, and may continue to ~500 m depth and at temperatures up to ~110 °C (Furnes and Staudigel, 1999). Potential energy sources include oxidation of ferrous and sulphide minerals, hydrogen formation from hydrolytic oxidation of ferrous iron in basaltic glass and mafic minerals (Bach and Edwards, 2003), and dissolved organic carbon (DOC) in circulating seawater. This may be reflected in the presence of prokaryotes in fluids from 3.5 Mya ocean crust (Cowen *et al.*, 2003), including both bacterial and archaeal sulphate-reducers that may have been responsible for the lowering of sulphate concentrations in these fluids. The interplay between basement rock and convective fluids and their relative importance in providing deep prokaryotic energy sources has still to be clarified, but considering the large concentration of sulphate in seawater-derived fluids SRB, should play an important role in this environment.

11.3.6 Quantitative importance of sulphate reduction in representative deep marine sediments

Even in low organic matter sub-seafloor sediments without gas hydrates, deep fluid flow, deep evaporites, or organic rich layers and with only an average thermal gradient ($\sim$30 °C km^{-1}), deep, sulphate reduction is still quantitatively important. Just how important depends on the depth of sulphate penetration (Wellsbury *et al.*, 2002). For example, in Woodlark Basin (ODP Leg 180, sites 1109 and 1115, water depths 2211 and 1150 respectively) the majority of the total prokaryotic population was in sediments below 20 mbsf ($\sim$77%, Wellsbury *et al.*, 2002). At site 1109 where maximum rates of sulphate reduction were highest (4.75 nmol cm^3 day^{-1}) and sulphate was removed by 107 mbsf, the upper 20 m of sediment accounted for the majority of sulphate reduction (65%). Although, this still means that a considerable $\sim$35% of total sediment sulphate reduction (as the complete sediment column was sampled down to a dolerite basement), occurs in deep sediments (down to 773 mbsf) at this site. In contrast, Site 1115 had lower maximum rates of sulphate reduction (1.26 nmol cm^3 day^{-1}) which resulted in sulphate being present much deeper ($\sim$199 mbsf) and enabled low rates of sulphate reduction to continue. At this site, which is representative of large areas of ocean sediment, the majority of sulphate reduction (72%) occurred below the upper 20 m of sediment.

11.4 SULPHATE-REDUCING BACTERIA IN THE MARINE SUBSURFACE

So far only a limited number of studies have applied cultivation-based methods for the quantification of microorganisms in subsurface environments. In general, viable counts decreased more rapidly with depth than total cell counts, and were several orders of magnitude lower than those obtained for near-surface environments (generally less than 0.01% of total cell count, Parkes *et al.*, 1994; 1995; Toffin *et al.*, 2004) (Table 11.2). One reason for this low culturability is assumed to be the "substrate shock" (Postgate and Hunter, 1963) to which the cells are exposed after transfer into relatively nutrient-rich media. Consistent with this view, the use of media with relatively low substrate concentrations increased the culturability of microorganisms from ancient Mediterranean sapropels to

Table 11.2. *Viable counts of sulphate-reducing bacteria in several marine subsurface environments*

Site	Sample type	Depth (mbsf)	Total cell counts ($\times 10^7$ cm^{-3})	Viable counts of SRB (cm^{-3})	SRB counts (% of TCC[b])	Reference
Peru Margin	Marine sediments (site 681C)	1–4	50–150	$10–1.2 \cdot 10^5$	<0.012	Parkes *et al.*, 1990
(ODP Leg 112)		4.5–80	20–30	b.d.l.[a]		
Japan Sea	Marine sediments	1–10	1.9–8.3	$0–10^3$	<0.003	Parkes *et al.*, 1994
(ODP Leg 128)		10–500	0.2–13	0–4	$< 8 \cdot 10^{-7}$	
Cascadia Margin	Marine sediments (site 888)	1–10	0.3–3.1	$2–10^3$	<0.004	Cragg *et al.*, 1996
(ODP Leg 146)		10–400	0.33–3.6	0–2	<0.00004	
	Marine sediments (site 889/890)	1–250	0.06–10	0–10	<0.00004	
Blake Ridge	Marine sediments (site 994)	1–640	0.22–3.2	0–10	<0.00003	Wellsbury *et al.*, 2000
(ODP Leg 164)	Marine sediments (site 995)	1–700	0.15–10	0–30	<0.00003	
Eastern	Sapropels	0.5–10	6.0–13	0–1000	<0.0008	Sass *et al.* (unpublished)
Mediterranean Sea	Hemipelagic sediments	1–10	1.8–2.7	0–200	<0.0007	

[a] b.d.l. below detection limit
[b] TCC: total cell counts

up to 3% of the total cell count (Süß *et al.*, 2004). The influence of limited in situ supply of organic substrates on the culturability of subsurface bacteria is supported by the finding that viable counts are elevated at contaminated ground water sites receiving input from landfills (Ludvigsen *et al.*, 1999) or oil spillage (Bekins *et al.*, 1999), and also in deep sediment layers with elevated organic carbon content, as for example, the Peru Margin (ODP Leg 112, Parkes *et al.*, 1990) and the Japan Sea (ODP Leg 128, Cragg *et al.*, 1992).

Although sulphate reduction is one of the most important processes in subsurface sediments, our knowledge of the community composition of sulphate-reducing prokaryotes (SRP) in these sediments is rather limited. So far only a limited number of sulphate-reducers from subsurface environments have been isolated. Most of these are affiliated with the deltaproteobacterial genera *Desulfovibrio* and *Desulfomicrobium*, or the sporeforming genera *Desulfosporosinus* and *Desulfotomaculum* within the *Firmicutes* (Table 11.3) (Bale *et al.*, 1997; Barnes *et al.*, 1998; D'Hondt *et al.*, 2004; Köpke *et al.*, 2005). Only recently, some *Desulfofrigus* sp. have been enriched and isolated from subsurface sediments (Süß *et al.*, 2004; Toffin *et al.*, 2004). Although, these genera are already well known from surface environments, the subsurface seems to harbour different species. In a cultivation-based study, Köpke *et al.* (2005) found almost no overlap between surface and subsurface (depth >1 m) sediment microbial communities in coastal sediments. From the upper 100 cm of the sediment *Desulfovibrio* spp. and some isolates related to the genera *Desulfotalea* and *Desulfuromonas* were obtained, while from the deeper layers mostly *Desulfosporosinus* and a few *Desulfovibrio* strains were isolated. The latter, however, were only distantly related to their surface relatives. The results of the cultivation-based studies are supported by several molecular studies. While members of the genus *Desulfotomaculum* were detected in the marine subsurface (Table 11.4) (Cowen *et al.*, 2003), *Desulfovibrio* and *Desulfosporosinus* spp. were also found in terrestrial (Boivin-Jahns *et al.*, 1996; Fry *et al.*, 1997; Moser *et al.*, 2003; Pedersen *et al.*, 1996) subsurface environments.

The subsurface seems to harbour only a limited diversity of SRB when compared with surface environments. Many phylotypes shown to dominate marine surface sediments such as *Desulfonema*, *Desulfobacter* or *Desulfosarcina* (Bidle *et al.*, 1999; Li *et al.*, 1999) have not yet been detected at greater depth. In addition, in several molecular surveys no sulphate-reducers were even detected (Kormas *et al.*, 2003; Reed *et al.*, 2002). Despite sulphate reduction resulting in sulphate depletion by $\sim$35 mbsf and being stimulated in deep sulphate brine in sediments of the Peru Margin

Table 11.3. *Physiological characteristics of sulphate-reducing bacteria isolated from brackish to saline subsurface environments*

	Desulfomic.	*Desulfovibrio*			*Desulfotomaculum*		
	macestense[a]	*'cavernae'*[b]	*indonesiensis*[b]	*profundus*[c]	*geothermicum*[d]	*kuznetsovii*[e]	sp. B2T[b]
Habitat	Ground-water	Triassic sandstone	Triassic sand-stone	Marine sedi-ment	Geothermal water	Geothermal water	Triassic sand-stone
Depth [m]	2500	580–900	580–900	80–520	2500	2800–3200	1060
Salinity [%]	0–2.5	0.5–22	0.5–22	0.2–10	0.2–5.0	0–3.0	0.2–16
Temperature [°C]	15–40	20–50	15–50	15–65	37–57	50–85	30–65
pH range	6.5–8.0	5.5–8.5	5.5–8.5	4.5–9.0	6.0–8.0	n.d.	5.7–8.2
Autotrophy	+	–	–	(+)	+	+	+
Electron donors							
H_2/CO_2	–	–	–	–	+	+	+
H_2	+	+	+	+	+	+	+
Formate	+	+	(+)	–	+	+	–
Acetate	–	–	–	–	–	+	–
n-Fatty acids	–	–	–	–	+	+	–
Lactate	+	+	+	+	+	+	+
Succinate	n.d.	+	+	n.d.	n.d.	+	+
Fumarate	n.d.	+	+	–	n.d.	+	+
Malate	–	+	+	–	n.d.	+	+
Methanol	–	(+)	(+)	–	–	+	–
Ethanol	+	+	+	–	+	+	+
n-Alcohols	–	+	+	n.d.	–	+	+

Glycerol	n.d.	+	+	n.d.	n.d.	n.d.	−
Amino acids	n.d.	(+)	−	−	n.d.	−	+
Manosac-charides	−	+	+	−	+	−	−
Electron acceptors							
Thiosulphate	+	+	+	(+)	n.d.	+	+
S°	n.d.	−	−	−	−	−	+
Nitrate	n.d.	+	+	+	−	−	+
Fe(III)	n.d.	−	(+)	+	n.d.	n.d.	+
Mn(IV)	n.d.	−	+	n.d.	n.d.	n.d.	+

Data from
[a] Gogotova and Vainshtein, 1989
[b] Sass and Cypionka, 2004
[c] Bale *et al.*, 1997
[d] Daumas *et al.*, 1988
[e] Nazina *et al.*, 1989
(+): activity weak or not present in all strains.

Table 11.4. *Environmental sequences of putative sulphate-reducing bacteria from different marine subsurface environments. Sequence similarities to most closely related species are given in brackets*

Sequence	Closest relative in GenBank	Phylum	Origin	References
Clone 1026B270	*Archaeoglobus veneficus* (97)	Euryarchaeota	Juan de Fuca Ridge	Cowen *et al.*, 2003
Clone 1026B15	*Desulfonatronum lacustre* (91)	Deltaproteobacteria	Juan de Fuca Ridge	Cowen *et al.*, 2003
Clone OHKB4.91	*Desulfobulbus mediterraneus* (93)	Deltaproteobacteria	Sea of Okhotsk	Inagaki *et al.*, 2003
DGGE Band 16	*Desulfobulbus mediterraneus* (92)	Deltaproteobacteria	North Sea	Wilms *et al.*, 2006
Clone OHKB2.18	*Desulfobacterium anilini* (87)	Deltaproteobacteria	Sea of Okhotsk	Inagaki *et al.*, 2003
DGGE Band 160J24	*Desulfonema magnum* (92)	Deltaproteobacteria	North Sea	Wilms *et al.*, 2006
DGGE Band Seq4 (Z1)	*Pelobacter carbinolicus* (88)	Deltaproteobacteria	Eastern Mediterranean	Coolen *et al.*, 2002
Clone 1026B114	*Desulfotomaculum thermosapovorans* (93)	'Firmicutes'	Juan de Fuca Ridge	Cowen *et al.*, 2003
Clone 1026B236	*Desulfotomaculum thermosapovorans* (94)	'Firmicutes'	Juan de Fuca Ridge	Cowen *et al.*, 2003

(Figure 11.6, Parkes *et al.*, 2005), no sulphate-reducers were detected, either in clone libraries or in sequenced DGGE (denaturing gradient gel electrophoresis) bands. Calculations based on the sulphate reduction rates indicate that the maximum proportion of SRB in the total microbial community was 0.002 to 0.02%, even at the sulphate–methane interfaces, where sulphate reduction was stimulated. Therefore it is most unlikely that SRB would be present in 16S gene libraries (Kemp and Aller, 2004). The estimated low numbers of sulphate-reducers were consistent with the lack of amplification products even with the highly specific functional primer (*dsr*AB) for sulphate-reducers (e.g. Parkes *et al.*, 2005). Hence, prokaryotes directly involved in sulphate reduction might be low in numbers but are highly active.

These results, however, were in apparent direct contradiction to other results for the same Peru Margin site (Mauclaire *et al.*, 2004) obtained by CARD-FISH (catalyzed reporter deposition fluorescence in situ hybridization). In this study, *Desulfobacteraceae* and *Desulfovibrionaceae* were the dominating Bacteria, and at some depths even representing all bacteria binding to the general eubacterial probe. This is surprising as even in shallow coastal sediments with intense sulphate reduction, SRB represent less than 20% of the total microbial community (e.g. Nedwell *et al.*, 2004). The study by Mauclaire *at al.* (2004) probably greatly overestimates SRB numbers since the probes used were not sufficiently specific for SRB and targeted other bacterial groups also. Analysis of these probes using the Probe Match programme of the RDP-II release 9 (http://rdp.cme.msu.edu/index.jsp) demonstrates that they match between 2 and 2.5% of the non-target organisms, including so far uncultured *Chloroflexi* (Green-non-sulphur bacteria). These were shown to be the dominating bacterial group in Peru Margin sediments (Parkes *et al.*, 2005).

It is highly likely that SRB belonging to novel phylogenetic groups will be discovered in the subsurface. Many DNA sequences retrieved from subsurface environments, particularly within the *Deltaproteobacteria*, are only distantly related to known sulphate-reducers (Table 6.4) (Coolen *et al.*, 2002; Inagaki *et al.*, 2003; Wilms *et al.*, 2006), so that they cannot, at present, be interpreted as being SRB. However, this does exclude the possibility of them being uncultured SRB. For example, an investigation of sulphite reductase (*dsr*AB) genes in the subsurface layers of coastal sediments (Thomsen *et al.*, 2001) and in fracture waters in rocks of a deep gold mine (Baker *et al.*, 2003), revealed the presence of so far unknown phylogenetic lineages of SRB in subsurface environments.

11.5 ADAPTATIONS OF SULPHATE-REDUCING BACTERIA TO THEIR SUBSURFACE HABITAT

Sulphate-reducing bacterial pure cultures obtained from marine subsurface sediments exhibit a salinity spectrum for growth that is typical for marine bacteria, ranging from brackish to slightly hypersaline conditions (Table 11.3, Bale *et al.*, 1997; Barnes *et al.*, 1998). But, surprisingly, most sulphate-reducing isolates from the terrestrial subsurface are also able to grow under marine conditions and even in hypersaline brines as well (Daumas *et al.*, 1988; Gogotova and Vainshtein, 1989; Nazina *et al.*, 1989; Sass and Cypionka, 2004). This extraordinarily high-salt tolerance allows these bacteria to access a wide range of terrestrial environments from low-salt groundwater to subsurface brines which are present in several subsurface sites (Sass and Cypionka, 2004), or even to colonize the marine subsurface.

Some of the isolates from the deep subsurface have an extraordinary broad temperature range for growth, such as *Desulfovibrio profundus* (Bale *et al.*, 1997) covering a span of 15 to 65 °C (Table 11.3). This broad temperature range might be seen as a prerequisite to colonize large parts of the deep biosphere, since temperature increases with depth. Whilst from the upper sediment layers (<1000 mbsf) predominantly mesophilic and slightly thermophilic species of SRB have been isolated (Table 11.3), the deeper, and hence, hotter layers can be expected to harbour mainly thermophilic sulphate-reducers such as *Archaeoglobus*, *Desulfotomaculum* (Cowen *et al.*, 2003) or *Thermodesulfovibrio* spp., found already in the deep terrestrial subsurface (Colwell *et al.*, 1997; Daumas *et al.*, 1988; Kimura *et al.*, 2005; Nazina *et al.*, 1989).

Pressure is probably one of the most important environmental factors influencing microbial communities in the deep subsurface, but which has received little attention so far due to technological difficulties (Bale *et al.*, 1997; Barnes *et al.*, 1998; Mangelsdorf *et al.*, 2005; Parkes *et al.*, 1995). However, available data indicate that sulphate-reducers from deep subsurface habitats are more barotolerant than related strains from surface environments. Subsurface sulphate-reducers showed no decrease in activity when pressure was increased to 20 MPa and were still active at pressures between 30 and 40 MPa (Bale *et al.*, 1997; Bradbrook, 2000), pressures twice to four times as high as tolerated by reference strain SRB. *Desulfotomaculum* sp. B10 is the most pressure-tolerant sulphate-reducing bacterium known, so far, showing no sign of inhibition at pressures up to 40 MPa (Bradbrook, 2000), and hence, probably still active at even higher pressures. However, although the optimum pressures for activity reflects in situ pressures, and therefore, indicates that these organisms are well adapted to their environment (Bale *et al.*, 1997), it is not clear how far these characteristics are shared by the majority of the indigenous deep subsurface microorganisms. To date, all samples used for the enrichment and isolation of SRB from subsurface environments have been decompressed during sampling and isolates obtained under atmospheric pressure, thereby selecting for less barotolerant types. Although sampling devices can be recovered slowly (Süß *et al.*, 2004), to reduce the speed of pressure change, bacteria may still not be able to adapt quickly enough, and barophilic bacteria might still be damaged.

11.6 METABOLIC CAPACITIES OF SULPHATE-REDUCING BACTERIA FROM THE MARINE SUBSURFACE

The metabolism of most subsurface sulphate-reducers so far isolated seems to be relatively similar to that of their surface counterparts. Among the electron donors used are organic acids (e.g. lactate, malate and fumarate) and alcohols, but also some more unusual compounds such as sugars or the amino acids, glutamate and arginine (Table 11.3) (Bale *et al.*, 1997; Barnes *et al.*, 1998; Sass and Cypionka, 2004). However, some strains, such as *Desulfovibrio profundus* (Bale *et al.*, 1997) or *Desulfovibrio putealis* (Basso *et al.*, 2005) show rather restricted electron donor range, although they do utilize hydrogen, and therefore, may rely on syntrophic associations or H_2 derived from deep geosphere processes. This view is supported by the finding that several Gram-negative, sulphate-reducing bacteria from the subsurface are also able to grow autotrophically (Bale *et al.*, 1997; Gogotova and Vainshtein, 1989). The sporeforming Gram-positive *Desulfotomaculum* and *Desulfosporosinus* spp. are in general nutritionally more versatile than Gram-negative strains. Many of them are able to utilize compounds difficult to degrade under anoxic conditions, but which might still be present in subsurface environments, such as long-chain fatty acids or aromatic compounds (Daumas *et al.*, 1988; Tasaki *et al.*, 1991). In addition, these bacteria are able to grow as homoacetogens, enabling them to thrive also in sulphate-depleted deep environments.

Most sulphate-reducers from subsurface sites can grow using alternative electron acceptors such as nitrate, ferric iron or manganese oxides (Bale *et al.*, 1997; Sass and Cypionka, 2004). At first sight this might be surprising, but at several subsurface sites reasonable concentrations of nitrate or oxidised metals like iron(III) or manganese(IV) were present and utilized (D'Hondt *et al.*, 2004; Parkes *et al.*, 2005; Sass and Cypionka, 2004), and the use of these alternative electron acceptors results in higher energy yields and a wider habitat range.

11.7 CONNECTIONS BETWEEN MARINE AND TERRESTRIAL SUBSURFACE ENVIRONMENTS

It has been suggested that the presence of the thermophilic, sulphate-reducing archaeon, *Archaeoglobus fulgidus*, which has an obligate salt requirement, in the terrestrial Paris Basin oilfield demonstrates that it had been deposited with the original marine sediment and survived over geological time (Haridon *et al.*, 1995). This is consistent with the active sulphate reduction in marine Cretaceous shales, now on land, described previously in this chapter (Krumholz *et al.*, 1997). However, as *A. fulgidus* was able to grow at the much lower, sub-optimal, in situ chloride concentrations, this does open the possibility of its more recent introduction into the reservoir from a non-marine source. Presumably, this was also the source of the freshwater sulphate reducer *Desulfovibrio putealis*, which is also present (Basso *et al.*, 2005). However, *D. putealis* has the opposite problem to *A. fulgidus*, as it does not grow at NaCl concentrations above $6\,g\,l^{-1}$, whilst the NaCl concentrations in the Paris Basin range from ~5 to $12\,g\,l^{-1}$. This highlights a general problem with extrapolating physiological ranges defined on the laboratory timescales of days to weeks to subsurface habitats, where activity can be on geological timescales. Hence, activity under sub-optimal conditions too slow to be reasonably detected in the laboratory could be very significant over much longer timescales. In addition, the presence of sulphate-reducers, able to grow from freshwater to marine salt concentrations, may reflect adaptation to their subsurface habitat which can have varying salinities, whether this is terrestrial or marine. This may also be reflected in the close phylogenetic relationship of some SRB (e.g. *Desulfotomaculum* spp.) present in either subsurface environment and indicate a remarkable overlap between microbial communities in the two contrasting subsurface environments. Activity of prokaryotes over geological time and the source of prokaryotes in the subsurface, including the possible presence of a globally distributed subsurface biosphere, are fascinating topics that require much more research.

REFERENCES

Anderson, R., Hobart, M. and Langseth, M. (1979). Geothermal convection through oceanic crust and sediments in the Indian Ocean. *Science*, **204**, 828–32.

Bach, W. and Edwards, K. J. (2003). Iron and sulfide oxidation within the basaltic ocean crust: implications for chemolithoautotrophic microbial biomass production. *Geochim. et Cosmochim. Acta.*, **67**, 3871–87.

Baker, B. J., Moser, D. P., MacGregor, B. J. *et al.* (2003). Related assemblages of sulphate-reducing bacteria associated with ultradeep gold mines of South Africa and deep basalt aquifers of Washington State. *Environ. Microbiol.*, **5**, 267–77.

Bale, S. J., Goodman, K., Rochelle, P. A. *et al.* (1997). *Desulfovibrio profundus* sp. nov., a novel barophilic sulphate-reducing bacterium from deep sediment layers in the Japan Sea. *Int. J. Syst. Bacteriol.*, **47**, 515–21.

Barnes, S. P., Bradbrook, S. D., Cragg, B. A. *et al.* (1998). Isolation of sulphate-reducing bacteria from deep sediment layers of the Pacific Ocean. *Geomicrobiol. J.*, **15**, 67–83.

Basso, O., Caumette, P. and Magot, M. (2005). *Desulfovibrio putealis* sp. nov., a novel sulphate-reducing bacterium isolated from a deep subsurface aquifer. *Int. J. Syst. Evol. Microbiol.*, **55**, 101–4.

Bekins, B. A., Godsy, E. M. and Warren, E. (1999). Distribution of microbial physiologic types in an aquifer contaminated by crude oil. *Microb. Ecol.*, **37**, 263–75.

Bidle, K. A., Kastner, M. and Bartlett, D. H. (1999). A phylogenetic analysis of microbial communities associated with methane hydrate containing marine fluids and sediments in the Cascadia Margin (ODP site 892B). *FEMS Microbiol. Lett.*, **177**, 101–8.

Boetius, A., Ravenschlag, K., Schubert, C. J. *et al.* (2000). A marine microbial consortium apparently mediating anaerobic oxidation of methane. *Nature*, **407**, 623–6.

Boivin-Jahns, V., Ruimy, R., Bianchi, A., Daumas, S. and Christen, R. (1996). Bacterial diversity in a deep-subsurface clay environment. *Appl. Environ. Microbiol.*, **62**, 3405–12.

Bottrell, S. H., Parkes, R. J., Cragg, B. A. and Raiswell, R. (2000). Isotopic evidence for anoxic pyrite oxidation and stimulation of bacterial sulphate reduction in marine sediments. *J. Geol. Soc.*, **157**, 711–14.

Bradbrook, S. D. (2000). *Physiological, metabolic, and genetic characteristics of sulphate-reducing bacteria from deep-sediment layers of the Cascadia Margin (ODP Leg 146)*. PhD thesis, University of Bristol.

Canfield, D. E. (1991). Sulphate reduction in deep-sea sediments. *Am. J. Sci.*, **291**, 177–88.

Chapelle, F. H. and Bradley, P. M. (1996). Microbial acetogenesis as a source of organic acids in ancient Atlantic Coastal Plain sediments. *Geology*, **24**, 925–8.

Colwell, F. S., Onstott, T. C., Delwiche, M. E. *et al.* (1997). Microorganisms from deep, high temperature sandstones: Constraints on microbial colonization. *FEMS Microbiol. Rev.*, **20**, 425–35.

Coolen, M. J. L., Cypionka, H., Sass, A. M., Sass, H. and Overmann, J. (2002). Ongoing modification of Mediterranean Pleistocene sapropels mediated by prokaryotes. *Science*, **296**, 2407–10.

Cowen, J. P., Giovannoni, S. J., Kenig, F. *et al.* (2003). Fluids from ageing ocean crust that support microbial life. *Science*, **299**, 120–3.

Cragg, B. A., Harvey, S. M., Fry, J. C., Herbert, R. A. and Parkes, R. J. (1992). Bacterial biomass and acyivity in the deep sediment layers of the Japan Sea, Hole 798B. *Proc. ODP Sci. Res.*, **127/128**, 761–76.

Cragg, B. A., Law, K. M., Cramp, A. and Parkes, R. J. (1997). Bacterial profiles in Amazon Fan sediments (Sites 934, 940). *Proc. ODP Sci. Res.*, **155**, 565–71.

Cragg, B. A., Law, K. M., Cramp, A. and Parkes, R. J. (1998). The response of bacterial populations to sapropels in deep sediments of the Eastern Medierranean (Site 969). *Proc. ODP Sci. Res.*, **160**, 303–7.

Cragg, B. A., Parkes, R. J., Fry, J. C. *et al.* (1996). Bacterial populations and processes in sediments containing gas hydrates (ODP Leg 146: Cascadia Margin). *Earth Planet. Sci. Lett.*, **139**, 497–507.

Cragg, B. A., Parkes, R. J., Fry, J. C. *et al.* (1995). The impact of fluid and gas venting on bacterial populations and processes in sediments from the Cascadia Margin Accretionary System (Sites 888–892) and the geochemical consequences. *Proc. ODP Sci. Res.*, **146**, 399–411.

Cragg, B. A., Wellsbury, P., Murray, R. W. and Parkes, R. J. (2003). Bacterial populations in deep-water, low-sedimentation-rate marine sediments and evidence for subsurface bacterial manganese reduction (ODP Site 1149 Izu-Bonin Trench). *Proc. ODP Sci. Res.*, v. Vol. **185**, online, http://www-odp.tamu.edu/publications/185_SR/008/008.htm.

Daumas, S., Cord-Ruwisch, R. and Garcia, J. L. (1988). *Desulfotomaculum geothermicum* sp. nov., a thermophilic, fatty acid-degrading, sulphate-reducing bacterium isolated with H_2 from geothermal ground water. *Antonie van Leeuwenhoek*, **54**, 165–78.

D'Hondt, S., Jørgensen, B. B., Miller, D. J. *et al.* (2004). Distributions of microbial activities in deep subseafloor sediments. *Science*, **306**, 2216–21.

D'Hondt, S., Rutherford, S. and Spivack, A. J. (2002). Metabolic activity of subsurface life in deep-sea sediments. *Science*, **295**, 2067–70.

Dickens, G. R., Paull, C. K., Wallace, P. and the ODP Leg 164 Scientific Party. (1997). Direct measurement of *in situ* methane quantities in a large gas-hydrate reservoir. *Nature*, **385**, 426–8.

Egeberg, P. K. and Barth, T. (1998). Contribution of dissolved organic species to the carbon and energy budgets of hydrate bearing deep sea sediments (Ocean Drilling Program Site 997 Blake Ridge). *Chem. Geol.*, **149**, 25–35.

Fisk, M. R., Giovannoni, S. J. and Thorseth, I. H. (1998). Alteration of oceanic volcanic glass: textural evidence of microbial activity. *Science*, **281**, 978–80.

Fowler, C. M. R. (1990). *The solid earth, an introduction to global geophysics*. Cambridge: Cambridge University Press.

Fry, N. K., Frederikson, J. K., Fishbain, S., Wagner, M. and Stahl, D. A. (1997). Population structure of microbial communities associated with two deep, anaerobic alkaline aquifers. *Appl. Environ. Microbiol.*, **53**, 1498–504.

Furnes, H. and Staudigel, H. (1999). Biological mediation in ocean crust alteration: how deep is the deep biosphere? *Earth Planet. Sci. Lett.*, **166**, 97–103.

Gogotova, G. I. and Vainshtein, M. B. (1989). Description of a sulphate-reducing bacterium, *Desulfobacterium macestii* sp. nov., which is capable of autotrophic growth. *Microbiology*, **58**, 64–8.

Haridon, S. L., Reysenbach, A., Glenat, P., Prieur, D. and Jeanthon, C. (1995). Hot subterranean biosphere in continental oil reservoir. *Nature*, **377**, 223–4.

Hedges, J. I. and Keil, R. G. (1995). Marine chemistry discussion paper. Sedimentary organic matter preservation: an assessment and speculative synthesis. *Mar. Chem.*, **4**, 81–115.

Inagaki, F., Suzuki, M., Takai, K. *et al.* (2003). Microbial communities associated with geological horizons in coastal subseafloor sediment from the Sea of Okhotsk. *Appl. Environ. Microbiol.*, **69**, 7224–35.

Ingle, J. C., Jr., Suyehiro, K. and von Breymann, M. T. (1990). Initial Reports Sites 794, 798–799 Japan Sea. *Proceedings of the Ocean Drilling Program, Initial Reports, v 128*, College Station, TX.

Jørgensen, B. B. (1983). Processes at the sediment-water interface. In B. Bolin and R. B. Cook (eds.), *The Major Biogeochemical Cycles and their Interactions*, Chichester: John Wiley. pp. 477–515.

Kallmeyer, J. and Boetius, A. (2004). Effects of temperature and pressure on sulphate reduction and anaerobic oxidation of methane in hydrothermal sediments of Guaymas Basin. *Appl. Environ. Microbiol.*, **70**, 1231–3.

Kemp, P. F. and Aller, J. Y. (2004). Bacterial diversity in aquatic and other environments: what 16S rDNA libraries can tell us. *FEMS Microbiol. Ecol.*, **47**, 161–77.

Kennicutt, M. C., Brooks, J. M. and Cox, B. C. (1993). The origin and distribution of gas hydrates in marine sediments. In M. H. Engel and S. A. Macko (eds.), *Organic Geochemistry*. New York: Plenum Press. pp. 535–44.

Kimura, H., Sugihara, M., Yamamoto, H. *et al.* (2005). Microbial community in a geothermal aquifer associated with the subsurface of the Great Artesian Basin, Australia. *Extremophiles*, **9**, 407–14.

Knittel, K., Lösekann, T., Boetius, A., Kort, R. and Amann, R. (2005). Diversity and distribution of methanotrophic archaea at cold seeps. *Appl. Environ. Microbiol.*, **71**, 467–79.

Köpke, B., Wilms, R., Engelen, B., Cypionka, H. and Sass, H. (2005). Microbial diversity in coastal subsurface sediments – a cultivation approach using various electron acceptors and substrate gradients. *Appl. Environ. Microbiol.* **71**, 7819–30.

Kormas, K. A., Smith, D. C., Edgcomb, V. and Teske, A. (2003). Molecular analysis of deep subsurface microbial communities in Nankai Trough sediments (ODP Leg 190, Site 1176). *FEMS Microbiol. Ecol.*, **45**, 115–25.

Krumholz, L. R., McKinley, J. P., Ulrich, F. A. and Suflita, J. M. (1997). Confined subsurface microbial communities in Cretaceous rock. *Nature*, **386**, 64–6.

Li, L., Guenzennec, J., Nichols, P. *et al.* (1999). Microbial diversity in Nankai Trough sediments at a depth of 3,843 m. *J. Oceanogr.*, **55**, 635–42.

Ludvigsen, L., Albrechtsen, H. J., Ringelberg, D. B., Ekelund, F. and Christensen, T. H. (1999). Distribution and composition of microbial populations in landfill leachate contaminated aquifer (Grindsted, Denmark). *Microb. Ecol.* **37**, 197–207.

Mangelsdorf, K., Zink, K.-G., Birrien, J.-L. and Toffin, L. (2005). A quantitative assessment of pressure dependent adaptive changes in the membrane lipids of a piezosensitive deep sub-seafloor bacterium. *Org. Geochem.*, **36**, 1459–79.

Mather, I. D. and Parkes, R. J. (2000). Bacterial populations in sediments of the eastern flank of the Juan de Fuca Ridge, Sites 1026 and 1027. *Proc. ODP Sci. Res.* **168**, 161–5.

Mauclaire, L., Zepp, K., Meister, P. and Mckenzie, J. (2004). Direct in situ detection of cells in deep-sea sediment cores from the Peru Margin (ODP Leg 201, Site 1229). *Geobiology*, **2**, 217–23.

Morita, R. Y. and ZoBell, C. E. (1955). Occurence of bacteria in pelagic sediments collected during the Mid-Pacific Expedition. *Deep-Sea Res.*, **3**, 66–73.

Moser, D. P., Onstott, T. C., Fredrickson, J. K. *et al.* (2003). Temporal shifts in the geochemistry and microbial community structure of an ultradeep mine borehole following isolation. *Geomicrobiol. J.*, **20**, 517–48.

Nazina, T. N., Ivanova, A. E., Kanchveli, L. P. and Rozanova, E. P. (1989). A new spore-forming thermophilic methylotrophic sulphate-reducing bacterium, *Desulfotomaculum kuznetsovii*. Microbiology, **57**, 659–63.

Nedwell, D. B., Embley, T. M. and Purdy, K. J. (2004). Sulphate reduction, methanogenesis and phylogenetics of the sulphate reducing bacterial communities along an estuarine gradient. *Aquat. Microb. Ecol.*, **37**, 209–17.

Parkes, R. J., Cragg, B. A., Bale, S. J. *et al.* (1994). Deep bacterial biosphere in Pacific Ocean sediments. *Nature*, **371**, 410–13.

Parkes, R. J., Cragg, B. A., Bale, S. J., Goodman, K. and Fry, J. C. (1995). A combined ecological and physiological approach to studying sulphate reduction within deep marine sediment layers. *J. Microbiol. Meth.*, **23**, 235–49.

Parkes, R. J., Cragg, B. A., Fry, J. C., Herbert, R. A. and Wimpenny, J. W. T. (1990). Bacterial biomass and activity in deep sediment layers from the Peru Margin. *Phil. Trans. R. Soc. Lond. A.*, **331**,139–53.

Parkes, R. J., Cragg, B. A. and Wellsbury, P. (2000). Recent studies on bacterial populations and processes in subseafloor sediments: a review. *Hydrogeol. J.*, **8**, 11–28.

Parkes, R. J., Webster, G., Cragg, B. A. *et al.* (2005). Deep sub-seafloor prokaryotes stimulated at interfaces over geological time. *Nature*, **436**, 390–4.

Paull, C. K., Buelow, W. J., Ussler, W. and Borowski, W. S. (1996). Increased continental-margin slumping frequency during sea-level lowstands above gas hydrate-bearing sediments. *Geology*, **24**, 143–6.

Pedersen, K., Arlinger, J., Ekendahl, S. and Hallbeck, L. (1996). 16S rRNA gene diversity of attached and unattached bacteria in boreholes along the access tunnel to the Äspö hard rock laboratory, Sweden. *FEMS Microbiol. Ecol.*, **19**, 249–62.

Postgate, J. R. and Hunter, J. R. (1963). Acceleration of bacterial death by growth substrate. *Nature*, **198**, 273.

Quigley, T. M. and Mackenzie, A. S. (1988). The temperatures of oil and gas formation in the sub-surface. *Nature*, **333**, 549–52.

Reed, D. W., Fujita, Y., Delwiche, M. E. *et al.* (2002). Microbial communities from methane hydrate-bearing deep marine sediments in a forearc basin. *Appl. Environ. Microbiol.*, **68**, 3759–70.

Sass, H. and Cypionka, H. (2004). Isolation of sulphate-reducing bacteria from the terrestrial deep subsurface and description of *Desulfovibrio cavernae* sp. nov. *System. Appl. Microbiol.*, **27**, 541–8.

Schippers, A., Neretin, L. N., Kallmeyer, J. *et al.* (2005). Prokaryotic cells of the deep sub-seafloor biosphere identified as living bacteria. *Nature*, **433**, 861–4.

Süß, J., Engelen, B., Cypionka, H. and Sass, H. (2004). Quantitative analysis of bacterial communities from Mediterranean sapropels based on cultivation-dependent methods. *FEMS Microbiol. Ecol.*, **51**, 109–21.

Tasaki, M., Kamagata, Y., Nakamura, K. and Mikami, E. (1991). Isolation and characterization of a thermophilic benzoate-degrading, sulphate-reducing bacterium, *Desulfotomaculum thermobenzoicum* sp. nov. *Arch. Microbiol.*, **155**, 348–52.

Thomsen, T. R., Finster, K. and Ramsing, N. B. (2001). Biogeochemical and molecular signatures of anaerobic methane oxidation in a marine sediment. *Appl. Environ. Microbiol.*, **67**, 1646–56.

Toffin, L., Webster, G., Weightman, A. J., Fry, J. C. and Prieur, D. (2004). Molecular monitoring of culturable bacteria from deep-sea sediment of the Nankai Trough, Leg 190 Ocean Drilling Program. *FEMS Microbiol. Ecol.*, **48**, 357–67.

Wellsbury, P., Goodman, K., Barth, T. *et al.* (1997). Deep marine biosphere fuelled by increasing organic matter availability during burial and heating. *Nature*, **388**, 573–6.

Wellsbury, P., Goodman, K., Cragg, B. A. and Parkes, R. J. (2000). The geomicrobiology of deep marine sediments from Blake Ridge containing methane hydrate (Sites 994, 995 and 997). *Proc. ODP Sci. Res.*, **164**, 379–91.

Wellsbury, P., Mather, I. and Parkes, R. J. (2002). Geomicrobiology of deep, low organic carbon sediments in the Woodlark Basin, Pacific Ocean. *FEMS Microbiol. Ecol.*, **42**, 59–70.

Wellsbury, P. and Parkes, R. J. (2000). Deep biosphere: source of methane for oceanic hydrate. In M. D. Max (ed.), *Natural Gas Hydrate in Oceanic and Permafrost Environments*. Dordrecht: Kluwer. pp. 91–104.

Whitman, W. B., Coleman, D. C. and Wiebe, W. J. (1998). Prokaryotes: the unseen majority: *Proc. Natl. Acad. Sci. USA.*, **95**, 6578–83.

Wilms, R., Köpke, B., Sass, H. *et al.* (2006). Deep-biosphere bacteria within the subsurface of tidal flat sediments. *Environ. Microbiol.*, 8, 709–19.

ZoBell, C. E. (1938). Studies on the bacterial flora of marine bottom sediments. *J. Sed. Petrol.*, **8**, 10–18.

CHAPTER 12

Ecophysiology of sulphate-reducing bacteria in environmental biofilms

Satoshi Okabe

12.1 INTRODUCTION

Sulphate-reducing bacteria (SRB) are a phylogenetically and physiologically diverse group of bacteria, characterized by their versatile metabolic capability to use various electron acceptors and donors (Widdel, 1988). SRB are therefore universally distributed in diverse environments and play significant ecophysiological roles in anaerobic biomineralization pathways. The degradation of organic matter by a complex microbial community is governed to a large extent by available electron acceptors. The terminal stages of the anaerobic mineralization of organic matter is catalyzed by SRB and methanogens, and their competitive and cooperative interactions have been described previously (Oude Elferink *et al.*, 1994).

Typical domestic wastewaters contain sulphate concentrations of 100–1000 μM and relatively low dissolved oxygen due to the lower solubility and rapid depletion of this gas by biological activity. Thus, sulphate reduction can be the dominant terminal electron accepting process and account for up to 50% of mineralization of organic matter in wastewater biofilms (Kühl and Jorgensen, 1992; Okabe *et al.*, 2003a). Multiple electron donors and electron acceptors are present in the wastewaters. As a result, wastewater biofilms are very complex multispecies biofilms, displaying considerable heterogeneity, with regard to both the microorganisms present and their physicochemical microenvironments. Sulphate reduction is anticipated to take place in the deeper anoxic biofilm strata even though the bulk liquid is oxygenated. It is, therefore, thought that a successive vertical zonation of respiratory processes can be found in aerobic wastewater biofilms with a typical thickness of only a few millimeters (Ito *et al.*, 2002b; Kühl and Jorgensen, 1992; Okabe *et al.*, 1999a; 2003a; Ramsing *et al.*, 1993).

The SRB are the only bacteria known to reduce sulphate and therefore the sulphur cycle in wastewater biofilm systems can only be initiated by reduction of SO_4^{2-} to S^{2-} by the SRB. Once sulphate reduction occurs in biofilms, internal biological and chemical reoxidation of reduced sulphur species begins, which is calculated to account for a substantial part of oxygen consumption in the biofilms (approximately up to 70%) (Kühl and Jorgensen, 1992, Norsker *et al.*, 1995, Okabe *et al.*, 1999a; 2003a). Reoxidation of the produced sulphide with oxygen and/or nitrate accordingly will take place in the strata close to the sulphate reduction zone, depending on the oxygen and nitrate penetration depths.

A major drawback of sulphate reduction in wastewater treatments is the production of toxic H_2S, which is also a possible precursor of odorants. Furthermore, the activity of SRB in the biofilms is believed to cause serious corrosion of metals in industrial water systems and concrete sewer pipes in wastewater treatment facilities, which is a universal problem (Hamilton, 1985; 2003; Hamilton and Lee, 1995; Lee *et al.*, 1994; Nielsen *et al.*, 1993; Postgate, 1984; and Chapter 16 of this book). However, one clear mechanism of biocorrosion does not exist presently. The complete understanding of the role of SRB in accelerating corrosion depends upon a better understanding of the relation of in situ microbial community structure of microorganisms, including not only the identification of organisms present but also identifying in situ activities, and their spatial relationships with others in the biofilm. Better knowledge of the ecophysiology of SRB communities and their metabolic output can, therefore, be of great practical and scientific relevance. However, such biofilm systems have been treated as "a black box" in most of the previous studies due to lack of analytical techniques and a limitation of their resolutions, as represented by traditional culture-dependent techniques (e.g. most probable number method (MPN)). In addition, the sulphur cycle in microaerophilic wastewater biofilms is very complex and mass balances of sulphide or sulphate flux across a biofilm–liquid interface cannot describe sulphur transformation within the biofilm due to the presence of an internal sulphur cycle. Therefore, it is necessary to explore analytical tools to overcome this problem.

12.2 POLYPHASIC APPROACH

Understanding the structure and function of complex microbial communities is a central theme in microbial ecology. However, traditional cultivation-dependent techniques are inadequate to fulfil this task because most members of microbial communities in natural and engineered systems cannot be cultured (Amann *et al.*, 1995). New molecular biological techniques (virtually based on the 16S ribosomal RNA sequences) provide comprehensive phylogenetic information on molecular diversity of microbial communities, including heretofore unknown microorganisms, without the need of cultivation. In particular, the fluorescence in situ hybridization (FISH) technique allows for in situ detection, localization, and quantification of target microorganisms in complex heterogeneous microbial communities such as wastewater biofilms. Several recent reviews provide technical details, applications, and limitation of FISH and the rRNA approach (Amann and Ludwig, 2000; Head *et al.*, 1998).

Since analysis of microbial community function (activity) in biofilms is difficult, studies relating community structure to community function are scarce. For biofilms, substrate transfer to the biofilms often limits microbial conversion rates. Thus, the biofilms develop various complex microenvironments due to this substrate transfer resistance. Microelectrode measurements are, therefore, the most reliable and direct way of determining microenvironments in the biofilm with high spatial and temporal resolution (Amann and Kühl, 1998; Schramm, 2003). One advantage of the use of microelectrodes is their ability to detect, directly and with minimal disturbance, in situ microbial activities, which are derived from the substrate profiles if the transport process (usually molecular diffusion) is known. Microelectrodes, however, only measure net chemical profiles. The spatial resolution of microelectrodes is at "community level", which is good enough to characterize the concentration gradients across the biofilms and to calculate the net rates of production and consumption at a certain depth within the biofilm. However, it is not high enough to determine the activity of individual microbial cells. In addition, when the substrates used by unidentified microorganisms are not known, or the abundance of the targeted microorganisms is low, the chemical profiles and fluxes cannot be correlated with the abundance of specific bacterial populations.

To overcome this technical problem, microautoradiography (MAR) combined with FISH (MAR-FISH) has been successfully developed (Lee *et al.*, 1999; Ouverney and Fuhrman, 1999). This MAR-FISH technique provides a direct link between 16S rRNA phylogeny and their in situ physiology (i.e. substrate uptake patterns) of both cultivated and uncultivated SRB on a single-cell level, leading to further insight into the ecophysiological importance of SRB in various environments.

In this chapter, ecophysiology of SRB in complex microbial communities such as wastewater biofilms will be described, based on the experimental results obtained by the combined use of the culture-dependent techniques; the molecular biological techniques (FISH with 16S rRNA-targeted oligonucleotide probes); and microelectrode measurements. The combination of these advanced techniques will provide more reliable and direct information about relationships between in situ spatial organization of SRB and their in situ activity in wastewater biofilms.

12.3 SRB COMMUNITY STRUCTURE

The community structure of SRB inhabiting wastewater biofilms growing under microaerophilic conditions were analyzed by 16S ribosomal RNA (rRNA) gene cloning analysis. The analysis revealed that SRB clones were phylogenetically diverse and affiliated with at least six major SRB genera in the *Deltaproteobacteria*: *Desulfomicrobium, Desulfovibrio, Desulfonema, Desulforegula, Desulfobacterium* and *Desulfobulbus*. SRB communities are well suited for rRNA-based studies, since the classical physiologically based taxonomy is in good agreement with the rRNA-derived phylogeny (Devereux *et al.*, 1989; 1992). Therefore, phylogenetic information gained about SRB allows an estimation of the physiological properties and metabolic activities of the corresponding population. A set of rRNA probes was developed to detect and identify Gram-negative SRB (Devereux *et al.*, 1992; Devereux and Stahl, 1993). Recently, an oligonucleotide microarray consisting of 132 16S rRNA gene-targeted oligonucleotide probes (18 mers) was developed for simultaneous detection of all recognized lineages of SRB (Loy *et al.*, 2002). This type of high-throughput molecular tool for community analysis provides relevant information about important SRB members in wastewater biofilms in a few hours.

12.3.1 Spatial distribution of SRB

Figure 12.1 shows a composite cross-section (20-μm-thick) image of a mature wastewater biofilm (biofilm thickness, approximately 1200 μm). The wastewater biofilm displayed a complex heterogeneous structure consisting of discrete biomass and interstitial voids, which connect the bulk water to the bottom part of the biofilm. The spatial distribution of SRB population was investigated using the FISH technique with family- and genus-specific fluorescence oligonucleotide probes. The probe sequences, their specificity, and hybridization conditions are given in Table 12.1. In situ hybridization of the vertical biofilm sections revealed that the fluorescent signals derived from probe 660-stained *Desulfobulbus* cells were found at all depths and in all

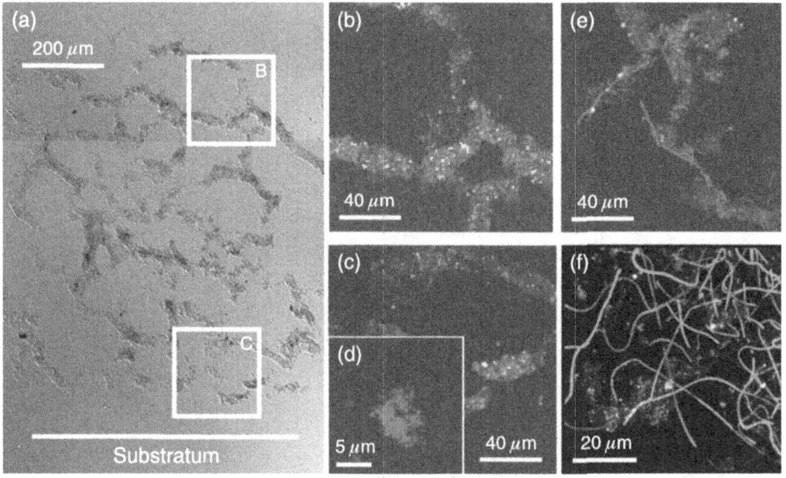

Figure 12.1. A composite differential interference contrast (DIC) image of a vertical section (20-μm-thick) of a wastewater biofilm (a). The biofilm thickness is about 1200 μm. In situ detection of *Desulfobulbus* hybridized with tetramethylrhodamine-5-isothiocyanate (TRITC)-labelled 660 probe (b) and (c) and a close-up view of SRB385 probe stained cell clusters (d). *Desulfonema* hybridized with TRITC-labelled DNMA657 probe (e). Coexistence of *Desulfobulbus* hybridized with TRITC-labelled 660 probe and *Thiothrix* hybridized with fluorescein isothiocyanate (FITC)-labelled G123T probe at the oxic/anoxic interface in the biofilm (f). (For a colour version of this figure, please refer to colour plate section.)

states from single scattered cells to clustered cells (Figures 12.1a–12.1d). Some of these lemon-shaped cells were linked together: these seem to be typical features of *Desulfobulbus*. The probe DNMA657-stained filamentous *Desulfonema* was also detected mainly in the surface of the biofilm (Figure 12.1e). The probe DSV698-stained vibrio-shaped cells were also detected, but the abundance was low.

These SRB species present in the surface biofilm closely coexisted with some sulphur-oxidizing bacteria (SOB), especially a newly isolated chemolithoautotrophic SOB, *Thiovirga sulfuroxydans* (Ito *et al.*, 2004; 2005) and *Thiothrix* (Figure 12.1f). This might suggest that SOB efficiently utilize sulphide produced by SRB with oxygen and nitrate as sole electron acceptor. At the oxic/anoxic interface, since a high concentration of S^0 (up to 22 μmol cm^{-3}) accumulated (Okabe *et al.*, 1999b; 2005), the anaerobic oxidation of S^0 to sulphate with oxidized metals as the electron acceptor, and S^0 disproportionation in the absence of an electron acceptor by *Desulfobulbus propionicus* might also be occurring (Lovly and Phillip, 1994).

Table 12.1. *A list of 16S rRNA-targeted oligonucleotide probes used in this study*

Probe	Specificity	Sequence of probe (5′–3′)	Target site[a]	FA[b] (%)	NaCl[c] [mM]	Ref.
EUB338	domain *Bacteria*	GCTGCCTCCCGTAGGAGT	338–355	20	0.166	(Amann *et al.*, 1992)
SRB385	SRB of the delta-proteobacteria plus several Gram-positive bacteria (e.g. *Clostridium*)	CGGCGTCGCTGCGTCAGG	385–402	30	0.071	(Amann *et al.*, 1992)
SRB385Db	*Desulfobacteriaceae* (except for *Desulfobulbus* spp.) plus some non-SRB (e.g. *Myxococcus xanthus* and *Pelobacter accetylenicus*)	CGGCGTTGCTGCGTCAGG	385–402	30	0.071	(Rabus *et al.*, 1996)

DSV698	Most *Desulfovibrio* spp. plus some non-SRB (e.g. *Lawsonia intracellularis*)	GTTCCTCCAGATATCTACGG	698–717	35	0.047	(Manz *et al.*, 1998)
660	*Desulfobulbus* spp.	GAATTCCACTTTCCCCTCTG	660–679	30	0.071	(Devereux *et al.*, 1992)
DNMA657	*Desulfonema* spp.	TTCCG(C/T)TTCCCTCTCCCATA	657–676	20	0.166	(Fukui *et al.*, 1999)

[a] 16S rRNA position according to *Escherichia coli* numbering.
[b] formamide concentration in the hybridization buffer.
[c] sodium chloride concentration in the washing buffer.

12.3.2 Quantification of SRB

The relative abundance of domain *Bacteria* detected with probe EUB338, and of SRB detected with family- or genus-specific probes in a typical microaerophilic wastewater biofilm, was determined by FISH (Ito *et al.*, 2002a). The total DAPI (4',6-diamodo-2-phenylindole) count was 6.7×10^{10} cells cm^{-3} biofilm, or 3.0×10^{10} cells g-VSS^{-1} (volatile suspended solids). The abundance of *Bacteria* detected with probe EUB338 was $93.5 \pm 3.4\%$ of the total DAPI-stained cells. The number of SRB detected with probes SRB385 and SRB385Db (defined as total SRB) accounted for only 4.8% of the total DAPI-stained cells. A similar number was found in the activated sludge process (Manz *et al.*, 1998). *Desulfobulbus* hybridized with probe 660 and *Desulfovibrio* hybridized with probe DSV698, were predominant SRB species and accounted for 23% and 9.4% of total SRB, respectively. *Desulfobulbus* is a propionate-utilizing SRB, mainly involved in the degradation of propionate to acetate in the microaerophilic wastewater. *Desulfovibrio* is generally recognized as an important member of the H_2-utilizing SRB. Both SRB species are known to be oxygen tolerant to some extent and to be the numerically important SRB in wastewater biofilms (Okabe *et al.*, 1999a; Santegoeds *et al.*, 1999) and in activated sludge flocs (Manz *et al.*, 1998; Schramm *et al.*, 1999). In addition, filamentous *Desulfonema* hybridized with probe DNMA 657 accounted for ca. 2.0% of total SRB. *Desulfonema* is a nutritionally versatile complete oxidizer that may grow on fatty acids up to chain lengths of 10 to 16 carbon atoms and is tolerant to oxygen to some extent (Fukui *et al.*, 1999). More than 65% of the total SRB, however, could not be identified by using FISH technique, indicating that many unidentified SRB species likely exist within the currently identified phylogenetic SRB groups. Thus, further development of FISH probes is needed.

12.3.3 Vertical distributions of SRB

To determine quantitatively the vertical distributions of SRB385-, SRB385Db-, 660-, DNMA657-, and DSV698-stained SRB cells with FISH technique, the probe-stained cells were directly counted along vertical transects across the biofilm sections (Okabe *et al.*, 2003a). The abundance of probe SRB385-hybridized cells was highest (ca. 8×10^9 cells cm^{-3}) in the surface region of the biofilm and gradually decreased with depth (Figure 12.2a). The SRB population detected with SRB385Db (mainly *Desulfobacteriaceae*) was approximately 6×10^8 cells cm^{-3} throughout the biofilm. A high abundance (ca. 4×10^9 cells cm^{-3}) of the probe 660-stained

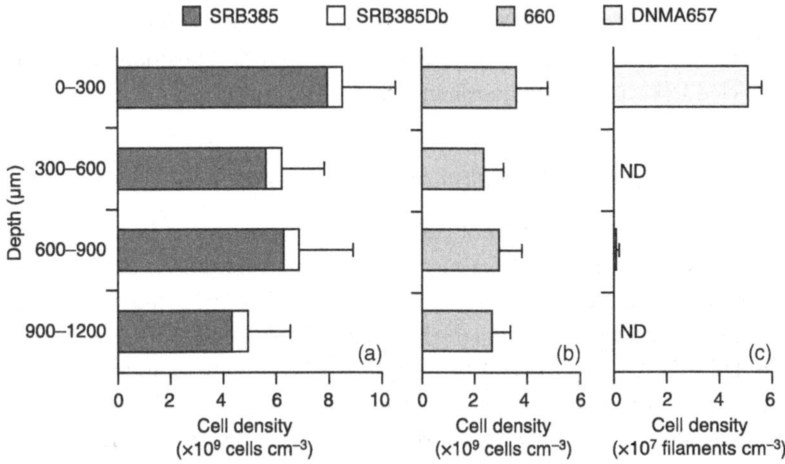

Figure 12.2. Vertical distributions of total SRB cells that are hybridized with probe SRB385 and SRB385 Db (a), *Desulfobulbus* hybridized with probe 660 (b), and *Desulfonema* hybridized with probe DNMA657 (c). The numbers of cells were determined at interval of 300 μm by FISH direct counting. The biofilm surface is at a depth of 0 μm. Error bars indicate the standard deviations ($n = 3$). ND: not determined. Figures are reproduced from Okabe *et al.* (2003a).

Desulfobulbus cells was also detected in the surface of the biofilm (Figure 12.2b). Approximately 200 μm long and 3–5 μm wide, filamentous *Desulfonema* hybridized with probe DNMA657 was also mainly detected in the surface of the biofilm (Figure 12.2c). One filament typically contained about 30–100 cells. The abundance of *Desulfovibrio* hybridized with probe DSV698 was about two orders of magnitude lower than that of 660-hybridized cells. When the biofilm was stained with other SRB group- and genus-specific probes, no clear hybridization signal was detected, suggesting that their abundances were low in the biofilm. When the biofilm was stained with the general bacterial probe, EUB338, the abundance and fluorescent intensity were clearly higher at the surface of the biofilm (Ito *et al.*, 2002b). This indicates that more metabolically active bacteria, including SRB, were present in the surface of the biofilm than in the deeper part of the biofilm, possibly due to substrate limitation. This FISH direct counting revealed that the presence of oxygen does not appear to restrict the distribution of recognized lineages of SRB. A preferential localization of SRB, especially *Desulfobulbus* and *Desulfonema*, to the oxic/anoxic chemocline was usually detected in wastewater biofilms (Ito *et al.*, 2002b; Okabe *et al.*, 1999a; 2003a) and in cyanobacterial mats (Minz *et al.*, 1999a; 1999b). This is not surprising because, considering avoidance of oxygen and positioning as close as possible to organic sources and sulphate, the optimum position for SRB is expected to be at the oxic/anoxic interface.

12.3.4 Spatial distribution of culturable SRB

The vertical distribution of culturable SRB populations in this biofilm can also be determined by the most probable number (MPN) method after the biofilms are sliced, without any pretreatment, into 100-μm-thick sections parallel to the substratum with the Microslicer® (Model TK-1000, Tosaka EM, Co. Ltd) (Okabe *et al.*, 1999a) (Figure 12.3). The slightly modified Postgate medium B containing either sodium propionate (500 mg-C/L) or sodium acetate (500 mg-C/L) as the sole carbon source was used for the MPN count. Interestingly, the culturable propionate-utilizing SRB were one order of magnitude higher than the acetate-utilizing SRB at the surface, which agreed with the FISH result. However, the propionate- and acetate-utilizing SRB were several orders of magnitude (10^2 to 10^4) lower than the FISH counting. This indicated the typical limitation of culture-dependent enumeration methods. Furthermore, the MPN counts decreased with depth, whereas the

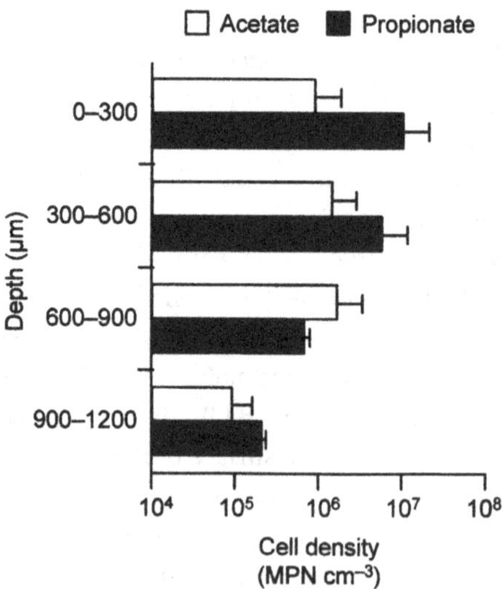

Figure 12.3. Vertical distributions of culturable SRB in the biofilm. The numbers of cells were determined by the five-tube multiple-dilution MPN method using a modified Postgate medium B containing either 500 mg-C/L of propionate or acetate, respectively. The biofilm surface is at a depth of 0 μm. Error bars indicate the standard deviations (n = 3).

FISH counts were relatively constant throughout the biofilm. This might indicate that more active SRB were present near the biofilm surface where both organic carbon and sulphate were more available, and oxygen stress was tolerable.

12.4 MICROELECTRODE MEASUREMENTS

Microelectrode measurements make it possible to analyze several metabolic processes with high spatial and temporal resolution, and have been used for the study of nitrogen cycle (Okabe *et al.*, 1999b; Santegoeds *et al.*, 1998; Schramm *et al.*, 1996) and sulphur cycles (Ito *et al.*, 2002a; Kühl and Jørgensen, 1992; Ramsing *et al.*, 1993; Okabe *et al.*, 1999a; 2003a) in various environmental biofilm samples. Ramsing *et al.* (1993) were the first to use both microelectrodes and molecular techniques to study sulphate reduction in a trickling-filter biofilm.

Figure 12.4 shows the typical steady-state microprofiles of O_2, pH, NO_3^-, and H_2S in a wastewater biofilm (Okabe *et al.*, 2005). Oxygen and nitrate

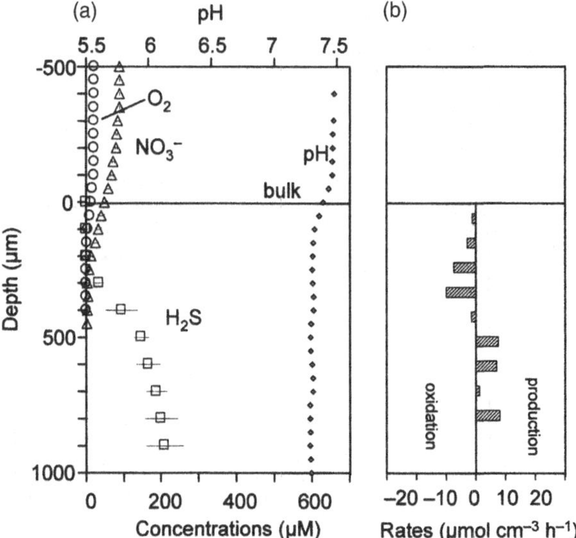

Figure 12.4. Steady-state concentration profiles for O_2, H_2S, NO_3^-, and pH in a wastewater biofilm. The biofilm was incubated in the synthetic medium with (90 μM NO_3^-) (a). The error bars indicate the standard deviations for three measurements at different positions. Spatial distribution of the specific H_2S production and oxidation rates was calculated based on the corresponding microprofiles (b). The biofilm surface is at a depth of 0 μm. Figures are reproduced from Okabe *et al.* (2005).

concentrations were maintained at ca. 10 and 90 μM, repectively, during microelectrode measurement. Sodium propionate (900 μM) was used as the sole carbon source because propionate-utilizing *Desulfobulbus* was the predominant SRB in this microaerophilic wastewater biofilms.

Oxygen and NO_3^- penetrated to about 200 μm and 500 μm, respectively, from the surface. An H_2S concentration of 220 μM was detected at the bottom of the biofilm. Sulphide was produced just below the oxic/anoxic interface (ca. 500 μm below the surface) with a production rate of 0.47 μmol H_2S cm^{-2} h^{-1}. The H_2S profile barely overlapped the O_2 profile, but did overlap the NO_3^- profile, indicating that H_2S was mainly oxidized anaerobically. These profiles clearly revealed the vertical microzonation of O_2 respiration, NO_3^- respiration, H_2S oxidation and SO_4^{2-} reduction occurring in the 1000-μm-thick biofilm. When NO_3^- was removed, the H_2S concentration rapidly increased and reached 430 μM at a depth of 500 μm and the sulphide production zone moved toward the biofilm surface (Figure 12.5). The H_2S produced was completely oxidized at the biofilm surface, indicating high sulphide oxidation capacity in this biofilm.

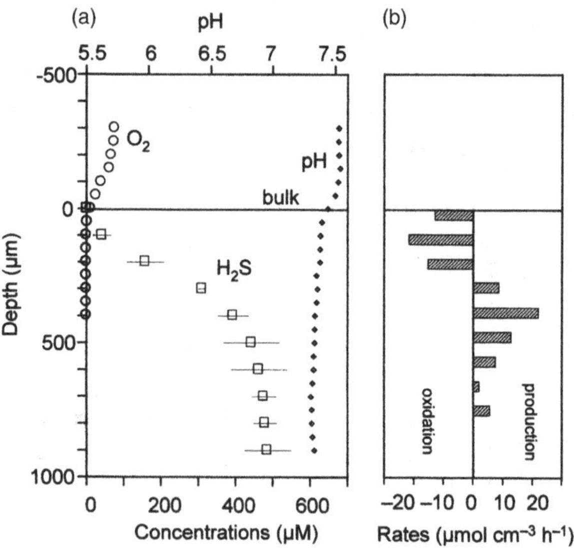

Figure 12.5. Steady-state concentration profiles for O_2, H_2S, NO_3^-, and pH in a wastewater biofilm. The biofilm was incubated in the synthetic medium without NO_3^- (a). The error bars indicate the standard deviations for three measurements at different positions. Spatial distribution of the specific H_2S production and oxidation rates were calculated based on the corresponding microprofiles (b). The biofilm surface is at a depth of 0 μm. Figures are reproduced from Okabe *et al.* (2005).

The specific sulphate reduction rate per cell in this biofilm was calculated to be on the order of 10^{-14} to 10^{-15} mol $SO_4{}^{2-}$ $cell^{-1}$ d^{-1}. This rate is in the range of previously reported values of pure cultures on H_2, lactate, or pyruvate: 2×10^{-16} to 5×10^{-14} mol SO_4^{2-} $cell^{-1}$ d^{-1} (Jørgensen, 1978). The in situ sulphate reduction rate was in the range of 5 to 23 μmol H_2S cm^{-3} h^{-1}, which is higher than the value reported for other wastewater biofilms (Kühl and Jorgensen, 1992; Okabe *et al.*, 1999a; Ramsing *et al.*, 1993; Santegoeds *et al.*, 1998), but is in the same range as anaerobic sulphidogenic granules (3.6 to 21.6 μmol H_2S cm^{-3} h^{-1}) (Santegoeds *et al.*, 1999). This is probably because a high abundance of SRB (ca. 8×10^9 cells cm^{-3}) was present due to the higher organic load.

12.4.1 Effect of nitrate on sulphate reduction

The toxicity and corrosive properties of H_2S dictate stringent control of the H_2S production in sewer systems. Inhibiting the growth of SRB can directly or indirectly prevent sulphide production. Aeration and biocides are common approaches. However, since SRB are generally present in biofilms, and oxygen and biocides do not effectively penetrate through the biofilm. Thus, these measures are of limited effectiveness. The use of nitrate (NO_3^-) to control biogenic sulphide production in sewer systems was recognized as early as 1929 (Allen, 1929; Jenneman *et al.*, 1986). However, it is not clear whether nitrate addition decreases or inhibits the sulphide production in the biofilm, by oxidizing produced sulphide or by inhibiting sulphate reduction directly. The effects of nitrate addition on sulphide production in the biofilm could be evaluated in details with microelectrodes.

An addition of nitrate forced the sulphate reduction zone deeper into the biofilm, and also reduced the specific sulphate reduction rate (see Figures 12.4 and 12.5) (Okabe *et al.*, 2003c). The sulphate reduction zone was consequently separated from O_2 and NO_3^- respiration zones. Anaerobic H_2S oxidation with NO_3^- was also induced by addition of nitrate to the medium. The NO_3^- treatment had only a transient effect on sulphide production, however, and the in situ sulphide production quickly recovered (Okabe *et al.*, 2003c). On the basis of these data, the addition of NO_3^- did not kill SRB, but induced an interspecies competition for common carbon source between nitrate-reducing bacteria and SRB, and enhanced the anaerobic oxidation of sulphide, which were therefore the main causes of the suppression of in situ sulphide production in the biofilm (Okabe *et al.*, 2003c).

This result is consistent with the generally accepted paradigm that electron acceptors (e.g., O_2, NO_3^-, and SO_4^{2-}) are used sequentially according to the thermodynamically predicted order, but is inconsistent with the earlier reports of high sulphate-reducing rates in the oxygenated surface regions of some microbial mats (Canfield and Des Marais, 1991; Frund and Cohen, 1992).

12.5 CONTRIBUTION OF SRB TO CARBON MINERALIZATION IN WASTEWATER BIOFILMS

Although microbial sulphate reduction is an important carbon mineralization process in wastewater biofilms, little is known about the specific contribution of SRB to carbon mineralization. A combined approach of molecular techniques and culture-based batch experiments with specific inhibitor (molybdate) indicated that the addition of molybdate inhibited propionate-utilizing *Desulfobulbus* and acetate-utilizing *Desulfobacterium* (Okabe *et al.*, 2003a). In particularly, *Desulfobulbus* is a numerically important member of SRB and the main contributor to the oxidation of propionate to acetate in wastewater biofilms. On the other hand, acetate was most likely utilized by nitrate-reducing bacteria but not by acetate-utilizing SRB.

SRB and methanogens are both capable of using acetate and H_2 as substrates. Kinetic studies have shown that SRB generally have the higher maximum specific growth rates and higher affinity for substrates (i.e. H_2 and acetate) (Oude Elferink *et al.*, 1994). Thus, SRB may compete with methanogens for common substrates (i.e. H_2, formate and acetate) under methanogenic conditions if sulphate is not limiting (Oude Elferink *et al.*, 1994; 1998). SRB also compete with syntrophic bacteria involved in the degradation of propionate and butyrate (Harmsen *et al.*, 1996). However, despite their kinetic advantages, SRB rarely predominate in anaerobic wastewater treatment processes (e.g. anaerobic digester) due to relatively low sulphate concentrations. In a full-scale anaerobic reactor treating paper mill wastewater that had a chemical oxygen demand/sulphate ratio of 9.5, acetate was mainly degraded by *Methanosaeta*-like microorganisms, although propionate was utilized by the *Desulfobulbus*-like propionate-degrading SRB, which competed with *Syntrophobacter*-like bacteria for available propionate (Oude Elferink *et al.*, 1998). Apparently, propionate oxidation via SRB alone is faster and more effective than via a syntrophy. Syntrophic degradation of propionate and butyrate, coupled to H_2 removal via sulphate reduction rather than via methanogenesis, is also conceivable at relatively low sulphate concentrations. In this way, SRB would keep the hydrogen partial pressure lower than methanogens, which leads to that syntrophic propionate-degrading acetogens grow much faster than in co-culture with hydrogen-consuming methanogens (Boone and Bryant, 1980). Therefore, SRB can act as a hydrogen consumer rather than a direct propionate oxidizer. Furthermore, *Desulfobulbus* may be able to oxidize propionate syntrophically with H_2-consuming anaerobic bacteria in the absence of sulphate or at low sulphate concentrations (Heppner *et al.*, 1992; Wu *et al.*, 1992).

12.6 ECOPHYSIOLOGY OF SRB IN ENVIRONMENTAL BIOFILMS

Although SRB are generally considered to be obligate anaerobic bacteria, high abundances of SRB, in particular *Desulfobulbus* and *Desulfonema*, were frequently detected in oxic surface region of the biofilm (see Figures 12.1–12.3). It is not clear whether these SRB remain or proliferate there via respiration with oxygen, nitrate, or other electron acceptors. To answer this question, MAR-FISH was used to simultaneously determine the phylogenetic affiliation and substrate uptake patterns of SRB inhabiting the wastewater biofilm with oxygen, nitrate, or sulphate as electron acceptor on a single-cell level (Ito *et al.*, 2002a).

The MAR-FISH analysis revealed that more than 90% of *Desulfobulbus* hybridized with probe 660 took up [^{14}C]propionate with sulphate as an electron acceptor (Figure 12.6), demonstrating the high substrate specificity of *Desulfobulbus* for propionate. Interestingly, approximately 9 and 27% of them could take up [^{14}C]propionate with oxygen and nitrate, respectively, as an electron acceptor. This result supports the previously reported ability of the versatile metabolism of *Desulfobulbus* to utilize propionate with SO_4^{2-}, NO_3^-, or even O_2 as an electron acceptor (Dannenberg *et al.*, 1992; Widdel and Pfenning, 1982). Furthermore, more than 40% of *Desulfobulbus* could incorporate [^{14}C]acetate into cells with nitrate and sulphate as the electron acceptor. It is conceivable that *Desulfobulbus* select sulphate, nitrate, or oxygen under microaerophilic conditions. Furthermore, *Desulfonema* hybridized with probe DNMA657 could utilize [^{14}C]acetate with nitrate as electron acceptor. This evidence clearly explains the high abundance of these species in the oxic surface zone of the biofilm (Okabe *et al.*, 1999a; 2003a; Risatti *et al.*, 1994). In addition, SRB were numerically important members of H_2-utilizing microbial populations in this microaerophilic wastewater biofilm if sulphate is present. Approximately 42% of total H_2-utilizing bacteria were SRB in this biofilm. *Desulfovibrio* hybridized with probe DSV698 and *Desulfobulbus* were thought to be main groups of H_2-utilizing SRB. These experimental results suggested the possible role of SRB in the organic carbon degradation in wastewater. At relatively low sulphate

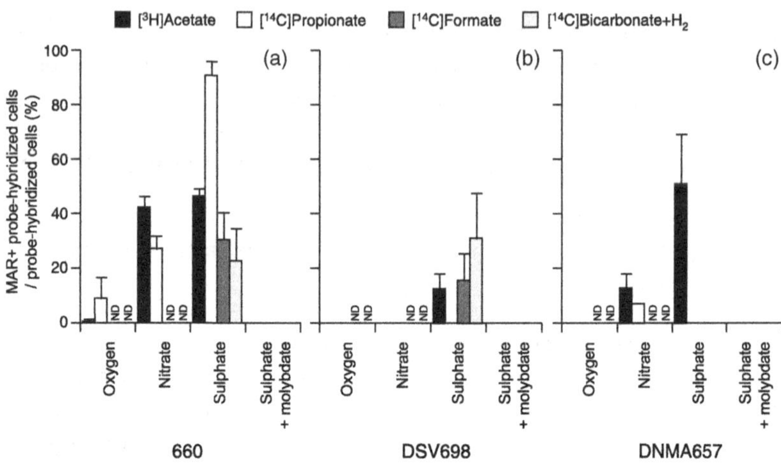

Figure 12.6. Substrate uptake patterns with oxygen, nitrate, and sulphate as electron acceptors by probe 660-hybridized *Desulfobulbus* (a), DVS698-hybridized *Desulfovibrio* (b), and DNMA657-hybridized *Desulfonema* (c), respectively. The fractions of probe-hybridized cells that simultaneously took up different radioactive substrates (MAR-positive cells) are shown as percentages of each genus-specific probe-hybridized cells. The error bars indicate standard errors. ND., not determined. Figures are reproduced from Ito *et al.* (2002a).

concentrations, as in wastewater conditions, part of the fatty acids may be directly oxidized by SRB, while the remainder is oxidized by syntrophic acetogenic bacteria, in which acetate and hydrogen are produced. The produced hydrogen would be partly utilized by SRB. In this way, hydrogen partial pressure can be maintained at low levels.

These results provide further insight into the correlation between the 16S rRNA phylogenetic diversity and the in situ physiological diversity of culturable and unculturable SRB populations inhabiting wastewater biofilms. It appears that SRB have the capacity to use a broad variety of electron acceptors including nitrate or even oxygen. The metabolic diversity of the SRB group has been revised repeatedly in recent years, suggesting a more general participation in the flow of carbon and electrons in microaerophilic and anoxic habitats than was earlier thought.

12.7 DEVELOPMENT OF SULPHUR CYCLE IN BIOFILMS

The internal sulphur cycle (consisting of sulphate reduction and subsequent sulphide oxidation) in wastewater biofilm systems can only be initiated by reduction of SO_4^{2-} to S^{2-} by the SRB. This is an important process for carrying electrons from the deeper anoxic zone to the oxic surface zone. The reductive side of this sulphur cycle (i.e. sulphate reduction) occurs only biologically by SRB. Thus, the population dynamics, biodiversity and in situ ecophysiology of SRB in wastewater biofilms have been extensively investigated (Ito *et al.*, 2002a; 2002b; Okabe *et al.*, 1999a; 2003a). The SRB population quickly developed in the wastewater biofilm within a week, and the first sulphide production was detected by microelectrode in the third week of growth (Ito *et al.*, 2002b). The active sulphide production zone gradually shifted upward and intensified with time. This indicates that SRB have the capability of adapting and proliferating in new environments. In contrast, the oxidative side of the sulphur cycle (i.e. sulphide oxidation) occurs both biologically and chemically, and thus is less understood due to its complexity. The chemical oxidation reaction is rather slow at neutral pH values and ambient temperature (e.g. in the range of minutes to several hours) (Chen and Morris, 1972; Eary and Schramke, 1990), and proceeds via $S_2O_3^{2-}$ as a major intermediate. In addition, the produced H_2S quickly reacts with metal ions, forms insoluble metal sulphides (i.e. FeS, FeS_2), and accumulates in the biofilm as an electron sink. The biological oxidation of H_2S occurs via S^0 (Buisman *et al.*, 1990; Janssen *et al.*, 1995; Jørgensen, 1982). S^0 accumulation has often been observed in wastewater biofilms (Okabe *et al.*, 1999a; 2005), indicating that biological oxidation of H_2S is dominant.

A complementary analysis for the accumulation of reduced inorganic sulphur compounds (i.e., S^0, FeS, and FeS_2) in the biofilm was performed to evaluate the importance and contribution of an internal iron-sulphur cycle in the overall sulphur cycle, which would provide more comprehensive information on a complex sulphur cycle occurring in microaerophilic wastewater biofilms (Okabe *et al.*, 1999a; 2005). The average accumulation rate of total reduced inorganic sulphur compounds (i.e. S^0, FeS, and FeS_2) was 0.026 $\mu mol\,cm^{-3}\,h^{-1}$ in the biofilm (unpublished data). Based on this average accumulation rate, and the in situ sulphate reduction rate that is determined with microelectrode, it can be estimated that less than 1% of all the produced sulphide was transformed to various reduced sulphur compounds and accumulated as an electron sink in this biofilm. Thus, the contribution of the iron-sulphur cycle to the overall sulphur cycle is negligible. This result highlights the importance of biological sulphur oxidation.

A complete sulphur cycle in the biofilm was established via S^0 accumulation during the biofilm development. This development was generally split into two phases; (i) a sulphur-accumulating phase and (ii) a sulphate-producing phase (Okabe *et al.*, 2005). The accumulation of S^0 is mainly dependent on the rate of sulphide production by SRB, the availability of electron acceptors (i.e. O_2 and NO_3^-), and the abundance and type of sulphide-oxidizing bacteria in the biofilm.

12.8 SUMMARY

In typical SRB habitats such as biofilms, substrates and physicochemical conditions are dynamically changing with time and across even very tiny distances because of metabolic activities and substrate transport limitation. Our major goal is to understand the ecophysiological roles that SRB play in such dynamic and complex ecosystems. To accomplish this goal, microbial identity (16S rRNA-based phylogeny) must be directly correlated to the in situ specific metabolic functions in the habitats. The combination of different molecular techniques and microelectrode measurement constitutes a powerful research tool and provides significant new insights into the ecophysiology of SRB and the internal sulphur cycle within complex wastewater biofilms with a typical thickness of only a few millimetres. The results demonstrated the correlation between the 16S rRNA phylogenetic and the physiological diversity of SRB inhabiting wastewater biofilms. There are, however, still many phylogenetically unidentified SRB in situ functions which are also unknown. Thus, further phylogenetic analysis followed by isolation and characterization of unidentified SRB is apparently required. In addition, sequential respiratory processes, O_2 respiration, NO_3^- respiration, and sulphate reduction, were vertically stratified in such thin biofilms and were highly sensitive to environmental changes. This dynamic nature of biofilm ecosystems consequently created the complex internal sulphur cycle in microaerophilic wastewater biofilms. Reductive and oxidative pathways of sulphur cycle occurring in microaerophilic biofilms are, however, still very complex, and further quantitative study is needed.

REFERENCES

Allen, L. A. (1929). The effect of nitro-compounds and some other substances on production of hydrogen sulphide by sulphate-reducing bacteria in sewage. *Proc Soc Appl Bacteriol*, **2**, 26–38.

Amann, R. and Ludwig, W. (2000). Ribosomal RNA-targeted nucleic acid probes for studies in microbial ecology. *FEMS Microbiol Rev*, **24**, 555–65.

Amann, R. I., Ludwig, W. and Schleifer, K.-H. (1995). Phylogenetic identification and *in situ* detection of individual microbial cells without cultivation. *Microbiol Rev*, **59**, 143–69.

Amann, R. and Kühl, M. (1998). *In situ* methods for assessment of microorganisms and their activities. *Curr Opin Microbiol*, **1**, 352–8.

Amann, R. I., Stomley, J., Devereux, R. K. and Stahl, D. A. (1992). Molecular and microscopic identification of sulphate-reducing bacteria in multispecies biofilms. *Appl Environ Microbiol*, **58**, 614–23.

Boone, D. R. and Bryant, M. P. (1980). Propionate-degrading bacterium, *Syntrophobacter wolinii* sp. nov. gen. nov., from methanogenic ecosystems. *Appl Environ Microbiol*, **40**, 626–32.

Buisman, C., Ijspeert, P., Janssen, A. and Lettinga, G. (1990). Kinetics of chemical and biological sulphide oxidation in aqueous solutions. *Wat Res*, **24**, 667–71.

Canfield, D. E. and Des Marais, D. J. (1991). Aerobic sulphate reduction in microbial mats. *Science*, **251**, 1471–3.

Chen, K. Y. and Morris, J. C. (1972). Kinetics of oxidation of aqueous sulfide by O_2. *Environ Sci Technol*, **6**, 529–37.

Dannenberg, S., Kroder, M., Dilling, W. and Cypionka, H. (1992). Oxidation of H_2, organic compounds and inorganic sulfur compounds coupled to reduction of O_2 or nitrate by sulphate-reducing bacteria. *Arch Microbiol*, **158**, 93–9.

Devereux, R., Kane, M. D., Winfrey, J. and Stahl, D. A. (1992). Genus- and group-specific hybridization probes for determinative and environmental studies of sulphate-reducing bacteria. *System Appl Microbiol*, **15**, 601–9.

Devereux, R., Delaney, M., Widdel, F. and Stahl, D. A. (1989). Natural relationships among sulphate-reducing eubacteria. *J Bacteriol*, **171**, 6689–95.

Devereux, R. and Stahl, D. A. (1993). Phylogeny of sulphate-reducing bacteria and a perspective for analyzing their natural communities. In J. M. Odom and R. Singleton Jr. (eds.), *The sulphate-reducing bacteria: contemporary perspectives*. New York: Springer-Verlag. pp. 131–60.

Eary, L. E. and J. A. Schramke. (1990). Rates of inorganic oxidation reactions involving dissolved oxygen. In D. C. Melchior and R. L. Basset (eds.), *Chemical modeling of aqueous systems II*. Washington, DC: American Chemical Society. pp. 379–96.

Frund, C. and Cohen, Y. (1992). Diurnal cycles of sulphate reduction under oxic conditions in cyanobacterial mats. *Appl Environ Microbiol*, **58**, 70–7.

Fukui, M., Teske, A., Assmus, B., Muyzer, G. and Widdel, F. (1999). Physiology, phylogenetic relationships, and ecology of filamentous sulphate-reducing bacteria (genus *Desulfonema*). *Arch Microbiol*, **172**, 193–203.

Hamilton, W. A. (1985). Sulphate-reducing bacteria and anaerobic corrosion. *Ann Rev Microbiol*, **35**, 195–217.

Hamilton, W. A. (2003). Microbially influenced corrosion as a model system for the study of metal microbe interactions: a unifying electron transfer hypothesis. *Biofouling*, **19**, 65–76.

Hamilton, W. A. and Lee, W. (1995). Biocorrosion. In L. L. Barton (ed.), *Sulphate-reducing bacteria. Biotechnology Handbooks 8*. New York: Plenum Press. pp. 243–62.

Harmsen, H. J. M., Akkermans, A. D. L., Stams, A. J. M. and de Vos, W. M. (1996). Population dynamics of propionate-oxidizing bacteria under methanogenic and sulfidogenic conditions in anaerobic granular sludge. *Appl Environ Microbiol*, **62**, 2163–8.

Head, I. M., Saunders, J. R. and Pickup, R. W. (1998). Microbial evolution, diversity and ecology: a decade of ribosomal RNA analysis of uncultured microorganisms. *Microb Ecol*, **35**, 1–21.

Heppner, B., Zellner, G. and Diekmann, H. (1992). Start-up and operation of a propionate-degrading fluidized-bed reactor. *Appl Microbiol Biotechnol*, **36**, 810–16.

Ito T., Nielsen, J. L., Okabe, S., Watanabe, Y. and Nielsen, P. H. (2002a). Phylogenetic identification and substrate uptake patterns of sulphate-reducing bacteria inhabiting an oxic-anoxic sewer biofilm determined by combining microautoradiography and fluorescent *in situ* hybridization. *Appl Environ Microbiol*, **68**, 356–64.

Ito, T., Okabe, S., Satoh, H. and Watanabe, Y. (2002b). Successional development of sulphate-reducing bacterial populations and their activities in a wastewater biofilm growing under microaerophilic conditions. *Appl Environ Microbiol*, **68**, 1392–402.

Ito T., Sugita, K. and Okabe, S. (2004). Isolation, characterization and *in situ* detection of a novel chemolithoautotrophic sulfur-oxidizing bacterium in wastewater biofilms growing under microaerophilic conditions. *Appl Environ Microbiol*, **70**, 3122–9.

Ito, T., Sugita, K., Yumoto, I., Nodasaka, Y. and Okabe, S. (2005). *Thiovirga sulfuroxydans* gen. nov., sp. nov., a chemolithotrophic sulfur-oxidizing bacterium isolated from a microaerophilic waste-water biofilm. *International Journal of Systematic and Evolutionary Microbiology*, **55**, 1059–64.

Janssen, A. J. H., Sleyster, R., van der Kaa, C. *et al.* (1995). Biological sulphide oxidation in a fed-batch reactor. *Biotechnol Bioeng*, **47**, 327–33.

Jenneman, G. E., McInerney, M. J. and Knapp, R. M. (1986). Effect of nitrate on biogenic sulfide production. *Appl Environ Microbiol*, **51**, 1205–11.

Jørgensen, B. B. (1978). A comparison of methods for the quantification of bacterial sulphate reduction in coastal marine sediments. III. Estimation from chemical and bacteriological field data. *Geomicrobiol J*, **1**, 49–64.

Jørgensen, B. B. (1982). Ecology of the bacteria of the sulphur cycle with special reference to anoxic–oxic interface environments. *Phil Trans R Soc Lond*, **298**, 543–61.

Kühl, M. and Jørgensen, B. B. (1992). Microsensor measurement of sulphate reduction and sulfide oxidation in compact microbial communities of aerobic biofilms. *Appl Environ Microbiol*, **58**, 1164–74.

Lee, N., Nielsen, P. H., Andreasen, K. H. *et al.* (1999). Combination of fluorescent *in situ* hybridization and microautoradiography – a new tool for structure-function analyses in microbial ecology. *Appl Environ Microbiol*, **65**, 1289–97.

Lee, W., Lewandowski, Z., Characklis, W. G. and Nielsen, P. H. (1994). Microbial corrosion of mild steel in a biofilm system. In G.G. Geesey, Z. Lewandowski and H.-C. Flemming (eds.), *Biofouling and biocorrosion in industrial water systems*. Lewis Publishers, Plenum Press, CRC Press, Inc., Florida, pp. 205–12.

Lovley, D. R. and Phillips, E. J. P. (1994). Novel processes for anaerobic sulphate production from elemental sulfur by sulphate-reducing bacteria. *Appl Environ Microbiol*, **60**, 2394–9.

Loy A., Lehner, A., Lee, N. *et al.* (2002). Oligonucleotide microarray for 16S rRNA gene-based detection of all recognized lineages of sulphate-reducing prokaryotes in the environment. *Appl Environ Microbiol*, **68**, 5064–81.

Manz, W., Eisenbrecher, M., Neu, T. R. and Szewzyk, U. (1998). Abundance and spatial organization of Gram-negative sulphate-reducing bacteria in activated sludge investigated by *in situ* probing with specific 16S rRNA targeted oligonucleotides. *FEMS Microbiology Ecology*, **25**, 43–61.

Minz, D., Fishbain, S., Green, S. J. *et al.* (1999a). Unexpected population distribution in a microbial mat community: sulphate-reducing bacteria localized to the highly oxic chemocline in contrast to a eukaryotic preference for anoxia. *Appl Environ Microbiol*, **65**, 4659–65.

Minz, D., Flax, J. L., Green, S. J. *et al.* (1999b). Diversity of sulphate-reducing bacteria in oxic and anoxic regions of a microbial mat characterized by comparative analysis of dissimilatory sulfite reductase genes. *Appl Environ Microbiol*, **65**: 4666–71.

Nielsen, P. H., Lee, W., Lewandowski, Z., Morison, M. and Characklis, W. G. (1993). Corrosion of mild steel in an alternating oxic and anoxic biofilm system. *Biofouling*, **7**, 267–84.

Norsker, N. H., Nielsen, P. H. and Hvitved-Jacobsen, T. (1995). Influence of oxygen on biofilm growth and potential sulphate reduction in gravity sewer biofilm. *Wat Sci Tech*, **31**(7), 159–67.

Okabe, S., Itoh, T., Satoh, H. and Watanabe, Y. (1999a). Analyses of spatial distributions of sulphate-reducing bacteria and their activity in aerobic wastewater biofilms. *Appl Environ Microbiol*, **65**, 5107–16.

Okabe, S., Ito T. and Satoh, H. (2003a). Sulphate-reducing bacterial community structure, function and their contribution to carbon mineralization in a wastewater biofilm growing microaerophilic conditions. *Appl Microbiol Biotechnol*, **63**, 322–34.

Okabe, S., Ito T., Satoh, H. and Watanabe, Y. (2003b). Effect of nitrite and nitrate on biogenic sulfide production in sewer biofilms as determined by use of microelectrodes. *Water Science and Technology*, **47**, 281–8.

Okabe, S., Ito T., Sugita, K. and Satoh, H. (2005). Succession of internal sulfur cycle and sulfide-oxidizing bacterial community in microaerophilic wastewater biofilms. *Appl Environ Microbiol*, **71**, 2520–9.

Okabe, S., Santegoeds, C. and de Beer, D. (2003c). Effect of nitrite and nitrate on *in situ* sulfide production in an activated sludge immobilized agar film as determined by use of microelectrodes. *Biotechnol Bioeng*, **81**, 570–7.

Okabe, S., Satoh, H. and Watanabe, Y. (1999b). *In situ* analysis of nitrifying biofilms as determined by *in situ* hybridization and the use of microelectrodes. *Appl Environ Microbiol*, **65**, 3182–91.

Oude Elferink, S. J. W. H., Visser, A., Hulshoff Pol, L. W. and Stams, A. J. M. (1994). Sulphate reduction in methanogenic bioreactors. *FEMS Microbiol Rev*, **15**, 119–36.

Oude Elferink, S. J. W. H., Vorstman, W. J. C., Sopjes, A. and Stams, A. J. M. (1998). Characterization of the sulphate-reducing and syntrophic population in granular sludge from a full-scale anaerobic reactor treating papermill wastewater. *FEMS Microbiol Ecology*, **27**, 185–94.

Ouverney, C. C. and Fuhrman, J. A. (1999). Combined microautoradiography-16S rRNA probe technique for determination of radioisotope uptake by specific microbial cell types *in situ*. *Appl Environ Microbiol*, **65**, 1746–52.

Postgate, J. R. (1984). The sulphate-reducing bactera, 2nd edn. Cambridge, UK: Cambridge University Press.

Ramsing, N. B., Kühl, M. and Jørgensen, B. B. (1993). Distribution of sulphate-reducing bacteria, O_2, and H_2S in photosynthetic biofilms determined by oligonucleotide probe and microelectrodes. *Appl Environ Microbiol*, **59**, 3840–9.

Rabus, R., Fukui, M., Wilkes, H. and Widdel, F. (1996). Degradative capacities and 16S rRNA-targeted whole cell hybridization of sulphate-reducing bacteria in an anaerobic environment culture utilizing alkylbenzenes from crude oil. *Appl Environ Microbiol*, **62**, 3605–13.

Risatti, J. B., Capman, W. C. and Stahl, D. A. (1994). Community structure of a microbial mat: the phylogenetic dimension. *Proc Natl Acad Sci USA*, **91**, 10173–7.

Santegoeds, C. M., Damgaard, L. R., Hesselink, G. *et al.* (1999). Distribution of sulphate-reducing and methanogenic bacteria in anaerobic aggregates determined by microsensor and molecular analyses. *Appl Environ Microbiol*, **65**, 4618–29.

Santegoeds, C. M., Ferdelman, T. G., Muyzer, G. and de Beer, D. (1998). Structural and functional dynamics of sulphate-reducing populations in bacterial biofilms. *Appl Environ Microbiol*, **64**, 3731–9.

Schramm, A. (2003). *In situ* analysis of structure and activity of the nitrifying community in biofilms, aggregate, and sediments. *Geomicrobiol J*, **20**, 313–33.

Schramm, A., Larsen, L. H., Revsbech, N. P., Amann, R. and Schleifer, K.-H. (1996). Structure and function of a nitrifying biofilm as determined by in situ hybridization and the use of microelectrodes. *Appl Environ Microbiol*, **62**, 4641–7.

Schramm, A., Santegoeds, C. M., Nielsen, H. K. *et al.* (1999). On the occurrence of anoxic microniches, denitrification, and sulphate reduction in aerated activated sludge. *Appl Environ Microbiol*, **65**, 4189–96.

Widdel, F. (1988). Microbiology and ecology of sulphate- and sulfur-reducing bacteria. In A. J. B. Zehnder (ed.), *Biology of anaerobic microorganisms*. John Wiley & Sons Inc., New York, pp. 469–585.

Widdel, F. and Pfenning, N. (1982). Studies on dissimilatory sulphate-reducing bacteria that decompose fatty acids II. Incomplete oxidation of propionate by *Desulfobulbus propionicus* gen. nov., sp. nov. *Arch. Microbiol*, **131**, 360–5.

Wu, W.-M., Jain, M. K., Conway de Macario, E., Thiele, J. H. and Zeikus, J. G. (1992). Microbial composition and characterization of prevalent methanogens and acetogens isolated from syntrophic methanogenic granules. *Appl Microbiol Biotechnol*, **38**, 282–90.

CHAPTER 13

Bioprocess engineering of sulphate reduction for environmental technology

Piet N. L. Lens, Marcus Vallero and Giovanni Esposito

13.1 INTRODUCTION

The microbiota present in the sulphur cycle have been studied since the end of the nineteenth century when the pioneering work of the famous microbiologists Winogradsky and Beijerinck took place. Sulphur conversions involve the metabolism of several different specific groups of bacteria, e.g. sulphate-reducing bacteria (SRB), phototrophic sulphur bacteria and thiobacilli, specialized to use these sulphur compounds in their different redox states (Lens and Kuenen, 2001). Many of these microorganisms possess unique metabolic and ecophysiological features, and to date there are still regular reports of novel microorganisms with extraordinary properties. Several of the microbial conversions of the sulphur cycle can be implemented for pollution control (Table 13.1). This chapter overviews the applications in environmental technology, which utilize the metabolism of SRB as the key process.

Technological utilization of SRB sounds at first somewhat controversial, as sulphate reduction has been considered unwanted for many years in anaerobic wastewater treatment (Hulshoff Pol *et al.*, 1998). Emphasis of the research in the 1970s–1980s was mainly on the prevention or minimalization of sulphate reduction during methanogenic wastewater treatment (Colleran *et al.*, 1995). From the 1990s onwards, interest has grown in applying sulphate reduction for the treatment of specific wastestreams, e.g. inorganic sulphate-rich wastewaters such as acid mine drainage, metal polluted groundwater and flue-gas scrubbing waters. Nowadays, sulphur-cycle-based technologies are not solely considered as "end-of-pipe" applications, but their potential for pollution prevention as well as for sulphur, metal or water recovery and re-use are now fully recognized.

Table 13.1. *Overview of applications in environmental biotechnology that mainly utilize conversions from the microbial sulphur cycle*

Application	Sulphur conversion utilized	Typical waste-stream
Wastewater treatment		
Removal of oxidized sulphurous compounds (sulphate, sulphite and thiosulphate)	S-oxyanion reduction to S^{2-}, followed by sulphide removal step	Industrial wastewaters, acid mine drainage and spent sulphuric acid
Sulphide removal	Partial S^{2-} oxidation to S^0	Industrial wastewaters
Heavy metal removal	SO_4^{2-} reduction	Extensive treatment in wetlands or anaerobic ponds High rate reactors for process water, acid mine drainage and groundwater
Nitrogen removal	S^{2-}, S^0 and $S_2O_3^{2-}$ oxidation	Domestic wastewater
Removal of xenobiotics	SO_4^{2-} reduction	Textile wastewaters
Microaerobic treatment	Internal sulphur cycle in a biofilm	Domestic sewage
Off-gas treatment		
Biofiltration of gases	Oxidation of S^{2-} and organosulphur compounds	Biogas, malodorous gases from composting and farming
Treatment of scrubbing waters	SO_4^{2-} and/or SO_3^{2-}reduction, plus partial S^{2-} oxidation to S^0	Scrubbing waters of SO_{2-} rich gases
Solid waste treatment		
Reduction of waste sludge production	Internal sulphur cycle in a biofilm	Sulphur cycle in biofilms
Desulphurization of resources	Organo-sulphur oxidation	Waste rubber, coal, oil, LPG, spent caustic

Bioleaching of metals	S^{2-} oxidation	Sewage sludge, compost
Gypsum processing	SO_4^{2-} reduction	Waste gypsum depots
Treatment of soils and sediments		
Bioleaching of metals	S^{2-} oxidation	Dredged sediments and spoils
Phytoextraction	SO_4^{2-} uptake by plants	Dredged sediments and spoils
Degradation of xenobiotics	SO_4^{2-} reduction	PCB-contaminated soil slurries

13.2 SULPHATE REDUCTION IN METHANOGENIC WASTEWATER TREATMENT

In anaerobic wastewater treatment, sulphate reduction is unwanted, since the production of H_2S causes a multitude of problems, such as toxicity, corrosion, odour, increase of the liquid effluent COD, as well as reduced quality and amount of biogas (Lens *et al.*, 1998a). Methanogenic treatment of sulphate-rich wastewater is still possible, if adequate measures which allow the integration of sulphate reduction with methanogenesis are applied (Table 13.2).

Trends in industries to close water cycles lead to the accumulation of salts (including sulphates) and heat in the wastewaters, and thus impose the need for the methanogenic treatment of hot and saline wastewaters that also contain moderate (1–3 g/l) sulphate levels. Several studies have reported the feasibility of thermophilic treatment of sulphate-rich wastewaters containing low energy substrates, e.g. a 1:1:1 mixture of acetate:propionate:butyrate at 55 °C (Visser *et al.*, 1992), acetate at 70 °C (Rintala *et al.*, 1993), methanol at 65 °C (Weijma *et al.*, 2000) and at 70 °C (Vallero *et al.*, 2007), and formate at 75 °C (Vallero *et al.*, 2005). For wastewaters rich in unacidified organic matter, additional precautions are needed to cope with the potential reactor acidification or deterioration of the granular sludge quality due to the excessive growth of acidifiers (Verstraete *et al.*, 1996). One way to overcome these problems is to separate the acidifying and methanogenic activities in phased (Rebac *et al.*, 1998) or staged (Van Lier *et al.*, 1994) reactor designs. If sulphate is present in the wastewater, sulphate reduction will occur together with acidification in the acidification phase (Reis *et al.*, 1995) or in the first stages of Upflow Staged Sludge Bed (USSB) reactors (Lens *et al.*, 1998b;

Table 13.2. *Process technological measures to reduce the reactor sulphide concentration, thus allowing the integration of methanogenesis and sulphate reduction in anaerobic bioreactors*

Measure	Reference
A. Dilution of the influent	
Non-sulphate containing process water	Rinzema and Lettinga, 1988
Recycle of effluent after a sulphide removal step by:	
– Sulphide stripping	Jensen and Webb, 1995
– Sulphide precipitation	Särner, 1990
– Biological sulphide oxidation to elemental sulphur	Buisman *et al.*, 1990
Thiobacillus sp., oxygen	Sublette and Sylvester, 1987
Thiobacillus denitrificans, nitrate	
Chlorobium limicola, sunlight	Kim *et al.*, 1993
– Chemical oxidation to elemental sulphur	
Ferric sulphate/silicone supported reactor	De Smul and Verstraete, 1999
B. Decrease of the unionized sulphide concentration	
Elevation of the reactor pH	Rinzema and Lettinga, 1988
Elevation of the reactor temperature	Rintala *et al.*, 1993
Precipitation of sulphide, e.g. with iron salts	McFarland and Jewell, 1989
Stripping of the reactor liquid using	
– High degree of mixing inside the reactor	
– Recirculation of biogas after scrubbing	Särner, 1990
– Other stripping gas (e.g. N_2)	
C. Separation of sulphide production and methanogenesis	Sipma *et al.*, 2000

Two-stage anaerobic digestion with SRB in acidifying stage	Lens *et al.*, 1998*b*
Upflow-staged sludge bed with MB in bottom and SRB in top compartment	
D. Selective inhibition of SRB	
Sulphate analogues (e.g. Molybdate)	Yadav and Archer, 1989
Transition elements (e.g. Copper addition)	Clancy *et al.*, 1992
Antibiotics	Tanimoto *et al.*, 1989

Sipma *et al.*, 2000). A complete sulphate reduction in the first stage or phase, together with high gas (CO_2) production rates during acidification may result in high hydrogen sulphide (H_2S) stripping efficiencies, and thus in high sulphur removal efficiencies in the acidifying reactor or compartment. Studies on thermophilic (55 °C) granular sludge reactors operated under acidifying (pH 6) conditions showed that SRB can coexist with acidifiers during the treatment of a sucrose:propionate:butyrate mixture (ratio 2:1:1 on COD basis) with a COD/sulphate ratio of 6.7 (Sipma *et al.*, 2000), or in synthetic cardboard production wastewater with a COD/sulphate ratio of 10 (Lens *et al.*, 2001; 2002) at organic loading rates up to, respectively, 46 and 35 gCOD·litre^{-1} reactor·d^{-1}.

13.3 HIGH RATE SULPHATE-REDUCING BIOREACTORS

Initially, the experience that sulphate reduction develops spontaneously during anaerobic wastewater treatment supported the adoption of bioreactor configurations commonly used in methanogenic wastewater treatment, i.e. upflow anaerobic sludge bed (UASB) reactors, for high rate sulphate-reducing bioreactors. In UASB reactors, sulphidogenic granules can be obtained by feeding methanogenic granular sludge with a sulphate-rich wastewater. However, it can take a very long time before the sulphate reducers outcompete the methanogens. Using a mathematical model, Omil *et al.* (1998) showed that the competition between acetate-utilizing SRB and methanogenic bacteria (MB) is very time-consuming. For a granular sludge with an inoculum size of 10^3 and 10^9 cells of, respectively, acetotrophic SRB and MB, it was calculated that it will take over 1000 days before the size of both populations is equal. This time period can be shortened by manipulating the population size of SRB and MB in the inoculum sludge, i.e. deactivating methanogens with chloroform (Visser *et al.*, 1993a), or temperature shocks (Visser *et al.*, 1993b), or by addition of pure cultures of SRB (Omil *et al.*, 1997b).

In sulphidogenic reactors, wastewater purification is mainly accomplished by SRB, which convert sulphate to dissolved sulphide (Visser *et al.*, 1992; 1993b). As methanogenic bacteria are not involved in the degradation of organic matter, less gaseous end products (e.g. no biogas formation) are formed. Biogas evolution has been reported to reduce the external diffusion resistance of granular sludge (Huisman *et al.*, 1990), imposed by the surrounding stagnant liquid layer of some 200 μm (Lens *et al.*, 1993). Consequently, sulphidogenic reactors operate at lower mass transfer rates. Several ways to accelerate mass transfer in sulphidogenic reactors have been studied: (i) elevated (up to 6 $m.h^{-1}$) superfical upflow velocity of the reactor liquid (Omil *et al.*, 1996); (ii) gas evolution in the granular sludge bed (Lens *et al.*, 2002); or (iii) the breathing granule principle (van den Heuvel *et al.*, 1995). Unfortunately, all these methods yielded lower sulphate-reduction efficiencies compared to control reactors because of disrupture and/or shearing off of the outer granule layers which contain high numbers of SRB, thus leading to selective washout of SRB from the reactor system.

For the treatment of inorganic wastewaters, the choice of electron donor is an important design parameter. One can, for example, supply an organic substrate (e.g. molasses) as electron donor, although this increases the risk of "residual" pollutants. For high rate sulphate-reducing bioreactors supplied with a hydrogen/carbon dioxide (H_2/CO_2) mixture, high conversion rates can be obtained in mesophilic (30 °C; van Houten *et al.*, 1994) or thermophilic (55 °C; van Houten *et al.*, 1997) gas-lift reactors in a short (10 days) start-up period. In H_2/CO_2-fed reactor systems, a consortium of SRB (*Desulfovibrio* sp.) and homoacetogens (*Acetobacterium* sp.) develop (van Houten *et al.*, 1995). In cases where pure hydrogen gas is not available, one can use synthesis gas (a mixture of H_2, CO_2 and CO), either directly (van Houten *et al.*, 1996) or after enriching its H_2 content by means of a water–gas–shift reaction, either chemically or biologically with anaerobic granular sludge (Sipma *et al.*, 2004).

The main drawback of gas lift bioreactors is the high pressure drop of the water column that needs to be overcome when supplying the gaseous substrate (H_2). Cell suspension bioreactors (Lens *et al.*, 2003) or bubbleless H_2 supply by hydrophobic membranes (Fedorovich *et al.*, 2000) might be elegant alternative reactor designs. Cell suspension bioreactors also allow, via the dilution rate, control of the competition between SRB and MB based on their growth kinetics (Paulo *et al.*, 2004). Alternatively, under thermophilic (55–65 °C) conditions, methanol can be supplied, as this substrate is converted to H_2/CO_2 at these high temperatures (Vallero *et al.* 2003a). As the formed H_2 is readily consumed by SRB, the H_2 partial pressure is kept below the threshold of H_2-utilizing methanogens, resulting in the outcompetition of methanogens (Visser *et al.*, 1996; 2003b). In cases where soluble substrates (such as methanol) are supplied, no H_2S stripping occurs and therefore H_2S removal needs to be adopted to prevent its accumulation to toxic concentrations. This can be done by stripping using an external gas stream (e.g. N_2), or via extractive H_2S membranes (De Smul and Verstraete, 1999).

Accumulation of acetate in the effluent is the limiting factor of high-rate sulphate-reducing systems fed with H_2/CO_2, ethanol, methanol or volatile fatty acids (Omil *et al.*, 1997a; Muthumbi *et al.*, 2001). Surprisingly, acetate degradation does not commence, even if sulphate is present in excess. Addition of nitrate to the last compartment of a baffled reactor is an elegant way to remove acetate from sulphidogenic reactor systems (Lens *et al.*, 2000). Further research must be oriented to understanding the reasons for the acetate accumulation, and establishing why acetate-consuming SRB do not develop in these reactors.

13.4 LOW RATE SULPHATE-REDUCING SYSTEMS

Microbial sulphate reduction is also regarded as an effective basic mechanism for treating acid or neutral waters contaminated with heavy metals and sulphate, which might simultaneously remove acidity and metals due to, respectively, the alkalinity produced during sulphate reduction and the very low solubility of metal sulphides. Application of high rate sulphate reduction systems for the treatment of mine waters is hampered by too high investment and operating costs. Attempts to overcome these problems have essentially focused on two strategies. Firstly, it has been attempted to adapt established industrial technologies for the purpose of mine drainage treatment, e.g. by choosing particular low-cost substrates such as whey, molasses or even wastewaters (Rose *et al.*, 1998). The second approach attempts to use SRB in passive processes, e.g. constructed wetlands (Gibert *et al.*, 2004; Markewitz *et al.*, 2004) and reactive walls (Waybrant *et al.*, 1998; Benner *et al.*, 2002).

Passive processes were developed on the basis of natural habitats such as marshes and wetlands and use both chemical and biological processes, thereby reducing the need for sophisticated process technology. Ideally passive processes are designed to be self-regulating long-term stable systems. Despite these applications being based on seemingly rather simple technical processes, the underlying chemical and microbiological mechanisms are multiple and often poorly understood. Therefore, long-term stability is often found to be a serious problem within various applications (Barton and Karathanasis, 1999).

The supply of dissolved organic carbon from the solid substrates is a time dependent process, since solid substrates are composed of different organic fractions, (carbohydrates, proteins, fats, etc.) of varying biodegradability. The resulting variations in the concentration and spectrum of the dissolved organic carbon result in unstable SRB rates. Moreover, there is a danger of process failure due to decreasing pH as a result of acidogenesis. This is of particular interest since acidogenic degradation of organic matter is needed (as the SRB cannot degrade solid substrates themselves) to form low-weight organic compounds such as volatile fatty acids or alcohols, which are the substrates of the SRB. But, at the same time, if the acidification proceeds too fast, the SRB will be inhibited by the low pH.

Passive processes apply low-cost solid substrates to supply carbon and energy for SRB and the continuous supply of sufficient amounts of dissolved organic carbon from solid substrates is crucial for maintaining long-term stability. The release of dissolved organic carbon is fundamentally a result of microbial degradation of the polymeric constituents of the solid substrates, which is a complex biochemical process catalyzed by various microorganisms. Hence, the long-term stability of the process depends on the stable interaction of these organisms. So far, a variety of compounds, e.g. cow manure and rice stalks (Cheong *et al.*, 1998), straw (Bechard *et al.*, 1994), molasses (Janssen and Teminghoff, 2004), sewage sludge and ryegrass (Harris and Ragusa, 2001), oak chips, spent oak from shiitake farms, spent mushroom compost, sewage sludge and organic-rich soil (Chang *et al.*, 2000), activated sludge, rabbit pellets and digested sludge (Prasard *et al.*, 1999) municipal compost, leaf mulch and wood chips (Benner *et al.*, 2002), mixtures of poultry manure, wood chips and leaf compost (Cocos *et al.*, 2002) have been tested as solid substrates for passive sulphate-reducing processes. Also more defined compounds such as polymers from lactic acid were investigated (Edenborn, 2004). Most solid substrates were only tested in short term experiments. Therefore, there is generally little information about the long-term behaviour of solid substrates or the release of organic carbon from solid substrates which is available for SRB in a continuous process.

13.5 FUNCTION AND STRUCTURE OF SULPHATE-REDUCING GRANULES AND BIOFILMS

13.5.1 Microbial ecology

Specific analytical tools are currently available for the study of the structure and functioning of sulphate-reducing granules and biofilms. Molecular techniques such as denaturing or thermal gradient gel electrophoresis (DGGE and TGGE), single strand conformation polymorphism (SSCP) and ribosomal RNA (mostly 16S rRNA) probes can be used to characterize the microorganisms present (Dabert *et al.* 2002). In general, these molecular techniques give more information than chemical methods, as, for example, quinone profiles (Kurisu *et al.*, 2002) and polar-lipid fatty acid biomarkers (Oude Elferink *et al.*, 1998). Using rRNA-based probes, one can detect the presence of a specific species or a group of bacterial species in a particular type of granule or biofilm. Moreover, when using fluorescently labelled 16S rRNA probes, bacteria can be spatially localized within the granule or biofilm (Figure 13.1). However, the presence of a bacterium, as detected by e.g. 16S rRNA probes, does not always reflect its metabolic activity. Therefore, one would need either to use mRNA probes or to determine the local concentration of reaction products in situ. By using miniaturized electrodes, one can determine the local concentration of a chemical compound with a 1 μm spatial resolution. A whole range of microelectrodes is currently available, e.g. for measurement of pH, sulphide, glucose or methane (Santegoeds *et al.*, 1998). Upon penetrating a granule or biofilm with such a microelectrode, a microprofile is recorded, from which a localized activity distribution can be calculated (Figure 13.1). For direct linking of microbial identification and also microorganism function, fluorescent in situ hybridization (FISH), coupled with in situ microautoradiography (Gray *et al.*, 2000) or chemical staining for storage polymers (Crocetti *et al.*, 2000), can be used (see also Chapter 12 in this book).

13.5.2 Transport

Transport of substrates and reaction products is an important factor in determining the overall process kinetics of bioreactors. In sulphidogenic granular sludge, diffusional transport is the mean transport process. The diffusional properties of granular sludges or biofilms can be determined non-destructively using pulsed field gradient nuclear magnetic resonance (PFG NMR) at low magnetic fields (0.47 T). Paramagnetic inclusions

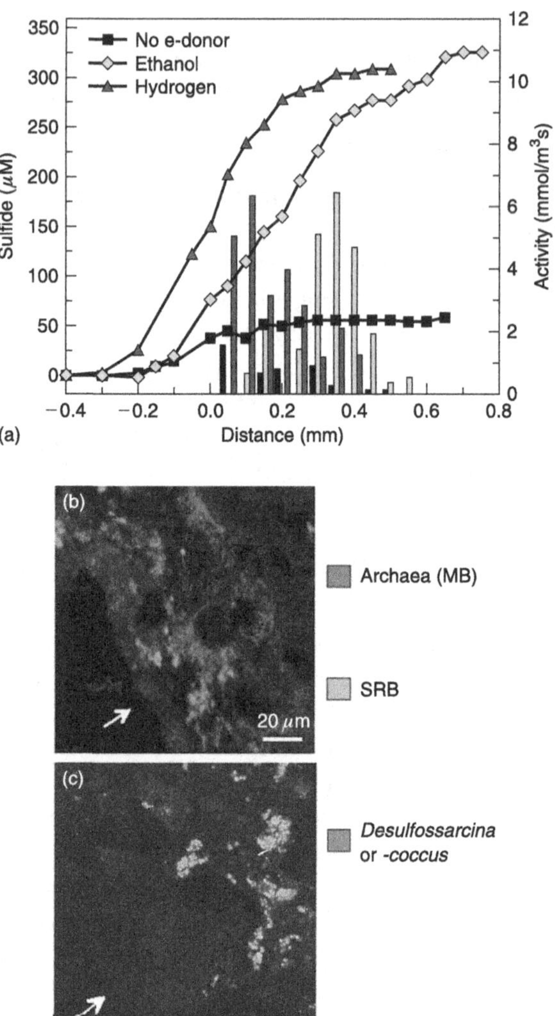

Figure 13.1. Steady state microsensor profiles and FISH analysis to localize specific SRB populations within a methanogenic-sulphidogenic aggregate (Adapted from Santegoeds *et al.*, 1999). (a) Microprofiles (lines) and activity values (bars) of sulphide and methane in aggregates in the presence of 10mM SO_4^{2-}. Sulphidogenic aggregate after one night without electron donor (■, closed bars) and with addition of 7 mM ethanol (□, open bars). (b) FISH analysis with general probes for sulphate reducers (SRB385, artificial colour green) and methanogens (ARC915, artificial colour red). (c) Hybridization with probe 660 (artificial colour purple). Microelectrode studies and in situ hybridizations were done by Dr. Cecilia M. Santegoeds and Dr. Dirk de Beer of the microsensor group of the Max Planck Institute of Marine Microbiology, Bremen (Germany) and Dr. Lars Damgaard from the University of Aarhus (Denmark). (For a colour version of this figure, please refer to the colour plate section.)

(e.g. FeS) in granular sludge hamper PFG NMR at high magnetic fields. Using PFG NMR, diffusion coefficients can be determined as an average value of a sludge sample (Beuling *et al.*, 1998; Lens *et al.*, 1999), or spatially resolved within a single granule with a resolution up to 80 μm (Gonzalez-Gil *et al.*, 2001).

13.5.3 Metal sorption and precipitation

The main mechanisms involved in metal accumulation within biofilms are complex formation, chelation, ion exchange, adsorption, inorganic microprecipitation and translocation of metals into the cells (van Hullebusch *et al.*, 2003). van Hullebusch *et al.* (2004; 2005) studied the kinetics and capacity of cobalt and nickel sorption onto granular sludge for UASB reactors. The sorption capacity, expressed in terms of q_m (Langmuir saturation constant), was generally quite low compared to that of other sorbents (Table 13.3).

Precipitation of metals, e.g. as carbonates and especially as sulphides, is important for the accumulation of metals in anaerobic sludges. Sulphide is ubiquitously present in anaerobic bioreactors because of the occurrence of sulphate reduction or organic matter mineralization (see also Chapter 14). The solubility product of metal sulphide precipitates are extremely low, and this property of sulphides is extensively used for the removal of metals from wastewaters (Kaksonen *et al.*, 2003; Jong and Perry, 2003). Although the metals are better retained in the sludge due to sulphide precipitation, it may also influence their bioavailability, e.g. making them no longer bioavailable (Gonzalez-Gil *et al.*, 2003). This may be especially the case when the sulphide precipitates age with time from amorphous to more crystalline forms (Jansen *et al.*, 2005).

13.6 SULPHUR AND METAL RECOVERY/RE-USE

13.6.1 Metal recovery and re-use

Heavy metals such as Cu, Zn, Cd, Pb, Ni, and Fe precipitate with biogenic sulphide to form sparingly soluble metal sulphides, thereby concentrating the metals into an easily separable and sometimes valuable form. In engineered systems, metal sulphide precipitation can be optimized with respect to the rate of biogenic sulphide production, metal precipitate product quality and selective precipitation of metal sulphides. Bench- and pilot-scale studies have shown that metal precipitation with

Table 13.3. *Comparison of the Langmuir constant (q_m) for nickel and cobalt with different sorbents*

Metal	Sorbent	q_m (mg·g^{-1})	T(°C)	pH	Reference
Nickel	Granular	7.9	30	6	van Hullebusch *et al.*, 2005
		9.4	30	7	
		11.5	30	8	
Nickel	Anaerobic digested sludge	25.2		7.2	Artola *et al.*, 2000
Nickel	Sphagnum moss peat	9.7	4.5		Ho *et al.*, 1996
Nickel	Anaerobic dead biomass	227			Haytoglu *et al.*, 2001
	Dried *Chlorella vulgaris*	54.8			Aksu, 2002
Nickel	*Ps.aeruginosa* biomass	62.6	35	7.2	
Nickel	Free cells	145	30	8	Lopez *et al.*, 2000
	Immobilized cells	37	30	8	
Cobalt	Anaerobic granules	8.4	30	6	van Hullebusch *et al.*, 2005
		8.9	30	7	
		9.5	30	8	
Cobalt	Ion exchange resin	60.0	25	5.3	Rengaraj and Moon, 2002
		75.6	25	5.3	
Cobalt	Carbon sorbent	17.3	25	6	El-Shafey *et al.*, 2002
Cobalt	*Oscillatoria anguitissima*	131.5	25	5	Ahuja *et al.*, 1999
	Biomass	150.0	25	6	

biogenic sulphide is a "working biotechnology", but some aspects are still poorly understood, especially the metal precipitation kinetics and biomass–precipitate interactions.

Technologies based on metal precipitation with sulphide have some fundamental advantages over hydroxide precipitation:

1. effluent concentrations are orders of magnitude lower: $\mu g.l^{-1}$ vs. $mg.l^{-1}$
2. the interference of chelating agents in the wastewater is less problematic
3. selective metal removal gives better opportunities for metal re-use
4. metal sulphide sludges have better settling, thickening and de-watering characteristics compared with hydroxide sludges
5. existing smelters can process sulphide precipitates, thus enabling metal recovery and eliminating the need for sludge disposal.

Earlier objections against the use of sulphide, i.e. that it is toxic, malodorous and corrosive, can today be overcome by adequate safety measures and the use of modern corrosion-resistant construction materials. Chemical forms of sulphide such as Na_2S, NaHS, CaS, FeS and H_2S can be used, but these need to be transported to the treatment site. In general, these sulphide sources are more expensive than lime or limestone. Moreover, the hazards that accompany transport, handling and storage of the chemical sulphides lead to additional costs for safety measures. These drawbacks can be overcome by the on-site production of biogenic sulphide in bioreactors described in Section 13.3. The formed metal sulphides are highly insoluble at neutral pH, while some metal sulphides (e.g. CuS) are highly insoluble at pH-values as low as 2.

The success of the precipitation process not only depends on removal of metal ions from the soluble phase, but also on the separation of the solid phase (metal sulphide precipitate) from the liquid phase. Therefore, solid–liquid separation processes such as sedimentation or filtration are of key importance in efficient metal removal processes. The settling and de-watering characteristics of metal precipitates are directly related to the morphology, density and particle size distribution of the precipitate. The particle size distribution is determined by the kinetics of precipitation, i.e. the competition between nucleation and crystal growth and agglomeration (Veeken *et al.*, 2003). Nucleation produces only very small particles which are difficult to separate from the liquid phase, whereas crystal growth and agglomeration of crystals result in larger particles. For precipitates with low solubility products such as heavy metal sulphides, very small particles normally prevail as (i) the supersaturation level cannot be controlled at low levels (Mersmann, 1999) and (ii) local supersaturation at the feed-points cannot be prevented due to micromixing (Esposito *et al.*, 2006). By adopting a membrane reactor with a high solids concentration of the precipitate, in combination with a pS-pH control (Konig *et al.*, 2006), a large specific surface area and low supersaturation levels can be created which promote crystal growth and suppress nucleation.

13.6.2 Sulphur recovery and re-use

Another option to minimize pollution is to recover sulphur from waste-streams and recycle it. Combination of sulphate reduction with biological or chemical sulphide oxidation (Table 13.1) allows a complete removal of sulphur from waste-streams by its conversion to insoluble elemental sulphur. The produced sulphur can be separated from the liquid stream and re-utilized as fertilizer or as raw material for sulphuric acid production.

Many industrial waste-streams and also H_2S-containing gas-streams have low amounts of organic material, but high concentrations of inorganic reduced-sulphur compounds, and are therefore selective towards obligate chemolithotrophs. Non-sterile laboratory and pilot plant studies, inoculated with obligately autotrophic species from the genus *Thiobacillus* form stable microbial communities when supplied with high sulphide and low organic influent. Biotechnological sulphide removal methods based on such cultures have been described using *Thiobacillus thioparus, Thiobacillus denitrificans* and *Thiobacillus ferrooxidans.* An aerobic biotechnological sulphide-removing method was developed, based on the ability of colourless bacteria to partially oxidize sulphide to elemental sulphur (H_2S + 0.5 $O_2 \rightarrow S^0 + H_20$) (Buisman *et al.*, 1990). The process has been applied in biorotors (Buisman *et al.*, 1990), completely stirred tank reactors (Janssen *et al.*, 1995), gas-lift reactors (Janssen *et al.*, 1997) and reversed fluidized bed reactors (Haridas *et al.*, 2000). Usually, colourless sulphur bacteria completely oxidize sulphide to sulphate ($H_2S + 2O_2 \rightarrow SO_4^{2-} + 2H^+$), generating more metabolically useful energy compared with the partial oxidation (Visser *et al.*, 1997). In order to obtain S^0 as a product, sulphide oxidation must be forced in the direction of sulphur production, for example by high sulphide loads or low oxygen concentrations (Stefess *et al.*, 1996). In bioreactors, sulphur formation can be regulated by dosing stoichiometric amounts of oxygen (Janssen *et al.*, 1995), although more rigid control mechanisms are based on monitoring the redox potential (Janssen *et al.*, 1998).

13.7 CONCLUSION

The presence of sulphate in a wastewater has to be considered when designing a treatment system. For sulphate-rich wastewaters, precautions are required to prevent sulfide toxicity in methanogenic or sulphate-reducing reactors. On the other hand, sulphate reduction also allows a number of unique process applications for environmental technology, e.g. organic matter removal by only SRB, or by SRB and sulphide oxidizers, utilizing the biological sulphur cycle; heavy metal and nitrogen removal; reduction of the waste sludge production.

REFERENCES

Ahuja, P., Gupta R. and Saxena, R. K. (1999). Sorption and desorption of cobalt by *Oscillatoria anguitissima*. *Curr. Microbiol.*, **39**, 49–52.

Aksu, Z. (2002). Determination of the equilibrium, kinetics and thermodynamic parameters of the batch sorption of nickel (II) ions onto *Chlorella vulgaris*. *Process Biochem.*, **38**, 89–99.

Artola, A., Martin M., Balaguer D. M. and Rigola M. (2000). Isotherm model analyses for the adsorption of Cd(II), Cu(II), Ni(II) and Zn(II) on anaerobically digested sludge. *J. Colloid. Inter. Sci.*, **232**, 64–70.

Barton, C. D. and Karathanasis, A. D. (1999). Renovation of a failed constructed wetland treating acid mine drainage. *Environmental Geology*, **39**, 39–50.

Bechard, G., Yamazaki, H., Gould, W. D. and Bedard, P. (1994). Use of cellulosic substrates for the microbial treatment of acid mine drainage. *J. Environmental Qual.*, **23**, 111–16.

Benner, S. G., Blowes, D. W., Ptacek, C. J. and Mayer, K. U. (2002). Rates of sulphate reduction and metal sulfide precipitation in a permeable reactive barrier. *Applied Geochemistry*, **17**, 301–20.

Beuling, E. E., van Dusschoten, D., Lens, P. *et al.* (1998). Characterization of the diffusive properties of biofilms using pulsed field gradient nuclear magnetic resonance. *Biotech. Bioeng.*, **60**, 283–91.

Buisman, C. N. J., Geraats, B. G., Ijspeert, P. and Lettinga, G. (1990). Optimization of sulphur production in a biotechnological sulphide-removing reactor. *Biotech. Bioeng.*, **35**, 50–6.

Chang, I. S., Shin, P. K. and Kim, B. H. (2000). Biological treatment of acid mine drainage under sulphate-reducing conditions with solid waste materials as substrate. *Wat. Res.*, **34**, 1269–77.

Cheong, Y.-W., Min, J.-S. and Kwon, K.-S. (1998). Metal removal efficiencies of substrates for treating acid mine drainage of the Dalsung mine, South Korea. *Journal of Geochemical Exploration*, **64**, 147–52.

Clancy, P. B., Venkataraman, N. and Lynd, L. R. (1992). Biochemical inhibition of sulphate reduction in batch and continuous anaerobic digesters. *Wat. Sci. Tech.*, **25**, 51–60.

Cocos, I. A., Zagury, G. J., Clement, B. and Samson, R. (2002). Multiple factor design for reactive mixture selection for use in reactive walls in mine drainage treatment. *Wat. Res.*, **32**, 167–77.

Colleran, E., Finnegan, S. and Lens, P. (1995). Anaerobic treatment of sulphate-containing waste streams. *Antonie van Leeuwenhoek*, **67**, 29–46.

Crocetti, G. R., Hugenholtz, P., Bond, P. L. *et al.* (2000). Identification of polyphosphate-accumulating organisms and design of 16S rRNA-directed probes for their detection and quantitation. *Appl. Environ. Microbiol.*, **66**, 1175–82.

Dabert, P., Delgenes, J.-P., Moletta, R. and Godon, J.-J. (2002). Contribution of molecular microbiology to the study in water pollution removal of microbial community dynamics. *Reviews in Environmental Science and Bio/Technology*, **1**, 39–49.

De Smul, A. and Verstraete, W. (1999). The phenomenology and the mathematical modeling of the silicone-supported chemical oxidation of aqueous sulfide to elemental sulfur with ferric sulphate. *J. Chem. Technol. Biotechnol.*, **74**, 456–66.

Edenborn, H. M. (2004). Use of poly(lactic acid) amendments to promote the bacterial fixation of metals in zinc smelter tailings. *Bioresource Technology*, **92**, 111–19.

El-Shafey, E., Cox, M., Pichugin, A. A. and Appleton, Q. (2002). Application of a carbon sorbent for the removal of cadmium and other heavy metal ions from aqeous solution. *J. Chem. Technol. Biotechnol.*, **77**, 429–36.

Esposito, G., Veeken, A., Weijma, J. and Lens, P. N. L. (2006). Effect of the use of biogenic sulphide on ZnS precipitation under different process conditions. *Separation and Purification Technology*, **51**, 31–9.

Fedorovich, V., Greben, M., Kalyuzhnyi, S. *et al.* (2000). Use of membranes for hydrogen supply in a sulphate reducing reactor. *Biodegradation*, **11**, 295–303.

Fedorovich, V., Lens, P. and Kalyuzhnyi, S. (2003). Extension of anaerobic digestion model no. 1 with the processes of sulphate reduction. *Applied Biochemistry and Biotechnology*, **109**, 33–46.

Gibert, O., de Pablo, J., Cortina, J. L. and Ayora, C. (2004). Chemical characterisation of natural organic substrates for biological mitigation of acid mine drainage. *Wat. Res.*, **38**, 4186–96.

Gonzalez-Gil, G., Lens, P., Van Aelst, A. *et al.* (2001). Cluster structure of anaerobic aggregates of an expanded granular sludge bed reactor. *Appl. Environ. Microbiol.*, **67**, 3683–92.

Gonzalez-Gil, G., Jansen, S., Zandvoort, M. H. and van Leeuwen, H. P. (2003). Effect of yeast extract on speciation and bioavailability in nickel and cobalt in anaerobic bioreactors. *Biotech. Bioeng.*, **82**, 134–42.

Gray, N. D., Howarth, R., Pickup, R. W., Gwyn Jones, J. and Head, I. M. (2000). Use of combined microautoradiography and fluorescence in situ hybridization to determine carbon metabolism in mixed natural communities of uncultured bacteria from the genus *Achromatium*. *Appl. Environ. Microbiol.*, **66**, 4518–22.

Haridas, A., Majumdar, S. and Kumar, K. (2000). Reverse fluidised loop reactor for oxidation of sulphide. In *Workshop on Anaerobic Processes in Wastewater Management.*, MHO-cooperation Cochin University of Science and Technology, Technical University Delft and Wageningen University. 9–15 October, Cochin, India.

Haytoglu, B., Demerir, G. N. and Yetis, U. (2001). Effectiveness of anaerobic biomass in adsorbing heavy metals. *Wat. Sci. Technol.*, **44**, 245–52.

Ho, Y. S., Wase, D. A. J. and Forster, C. F. (1996). Kinetic studies of competitive heavy metal sorption by sphagnum moss peat. *Environ. Technol.*, **17**, 71–6.

Huisman, J. W., Van den Heuvel, J. C. and Ottengraf, S. P. P. (1990). Enhancement of external mass transfer by gaseous end products. *Biotechnol. Progr.*, **6**, 425–9.

Hulshoff Pol, L., Lens, P., Stams, A. J. M. and Lettinga, G. (1998). Anaerobic treatment of sulphate-rich wastewaters. *Biodegradation*, **9**, 213–24.

Jensen, A. B. and Webb, C. (1995). Treatment of H_2S-containing gases: a review of microbiological alternatives. *Enzyme Microbiol. Technol.*, **17**, 2–10.

Janssen, G. M. C. M. and Teminghoff, E. J. M. (2004). *In situ* metal precipitation in a zinc contaminated aerobic sandy aquifer by means of biological sulphate reduction. *Environ. Sci. Technol.*, **38**, 4002–11.

Jansen, S., Steffen, F., Threels, W. F. and Van Leeuwen, H. P. (2005). Speciation of Co(II) and Ni(II) in anaerobic bioreactors measured by competitive ligand exchange-adsorptive stripping voltammetry. *Environ. Sci. Technol.*, **39**, 9493–9.

Janssen, A. J. H., Sleyster, R., van der Kaa, C. *et al.* (1995). Biological sulphide oxidation in a fed-batch reactor. *Biotech. Bioeng.*, **47**, 327–33.

Janssen, A. J. H., Ma, S. C., Lens, P. and Lettinga, G. (1997). Performance of a sulphide-oxidizing expanded-bed reactor supplied with dissolved oxygen. *Biotech. Bioeng.*, **53**, 32–40.

Janssen, A. J. H., Meijer, S., Bontsema, J. and Lettinga, G. (1998). Application of the redox potential for controlling a sulfide oxidizing bioreactor. *Biotech. Bioeng.*, **60**, 147–55.

Jong, T. and Perry, L. (2003). Removal of sulphate and heavy metals by sulphate reducing bacteria in short-term bench scale upflow anaerobic packed bed reactors. *Wat. Res.*, **37**, 3379–89.

Kaksonen, A. H., Riekkola-Vanhanen, M. L. and Puhakka, J. A. (2003). Optimization of metal sulfide precipitation in fluidized-bed treatment of acidic wastewater. *Wat. Res.*, **37**, 255–66.

Harris, M. A. and Ragusa, S. (2001). Bioremediation of acid mine drainage using decompostable plant material in a constant flow bioreactor. *Environmental Geology*, **40**, 1192–204.

Kim, B. W., Kim, E. H., Lee, S. C. and Chang, H. N. (1993). Model-based control of feed rate and illuminance in a photosynthetic fed-batch reactor for H_2S removal. *Bioprocess Eng.*, **8**, 263–9.

König, J., Keesman, K. J., Veeken, A. and Lens, P. N. L. (2006). Dynamic modelling and process control of ZnS precipitation. *Separation Science Technology*. **41**(6), 1025–42.

Kurisu, F., Satoh, H., Mino, T. and Matsuo, T. (2002). Microbial community analysis of thermophilic contact oxidation process by using ribosomal RNA and the quinone profile method. *Wat. Res.*, **36**, 429–38.

Lens, P., de Beer, D., Cronenberg, C. *et al.* (1993). Inhomogenic distribution of microbial activity in UASB aggregates: pH and glucose microprofiles. *Appl. Environ. Microbiol.*, **59**, 3803–15.

Lens, P., Gastesi, R., Hulshoff Pol, L. and Lettinga, G. (2003). Use of sulphate reducing cell suspension bioreactors for the treatment of SO_2 rich flue gases. *Biodegradation*, **14**, 229–40.

Lens, P., Sipma, J., Hulshof Pol, L. and Lettinga, G. (2000). Effect of staging and nitrate addition on sulfidogenic acetate removal. *Wat. Res.*, **34**, 31–42.

Lens, P., van den Bosch, M., Hulshoff Pol, L. and Lettinga, G. (1998b). Effect of staging on volatile fatty acid degradation in a sulfidogenic granular sludge reactor. *Wat. Res.*, **32**, 1178–92.

Lens, P., Vergeldt, F., Lettinga, G. and van As, H. (1999). ^{1}H-NMR study of the diffusional properties of methanogenic aggregates. *Wat. Sci. Tech.*, **39**, 187–94.

Lens, P., Visser, A., Janssen, A., Hulshoff Pol, L. and Lettinga, G. (1998a). Biotechnological treatment of sulphate rich wastewaters. *Crit. Rev. Env. Sci. Technol.*, **28**, 41–88.

Lens, P. N. L., Klijn, R., van Lier, J. B., Hulshoff Pol, L. W. and Lettinga, G. (2002). Effect of specific gas loading rate on thermofilic sulphate reduction under acidifying conditons. *Wat. Res.*, **37**, 1033–47.

Lens, P. N. L., Korthout, D., van Lier, J. B., Hulshoff Pol, L. W. and Lettinga, G. (2001). Effect of upflow velocity on thermofilic sulphate reduction under acidifying conditons. *Environ. Technol.*, **22**, 183–93.

Lens, P. N. L. and Kuenen, J. G. (2001). The biological sulfur cycle: novel opportunities for environmental biotechnology. *Wat. Sci. Tech.*, **44**, 57–66.

Lopez, A., Lazaro, N., Priego, J. M. and Marques, A. M. (2000). Effect of pH on the biosorption of nickel and other metals by *Pseodomonas fluorescens* 4F39. *J. Ind. Microbiol. Bioetechnol.*, **24**, 146–51.

Markewitz, K., Cabral, A. R., Panarotto, C. T. and Lefebvre, G. (2004). Anaerobic biodegradation of an organic by-products leachate by interaction with different mine tailings. *Journal of Hazardous Materials*, **110**, 93–104.

McFarland, M. J. and Jewell, W. J. (1989). In situ control of sulfide emission during thermophilic anaerobic digestion process. *Wat. Res.*, **23**, 1571–7.

Mersmann, A. (1999). Crystallization and precipitation. *Chem. Eng. Process*, **38**, 345–53.

Muthumbi, W., Boon, N., Boterdaele, R. *et al.* (2001). Microbial sulphate reduction with acetate: process performance and composition of the bacterial communities in the reactor at different salinity levels. *Appl. Microbiol. Biotechol.*, **55**, 787–93.

Omil, F., Lens, P., Hulshoff Pol, L. and Lettinga, G. (1996). Effect of upward velocity and sulphide concentration on volatile fatty acid degradation in a sulphidogenic granular sludge reactor. *Process Biochem.*, **31**, 699–710.

Omil, F., Lens, P., Hulshoff Pol, L. and Lettinga, G. (1997a). Characterization of biomass from a sulphidogenic, volatile fatty acid-degrading granular sludge reactor. *Enzyme Microb. Technol.*, **20**, 229–36.

Omil, F., Lens, P., Visser, A., Hulshoff Pol, L. W. and Lettinga, G. (1998). Long term competition between sulphate reducing and methanogenic bacteria in UASB reactors treating volatile fatty acids. *Biotech. Bioeng.*, **57**, 676–85.

Omil, F., Oude Elferink, S. J. W. H., Lens, P., Hulshoff Pol, L. and Lettinga, G. (1997b). Effect of the inoculation with *Desulforhabdus amnigenus* and pH or O_2 shocks on the competition between sulphate reducing and methanogenic bacteria in an acetate fed UASB reactor. *Biores. Technol.*, **60**, 113–22.

Oude Elferink, S. J. W. H., Boschker, H. T. S. and Stams, A. J. M. (1998). Identification of sulphate reducers and *Syntrophobacter* sp. in anaerobic granular sludge by fatty-acid biomarkers and 16S rRNA probing. *Geomicrobial J.*, **15**, 3–18.

Paulo, P., Kleerebezem, R., Lettinga, G. and Lens, P. N. L. (2005). Cultivation of high-rate sulphate reducing sludge by pH-based electron donor dosage. *Journal of Biotechnology*, **118**, 107–16.

Prasad, D., Wai, M., Berube, P. and Henry, J. G. (1999). Evaluating substrates in the biological treatment of acid mine drainage. *Environmental Technology*, **20**, 449–58.

Rebac, S., van Lier, J. B., Lens, P. *et al.* (1998). Psychrophilic (6–15 °C) high-rate treatment of malting waste water in a two module EGSB system. *Biotechnol. Progr.*, **14**, 856–64.

Reis, M. A. M., Lemos, P. C. and Carrondo, M. J. T. (1995). Biological sulphate removal of industrial effluents using the anaerobic digestion. Med. Fac. Landbouww. Univ. *Gent.*, **60**, 2701–7.

Rengaraj, S. and Moon, S. H. (2002). Kinetics of adsorption of Co (II) removal from water and and wastewater by ion exchange resin. *Wat. Res.*, **36**, 1783–93.

Rintala, J., Lepisto, S. and Ahring, B. (1993). Acetate degradation at 70 °C in upflow anaerobic sludge blanket reactors and temperature response of granules grown at 70 °C. *Appl. Environ. Microbiol.*, **59**, 1742–6.

Rinzema, A. and Lettinga, G. (1988). Anaerobic treatment of sulphate containing waste water. In D. L. Wise (ed.), *Biotreatment systems*, Vol III, pp. 65–109. Boca Raton, FL: CRC Press Inc.

Rose, P. D., Boshoff, G. A., van Hille, R. P. *et al.* (1998). An integrated algal sulphate reducing high rate ponding process for the treatment of acid mine drainage wastewaters. *Biodegradation*, **9**, 247–57.

Santegoeds, C. M., Damgaard, L. R., Hesselink, G. *et al.* (1999). Distribution of sulphate reducing and methanogenic bacteria in UASB aggregates determined by microsensors and molecular techniques. *Appl. Environ. Microbiol.*, **65**, 4618–29.

Santegoeds, C. M., Schramm, A. and de Beer, D. (1998). Microsensors as a tool to determine chemical microgradients and bacterial activity in wastewater biofilms and flocs. *Biodegradation*, **9**, 159–67.

Särner, E. (1990). Removal of sulphate and sulphite in an anaerobic trickling (ANTRIC) filter. *Wat. Sci. Tech.*, **22**, 395–404.

Sipma, J., Lens, P. N. L., Vieira, A. *et al.* (2000). Thermofilic sulphate reduction in UASB reactors under acidifying conditons. *Process Biochem.*, **35**, 509–22.

Sipma, J., Meulepas, R. J. W., Parshina, S. N. *et al.* (2004). Effect of carbon monoxide, hydrogen and sulphate on thermophilic (55 °C) hydrogenogenic carbon monoxide conversion in two anaerobic bioreactor sludges. *Applied Microbiology and Biotechnology*, **64**, 421–8.

Stefess, G. C., Torremans, R. A. M., De Schrijver, R., Robertson, L. A. and Kuenen, J. G. (1996). Quantitative measurement of sulphur formation by steady-state and transient-state continuous cultures of autotrophic *Thiobacillus* species. *Appl. Microbiol. Biotechnol.*, **45**, 169–75.

Sublette, K. L. and Sylvester, N. D. (1987). Oxidation of hydrogen sulfide by continuous cultures of *Thiobacillus denitrificans. Biotech. Bioeng.*, **29**, 753–8.

Tanimoto, Y., Tasaki, M., Okamura, K., Yamaguchi, M. and Minami, K. (1989). Screening growth inhibitors of sulphate-reducing bacteria and their effects on methane fermentation. *J. Ferment. Bioeng.*, **68**, 353–9.

Vallero, M. V. G., Camarero, E., Lettinga, G. and Lens, P. N. L. (2007). Hyperthermophilic sulphate reduction in methanol and formate fed UASB reactors. *Appl. Environ. Microbiol.* Submitted.

Vallero, M. V. G., Lens, P. N. L., Hulshoff Pol, L. W. and Lettinga, G. (2003a). Effect of NaCl on thermophilic (55 °C) methanol degradation in sulphate reducing reactors. *Wat. Res.*, **37**, 2269–80.

Vallero, M. V. G., Paulo, P. L., Trevino, R. H. M., Lettinga, G. and Lens, P. N. L. (2003b). Effect of sulphate on methanol degradation in thermophilic (55 °C) methanogenic UASB reactors. *Enzyme Microb. Technol.*, **32**, 676–87.

van den Heuvel, J. C., Vredenbregt, L. H. J., Portegies-Zwart, I. and Ottengraf, S. P. P. (1995). Acceleration of mass transfer in methane-producing loop reactors. *Antonie van Leeuwenhoek*, **67**, 125–30.

van Houten, R. T., Hulshoff Pol, L. W. and Lettinga, G. (1994). Biological sulphate reduction using gas-lift reactors fed with hydrogen and carbon dioxide as energy and carbon source. *Biotech. Bioeng.*, **44**, 586–94.

van Houten, R. T., Oude Elferink, S. J. W. H., van Hamel, S. E. *et al.* (1995). Sulphate reduction by aggregates of sulphate-reducing bacteria and homo-acetogenic bacteria in a lab-scale gas-lift reactor. *Biores. Technol.*, **54**, 73–9.

van Houten, R. T., van der Spoel, H., van Aelst, A. C., Hulshoff Pol, L. W. and Lettinga, G. (1996). Biological sulphate reduction using synthesis gas as energy and carbon source. *Biotech. Bioeng.*, **50**, 136–44.

van Houten, R. T., Yun, S. Y. and Lettinga, G. (1997). Thermophilic sulphate and sulfite reduction in lab-scale gas-lift reactors using H_2 and CO_2 as energy and carbon source. *Biotech. Bioeng.*, **55**, 807–14.

van Hullebusch, E. D., Zandvoort, M. H. and Lens, P. N. L. (2003). Metal immobilisation in biofilms: mechanisms and analytical tools. *Re/view Environ. Sci. Bio/Technol.*, **2**, 9–33.

van Hullebusch, E. D., Peerbolte, A., Zandvoort, M. H. and Lens, P. N. L. (2005). Sorption of cobalt and nickel on anaerobic granular sludges: isotherms and sequential extraction. *Chemosphere*, **58**, 493–505.

van Hullebusch, E., Zandvoort, M. H. and Lens, P. N. L. (2004). Nickel and cobalt sorption on anaerobic granular sludges: kinetic and equilibrium studies. *J. Chem. Technol. Biotechnol.*, **79**, 1219–27.

Van Lier, J. B., Boersma, F., Debets, M. M. W. H. and Lettinga, G. (1994). High-rate thermophilic anaerobic wastewater treatment in compartmentalized upflow reactors. *Wat. Sci. Tech.*, **30**, 251–61.

Veeken, A. H. M., Vries, S. de, Mark, A van der, Rulkens, W. H. (2003). Selective precipitation of heavy metals as controlled by a sulfide-selective electrode. *Sep. Sci. Tech.*, **38**, 1–19.

Visser, A., Beeksma, I., van der Zee, F., Stams, A. J. M. and Lettinga, G. (1993a). Anaerobic degradation of volatile fatty acids at different sulphate concentrations. *Appl. Microbiol. Biotechnol.*, **40**, 549–56.

Verstraete, W., de Beer, D., Pena, M., Lettinga, G. and Lens, P. (1996). Anaerobic bioprocessing of waste. *World J. Microbiol. Biotechnol.*, **12**, 221–38.

Visser, A., Gao, Y. and Lettinga, G. (1992). The anaerobic treatment of a synthetic sulphate containing wastewater under thermophilic (55 °C) conditions. *Wat. Sci. Tech.*, **25**, 193–202.

Visser, A., Gao, Y. and Lettinga, G. (1993b). Effects of short-term temperature increases on the mesophilic anaerobic breakdown of sulphate containing synthetic wastewater. *Wat. Res.*, **27**, 541–50.

Visser, A., Hulshoff Pol, L. W. and Lettinga, G. (1996). Competition of methanogenic and sulfidogenic bacteria. *Wat. Sci. Tech.*, **33**, 99–110.

Visser, J. M., Robertson, L. A., Van Verseveld, H. W. and Kuenen, J. G. (1997). Sulfur production by obligately chemolithoautotrophic *Thiobacillus* species. *Appl. Environ. Microbiol.*, **63**, 2300–5.

Waybrant, K. R., Blowes, D. W. and Ptacek, C. J. (1998). Selection of reactive mixtures for use in permeable reactive walls for treatment of acid mine drainage. *Environ. Sci. Tech.*, **32**, 1972–9.

Weijma, J., Stams, A. J. M., Hulshoff Pol, L. W. and Lettinga, G. (2000). Thermophilic sulphate reduction and methanogenesis with methanol in a high rate anaerobic reactor. *Biotech. Bioeng.*, **67**, 354–63.

Yadav, V. K. and Archer, D. B. (1989). Sodium molybdate inhibits sulphate reduction in the anaerobic treatment of high sulphate molasses wastewater. *Appl. Microbiol. Biotechnol.*, **31**, 103–6.

CHAPTER 14

Bioremediation of metals and metalloids by precipitation and cellular binding

Simon L. Hockin and Geoffrey M. Gadd

14.1 INTRODUCTION

Interactions between dissimilatory sulphate-reducing bacteria (SRB) and metal(loid) ions have been studied since the first half of the twentieth century and the ability of SRB to bring about changes in the speciation of metal(loid)s has been recognized for much of this time. Early work focused on the role of SRB as nuisance organisms and metal(loid) interactions with SRB were often studied in the context of their use as metabolic poisons to control SRB activity (Postgate, 1952; Newport and Nedwell, 1988; Nemati *et al.*, 2001). With growing awareness of the importance of microorganisms in biogeochemical cycling, the emphasis of research has shifted to the environmental roles of SRB. Their capacity to control the mobility of metals in aqueous sediments by the formation of poorly-soluble metal sulphides is now apparent and SRB-generated metal sulphides constitute an important environmental sink for many metals (Morse *et al.*, 1987; see also Chapter 13, this volume). Bioremediation of dissolved metal(loid)s is an application for which SRB may be particularly suitable, given that sulphate frequently co-occurs with toxic metal ions in, e.g. metal-processing wastes and acid mine-drainage waters. SRB have the apparently unique potential to simultaneously remove metals, sulphate and acidity through the bioprecipitation of metal sulphides – a phenomenon that still represents one of the most successful biological approaches to metal removal from aqueous media. Other abiotic chemical reactions that can effect the removal of metal(loid) ions from solution take place during active sulphate-reduction. These include direct chemical reduction of ions by SRB-produced sulphides and the precipitation of metal oxides and carbonates following the rise in pH that results from sulphate removal (e.g. White and Gadd, 1998; Gadd, 2002).

Contemporary work on the nature of SRB enzyme systems has revealed that common SRB genera are rich in broad-specificity enzymes which have the capacity to change the oxidation state of metal(loid)s and reduce their solubility (Lovley *et al.*, 1993a; 1993b; Lovley and Phillips, 1994; Lloyd *et al.*, 1998a,b; Macy *et al.*, 2000; De Luca *et al.*, 2001; Naz *et al.*, 2005; Hockin and Gadd, 2003; 2006). This greatly extends the range of metal-precipitation mechanisms now recognized in the SRB and has opened new research directions in the application of direct enzymatic reduction of inorganic pollutants by SRB, with possibilities for *ex situ* cell-based and metabolically engineered approaches.

The capacity for sorption of metal(loid) ions by SRB cells and extracellular polymers has also attracted attention, particularly for SRB that exist as biofilms or flocs. The biosorption and bioconcentration of soluble ions, nucleation of precipitate at biopolymer surfaces, and electrostatic entrapment of particulates may all be significant mechanisms of metal(loid) immobilization, as well as having the potential to enhance the efficiency of other removal mechanisms as the primary focus (Beech and Cheung, 1995; Gadd, 2001; Hockin and Gadd, 2003; 2006; White and Gadd, 1998; 2000; White *et al.*, 2003; Glasauer *et al.*, 2004).

14.2 METAL REMOVAL BY SULPHIDE GENERATION

While dissimilatory sulphate-reducing bacteria are a metabolically and taxanomically diverse group, they are united by the common ability to utilize sulphate as a terminal electron acceptor to conserve energy for growth:

$$SO_4^{2-} + 9H^+ + 8e^- \rightarrow HS^- + 4H_2O \qquad \Delta G = -191.8\,\text{kJ mol}^{-1} \tag{14.1}$$

Sulphur in the S(VI) oxidation state is stoichiometrically reduced to S(-II) and, under the circumneutral conditions in which SRB are generally encountered, the main product is bisulphide (HS^-), with a small proportion of volatile H_2S (Stumm and Morgan, 1996). Bisulphide is a highly reactive species, with the propensity to bond with metal cations in solution forming metal sulphide solids. Metal sulphides are chemically characterized as sparingly soluble, in equilibrium between the solid and aqueous phases. The degree of solubilization is represented by the solubility product constant, K_s (usually expressed as the log) and given by the equations below: where Me represents a divalent metal cation:

$$MeS_{(s)} + H^+ \rightleftharpoons Me^{2+}_{(aq)} + HS^-_{(aq)} \tag{14.2}$$

$$Me_2S_{(s)} + H^+ \rightleftharpoons 2Me^{2+}_{(aq)} + HS^-_{(aq)} \qquad (14.3)$$

K_s is generally calculated using the form of a direct reaction with bisulphide and assuming a dilute solute with a hydrogen (or hydronium) ion concentration equal to one. Certain modifications to the above equations are applied for non-stoichiometric sulphide phases and for diagenetic MeS_2 forms, such as the iron sulphides greigite and pyrite (Stumm and Morgan, 1996). The solubility of most metal sulphides is extremely low (Table 14.1), meaning that most of these are virtually insoluble under anaerobic conditions at circumneutral pH. Nevertheless, variations of many orders of magnitude in the value of K_s exist, even between sulphide mineral forms of the same metal counterion, while some alkali metals and species such as MnS, with unusually complex chemistry, do have more soluble sulphides.

As the equilibrium equations (14.2) and (14.3) show, the concentration of the metal remaining in solution is an inverse function of the sulphide concentration. As sulphate is frequently present at millimolar concentrations while toxic metal ions generally occur in the micromolar range, even moderate sulphate-reducing activity can result in effective metal sulphide deposition. Furthermore, the extremely low solubility products and wide stability fields of many transition metal sulphides mean that they can be precipitated at pH values well below neutral when sulphide is available in stoichiometric excess. In fact, equations (14.2) and (14.3) are

Table 14.1. *Solubility products of selected metal sulphides at 18 °C*[a]

Counterion	Mineral form	lg K_s
Cadmium	(CdS greigite)	−14.36
Cobalt	(β CoS)	−11.07
Copper	(CuS)	−22.30
Copper	(Cu_2S)	−34.65
Iron	(FeS mackinawite)	−3.60
Iron	(FeS_2 pyrite)	−16.40
Lead	(PbS)	−13.97
Manganese	(MnS pink)	3.34
Mercury	(HgS black)	−38.80
Nickel	(β-NiS)	−11.10
Zinc	(α-ZnS)	−10.93

[a] Values taken from Stumm & Morgan (1996) and references therein.

simplifications of metal sulphide formation, which is arguably a multi-stage process and frequently results in the primary deposition of non-stoichiometric phases, which then undergo diagenetic reorganization to more crystalline minerals. Nevertheless, most authigenic metal sulphides are (meta)stable under anaerobic conditions once formed, but decomposition of precipitated sulphides at the oxic–anoxic interface can lead to dissolution and release of covalently bound metals, as well as co-precipitated or adsorbed metal(loid) ions (Morse and Arakari, 1993; Morse, 1994; Cooper and Morse, 1998).

The efficiency of metal sulphide deposition (and the relative efficiency for metals with relatively soluble sulphides) was demonstrated by White and Gadd (1996a) who studied the effects of process variables on the rate of sulphate reduction and metal removal (Cd, Co, Cr, Cu, Mn, Ni, Zn) from a simulated acid soil leachate, using a mixed SRB culture. Copper was almost entirely removed but, for other metals, removal varied with the amount of sulphide produced, although high percentage removals ($>95\%$) were generally achieved, except for Mn. In other work, using a laboratory-scale internal sedimentation bioreactor, $>95\%$ removal was successfully achieved for all metals tested, meeting (or marginally exceeding) European Water Standards. However, Mn^{2+} removal was again problematic with less than 45% removal from an initial metal concentration of 500 μM (White and Gadd, 1997). In both sets of experiments the reductions in aqueous concentration achieved were less than would be predicted by the sulphide chemistry alone. It is likely that other factors, such as the formation of hydrated sulphide ions, or of a colloidal sulphide phase, may have inhibited efficient metal precipitation. Kinetic modelling of manganese sulphide/carbonate precipitation by SRB gave a closer fit to experimental data when terms for hydrated aqueous polysulphides or dissolved colloids were included (Hockin and Gadd, unpublished data). Subsequent studies with cadmium confirmed that a limiting factor for metal removal in laboratory SRB systems was the formation of fine, suspended particulates which are slow to precipitate from suspension (White and Gadd, 1998; White *et al.*, 2003).

The best known commercial application involving metal sulphide precipitation is the THIOPAQ® technology, developed and marketed by PAQUES BV (http://www.paques.nl/paques/) in Balk, Netherlands, and first applied in 1992 for the treatment of contaminated groundwater at the Budelco zinc refinery in the Netherlands. The basic THIOPAQ® system consisted of two biological process stages in series: anaerobic sulphate reduction to sulphide followed by aerobic sulphide oxidation to elemental sulphur. Since the solubilities of most metal sulphides are much lower than those of their hydroxides, an advantage of the THIOPAQ® system is that considerably lower effluent metal concentrations can be achieved than in neutralization processes which immobilize metals by hydroxide precipitation. In addition, the metal sulphide precipitate formed may be reprocessed in a smelter or a refinery.

Suitable electron donors for small-scale THIOPAQ® installations were ethanol, various fatty acids and organic waste-streams. For large-scale applications, where more than 2.5 tonnes of hydrogen sulphide are produced per day, hydrogen gas is preferentially used as the reductant. Hydrogen gas can be produced on-site by cracking methanol or by steam-reforming natural gas or LPG. If ethanol is used, the chemical costs are somewhat higher than in the case of hydrogen but the investment is lower since no reformer is needed. The main reaction that occurs in a reactor operated with H_2 is:

$$H_2SO_4 + 4H_2 \rightarrow H_2S + 4H_2O \qquad (14.4)$$

The hydrogen sulphide in the reactor gas (3–15% v/v) can be employed for metal precipitation by contacting it with the solution to be treated. Compared with the addition of a NaHS or Na_2S solution, an advantage with the use of H_2S is that sodium is not introduced into the system. It is also possible that careful control of the pH and the redox potential of the process liquid may allow selective recovery of metals. Thus, sulphide precipitation makes it possible to separate copper from zinc, arsenic from copper, iron from nickel, etc. in multiple reaction stages at different pH values. Alternatively, metals may also be precipitated as sulphides inside the anaerobic bioreactor.

At the Budelco zinc refinery in Budel-Dorplein, Netherlands, a THIOPAQ® system processing approximately 300 m^3/h of polluted groundwater has been in operation since 1992. Its products, a metal sulphide sludge (mainly ZnS) and a sulphur slurry, are fed back to the roasters in the refinery. The capacity of the installation was increased to 400 m^3/h (in 1998) with the feed also including a mixture of groundwater and process water. The most recent THIOPAQ® installation (called Budelco II, and in operation from 1999) treats several bleed streams and process water: sulphate reduction occurs in a 500 m^3 bioreactor where hydrogen is used as the electron donor.

Precipitation of metalloid-sulphide phases has been less widely reported, but the occurrence of technetium sulphides in sediments has been attributed to SRB (Pignolet *et al.*, 1989). Lloyd *et al.* (1998a) reported the indirect precipitation of Tc(VII) by resting cells of SRB supplied with an electron donor and sulphate as the electron acceptor. Tc precipitated as a sulphidic phase with S in stoichiometric excess. There was no indication of prior reduction of technetium to a lower valence state. Arsenical sulphides are predominant among the source minerals associated with the contamination of groundwaters by arsenic, which constitutes a major problem in parts of the world, notably Bangladesh and the USA. Attention has been paid to the bacterial leaching mechanisms responsible for the solubilization of these As-bearing minerals, but there has been less investigation of the controls on arsenic solubilization. While bacterial arsenate reduction has been proposed as the dominant control under acidic conditions, this does not appear to be the case in circumneutral groundwaters. Chemical precipitation of As(III)sulphides is, however, rapid under laboratory circumstances and convincing evidence of the role of sulphate-reducers in natural environments exists (Moore *et al.*, 1988; Rittle *et al.*, 1995; McReadie *et al.*, 2000; Macy *et al.*, 2000; Kirk *et al.*, 2004). Bacterial sulphate reduction has also been proposed as a means of remediating arsenic-contaminated aquifer water (Kirk *et al.*, 2004). While there are obvious problems in using sulphide to treat potential drinking water supplies, arsenic mobilization in Bangladesh is associated with surface recharge and recent addition of dissolved carbon electron donors to the aquifer (Harvey *et al.*, 2002). This appears to stimulate methanogenesis and iron oxidation, which lead to arsenic solubilization. The addition of sulphate as an alternative electron acceptor might promote SRB activity, simultaneously removing carbon and arsenic and restoring stable, oligotrophic conditions. SRB-based bioremediation would therefore seem to offer at least a technically feasible option, although recharge of aquifers to historical levels would be a prerequisite to prevent further draw-down of surface water.

14.3 OTHER INDIRECT CHEMICAL EFFECTS

The use of chemical treatments to stimulate the alkaline precipitation of heavy metals from acidic wastewaters is well established, but while such processes may be highly efficient in terms of metal removal abilities, they require the continuous addition of alkaline reactants and are thus expensive. The use of biologically generated alkalinity, based on cheap and readily available substrates, is therefore a potentially attractive alternative. SRB are able to remove acidity from local and bulk environments and some SRB strains are reported to maintain sulphate-reducing activity in waters with an initial bulk acidity of pH 2–4. Fortin and Beveridge (1997) found that SRB were active and important in the localized cycling of iron, copper and zinc in anoxic sediments of sulphidic mine tailings, despite a bulk porewater pH of 3–4.

In confined systems, SRB can also bring about significant rises in the bulk pH, which can enhance sulphide precipitation and lead to precipitation of hydroxides and carbonates of transition metals. The effect is attributable to the removal from solution of sulphate – as the strong sulphuric acid ($-\lg pK\alpha\ H_2SO_4 = 3$) – and its replacement in solution by the weak acid bisulphide ($-\lg pK\alpha\ HS^- \approx 17$), so removing free hydrogen ions (Stumm and Morgan, 1996). Where an organic substrate acts as the electron donor, bicarbonate is also generated:

$$2CH_3CHOHCOO^- + SO_4^{2-} \rightarrow 2CH_3COO^- + 2HCO_3^- + HS^- + H^+ \quad (14.5)$$

This also has useful implications for the use of SRB in the remediation of acidic metal-processing waters and mine wastes, particularly where they are active within suspended, or surface-attached mesophilic biofilms. SRB are reported to contribute significantly to metal removal in constructed wetlands that have successfully been used for metals removal and alkalization of acidic mine wastes. Both sulphide generation and pH-related precipitation appear to be important (Webb *et al.*, 1998; Fortin *et al.*, 2000; Gadd, 2002). However, some studies have questioned the contribution of SRB in these broad-scale systems, arguing that Fe(III)-reducing bacteria make a greater contribution where carbon is limiting, both in terms of metal removal and in ameliorating low pH (Vile and Weider, 1993; Fortin *et al.*, 1996).

It is clear that pH effects are important in enhancing metal(loid) removal in *ex situ* SRB-based systems. Chang *et al.* (2000) showed that a mixed-species SRB culture incubated in laboratory-scale through-flow reactors was able to generate sufficient alkalinity to efficiently remove Zn and Cu, as sulphides, from acid mine wastewater when a suitable organic substrate was provided. Iron and manganese were less efficiently removed and this was probably related to the difficulty in maintaining reactor pH over the operational cycle, even though mean residence times of up to 20 d were used. A straightforward solution to this problem involves the addition of an alkaline mineral, such as limestone, to achieve chemical neutralization. Dvorak *et al.* (1991) used a simple SRB fixed-bed reactor with spent mushroom compost (containing finely crushed limestone) as the nutrient source to treat metal-contaminated water in an underground coal mine and at a smelting residues dump in Pennsylvania. Approximately 95% of supplied Al, Cd, Fe, Mn, Ni, and Zn was removed using a flow-through process. Cd, Fe and Ni were deposited primarily as sulphides, while

Al, Mn and Zn were precipitated as hydroxides, following a bulk pH rise. Hammack and Edenborn (1992) developed a pilot fixed-bed system for the removal of nickel from simulated mine waters. This system, which used spent mushroom compost as a matrix, was able to remove 75% of the nickel, and 90% of the sulphate and brought the pH to >6. The addition of lactate as a supplementary carbon/energy source increased overall sulphate reduction, raised the effluent pH and increased metal removal to $>95\%$.

Both the acidity and metal loading of mine waters are associated with the (bio)chemical leaching activities of iron-oxidizing bacteria (IOB) and sulphur-oxidizing bacteria (SOB) (Ewart and Hughes, 1991). This capacity of SOB to acidify waters and to bring metals into solution by the production of sulphuric acid has been applied to contaminant remediation at laboratory and pilot scale, by integrating the activities of SOB and SRB in a single process for the remediation of mixed metal-contaminated soils. An internal sedimentation SRB reactor was developed for the removal of metals from a SOB-generated metals-laden acid leachate (White and Gadd, 1997; White *et al.*, 1998). The reactor was able to achieve high internal biomass production and almost complete sulphate reduction for influent concentrations of up to 50 mM, while raising the reactor bulk pH to near neutrality. Precipitation efficiency was further increased by the addition of flocculating agents. All metals were removed with $>95\%$ efficiency, with the exception of Mn and Ni, as described above.

Notwithstanding the above, the pH sensitivity of SRB remains an impediment to their wider use in bioremediation. Active sulphate reduction has nevertheless been recorded in extremely acidic environments (e.g. Gyure *et al.*, 1990), leading researchers to seek to identify acidophilic, or acid-tolerant SRB strains. Although SRB have been isolated from such environments, this apparent acid tolerance seems to be the result of the high pH microclimates maintained by the activity of the SRB (Fortin *et al.*, 1996; Hard and Babel, 1997). While this can be enhanced during biofilm growth (see below), there are clearly limits to the abilities of mesophilic SRB to maintain active metabolism in hostile acid environments. It seems a truly acidophilic dissimilatory SRB remains to be discovered.

A possible alternative approach is to physically separate metabolizing cells from the effluent to be treated. For *ex situ* engineered applications it is relatively straightforward to achieve this using gas-stripping or gas-permeable membranes to draw off the H_2S. This also offers the possiblity of controlling the chemical environment in which H_2S-mediated precipitation takes place, with the ability to alter pH to separate and selectively recover precipitated elements, according to the solubility product of their sulphides.

The above approach has been successfully applied at both demonstration and full commercial level, as in the Paques B.V. THIOPAQ® system (Barnes *et al.*, 1994; Rowley *et al.*, 1997; Battelle Bioprocessing, 2001 and references therein).

The extremely reducing conditions that SRB create can also result in indirect chemical reduction of metal(loid) and radionuclide species. This appears to be the case for U(VI) which is reduced chemically to U(IV) under highly reducing, sulphidic conditions. The ability of SRB to enzymatically reduce uranium is now established (see below) and there has been some debate as to the relative contribution of chemical and enzymatic uranium reduction, with chemical reduction rates appearing relatively low (Lovley and Phillips, 1992; Lovley *et al.*, 1993a; Tucker *et al.*, 1998). Nevertheless, while enzymatic reduction is effective in non-metabolizing cells, some studies appear to support a role for chemical reduction in the presence of growing cells. Barton *et al.* (1996) examined the use of mixed sulphate-reducing cultures with cellulose substrates. In laboratory-scale batch reactors, 1.0 mM uranium was reduced to <0.1 mM in <40 days. In through-flow column reactors, $>99.9\%$ uranium was removed from a 1.0 mM starting solution. In the latter experiments initial removal rates were low and increased rapidly, following sequential reduction of the medium and the onset of sulphate-reducing activity. Uranium was precipitated extracellularly as microcrystalline UO_2, implying that indirect chemical reduction exterior to the cells was the primary mechanism responsible. Although the removal rates recorded would be considered too slow for *ex situ* applications, the work aimed to establish the viability of long-term, "passive" remediation. This was proposed as a suitable basis for the use of underground "permeable-reactive biobarriers" in the remediation of uranium mine tailings in the Western USA.

Technetium (which, as ^{99}Tc, is a long half-life product of the nuclear fuel cycle) is present in many environments as Tc(VII) in the form of the highly mobile pertechnetate ion (TcO_4^-). Both chemical and enzymatic reductive precipitation of Tc has been demonstrated in SRB and chemical precipitation appears to be more efficient than for uranium, with sulphide as the reductant. Lloyd *et al.* (1998a) found that under sulphidogenic conditions, chemical precipitation operated in preference to enzymatic reduction and *Dv. desulfuricans* was able to precipitate technetium extracellularly, probably as sulphide.

Selenium contamination from industrial and agricultural activities is a significant problem in some regions and ^{79}Se is also a long half-life β-emitting fission product of ^{235}U, occurring in spent nuclear fuel and reprocessing wastes. Although it is likely that other organisms offer greater potential than SRB for the remediation of selenium as a sole contaminant (Oremland *et al.*, 1989; 1994; 1999), SRB may be suitable for treatment of selenium where it is present in mixed-metal polluted wastes and can have a disproportionate impact due to its extreme chemical toxicity (Jacobs, 1989). SRB reduce soluble selenium oxyanions by a several means, including the reductive precipitation of Se(IV) in the form of biselenite ($HSeO_4^-$) by SRB-generated sulphide. This reaction can rapidly remove selenite from solution with only moderate sulphide production and in the presence of significant concentrations of sulphate and metal cations. Under laboratory conditions, the reaction results in co-precipitation of both elemental selenium and elemental sulphur (Hockin and Gadd, 2003).

There is also evidence that the extremely reducing conditions that develop during sulphate reduction can lead to chemical conversion of oxyanions to cationic species that are more easily precipitated, or biosorbed. The indirect chemical reduction of Cr(VI), as soluble chromate (CrO_4^{2-}), to much less soluble Cr(III) cationic species by sulphide and/or Fe^{2+} in SRB culture appears to be at least partially responsible for the removal of chromate from solution by SRB (Fude *et al.*, 1994; Lloyd *et al.*, 2000).

14.4 ENZYMATIC REDUCTION OF METALS AND METALLOIDS

Sulphate-reducing bacteria have been subject to considerable attention for their apparent capacity for the direct, enzymatically mediated reductive precipitation of a wide range of toxic metal(loid)s (see Lloyd *et al.*, 2004; Chapter 15, this volume). It has been suggested that enzymatic metal(loid) reduction is more characteristic of spore-forming Gram-positive genera than of SRB of the delta-proteobacteria, but a growing body of work appears to contradict this view and the Gram-negative *Desulfovibrio* spp. have shown high metal reductase activity, with broad metal(loid) specificity (Lovley *et al.*, 1993a; Lovley and Phillips, 1994; Tucker *et al.*, 1998; Lloyd *et al.*, 1999a,b). Widespread as this property seems to be, the ability of SRB to conserve energy for growth by the dissimilatory reduction of metal(loid)s appears to be rare (Lovley and Phillips, 1994; Tebo and Obraztsava, 1998; Elias *et al.*, 2004).

Hydrogenases and/or cytochromes have been implicated as the active complexes in the reduction of a range of metal(loid)s that may be precipitated by SRB. Cytochrome c_3 was confirmed as the enzymatic pathway for the reduction of Cr(VI) to Cr(III) by non-sulphidogenic cells of

Desulfovibrio vulgaris. The reductive pathway operated via hydrogenase when H_2 was supplied directly as the electron donor (Lovley and Phillips, 1994). The authors acknowledged that enzymatic reduction is unlikely to be significant in natural environments (where chemical reduction of chromate by SRB-produced sulphide, Fe(II), or organic moieties will predominate), but that enzymatic precipitation may offer a useful approach for *ex situ* remedial systems. Unlike other bacterial dissimilatory metal(loid) reducers such as *Enterobacter cloacae*, the presence of other toxic metals and of sulphate does not appear to inhibit metalloid reduction by *Desulfovibrio* spp., which may thus offer distinct advantages in this respect.

Uranium (as the radionuclide ^{235}U) is clearly an element of great environmental concern. While U(VI) species are readily soluble, U(IV) compounds are often poorly soluble in the circumneutral pH range and thus partial enzymatic reduction of U(VI) to U(IV) can be an effective removal strategy. Enzymatic U(VI) reduction appears to be widespread among SRB. In *Dv. vulgaris*, the cytochrome c_3 complex appears again to be the key cellular mechanism responsible for partial uranium reduction (Lovley and Phillips, 1992; Lovley *et al.*, 1993b). *Desulfovibrio* group SRB have also been shown to enzymatically reduce Fe(III) to Fe(II) (Lovley *et al.*, 1993a) and the effectiveness of enzymatic U(VI) removal (and that of other metalloids) by *Desulfovibrio spp.* may therefore be further enhanced by the simultaneous enzymatic reduction of Fe(III) to Fe(II) and additional chemical reduction of U(VI) to U(IV).

During sulphidogenesis, SRB can bring about the precipitation of technetium as Tc(IV) sulphide, as described above, but SRB have also been shown to have the capacity for effective enzymatic technetium reduction. *Dv. desulfuricans* is reported to reduce Tc(VII) via a periplasmic hydrogenase, leading to precipitation at the cell margin, probably as a Tc(V) oxide (Lloyd *et al.*, 1998a; 1999a; De Luca *et al.*, 2001). Lloyd *et al.* (1999a) showed that immobilized, non-growing cells of *Dv. desulfuricans* removed $>85\%$ Tc from a solution containing 250 μM pertechnate in 1 h when hydrogen was the electron donor, although removal was less efficient when formate or lactate were used. The enzyme-mediated reduction of soluble palladium by cells of *Dv. desulfuricans* has also been reported (Lloyd *et al.*, 1998b).

Selenate (Se(VI), as SeO_4^{2-}) is known to be enzymatically reduced by SRB via a number of pathways, including assimilatory reduction, leading to the release of the volatile species dimethyl selenide (DMSe) and dimethyl diselenide (DMDSe) (Michalke *et al.*, 2000). Reduction to selenide in nanomolar amounts by the dissimilatory sulphate-reducing pathway has also been demonstrated, resulting in the formation of volatile hydrogen selenide (Zehr and Oremland, 1987). The existence of at least one separate pathway by which SRB enzymatically reduce both Se(VI) and Se(IV) oxyanions to elemental selenium has also been shown (Tomei *et al.*, 1995). Tucker *et al.* (1998) demonstrated effective enzymatic precipitation of selenium (and of chromium, molybdenum and uranium) by immobilized cells of *Dv. desulfuricans*, using a distilled water medium containing only lactate as electron donor and the metal(loid) target anion as potential terminal electron acceptor. The work described above used SRB that utilized non-sulphate-reducing metabolism, but other studies suggest that enzymatic reduction of selenate can also take place during active sulphate reduction. While Se(IV) is preferentially chemically precipitated under sulphate-reducing conditions, chemical reduction does not appear to be significant for Se(VI) when present in solution as the much more stable selenate oxyanion. Experiments carried out using a *Desulfomicrobium* strain found that, under sulphate-reducing conditions, selenate was precipitated as elemental Se in the cell periplasm, probably by direct enzymatic reduction (Hockin and Gadd, 2006).

In contrast to the reductive precipitation of metal(loid)s described above, arsenic reduction frequently increases the solubility of this toxic element. Microbial dissimilatory reduction of As(V) to As(III) has been identified as an important route for increased As toxicity in the environment. This capacity appeared for some time to be phylogenetically and metabolically separate from dissimilatory sulphate reduction (Laverman *et al.*, 1995; Dowdle *et al.*, 1996), but at least one *Desulfotomaculum* strain has been shown to have the capability for simultaneous arsenic and sulphate reduction and to stimulate the precipitation of As(III) sulphide (Newman *et al.*, 1997; Macy *et al.*, 2000). The capacity of SRB to partially reduce and solubilize As and for soluble As(III) to precipitate with sulphide has further potential for bioremediation applications. Rittle *et al.* (1995) proposed a novel process for arsenic remediation using a two-step process to avoid the need for simultaneous As and sulphate reduction. In the first stage, enzymatic reduction of As(V) to As(III) by SRB cells was used to bring arsenic into solution, which was then deposited by exposure to SRB-generated sulphide.

14.5 BIOSORPTION AND BIOPRECIPITATION BY CELLULAR AND EXTRACELLULAR COMPONENTS

Bacterial cell surface components and extracellular biomolecules can bind metal(loid) ions by a variety of (electro)chemical means (Gadd, 2000; 2001; 2005). SRB cell-surface polymers, soluble autolysis products and extracellular polymers have all been shown to bind metals (Gadd, 2002). Where hydrophilic ligand complexes are formed, this can bring metals into solution so contributing to the corrosion of metal surfaces, or retarding precipitate formation and inhibiting metal sulphide deposition (Fortin *et al.*, 1994; Beech and Cheung, 1995; Zinkevich *et al.*, 1996).

In the current context, the ability of surface-associated macromolecules to effect the immobilization of aqueous metal(loid) species may be of greater importance, particularly where SRB grow as surface-attached biofilms, enmeshed in a matrix of extracellular polymeric substances (EPS). The biofilm mode of growth is now widely accepted to be the predominant form in which natural SRB populations occur and it appears that natural mixed-species, SRB-containing biofilms can act as sinks for precipitated minerals, including potentially toxic metals, in aqueous environments (Brown *et al.*, 1994; Labrenz *et al.*, 2000).

The EPS matrix of bacterial biofilms can act as a direct adsorbent of dissolved metal ions, with the ionic state and charge density of EPS components principally determining the ionic binding and electrostatic immobilization properties (Geesey *et al.*, 1989). Bacterial EPS is dominated by polysaccharides, but secreted polymers also include proteins, nucleic acids, peptidoglycan, lipids, phospholipids and other molecules. This heterogeneous matrix is generally depicted as having a net negative charge, with predominantly polyanionic moieties effectively acting as an ion-exchange matrix for metal cations. Well-characterized examples include the propensity of uronic acid-containing polysaccharides to bond with carboxyl groups and so to bind metals, while neutral carbohydrates can bind metals by the formation of weak electrostatic bonds around hydroxyl groups (Geesey *et al.*, 1989). It should be noted that cross-linking of SRB extracellular polysaccharides by the metal ions themselves can alter both the mechanical and chemical properties of EPS (Stoodley *et al.*, 2001).

The biofilm growth mode appears to further enhance metal removal in various ways (Douglas and Beveridge, 1998; Langley and Beveridge, 1999). Biosorption and bioprecipitation are also interrelated phenomena, such that ionic concentration by sorption at low-energy cellular, or EPS surface sites within biofilms can initiate mineral formation and immobilization within the biofilms (Glasauer *et al.*, 2001; Lee and Beveridge, 2001). Mineral precipitates formed in the bulk solution may also be physically entrapped, or chemically adsorbed by the biofilm EPS matrix. This may be important in enhancing metal removal where the formation of a persistent colloidal

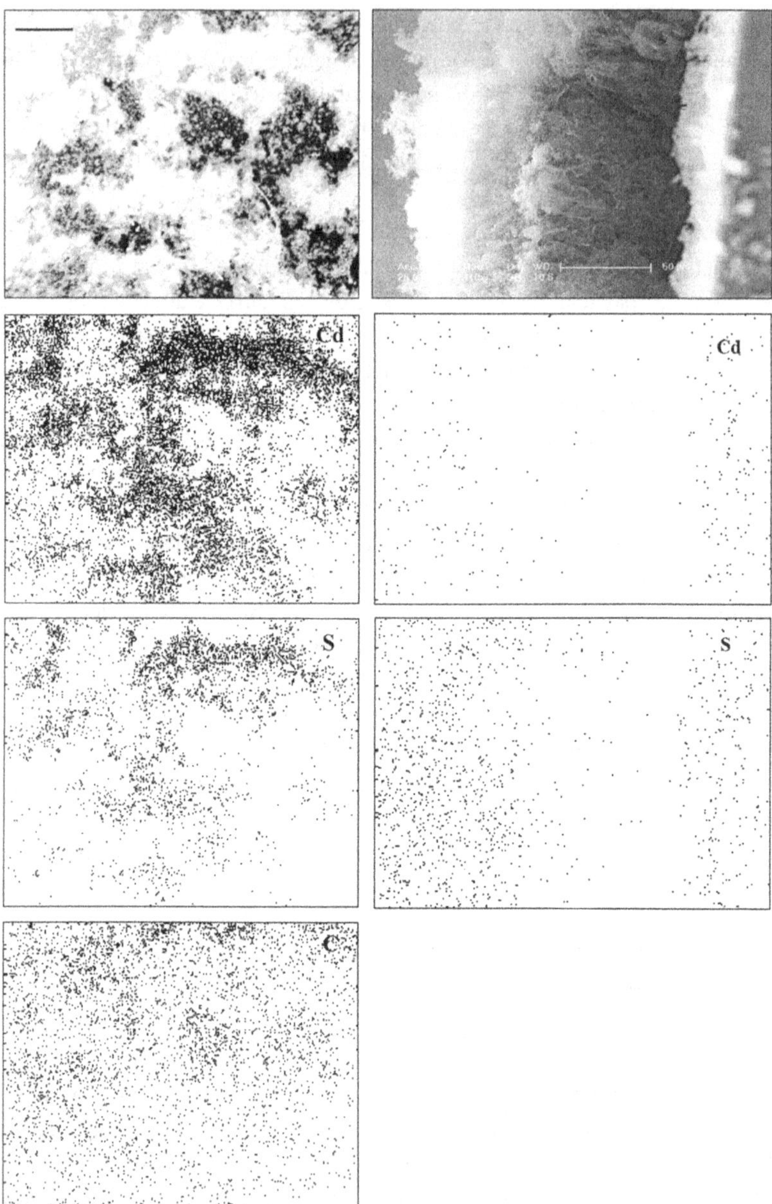

Figure 14.1. Electron micrographs and element-specific energy dispersive X-ray analysis (EDX) maps of a mixed-species SRB biofilm, dominated by a *Desulfomicrobium sp.*, grown in through-flow culture and exposed to 500 μM Cd^{2+}. EDX maps show the approximate distribution of elements in the areas shown in the micrographs. The left-hand column, generated using environmental scanning electron microscopy (ESEM), shows the biofilm

phase of hydrated metal sulphides is a limiting factor for metal removal in laboratory-scale SRB systems (White *et al.*, 2003). Precipitation, biosorption and particulate entrapment all appeared to contribute to metal cation removal by SRB in a constructed wetland system (Webb *et al.*, 1998). White and Gadd (2000) found that EPS production enhanced the removal of copper by SRB, proposing that entrapment of colloidal copper sulphide was the mechanism by which this occurred. Studies have also shown that mixed SRB consortia are more effective than pure cultures in the removal of toxic metals from solution and this may be attributable to greater metabolic efficiency and higher EPS production by mixed cultures (White and Gadd, 1998; 2000).

Precipitated metal sulphides accumulate preferentially in biofilm surface layers, supporting the view that entrapment of suspended sulphides takes place (Figure 14.1). Entrapment by the biofilm was also found to be important in the immobilization of colloidal selenium granules formed in the bulk solution by selenite-exposed biofilms (Figures 14.2 and 14.3) (Hockin and Gadd, 2003). However, in another study, mixed-culture SRB biofilm precipitated >88% of added chromate in 48 h from a medium containing a starting concentration of 500 μM and only around 10% of precipitated chromium, was associated with biofilm components (Smith and Gadd, 2000). Such differences suggest that secondary adsorption may depend on surface chemistry and the interaction between biological and non-biological components.

Secondary metal(loid) removal by adsorption on SRB-produced metal sulphides deposited within, or immobilized by, the biofilm matrix can also contribute to overall removal. SRB-generated metal sulphides can adsorb a range of cations and anions (Morse and Arakari, 1993; Morse, 1994; Cooper and Morse, 1998). Fortin and Beveridge (1997) attributed the removal of copper and zinc from porewaters in acid mine tailings to sorption on

Caption for Figure 14.1. (*cont.*) in plan view. Dense accumulations of cadmium were present at the biofilm surface, closely co-distributed with sulphur, suggesting the entrapment of cadmium sulphide. The densest accumulations were associated with regions of dense biofilm that were covered by copious EPS (Bar = 500 μm). The right-hand column, generated using cryo-field emission scanning (FESEM) electron microscopy, shows a fractured section of the same biofilm. The biofilm runs vertically with the support surface at right of shot. In this region of dense, confluent biofilm, cadmium and sulphur were present as a thick layer of very fine-grained deposits at the biofilm surface. The band of Cd/S on the right hand side is from debris lost from the biofilm during the fracture, which took place slightly behind the substrate surface (Bar = 50 μm).

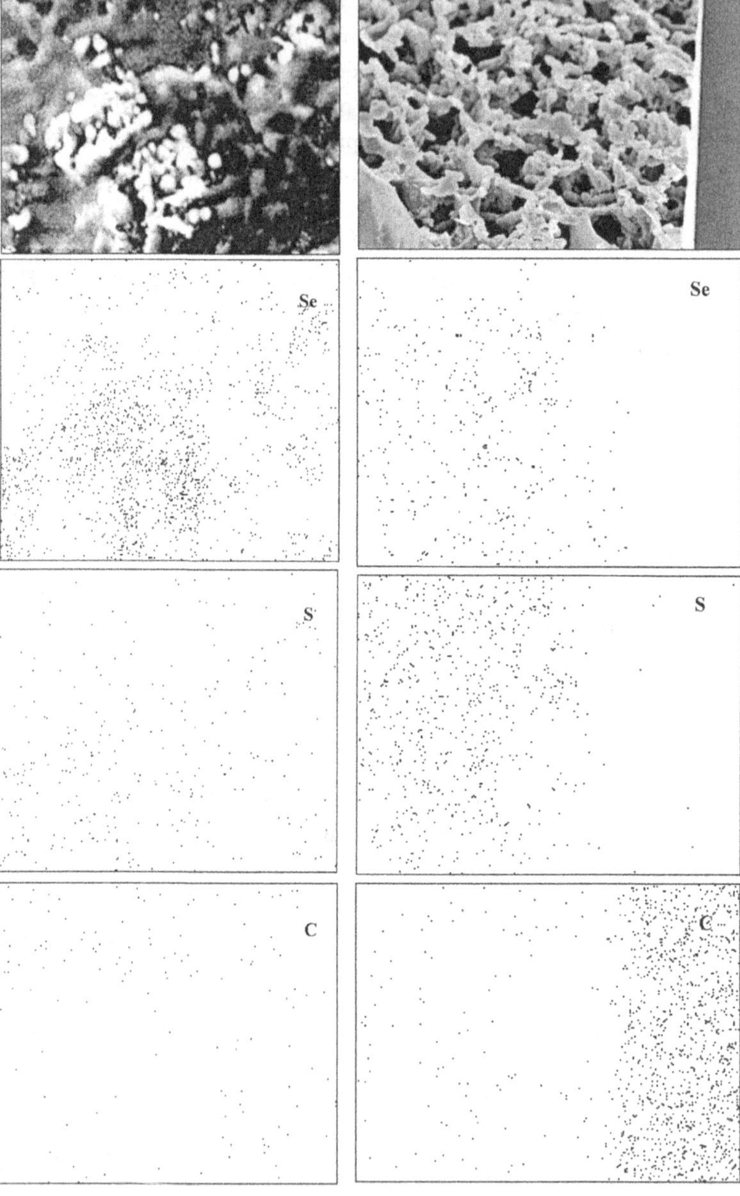

Figure 14.2. Electron micrographs and element-specific EDX maps, showing a hydrated mixed SRB biofilm exposed to 500 μM selenite. Wet-mode ESEM was used to generate a plan view of the fully hydrated biofilm (left-hand column). Selenium can be seen deposited at the biofilm surface, co-distributed with sulphur as abundant granular precipitates

SRB-produced iron sulphides. Watson *et al.* (1995; 2001) showed that Fe(II) sulphides bound a range of metals, allowing the level of metals in solution to be reduced from original concentrations in the order of $mg\,l^{-1}$ to $\mu g\,l^{-1}$. Significantly, metals having relatively soluble sulphides were also removed by this mechanism. Synchrotron-based X-ray studies of SRB-produced iron sulphide have found evidence that unreduced selenate was adsorbed to non-stoichiometric iron sulphides deposited in SRB biofilms, possibly at localized cationic sites (Hockin and Gadd, unpublished data).

14.6 CURRENT LIMITATIONS AND RESEARCH PROSPECTS

The broad spectrum of SRB action on metal(lloid) ions and the efficiency of metal sulphide precipitation makes them suitable candidates for the bioremediation of a range of metal-contaminated media. Nevertheless, while SRB are recognized as important components of constructed wetlands used to mitigate acid mine drainage, the most significant commercial applications of SRB have led from the use of anaerobic sludge-blanket reactors to treat zinc-contaminated groundwater in the THIOPAQ® technology developed by Paques B.V. (Buisman *et al.*, 1989; Barnes *et al.*, 1991; van Houten *et al.*, 1994). However, widespread commercial uptake of SRB bioremediation technology remains limited. While the efficacy of biogenic sulphide production for the precipitation of toxic metals is demonstrable, SRB are highly susceptible to metal(loid) toxicity, to fluctuations in pH and redox conditions and to competition from more energy-efficient organisms. The relatively slow growth rate of SRB and the large working volumes that this engenders also present engineering problems, as does the requirement for reducing conditions in the bulk feed. Hydrogen sulphide gas is itself flammable, highly toxic and requires careful process design and management.

Caption for Figure 14.2. (*cont.*) (light colour). Cells of *Desulfomicrobium* sp. are visible, partially obscured by the extracellular matrix, which is penetrated by frequent pores. A cryo-FESEM micrograph of fractured, hydrated biofilm sections (right-hand column) shows that selenium was present, not only at the surface, but also throughout the biofilm matrix. The heavy carbon line shows the position of the substrate. Some topographical effects are apparent, caused by the etched ice domains. Bars = 1 µm. (See also Hockin and Gadd, 2003.)

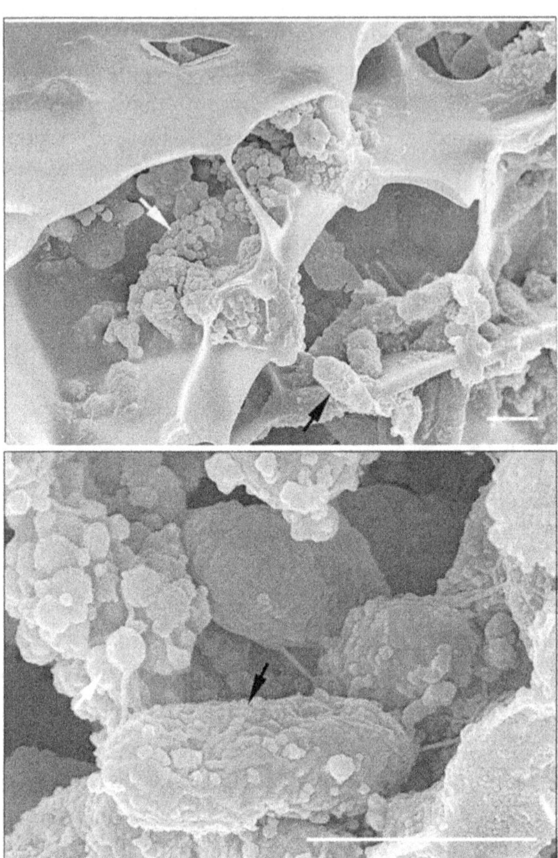

Figure 14.3. At higher magnification, the precipitation of selenium-sulphur granules within the biofilm matrix can be seen in chromium-coated cryo-FESEM examples. The cryo-sections show isometric biofilm sections, with the surface of the extracellular matrix top left. Individual cells of *Dm. norvegicum* (black arrows) form colonies within the matrix. Abundant Se/S granules (white arrows) are clearly seen precipitated beneath the biofilm canopy. The bacterium at bottom right has been sectioned during freeze-fracture. Bars = 1 μm. (From Hockin and Gadd, 2003.)

These drawbacks have been implicit in the focus of much research over the last few years, aimed at identifying and optimizing parameters to achieve greater process efficiency in the rate of metal removal and on the design of economic reactor conditions. The use of immobilized, surface-attached biofilms, or semi-suspended flocs is currently a favoured approach to further development of engineered SRB systems that use growing cells, offering the potential for increased process intensity and reduced working volumes.

Laboratory and pilot-scale studies have also increasingly sought to utilize metal-adapted SRB, often present as components of mixed cultures (Fude *et al.*, 1994; Hard and Babel, 1997; White and Gadd, 1998). Growth of SRB as biofilms may also confer further protection from metal toxicity (White and Gadd, 1998).

The procurement of plentiful, cheap substrates is also something of a problem in achieving economic efficiency of SRB bioreactors and, while there is a wide and varied capacity for organic carbon substrates among the SRB, utilization of simple sugars is rare. This presents both opportunities and difficulties for engineered applications: while the use of less widely metabolized substrates may enable SRB activity to be selected in mixed systems, the presence of simple sugars can mitigate against this (White and Gadd, 1996a). There may be a problem when complex organic wastes are used as carbon/nutrient sources in bulk feeds, resulting in a reduction in effectiveness during long-term operation (e.g. Chang *et al.*, 2000). White and Gadd (1997) concluded that ethanol, with the addition of a complex nitrogen source, offered the greatest potential for efficient sulphide production and the use of ethanol has proven also to be efficient in commercial operation (Barnes *et al.*, 1991).

The perceived strict requirement for low-redox, anaerobic conditions also has obvious drawbacks in the commercial application of SRB systems. It has long been assumed that oxygen is extremely toxic to SRB and that all SRB require established reducing conditions in the region of −200 mV in order to begin dissimilatory sulphate reduction. Redox reduction of reactor feeds has therefore been taken as essential, either by prior chemical reduction, or by sequential biological reduction in column-type reactors. However, evidence for oxygen tolerance and microaerophilicity among SRB genera is now compelling, as is the capacity for some SRB strains to enzymatically reduce oxygen and to upregulate sulphate reduction in the presence of oxygen (Dilling and Cypionka, 1990; Marschall *et al.*, 1993; Cypionka 2000; Lemos *et al.*, 2001). Dissimilatory reduction of nitrate among members of the *Desulfovibrio* group probably contributes to the ability of many SRB to remain active in aerobic environments where sulphate reduction appears to be thermodynamically impossible (Mitchell *et al.*, 1986), but this does not generate the large amounts of bisulphide required for the efficient removal of metal cations from solution. However, it is now clear that active dissimilatory sulphate reduction can take place in anoxic biofilm microniches with mixed-species biofilms, even in oxidized bulk environments (see Chapters 5 and 6, this volume). The biofilm mode of growth also appears to confer the ability for consortia to exert a degree of local control over parameters such as pH and E_h, while close proximity of metabolically complementary organisms benefits growth and metabolism of complex and/or toxic substrates (Kühl and Jørgensen, 1992; Lens *et al.*, 1995; Santegoeds *et al.*, 1998; White and Gadd, 1996a, b; 1998; see also Chapters 12 and 13, this volume).

There is, as yet, only a very limited body of knowledge with respect to the composition and function of the EPS produced by SRB. This may also be a promising avenue for improving the efficiency and selectivity of metal(loid) removal by SRB-based systems. Blenkinsop *et al.* (1992) found that a DC electrical field was able to induce a bioelectrical effect that apparently modified the surface characteristics of biofilm EPS to expose cationic charge sites. Such approaches may have interesting applications in altering the chemical binding characteristics of biofilms to enhance removal of anionic metal(loid) species.

Increased understanding of the propensity of bacterial cells to remove dissolved metal(loid)s by enzymatic reduction has also stimulated research using growth-decoupled SRB cells. Many key reductive enzyme systems appear to be constitutively expressed and have broad substrate and metal(loid) specificity, offering potential for remediation of mixed metal-liferous waters. A significant obstacle to the use of enzymatic reduction in non-growing cells has been the difficulty in achieving sustained reductase activity over time, probably due to the depletion of metabolic cofactors when organic electron donors requiring extended electron transport chains are used. However SRB of the *Desulfovibrio* group are able to utilize molecular hydrogen for energy generation and non-growing cells appear able to couple this directly to the reduction of toxic metal(loid)s (Lovley *et al.*, 1993b; Lovley and Phillips, 1994; De Luca *et al.*, 2001; Elias *et al.*, 2004). The possibility of further process intensification is also offered by the use of high-density, pregrown cells, immobilized on surfaces, or within permeable gels (Tucker *et al.*, 1998; Lloyd *et al.*, 1999a; 2004). The recent discovery that enzymatic precipitation of Pd by pregrown, immobilized cells of *Dv. desulfuricans* is possible in the presence of oxygen (Lloyd *et al.*, 1998b) also offers the possibility of using growth-decoupled enzymatic reduction of metal(loid)s by SRB to treat waste liquids. Further research into enzymatic metal(loid) reduction by SRB is clearly needed therefore, but understanding of the genetic and biomolecular bases of SRB reductase activity towards metal(loid)s has proven something of an obstacle to their detailed characterization. However, SRB genomic sequences are now rapidly becoming available (http://www.jgi.doe.gov; http://www.ncbi.nlm.nih.gov), including the full genome sequence of *Dv. vulgaris* Hildenborough

(Heidelberg *et al.*, 2004). This is generating insights into gene function, homology and regulation (Hemme and Wall, 2004) and appears to confirm the rich diversity and biotechnological potential of redox enzymes present in this organism (Heidelberg *et al.*, 2004; Valente *et al.*, 2005). Sequence availability has also enabled Rapp-Giles *et al.* (2000) to engineer cytochrome *c*3 insertion mutants of *Dv. desulfuricans* and Goenka *et al.* (2005) to produce [NiFe] hydrogenase deletion mutants of *Dv. vulgaris* Hildenborough.

The growing genomic database is generating research using transgenic approaches to biotechnological applications of sulphate reduction. Although it is difficult to see this achieving widespread adoption, transgenic cloning has been successfully applied to improving the robustness of biosulphide generation in the laboratory. Earlier work utilized the better-characterized sulphate-reduction pathways from non-SRB organisms. This was successful for expression of the *Salmonella enterica* thiosulphate reductase gene in *Escherichia coli* (Bang *et al.*, 2000). In another study, genes for assimilatory sulphate reduction were overexpressed in *E. coli*, resulting in the production of sulphide under aerobic conditions that achieved cadmium sulphide precipitation at the cell surface (Wang *et al.*, 2000). Transgenic approaches to SRB reductase expression systems have also yielded some promising results, particularly with respect to the *c*-type cytochromes which appear to have broad reductive ability. Aubert *et al.* (1998) successfully transferred the high-activity cytochrome c_7 of *Desulfuromonas acetoxidans* to *Dv. desulfuricans*. Transformed cells over-expressed c_7, engendering enhanced metal reductase activity. Ozawa *et al.* (2000) expressed a c_3 cytochrome from *Dv. vulgaris* in the facultative anaerobe *Shewanella oneidensis*, achieving reductive activity under anaerobic, microaerophilic and aerobic conditions. Sadeghi *et al.* (2000) used electron transport protein cassettes from *Desulfovibrio* group SRB in combination with those for redox proteins from other organisms, assembled in a "molecular Lego" approach to construct electron transport chains. Further advances in transgenic manipulation of electron transport pathways might be expected, and there is scope for exciting future developments as more molecular information becomes available.

ACKNOWLEDGEMENTS

SH gratefully acknowledges receipt of a BBSRC Industrial CASE postgraduate studentship with BNFL. Thanks are also due to Martin Kierans of the Centre for High Resolution Imaging and Processing, University of Dundee, for assistance with the electron microscopy.

REFERENCES

Aubert, C., Lojou, E., Bianco, P. *et al.* (1998). The *Desulfuromonas acetoxidans* tri-heme cytochrome *c*7 produced in *Desulfovibrio desulfuricans* retains its metal reductase activity. *Applied and Environmental Microbiology*, **64**, 1308–12.

Bang, S. W., Clark, D. S. and Keasling, J. D. (2000). Engineering hydrogen sulphide production and cadmium removal by expression of the thiosulphate reductase gene (phsABC) from *Salmonella enterica* serovar *typhimurium* in *Escherichia coli*. *Applied and Environmental Microbiology*, **66**, 3939–44.

Barnes, L. J., Janssen, F. J., Sherren, J. *et al.* (1991). A new process for the microbial removal of sulphate and heavy metals from contaminated waters extracted by a geohydrological control system. *Transactions of the Institute of Chemical Engineering*, **69**, 184–6.

Barnes, L. J., Scheeren, P. J. and Buisman, C. J. N. (1994). Microbial removal of heavy metals and sulphate from contaminated groundwaters. In J. L. Means and R. E. Hinchee (eds.), *Emerging technology for the bioremediation of metals*. Boca Raton, FL: Lewis Publishers. pp. 38–49.

Barton, L. L., Choudhury, K., Thomson, B. M. and Steenhoudt, K. (1996). Bacterial reduction of soluble uranium: the first step of *in situ* immobilization of uranium. *Radioactive Waste Management and Environmental Restoration*, **20**, 141–51.

Battelle Bioprocessing (2001). http://bioprocess.pnl.gov/sulfide.htm.

Beech, I. B. and Cheung, C. W. S. (1995). Interactions of exopolymers produced by sulphate-reducing bacteria with metal ions. *International Biodeterioration and Biodegradation*, **35**, 59–72.

Blenkinsop, S. A., Khoury, A. E. and Costerton, J. W. (1992). Electrical enhancement of biocide efficacy against *Pseudomonas aeruginosa* biofilms. *Applied and Environmental Microbiology*, **58**, 3770–3.

Brown, D. A., Choudari Kamineni, D., Sawicki, J. A. and Beveridge, T. J. (1994). Minerals associated with biofilms occurring on exposed rock in a granitic underground research laboratory. *Applied and Environmental Microbiology*, **60**, 3182–91.

Buisman, C., Post, R., Yspeert, P., Geraats, G. and Lettinga, G. (1989). Biotechnological process for sulfide removal with sulfur reclamation. *Acta Biotechnologica*, **9**, 255–67.

Chang, I. S., Shin, P. K. and Kim, B. H. (2000). Biological treatment of acid mine drainage under sulphate-reducing conditions with solid waste materials as substrate. *Water Research*, **34**, 1269–77.

Cooper, D. C. and Morse, J. W. (1998). Biogeochemical controls on trace metal cycling in anoxic marine sediments. *Environmental Science and Technology*, **32**, 327–30.

Cypionka, H. (2000). Oxygen respiration by *Desulfovibrio* species. *Annual Review of Microbiology*, **54**, 827–48.

DeLuca, G., de Philip, P., Dermoun, Z., Rousset, M. and Vermeglio, A. (2001). Reduction of technetium(VII) by *Desulfovibrio fructosovorans* is mediated by the nickel-iron hydrogenase. *Applied and Environmental Microbiology*, **67**, 4583–7.

Dilling, W. and Cypionka, H. (1990). Aerobic respiration in sulphate-reducing bacteria. *FEMS Microbiology Letters*, **71**, 123–8.

Douglas, S. and Beveridge, T. J. (1998). Mineral formation by bacteria in natural communities. *FEMS Microbial Ecology*, **26**, 79–88.

Dowdle, P. R., Laverman, A. M. and Oremland, R. S. (1996). Bacterial dissimilatory reduction of arsenic(V) to arsenic(III) in anoxic sediments. *Applied and Environmental Microbiology*, **62**, 1664–9.

Dvorak, D. H., Hedin, R. S., Edenborm, H. M. and McIntyre, P. E. (1991). Treatment of metal-contaminated water using bacterial sulphate-reduction: results from pilot-scale reactors. *Biotechnology and Bioengineering*, **40**, 609–16.

Elias, D. A., Suflita, J. M., McInerney, M. J. and Krumholz, L. R. (2004). Periplasmic cytochrome *c*3 of *Desulfovibrio vulgaris* is directly involved in H_2-mediated metal but not sulphate reduction. *Applied and Environmental Microbiology*, **70**, 413–20.

Ewart, D. K. and Hughes, M. N. (1991). The extraction of metals from ores using bacteria. *Advances in Inorganic Chemistry*, **36**, 103–35.

Flemming, H.-C. (1995). Sorption sites in biofilms. *Water Science and Technology*, **32**, 27–33.

Fortin, D. and Beveridge, T. J. (1997). Microbial sulphate reduction within sulphidic mine tailings: formation of diagenetic iron sulphides. *Geomicrobiology Journal*, **14**, 1–21.

Fortin, D., Davis, B. and Beveridge, T. J. (1996). Role of *Thiobacillus* and sulphate-reducing bacteria in iron biocycling in oxic and acidic mine tailings. *FEMS Microbiology Ecology*, **21**, 11–24.

Fortin, D., Goulet, R. and Roy, M. (2000). The effect of seasonal variations in sulphate-reducing bacterial populations on Fe and S cycling in a constructed wetland. *Geomicrobiology Journal*, **17**, 221–35.

Fortin, D., Souham, G. and Beveridge, T. J. (1994). Nickel sulfide, iron-nickel sulfide and iron sulfide precipitation by a newly-isolated *Desulfotomaculum* species and its relation to nickel resistance. *FEMS Microbiology Ecology*, **14**, 121–32.

Fude, L., Harris, B., Urrutia, M. M. and Beveridge, T. J. (1994). Reduction of Cr(VI) by a consortium of sulphate-reducing bacteria (SRB III). *Applied and Environmental Microbiology*, **60**, 1525–31.

Gadd, G. M. (2000). Heavy metal pollutants: environmental and biotechnological aspects. In J. Lederberg (ed.), *The Encyclopedia of Microbiology*, 2nd edn. San Diego: Academic Press, Inc. pp. 607–17.

Gadd, G. M. (2001). Accumulation and transformation of metals by microorganisms. In H.-J. Rehm, G. Reed, A. Puhler and P. Stadler (eds.), *Biotechnology, a Multi-Volume Comprehensive Treatise, Volume 10: Special Processes*. Weinheim, Germany: Wiley-VCH Verlag. pp. 225–64.

Gadd, G. M. (2002). Interactions between microorganisms and metals/radionuclides: the basis of bioremediation. In M. J. Keith-Roach and F. R. Livens (eds.), *Interactions of Microorganisms with Radionuclides*. Amsterdam: Elsevier. pp. 179–203.

Gadd, G. M. (2005). Microorganisms in toxic metal polluted soils. In F. Buscot and A. Varma (eds.), *Microorganisms in Soils: Roles in Genesis and Functions*. Berlin: Springer-Verlag. pp. 325–56.

Geesey, G. G., Lang, J., Jolly, J. G. *et al.* (1989). Binding of metal ions by extracellular polymers of biofilm bacteria. *Water Science and Technology*, **20**, 161–5.

Glasauer, S., Beveridge, T. J., Burford, E. P., Harper, F. A. and Gadd, G. M. (2004). Metals and metalloids, transformations by microorganisms. In D. Hillel, C. Rosenzweig, D. S. Powlson *et al.* (eds.), *Encyclopedia of Soils in the Environment*. Amsterdam: Elsevier. pp. 438–47.

Glasauer, S., Langley, S. and Beveridge, T. J. (2001). Sorption of Fe (hydr)oxides to the surface of *Shewanella putrefaciens*: cell-bound fine-grained minerals are not always formed *de novo*. *Applied and Environmental Microbiology*, **67**, 5544–50.

Goenka, A., Voordouw, J. K., Lubitz, W., Gartner, W. and Voordouw, G. (2005). Construction of a NiFe-hydrogenase deletion mutant of *Desulfovibrio vulgaris* Hildenborough. *Transactions of the Biochemical Society*, **33**, 59–60.

Gyure, R. A., Konpka, A., Brooks, A. and Doemel, W. (1990). Microbial sulphate reduction in acidic (pH 3) strip-mine lakes. *FEMS Microbiology Ecology*, **73**, 193–202.

Hammack, R. W. and Edenborm, H. M. (1992). The removal of nickel from mine waters using bacterial sulphate reduction. *Applied Microbiology and Biotechnology*, **37**, 674–8.

Hard, B. C. and Babel, F. W. (1997). Bioremediation of acid minewater, using facultatively methylotrophic metal-tolerant sulphate-reducing bacteria. *Microbiological Research*, **152**, 65–73.

Harvey, C. F., Swartz, C. H., Badruzzaman, A. B. M. *et al.* (2002). Arsenic mobility and groundwater extraction in Bangladesh. *Science*, **98**, 1602–6.

Heidelberg, J. F., Seshadri, R., Haveman, S. A. *et al.* (2004). The genome sequence of the anaerobic, sulphate-reducing bacterium *Desulfovibrio vulgaris* Hildenborough. *Nature Biotechnology*, **22**, 554–9.

Hemme, C. L. and Wall, J. D. (2004). Genomic insights into gene regulation of *Desulfovibrio* vulgaris Hildenborough. *OMICS: A Journal of Integrative Biology*, **8**, 43–55.

Hockin, S. L. and Gadd, G. M. (2003). Linked redox precipitation of sulfur and selenium under anaerobic conditions by sulphate-reducing bacterial biofilms. *Applied and Environmental Microbiology*, **69**, 7063–72.

Hockin, S. and Gadd, G. M. (2006). Removal of selenate from sulphate-containing media by sulphate-reducing bacterial biofilms. *Environmental Microbiology*, **8**, 816–26.

Jacobs, L. (1989). *Selenium in agriculture and the environment*. Madison, Wisconsin: American Society of Agronomy.

Kirk, M. F., Holm, T. R., Park, J. *et al.* (2004). Bacterial sulphate reduction limits natural arsenic contamination in groundwater. *Geology*, **32**, 953–6.

Kühl, M. and Jørgensen, B. B. (1992). Microsensor measurements of sulphate reduction and sulfide oxidation in compact microbial commuities of aerobic biofilms. *Applied and Environmental Microbiology*, **58**, 1164–74.

Labrenz, M., Druschel, G. K., Thompson-Ebert, T. K. *et al.* (2000). Formation of sphaelerite (ZnS) deposits in natural biofilms of sulphate-reducing bacteria. *Science*, **290**, 1744–6.

Langley, S. and Beveridge, T. J. (1999). Metal binding by *Pseudomonas aeruginosa* PAO1 is influenced by growth as a biofilm. *Canadian Journal of Microbiology*, **45**, 616–22.

Laverman, A. M., Switzer Blum, J., Schaefer, J. K. *et al.* (1995). Growth of strain SES-3 with arsenate and other diverse electron acceptors. *Applied and Environmental Microbiology*, **61**, 3556–61.

Lee, J.-U. and Beveridge, T. J. (2001). Interaction between iron and *Pseudomonas aeruginosa* biofilms attached to sepharose surfaces. *Chemical Geology*, **180**, 67–80.

Lens, P. N., DePoorter, M. P., Cronenberg, C. C. and Verstraete, W. H. (1995). Sulphate-reducing and methane-producing bacteria in aerobic wastewater treatment systems. *Water Research*, **29**, 857–70.

Lemos, R. S., Gomes, C. M., Santana, M. *et al.* (2001). The 'strict' anaerobe *Desulfovibrio gigas* contains a membrane-bound oxygen-reducing respiratory chain. *FEBS Letters*, **496**, 40–3.

Lloyd, J. R., Lovley, D. R. and Macaskie, L. E. (2004). Biotechnological applications of metal-reducing microorganisms. *Advances in Applied Microbiology*, **53**, 85–128.

Lloyd, J. R., Nolting, H. F., Sole, V. A., Bosecker, K. and Macaskie, L. E. (1998a). Technetium reduction and precipitation by sulphate-reducing bacteria. *Geomicrobiology Journal*, **15**, 45–58.

Lloyd, J. R., Yong, P. and Macaskie, L. E. (1998b). Enzymatic recovery of elemental palladium by using sulphate-reducing bacteria. *Applied and Environmental Microbiology*, **64**, 4607–9.

Lloyd, J. R., Ridley, J., Khizniak, T., Lyalikova, N. N. and Macaskie, L. E. (1999a). Reduction of technetium by *Desulfovibrio desulfuricans*: biocatalyst characterization and use in a flowthrough bioreactor. *Applied and Environmental Microbiology*, **65**, 2691–6.

Lloyd, J. R., Sole, V. A., Van Praagh, C. V. and Lovley, D. R. (2000). Direct and Fe(II)-mediated reduction of technetium by Fe(III)-reducing bacteria. *Applied and Environmental Microbiology*, **66**, 3743–9.

Lloyd, J. R., Thomas, G. H., Finlay, J. A., Cole, J. A. and Macaskie, L. E. (1999b). Microbial reduction of technetium by *Escherichia coli* and *Desulfovibrio desulfuricans*: enhancement by the use of high activity strains and effects of process parameters. *Biotechnology and Bioengineering*, **66**, 122–30.

Lovley, D. R. and Phillips, E. J. P. (1992). Reduction of uranium by *Desulfovibrio desulfuricans*. *Applied and Environmental Microbiology*, **58**, 850–6.

Lovley, D. R. and Phillips, E. J. P. (1994). Reduction of chromate by *Desulfovibrio desulfuricans* and its c_3 cytochrome. *Applied and Environmental Microbiology*, **60**, 726–8.

Lovley, D. R., Roden, E. E., Phillips, E. J. P. and Woodward, J. C. (1993a). Enzymatic iron and uranium reduction by sulphate-reducing bacteria. *Marine Geology*, **113**, 41–53.

Lovley, D. R., Widman, P. K., Woodward, J. C. and Phillips E. J. P. (1993b). Reduction of uranium by cytochrome *c*3 of *Desulfovibrio vulgaris*. *Applied and Environmental Microbiology*, **59**, 3572–6.

Macy, J. M., Santini, J. M., Pauling, B. V., O'Neill, A. H. and Sly, L. I. (2000). Two new arsenate/sulphate-reducing bacteria: mechanisms of arsenate reduction. *Archives of Microbiology*, **173**, 49–57.

Marschall, C., Frenzel, P. and Cypionka, H. (1993). Influence of oxygen on sulphate reduction and growth of sulphate-reducing bacteria. *Archives of Microbiology*, **159**, 168–73.

Michalke, K., Wickenheiser, E. B., Mehring, M., Hirner, A. V. and Hensel, R. (2000). Production of volatile derivatives of metal(loid)s by microflora involved in anaerobic digestion of sewage sludge. *Applied and Environmental Microbiology*, **66**, 2791–6.

McCreadie, H., Blowes, D. W., Ptacek, C. J. and Jambor, J. L. (2000). Influence of reduction reactions and solid-phase composition on porewater concentrations of arsenic. *Environmental Science and Technology*, **34**, 3159–66.

Mitchell, G. J., Jones, J. G. and Cole, J. A. (1986). Distribution and regulation of nitrate and nitrite reduction by *Desulfovibrio* and *Desulfotomaculum* species. *Archives of Microbiology*, **144**, 35–40.

Moore, J. N., Ficklin, W. H. and Johns, C. (1988). Partitioning of arsenic and metals in reducing sulfidic sediments. *Environmental Science and Technology*, **22**, 432–7.

Morse, J. W. (1994). Interactions of trace metals with authigenic sulfide minerals: implications for their bioavailability. *Marine Chemistry*, **46**, 1–6.

Morse, J. W. and Arakaki, T. (1993). Adsorption and coprecipitation of divalent metals with mackinawite (FeS). *Geochimica et Cosmochimica Acta*, **57**, 3635–40.

Morse, J. W., Millero, F. J., Cornwell, J. C. and Rickard, D. (1987). The chemistry of hydrogen sulfide and iron sulfide systems in natural waters. *Earth Science Reviews*, **24**, 1–42.

Naz, N., Young, H. K., Ahmed, N. and Gadd, G. M. (2005). Cadmium accumulation and homology with metal resistance genes in sulphate-reducing bacteria. *Applied and Environmental Microbiology*, **71**, 4610–18.

Nemati, M., Mazutinec, T. J., Jenneman, G. E. and Voordrouw, G. (2001). Control of biogenic H_2S production with nitrite and molybdate. *Journal of Industrial Microbiology and Biotechnology*, **26**, 350–5.

Newman, D. K., Beveridge, T. J. and Morel, F. M. M. (1997). Precipitation of arsenic trisulphide by *Desulfotomaculum auripigmentum*. *Applied and Environmental Microbiology*, **63**, 2022–8.

Newport, P. J. and Nedwell, D. B. (1988). The mechanism of inhibition of *Desulfovibrio* and *Desulfotomaculum* species by selenate and molybdate. *Journal of Applied Bacteriology*, **65**, 419–23.

NTBC (2000). http://www.direct.ca/ntbc/srb.htm.

Oremland, R., Hollibaugh, J. T., Maest, A. S. *et al.* (1989). Selenate reduction to elemental selenium by anaerobic bacteria in sediments and culture: biogeochemical significance of a novel, sulphate independent respiration. *Applied and Environmental Microbiology*, **55**, 2333–43.

Oremland, R. S., Switzer-Blum, J., Culbertson, C. W. *et al.* (1994). Isolation, growth and metabolism of an obligately anaerobic, selenate-respiring bacterium, strain SES-3. *Applied and Environmental Microbiology*, **60**, 3011–19.

Oremland, R., Switzer-Blum, J., Burns Bindi, A. *et al.* (1999). Simultaneous reduction of nitrate and selenate by cell suspensions of selenium-respiring bacteria. *Applied and Environmental Microbiology*, **65**, 4385–92.

Ozawa, K., Tsapin, A. I., Nealson, K. H., Cusanovich, M. A. and Akutsu, H. (2000). Expression of a tetraheme protein, *Desulfovibrio vulgaris* Miyazaki F cytochrome c(3), in *Shewanella oneidensis* MR-1. *Applied and Environmental Microbiology*, **66**, 4168–71.

Pignolet, L., Fonsy, K., Capot, F. and Moureau, Z. (1989). The role of various microorganisms on Tc behaviour in sediments. *Health Physics*, **57**, 791–800.

Postgate, J. R. (1952). Competitive and non-competitive inhibitors of bacterial sulphate reduction. *Journal of General Microbiology*, **6**, 128–42.

Rapp-Giles, B. J., Casalot, L., English, R. S. *et al.* (2000). Cytochrome c(3) mutants of *Desulfovibrio desulfuricans. Applied and Environmental Microbiology*, **66**, 671–7.

Rittle, K. A., Drever, J. I. and Colberg, P. J. S. (1995). Precipitation of arsenic during bacterial sulphate reduction. *Geomicrobiology Journal*, **13**, 1–12.

Rowley, M. V., Warkentin, D. D. and Sicotte, V. (1997). Site demonstration of the biosulphide process at the former Britannia mine. In *Proceedings of the Fourth International Conference of Acid Rock Drainage*. Vancouver, BC, Canada: Canadian Institute of Mining, Metallurgy and Petroleum. pp. 1533–48.

Sadeghi, S. J., Meharenna, Y. T., Fantuzzi, A., Valetti, F. and Gilardi, G. (2000). Engineering artificial redox chains by molecular 'Lego'. *Faraday Discussions*, **116**, 135–53 (discussion 171–90).

Santegoeds, C., Ferdelman, G. and Muyzer, G. (1998). Structural and functional dynamics of sulphate-reducing populations in bacterial biofilms. *Applied and Environmental Microbiology*, **64**, 3731–9.

Smith, W. L. and Gadd, G. M. (2000). Reduction and precipitation of chromate by mixed culture sulphate-reducing bacterial biofilms. *Journal of Applied Microbiology*, **88**, 983–91.

Stoodley, P., Jacobsen, A., Dunsmore, B. C. *et al.* (2001). The influence of fluid shear and $AlCl_3$ on the material properties of *Pseudomonas aeruginosa* PAO1 and *Desulfovibrio sp.* EX265 biofilms. *Water Science and Technology*, **43**, 113–20.

Stumm, W. and Morgan, J. J. (1996). *Aquatic Chemistry: Chemical Equilibria and Rates in Natural Waters*, 3rd edn. New York: John Wiley and Sons Inc.

Tebo, B. M. and Obraztsova, A. Y. (1998). Sulphate-reducing bacterium grows with Cr(IV), U(VI), Mn(IV) and Fe(III) as electron acceptors. *FEMS Microbiology Letters*, **162**, 193–8.

Tomei, F. A., Barton, L. L., Lemanski, C. L. *et al.* (1995). Transformation of selenate and selenite to elemental selenium by *Desulfovibrio desulfuricans*. *Journal of Industrial Microbiology and Biotechnology*, **14**, 329–36.

Tucker, M. D., Barton, L. L. and Thompson, B. M. (1998). Reduction of Cr, Mo, Se, and U by *Desulfovibrio desulfuricans* immobilised in polyacrylamide gels. *Journal of Industrial Microbiology and Biotechnology*, **20**, 13–19.

Valente, F. M., Oliveira, A. S., Gnadt, N. *et al.* (2005). Hydrogenases in *Desulfovibrio vulgaris* Hildenborough: structural and physiologic characterisation of the membrane-bound NiFeSe hydrogenase. *Journal of Biological Inorganic Chemistry*, **10**, 667–82.

van Houten, R. T., Pol, L. W. H. and Lettinga, G. (1994). Biological sulphate reduction using gas-lift reactors fed with hydrogen and carbon dioxide as energy and carbon source. *Biotechnology and Bioengineering*, **44**, 586–94.

Vile, M. A. and Wieder, R. K. (1993). Alkalinity generation by Fe(III) reduction versus sulphate reduction in wetlands constructed for acid mine drainage treatment. *Water, Air and Soil Pollution*, **69**, 425–41.

Wang, C. L., Maratukulam, P. D. L., Clark, D. S. and Keasling, J. D. (2000). Metabolic engineering of an aerobic sulphate reduction pathway and its application to precipitation of cadmium on the cell surface. *Applied and Environmental Microbiology*, **66**, 4497–502.

Watson, J. H. P., Ellwood, D. C., Quixi, D. *et al.* (1995). Heavy metal adsorption on bacterially produced FeS. *Minerals Engineering*, **8**, 1097–108.

Watson, J. H. P., Croudace, I. W., Warwick, P. E. *et al.* (2001). Adsorption of radioactive metals by strongly magnetic iron sulfide nanoparticles produced by sulphate-reducing bacteria. *Separation Science and Technology*, **36**, 2571–607.

Webb, J. S., McGinness, S. and Lappin-Scott, H. M. (1998). Metal removal by sulphate-reducing bacteria from natural and constructed wetlands. *Journal of Applied Microbiology*, **84**, 240–8.

White, C., Dennis, J. S. and Gadd, G. M. (2003). A mathematical process model for cadmium bioprecipitation by sulphate-reducing bacterial biofilms. *Biodegradation*, **14**, 139–51.

White, C. and Gadd, G. M. (1996a). Mixed sulphate-reducing cultures for the bioprecipitation of toxic metals: factorial and response-surface analysis of the effects of dilution rate, sulphate and substrate concentration. *Microbiology*, **142**, 2197–205.

White, C. and Gadd, G. M. (1996b). A comparison of carbon/energy and complex nitrogen sources for bacterial sulphate-reduction: potential applications to bioprecipitation of toxic metals as sulphides. *Journal of Industrial Microbiology*, **17**, 116–23.

White, C. and Gadd, G. M. (1997). An internal sedimentation bioreactor for laboratory-scale removal of toxic metals from soil leachates using biogenic sulphide precipitation. *Journal of Industrial Microbiology and Biotechnology*, **18**, 414–21.

White, C. and Gadd, G. M. (1998). Accumulation and effects of cadmium on sulphate-reducing bacterial biofilms. *Microbiology*, **144**, 1407–15.

White, C. and Gadd, G. M. (2000). Copper accumulation by sulphate-reducing bacterial biofilms. *FEMS Microbiology Letters*, **183**, 313–18.

White, C., Sharman, A. K. and Gadd, G. M. (1998). An integrated microbial process for the bioremediation of soil contaminated with toxic metals. *Nature Biotechnology*, **16**, 572–5.

Zehr, J. P. and Oremland, R. S. (1987). Reduction of selenate to selenide by sulphate-respiring bacteria: experiments with cell suspensions and estuarine sediments. *Applied and Environmental Microbiology*, **53**, 1365–9.

Zinkevich, V., Bogdarina, I. and Kang, H. (1996). Characterisation of exopolymers produced by different isolates of marine sulphate-reducing bacteria. *International Biodeterioration and Biodegradation*, **37**, 163–72.

CHAPTER 15

Enzymatic and genomic studies on the reduction of mercury and selected metallic oxyanions by sulphate-reducing bacteria

Mireille Bruschi, Larry L. Barton, Florence Goulhen and Richard M. Plunkett

15.1 INTRODUCTION

Toxic heavy metals and metalloids constitute an international pollution problem that not only impacts public health but also is of environmental and economic importance. Prokaryotes with the physiological activity of sulphate reduction are found in a number of environmental sites containing toxic metals and these microorganisms have developed several different strategies for resistance to toxic elements. Some bacteria have developed detoxification strategies that are potentially useful for bioremediation. Since sulphate-reducing bacteria (SRB) are found in a large number of contaminated sites containing toxic metals, it is apparent that these organisms have a functional defence system that enables them to persist and even grow under metal stress. The enzymatic metal reduction by SRB offers an alternative to chemical processes to remediate environments containing redox-active toxic metals and metalloids. While Hockin and Gadd discuss in Chapter 14 the bioremediation activities of sulphate-reducing bacteria, this chapter focuses on the enzymatic processes associated with metal reduction. We review results obtained with isolated proteins and discuss the potential of sulphate-reducers by reviewing putative proteins found in their genomes. Reference is made to putative genes present in *Desulfovibrio* (*D.*) *vulgaris* strain Hildenborough (Heidelberg *et al.*, 2004), *D. desulfuricans* strain G20 (http://www.jgi.doe.gov), *Desulfotalea* (*Des.*) *psychrophila* (Rabus *et al.*, 2004) and *Archaeoglobus* (*A.*) *fulgidus* (Klenk *et al.*, 1997).

15.2 ENZYMATIC ACTIVITIES INVOLVING REDOX-ACTIVE ELEMENTS

The detoxification of an environment arising from SRB reductions is considered by many as an important event for bioremediation of various polluted environments In addition to precipitation of metals by biogenic hydrogen sulfide, the SRB are highly capable of reducing many soluble redox-active elements. While earlier reviews have enumerated the various toxic metals and metalloids reduced by different anaerobic bacteria (Barton *et al.*, 2003; Hobman *et al.*, 2000; Oremland and Stolz, 2000; Wang, 2000; Lloyd and Macaskie, 2000), we focus on the enzymatic reactions of SRB in the reduction of divalent mercury and the oxyanions of chromium, selenium, arsenic and uranium.

The initial observations concerning the reduction of elemental sulphur by cytochrome c_3 from *D. vulgaris* Miyazaki were considered to be nonphysiological (Ishimoto *et al.*, 1958); however, subsequently it was shown by Fauque and colleagues (1979) that the reduction of elemental sulphur by cytochrome c_3 from several different *desulfovibrio* displayed enzymatic characteristics. The reports by Lovley *et al.* (1993b) describing the reduction of U(VI) and Cr(VI) by cytochrome c_3 rekindled an interest in cytochrome c_3 as a metal reductase. To date, several oxidized elements are reduced by cytochromes from various different strains of sulphate-reducing bacteria and these are summarized in Table 15.1.

15.2.1 Cytochromes

Metal reduction has been reported due to an enzymatic process involving c_3-type cytochromes (Lovley and Philips, 1992; 1994; Lovley *et al.*, 1993a; 1993b). These cytochromes are indeed unique because electrochemistry experiments have demonstrated the direct reduction of various heavy metals (Lojou *et al.*, 1998b; Lojou and Bianco, 1999) by purified multihaemic cytochrome whereas mitochondrial *c*-type cytochrome did not exhibit comparable activity (Lojou *et al.*, 1998a). The amino-acid sequence and three dimensional structure comparisons of multihaemic cytochromes, characterized by bishistidinyl axial iron coordination and low redox potentials, lead to classify all these cytochromes to a cytochrome c_3 superfamily sharing a common ancestral origin (Bruschi *et al.*, 1992; 1994; Bruschi, 1994). Since all the tested cytochromes belonging to this family exhibit a metal reductase activity, we propose that other cytochromes with the

Table 15.1. *Reduction of diverse elements by cytochromes and hydrogenases from sulphate-reducing bacteria*

Organism	Protein	Element reduced	Reference
Dsf. acetoxidans	cytochrome c_7	Cr(VI)	Lojou *et al.*, 1998b
	cytochrome c_7	Mn(IV)	Lojou *et al.*, 1998b
	cytochrome c_7	Fe(III)	Lojou *et al.*, 1998a; 1998b
D. desulfuricans G20	hydrogenase	Tc(VII)	Lloyd *et al.*, 1999
	hydrogenase	Pd(II)	Lloyd *et al.*, 1998
D. desulfuricans Norway 4[a]	cytochrome c_3	S°	Fauque *et al.*, 1979
D. fructosovorans	[Fe] hydrogenase	Tc(VII)	De Luca *et al.*, 2001
	cytochrome c_3	Tc(VII)	De Luca *et al.*, 2001
	[NiFe] hydrogenase	Cr(VI)	Chardin *et al.*, 2003
D. gigas	cytochrome c_3	Fe(III)	Lojou *et al.*, 1998b
	cytochrome c_3	S°	Fauque *et al.*, 1979
Dsm. norvegicum	[NiFeSe] hydrogenase	Cr(VI)	Michel *et al.*, 2001
	cytochrome c_3	Cr(VI)	Michel *et al.*, 2001
	cytochrome c_3	Fe(III)	Lojou *et al.*, 1998a; 1998b
D. vulgaris Hildenborough	cytochrome c_3	Cr(VI)	Lovley and Philips, 1994; Michel *et al.*, 2001
	cytochrome c_3	Fe(III)	Lovley *et al.*, 1993b
	cytochrome c_3	U(VI)	Lojou *et al.*, 1998a; 1998b Lovley *et al.*, 1993b
	cytochrome c_3	S°	Fauque *et al.*, 1979
	[Fe] hydrogenase	Cr(VI)	Michel *et al.*, 2001
	cytochrome c_3	Se(VI)	Abdelous *et al.*, 2000
D. vulagris Miyazaki	cytochrome c_3	S°	Ishimoto *et al.*, 1958

[a] *D. desulfuricans* Norway 4 is now classified as *Desulfomicrobium baculatum* Norway 4.

tetrahaem motif will exhibit metal reductase activity (Czjzek *et al.*, 1996; 2002; Aubert *et al.*, 1998).

Physiological activity associated with sulphur reduction is important for cell energetics. For example the tetrahaeme cytochrome c_3 from *D. vulgaris* Hildenborough is not highly active in reduction of colloidal sulphur and is inhibited by hydrogen sulphide. Perhaps this explains why *D. vulgaris*

Hildenborough is unable to grow on elemental sulphur instead of sulphate as the terminal electron acceptor (Fauque, 1994).

15.2.2 Hydrogenases

The reduction of metals by hydrogenase has been reported by Chardin *et al.* (2003). In sulphate reducers there are [Fe] hydrogenases, [NiFe] hydrogenases and [NiFeSe] hydrogenases, containing [4Fe4S] or 3Fe4S] clusters and low potential redox proteins (Vignais *et al.*, 2001; Fauque *et al.*, 1988). The metal reduction activity results from the direct reduction by the [FeS] cluster closest to the protein surface, the electrons being transferred from the buried active site by the other [FeS] clusters. Hydrogenase from several different sulphate reducers will reduce numerous elements, see Table 15.1.

Other proteins from bacteria appear to be capable of reducing certain elements. The [Fe] hydrogenase from *Clostridium pasteuranium* reduces selenite to elemental selenium (Yanke *et al.*, 1995) while the hydrogenase from *Pyrococcus furiosus* reduces elemental sulphur (Kesen *et al.*, 1993). Additionally, *D. gigas* ferredoxin, a protein containing Fe-S clusters, has been found to reduce Cr(VI) (Chardin *et al.*, 2003).

15.3 AN EXAMPLE OF METAL REDUCTION BY SRB: CHROMIUM REDUCTION

15.3.1 Proteins and chemistry

Chromium is one of the most widely used metals in the leather tanning, wood preservation and alloy preparation industries and is considered as a priority pollutant by US Environmental Agency. Various species of SRB, which tolerate high Cr(VI) concentration (up to 500 μM for *Desulfomicrobium* (*Dsm.*) *norvegicum*) were tested to compare their potent efficacy to reduce chromium to a less toxic and soluble form Cr(III) and remove it from contaminated environments (Michel *et al.*, 2001). The best enzymatic activity for chromate reduction in intact cells was observed for *Dsm. norvegicum* (Michel *et al.*, 2001). Surprisingly, environmental strains isolated from mining sites or black smokers, did not exhibit a higher Cr(VI) reduction activity when compared to collection strains (Michel *et al.*, 2001).

Because metal reduction could be achieved enzymatically, the metal reduction activity of purified cytochrome c_3 and hydrogenase has been studied. It was concluded that only cytochromes of the cytochromes c_3 family showed a Cr(VI) reductase activity and among them, cytochrome c_3 from *Dsm. norvegicum* (redox potential −400 mV) presented the highest activity. Site-directed mutagenesis mutants of cytochromes c_3 have demonstrated that the metal reductase activity is directly linked to the lowest value of the redox potential of the cytochrome. The molecular mechanism of the metal reduction has been reported for the interaction between Cr(VI) and cytochrome c_7, a c_3-type cytochrome isolated from *Desulfuromonas* (*Dsf.*) *acetoxidans*. In studies using ^{1}H NMR (Assfalg *et al.*, 2002), the resolution of the three-dimensional structure of the complex (Figure 15.1) indicates a single binding site of Cr(III) around Haem IV already reported as the interacting site for the physiological partner, hydrogenase.

In addition, [Fe], [NiFe] and [NiFeSe] hydrogenases isolated from SRB of the genera *Desulfovibrio* and *Desulfomicrobium* were reported to be Cr(VI) reductases (Michel *et al.*, 2001; Chardin *et al.*, 2003). The highest rate for Cr(VI) reduction was observed using purified [Fe] hydrogenase from *D. vulgaris* Hildenborough and it was 10 times higher than when cytochrome c_3 was used from the same organism. The chromium reductase activity of hydrogenases from SRB results from the direct reduction of Cr(VI) species by the [Fe-S] clusters of the hydrogenases. Presumably, other proteins with [Fe-S clusters] would be capable of reducing metals since chromate reducing capacity has also been demonstrated for purified [Fe-S] ferredoxin II from *D. gigas* (Chardin *et al.*, 2003).

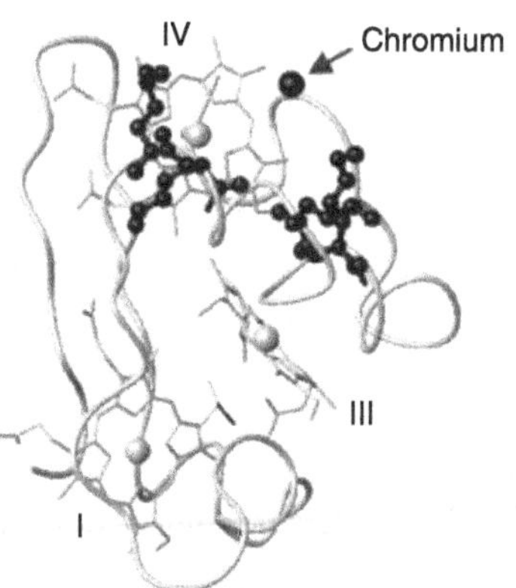

Figure 15.1. The Cr(III) binding site on cytochrome c_7 from *Desulfuromonas acetoxidans*. The Cr(III) ion is represented by a black sphere, and the haems are labelled by roman numbers (Assfalg *et al.*, 2002).

15.3.2 Physiological activities due to Cr(VI) exposure

It can be proposed that the enzymatic reduction of heavy metals by whole cells may be due to several different proteins with reductase activities. Chromate acts as a stressing agent for sulphate reducers signalling a physiological response which includes increased cell fragility and morphological changes. Numerous SRB can reduce Cr(VI) using several low redox potential enzymes involved in the electron chain transfer but the reduction of this metal does not support growth (Chardin *et al.*, 2002). If chromate (CrO_4^{2-}) enters SRB cells, it could be by the sulphate transport mechanism. Once inside the cell, chromate may be either reduced to Cr(III) by enzymatic processes or it is exported from the cell as Cr(VI). Chromate reduction occurs by haem proteins or proteins with FeS clusters; however, to date, a specific chromate reductase has not been identified in sulphate-reducing organisms. As previously discussed in this chapter, hydrogenase with [Fe-S] clusters and c_3-type cytochromes are effective in reducing Cr(VI). In SRB, the reduction of Cr(VI) (and other heavy metals) may occur at the expense of molecular hydrogen or organic substrates such as lactate (Lloyd *et al.*, 2001; Mabbett *et al.*, 2002). The involvement of low redox enzymes in reduction of chromate is suggested by the findings that the lack of hydrogenase gene in *D. vulgaris* Hildenborough did not result in a complete loss of the Cr(VI)-reducing activity. Similarily, a mutant devoid of a cytochrome gene has a decrease of one half of the U(VI) reduction (Payne *et al.*, 2002).

The genome of *D. vulgaris* Hildenborough contains two genes, one on the chromosome and one plasmid-borne, which have been identified as encoding chromate family transport proteins (Table 15.2). The translated amino acid sequences of these two genes are similar to chromate resistance family proteins that have been characterized by Nies *et al.* (1998), and products of these chromate transport genes may function in *D. vulgaris* as chromate efflux pumps. Chromate transport genes have not been identified in genomes of *D. desulfuricans* 20, *Des. psychrophila* or of *A. fulgidus*.

In order to evaluate the adaptation mechanisms to high chromate concentrations, the effects of Cr(VI) on bioenergetics metabolism were monitored using isothermal microcalorimetry (Chardin *et al.*, 2002). An extension of the lag growth phase and deep changes in the bacterial metabolism of the carbon source (lactate) were observed in the presence of high Cr(VI) concentration. The growth was inhibited with a concomitant energy production, suggesting that lactate is catabolized for lowering the redox potential to maintain survival conditions for SRB. The redox potential of the culture medium is increased by Cr(VI) (Eh = +135 mV, Cr(VI) 25 μM)

Table 15.2. *Putative genes associated with arsenic, chromium and mercury metabolism by sulphate reducers*

Enzymatic activity	Common name	Gene locus *D. vulgaris* Hildenborough	Gene locus *D. desulfuricans* G20	Gene locus *Des. psychrophila*	Genus locus *A. fulgidus*
Arsenic					
Arsenite transport protein					AF2308
Arsenate reductase	arsC	DVU1646	Dde2793	DP1879	AF1361
Putative arsenate reductase			Dde2792		
Transcriptional regulator ArsR family	arsR	DVU0606	Dde0747	DP1300	AF1270
		DVU1645	Dde3135		AF1544
		DVU2788	Dde3721		AF1853
Related to arsR			Dde2776		
Arsenite efflux pump	acr3		Dde2791	DP1778	
Chromium					
Chromate transport family protein		DVU0426			

Table 15.2. (*cont.*)

Enzymatic activity	Common name	Gene locus *D. vulgaris* Hildenborough	Gene locus *D. desulfuricans* G20	Gene locus *Des. psychrophila*	Genus locus *A. fulgidus*
Mercury		DVUA0093[a]			
Mercuric reductase	merA	DVU1037	Dde1463		
	related to merA			DP0504	
Mercuric resistance regulatory protein	merR				AF0673
Mercuric transport	merP	DVU2325	Dde1312		AF0346
Related to Hg^{2+} Binding protein				DP1470	
Probable heavy metal ATPase				DP1474	

[a] Found on plasmid.

Table 15.3. *Putative genes for proposed selenium metabolism present in sulphate reducers*

Enzymatic activity	Common name	Gene locus *D. vulgaris* Hildenborough[a]	Gene locus *D. desulfuricans* G20[b]	Gene locus *Des. psychrophila*[c]	Genus locus *A. fulgidus*[d]
Nitrate reductase	*narI*				AF0546
	γ-subunit	DVU1290	Dde2271	DP3075	AF0501
Thioredoxin	*trx*	DVU0037	Dde0464	NT01DP0963	
		DVU1839	Dde1202	NT01DP1108	
			Dde2067	DP0810	
			Dde2781	DP0948	
			Dde3416		
	trx-3				AF0711
					AF2144
					AF0769
					AF1284
Thioredoxin family		DVU1586	Dde2114		
Thioredoxin-like protein				NT01DP3585	
Thioredoxin/glutaredoxin-related protein				NT01DP3294	
				DP2832	
Thioredoxin reductase		DVU1457	Dde0465		
			Dde1203		
			Dde2151		
	trxB			NT01DP0962	AF1554
	trxB-1	DV00377			
	trxB-2	DVU1838			

Table 15.3. (*cont.*)

Enzymatic activity	Common name	Gene locus *D. vulgaris* Hildenborough[a]	Gene locus *D. desulfuricans* G20[b]	Gene locus *Des. psychrophila*[c]	Genus locus *A. fulgidus*[d]
Thioredoxin disulfide reductase			Dde2066		
Ferredoxin-thioredoxin reductase	*ftrB*			DP2155	AF1535
Cysteine synthase	*cysK*	DVU0663	Dde3080	NT01DP1455	
Cysteine desulfurase	*nifS*			DP2229	
	nifS-1				AF0186
	nifS-2				AF0564
	nifS-like protein	DVU0664			
L-seryl-trNA selenium transferase[e]	*selA*	DVU2883	Dde3059	DP11691	ni
Selenide, water dikinase[f]	*selD*	DVU1332	Dde2225	DP0969	ni[g]

[a] Heidelberg *et al.*, 2004
[b] www.jgi.doe.gov
[c] Rabus *et al.*, 2004
[d] Klenk *et al.*, 1997
[e] Selenocysteine synthetase
[f] Selenophosphate synthetase
[g] Not identified.

and it is lowered to its optimal value of about −150 mV. Indeed, Cr(VI) reduction is a protective escape to keep the bacterial environment favourable. The complete reduction of Cr(VI) to Cr(III) was observed by spectrophotometry and by speciation, using a combination of high performance liquid chromatography and inductively coupled plasma mass spectrometry.

Desulfovibrio vulgaris Hildenborough is an important strain for studying chromate detoxification because it is highly resistant to Cr(VI) (Michel *et al.*, 2001; Humphries and Macaskie, 2002), the biochemistry of the electron transport proteins is extensively studied, and the genome of this organism has been sequenced (Heidelberg *et al.*, 2004). Cells of *D. vulgaris* grown in the presence of 250 μM Cr(VI) exhibited electron dense particulate material near the cell surface (Figure 15.2). These metal precipitates of nanometer dimension are localized at the bacterial surface and energy electron loss spectrometry analysis (EELS) showed that Cr(III) is part of these amorphous precipitates. Moreover, these precipitates of Cr(III) also accumulate on inner and outer membranes of *D. vulgaris* grown in the presence of Cr(VI). The metal reductase activity which has been localized at the cell surface is consistent with a direct electron transfer to the metal by cytochrome and hydrogenase, which are periplasmic or membrane bound proteins (Goulhen *et al.*, 2006).

The microscopy analysis showed that the accumulation of trivalent chromium as chromium phosphate in the medium lowers the redox potential of the medium. It also shows the direct evidence of CrIII mineralization in biological membranes. In contrast to several bacterial groups in which the metal in used in metabolism as a terminal electron acceptor, Cr(VI) could penetrate the cells using probably the sulphate

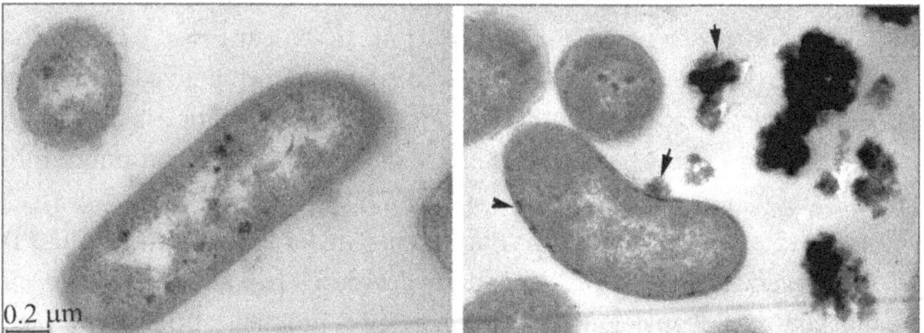

Figure 15.2. Subcellular localization of Cr(III) precipitates. Electron micrographs of ultrathin sections of early grown phase *D. vulgaris* Hildenborough in the presence (right panel) or absence (left panel) of 250 μM Cr(VI). Arrows indicate the precipitates of trivalent chromium in the medium, at the surface of the cell and at the membrane surfaces.

transport pathway and form precipitates in the periplasm and membrane compartments. This detoxification strategy helps the bacteria to keep the environment as favourable as possible. Nevertheless, the mineral precipitation reduces the cell surface for nutrient uptake. As the inhibited cells die, survival of the resistant bacteria is favoured. This strategy may be used by SRB in their metabolism of other toxic metals, including Tc(VII), U(VI), Pd(II), and V(V) (Goulhen *et al.*, 2006).

The exposure to chromium could be considered as an oxidative stress for the bacteria and in order to study the metal resistance and the detoxification mechanism, the proteomic pattern of periplasmic fractions of *D. vulgaris* Hildenborough grown in the presence of Cr(VI) has been examined. The proteomic profile of adapted bacteria showed a number of up- and down-regulated proteins when compared with bacteria grown in the absence of chromate. Preliminary results suggest that chromate stress influences production of proteins involved in energy metabolism, in redox regulation, cell surface biogenesis and sugar transport. One protein, superoxide dismutase, is up regulated in response to chromate stress (Bruschi and Goulhen, unpublished results).

15.4 URANYL REDUCTION

There has been considerable interest in examining the reduction of uranium salts by bacteria in natural environments, as well as in systems using cells or protein fractions. Reduction of the uranyl ion, UO_2^{2-}, to urananite, UO_2, has been reported for bacteria that are members of the physiological group referred to as dissimilatory metal reducers. The reduction of U(VI) by suspension cultures of *Clostridium, Geobacter, Veillonella, Cellulomonas, Desulfosporosinus,* and several different species of SRB has been reviewed earlier (Barton *et al.*, 2003; Sani *et al.*, 2002). The kinetics of U(VI) reduction have been studied using immobilized cultures *of D. desulfuricans* (Tucker *et al.*, 1998). From a bioremediation perspective, sulphate reducers have been characterized in groundwater at a uranium mill tailings site (Chang *et al.*, 2001).

In an effort to isolate the enzyme responsible for this reduction of U(VI) by SRB, several avenues have been pursued with cell-free studies. It has been established that uranyl ion is reduced by cytochrome c_3 from *D. vulgaris* Hildenborough (Lovley *et al.*, 1993b; Lojou *et al.*, 1998a; 1998b). Cytochrome c_3 from *D. desulfuricans* G20 also reduces U(VI) to U(IV), uraninite, and this cytochome binds to uraninite and other metal oxides (Payne *et al.*, 2004). There is the possibility that there are multiple proteins with uranyl reductase activity in SRB and this is suggested by the evidence that a mutant devoid of cytochrome c_3 gene still retains significant U(VI) reduction (Payne *et al.*, 2002). Additionally, cell-free extracts of *Desulfosporosinus* species reduce U(VI), even though these strains of SRB do not produce cytochrome c_3 (Suzuki *et al.*, 2004). Preliminary results indicates that the periplasmic [Fe] hydrogenase from *D. vulgaris* Hildenborough reduces several metals including U(VI) (Barton, Bryant and Laishley, unpublished results).

15.5 ENZYMOLOGY AND TRANSPORTER FOR ARSENIC REDUCTION

Sulphate-reducing prokaryotes have been shown to play a key role in environmental arsenic transformations, and a recent review by Oremland and Stolz (2003) of the ecology of arsenic identifies microbial arsenate reduction as part of the environmental cycle of this rare but important element. Investigations of arsenic-contaminated ecosystems reveal that the activities of SRB may quite strongly influence the measured levels of arsenic. Kirk *et al.* (2004) found that varying levels of arsenic contamination in groundwater they studied were dependent on the presence of sulphate reducers; not on the natural arsenic supply. They propose that lowered arsenic levels in the presence of SRB are not only a result of microbial-mediated arsenic reduction, but also because of arsenic precipitating with sulphide, or co-precipitating with iron and sulfide.

Sulphate-reducing organisms certainly do carry out enzymatic transformations of arsenic. The genes of bacterial arsenic reduction and oxidation were reviewed (Silver and Phung, 2005), presenting a model of bacterial arsenic reduction. Arsenate reduction, As(III) produced from As(V), has been observed in sulphate-reducing prokaryotes via two mechanisms. The first activity is associated with cellular resistance to arsenic, and is characterized by the reduction of arsenate. A putative gene for a cytoplasmic arsenate reductase (*arsC*) has been identified in several SRB (Table 15.2). The second activity is the exporting of arsenite via an ATP-independent arsenic transporter (ArsA/ArsB). A putative *arsB* has been identified in *A. fulgidus* while an arsenite efflux pump gene (*acr3*) is present in the genomes of *D. desulfuricans* G20 and *Des. psychrophila*. An alternate route for arsenite export could be an ABC transporter. Entry of arsenate into the cell could be mediated by phosphate transport mechanisms. Arsenic resistance is controlled by the regulatory gene *arsR*; this gene or similar genes have been found in the genome of each of the four sulphate reducers sequenced to date.

Despite its toxicity, some prokaryotic organisms are able to take advantage of As(V) as a terminal electron acceptor for respiration, and sulphate reducers have been identified among them (Newman *et al.*, 1997; Macy *et al.*, 2000). Genes for a respiratory arsenate reductase system (*arrAB*) has been identified in bacteria (Saltikov and Newman, 2003), but homologs have not been identified in SRB to date. It is likely that a *c*-type cytochrome, or associated protein, is at work in sulphate reducers for the respiratory reduction of As(V), as has been proposed for *Desulfovibrio* strain Ben-RA and *Desulfomicrobium* strain Ben-RB (Macy *et al.*, 2000).

15.6 REDUCTION OF SELENATE AND SELENITE

Selenium is required at trace levels by the sulphate-reducers for the synthesis of seleno-enzymes where selenium is present as selenocysteine and selenomethionine. Although the metabolic pathway for selenate reduction in SRB is not established, we consider a possible pathway based on the one proposed for *Escherichia coli* (Turner *et al.*, 1998) and the presence of corresponding genes in the sequenced genomes of sulphate reducers. The metabolic pathway proposed for sulphate-reducers is presented in Figure 15.3. The uptake of selenate or selenite is presumed to occur by a sulphate transporter and enzymes such as nitrate reductase have been found to have selenate reductase activity. Of the nitrate reductase genes (*narGHIJ*), only *A. fulgidus* potentially has one of these genes (*narI*) but *D. vulgaris, D. desulfuricans* and *Des. psychrophila* have a putative gene for the γ subunit of nitrate reductase. These putative genes may be physiologically important in selenate reduction to selenite because nitrate reduction to ammonia occurs in SRB. In *E. coli*, selenite reduction to H_2Se is proposed to require reduced glutathione (GSH) with the transient production of GS-Se-S-G prior to HSe^- formation; however, glutathione is not present in sulphate reducers. It may be that thioredoxin substitutes for glutathione in the sulphate reducers and upon examination of the genomes, numerous genes for thioredoxin synthesis, thioredoxin reductase are present. Selenophosphate is an important intermediate in the formation of the appropriate tRNA and selenophosphate synthase is present as *selD*. For the charging of the serine-$tRNA^{Sec}$ with selenium, selenocysteine synthetase may be involved and this is present in all four genomes as *selA*. The production of elemental selenium from selenate by *D. desulfuricans* has been reported (Tomei *et al.*, 1995) but the enzymology has not been established for this reaction. While selenate reduction by *D. desulfuricans* may result from cysteine desulfurase (*nifS* gene), this gene is not present in all sulphate reducers. However, cytochrome

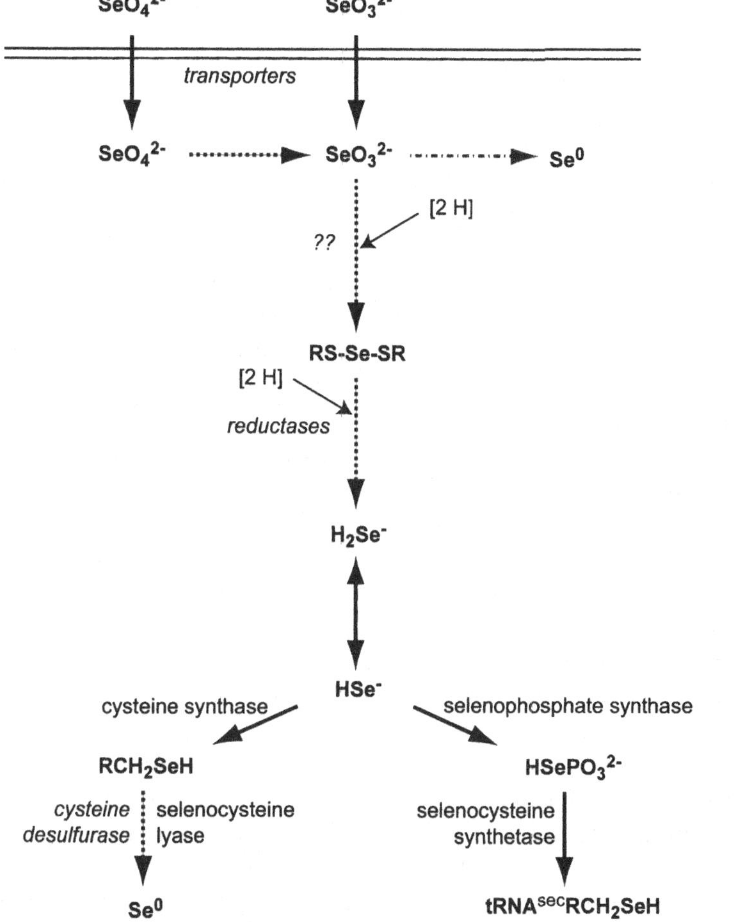

Figure 15.3. Proposed scheme for selenate reduction in sulphate-reducers. Solid lines are by enzymes discussed in this review, dashed lines are by enzymes not readily identified in the genomes, and the dot-dash line represents a reaction for which there is no evidence in the sulphate-reducing bacteria.

c_3 is also known to reduce selenate and this could produce elemental selenium by cells of *D. desulfuricans*.

15.7 ENZYMOLOGY OF MERCURY DETOXIFICATION

Many strains of bacteria are resistant to mercury and have the capability of detoxifying both organic and inorganic mercury compounds. Since the 1970s, it had been known that bacteria have the enzymatic capability of reducing water-soluble Hg^{2+} to lipid-soluble Hg^0. Some bacteria add methyl groups to Hg^{2+} with the formation of mono- and dimethylmercury. Dimethylmercury, as well as Hg^0, could exit the soil/aquatic environment and become dispersed in the atmosphere. Due to photochemical activities in the upper atmosphere, the mercury compounds are converted to Hg^{2+} and this heavy metal gradually settles over the earth's surface. Organomercury compounds are converted to Hg^{2+} in the soil by an organomercurial lyase which is produced by certain bacteria. Most of the research has been carrried out using plasmid-containing bacteria and, since genes for resistance and metabolism are laterally transferred in nature (Barkay and Smets, 2005), bacteria displaying enzymatic metabolism of mercury are relatively abundant.

The genes for Hg^{2+}-reduction are attributed to the *mer* operon which contains the following genes: *merA, merB, merC, merD, merF, merP, merT* and *merR* (Nies, 1999; Osborn *et al.*, 1997). Mercury resistance may be conferred by the presence of at least the components of mercuric reductase (MerA), which reduces Hg(II) to elemental Hg(0); MerT, a membrane-bound mercuric uptake protein; the *merA* and *merT* genes may be under the control of the mercury-sensitive regulator MerR. Several putative components of the mer system have been identified in the genomes of SRP (Table 15.2). Genes for merA, which encodes for mercuric reductase, have been identified in *D. vulgaris* strain Hildenborough and *D. desulfuricans* strain 20 and *Des. psychrophila*. The genome of the Achaeon *A. fulgidus* contains a *merR* gene, and other genes coding for mercury-associated proteins have been designated in SRB. With the absence of key mer operon genes, the traditional *mer* system for mercury reduction does not appear to be important in some of the SRB.

Another process for detoxification of the environment involves methylation of Hg^{2+} with the release of methyl mercury into the atmosphere. Various species of *Desulfovibrio, Desulfobulbus, Desulfococcus, Desulfobacter and Desulfobacterium* have been reported to methylate mercury (Pak and Bartha, 1998a; King *et al.*, 2000; Macalady *et al.*, 2000). *D. desulfuricans* produces methylmercury by an enzymatic process that proceeds with methyl transfer from methyltetrahydrofolate to cobalamin and ultimately to Hg^{2+} (Choi *et al.*, 1994). The enzyme for this methylation reduction has been proposed to be acetyl coenzyme A synthase. Pak and Bartha (1998b) have proposed that in SRB, the release of methane from methylmercury is by a mercuric lyase system.

In anoxic environments where SRB are active, many factors influence mercury methylation. Gilmore *et al.* (1992) report that the addition of sulphate promoted methylation of mercury and, if molybdate was added to a freshwater environment, it inhibited sulphide production from sulphate; methylation of mercury also ceased. King *et al.* (2000) indicated that accumulation of methylmercury in marine sediments was coupled to the rapid reduction of sulphate to sulphide. These reports have led some to propose that sulphate-containing fertilizers should not be applied to regions where methylmercury is being produced from Hg^{2+}. Gilmore *et al.* (1992) have reported that at <200 μM sulphate, the high concentration of hydrogen sulphide inhibits mercury methylation. In a hydrogen sulphide environment, Hg^{2+} combines with sulphide to produce insoluble HgS. The methylation of mercury as reported by Pak and Bartha (1998c) rapidly proceeds in sulphate-free environments where SRB are in co-culture with a methanogen and hydrogen is transferred by an interspecies process. Net mercury methylation rates associated with microbial communities containing SRB may vary widely. Reasons for these differences may be attributed to many different conditions, such as terminal electron accepting processes (Warner *et al.*, 2003), seasonal and spatial changes (Korthals and Winfrey, 1987; Marvin-DiPasquale and Agee, 2003), sulphate and sulphide concentrations, and specific species of sulphate-reducing microbes present (King *et al.*, 2000).

Recent studies of different ecosystems carried out by Marvin-DiPasquale *et al.* (2000) suggest that the degradation of methylmercury is effected by different enzymatic systems, depending on the concentration of mercury present in the environment. In sediments with high concentrations of mercury, methylmercury is reductively degraded (by the *mer* operon), while in less contaminated sediments, oxidative methylmercury degradation is dominant. This would not be unexpected, as many mercury-resistant organisms carry the *mer* operon.

15.8 PERSPECTIVE ON DISSIMILATORY METAL REDUCTION

From a physiological perspective, cellular mechanisms attributed to reduction of several toxic metals or metalloids remain unresolved. For detoxification of mercuric ion, specific enzymes from the mercury reduction and mercury methylation pathways appear to be involved and these systems are consistent with those found in other bacteria displaying mercury resistance. The reduction of chromate can be accomplished by c_3 cytochromes and by hydrogenases acting in the periplasmic region (Goulher *et al.*, 2006). The cellular region, where reduction of uranyl ions occurs, has not been fully established. Arsenate reduction to arsenite apparently occurs by enzymes present in other bacteria; however, the conversion of arsenite to arsenic sulfide has not been addressed. While selenium is required by the SRB as either selenate or selenite in trace levels, both selenate and selenite are highly toxic at elevated concentrations. There remains a great deal to be learned from the dissimilatory reduction capabilities of the sulphate-reducers.

From an environmental perspective, bacterial enzymatic metal reduction offers an alternative to chemical processes used to remediate toxic heavy metals. Sulphate-reducing bacteria are good candidates for bioremediation processes since growing cells or proteins from these cells could also be applied to treat contaminated soils or groundwater. The use of purified enzymes to develop amperometric biosensors for in situ measurement of metal bioavailability or monitoring remediation would be useful. A better understanding of the enzymatic activities associated with detoxification of metals can be used for monitoring the metabolic potential for such transformations and to enhance the activity of strains against various heavy metals using the molecular biology tools. The over-expression of target enzymes and improvement of the kinetic constants of the enzymes for the metals could undoubtedly make an impact on the arena of environmental biotechnology. With the increasing knowledge of genomic sequences for metal reducing bacteria, post-genomic and proteomic approaches are the new tools for environmental technological advances.

REFERENCES

Abdelous, A., Gong, W. L., Lutze, W. *et al.* (2000). Using cytochrome c_3 to make selenium wires. *Chem Mat*, **12**, 1510–12.

Assfalg, M., Bertini, I., Bruschi, M., Michel, C. and Turano, P. (2002). The metal reductase activity of some multiheme cytochromes *c*: NMR structural characterization of the reduction of chromium(VI) to chromium(III) by cytochrome c_7. *Proc Natl Acad Sci USA*, **99**, 9750–4.

Aubert, C., Lojou, E., Bianco, P. *et al.* (1998). The *Desulfuromonas acetoxidans* triheme cytochrome c_7 produced in *Desulfovibrio desulfuricans* retains its metal reductase activity. *Environ Microbiol*, **64**, 1308–12.

Barkay, T. and Smets, B. F. (2005). Horizontal gene flow in microbial communities. *ASM News*, **71**, 412–19.

Bruschi, M., Bertrand, P., More, C. *et al.* (1992). Biochemical and spectroscopic characterization of the high molecular weight cytochrome *c* from

Desulfovibrio vulgaris Hildenborough expressed in *Desulfovibrio desulfuricans* G200. *Biochem*, **31**, 3281–8.

Bruschi, M. (1994). Cytochrome c_3 (Mr26000) isolated from sulphate-reducing bacteria and its relationships to other polyhemic cytochromes from *Desulfovibrio*. *Meth Enzymol*, **243**, 140–55.

Bruschi, M., Leroy, G., Guerlesquin, F. and Bonicel, J. (1994). Amino-acid sequence of the cytochrome c_3 (M(r) 26,000) from *Desulfovibrio desulfuricans* Norway and a comparison with those of the other polyhemic cytochromes from *Desulfovibrio*. *Biochim Biophys Acta*, **1205**, 123–31.

Barton, L. L., Plunkett, R. M. and Thomson, B. M. (2003). Reduction of metals and nonessential elements by anaerobes. In L. G. Ljungdahl, M. W. Adams, L. L. Barton, J. G. Ferry and M. K. Johnson (eds.), *Biochemistry and Physiology of Anaerobic Bacteria*. New York: Springer-Verlag. pp. 220–34.

Chang, Y. J., Peacock, A. D., Long, P. E. *et al.* (2001). Diversity and characterization of sulphate-reducing bacteria in groundwater at a uranium mill tailing site. *Appl Environ Microbiol*, **67**, 3149–60.

Chardin, B., Dolla, A., Chaspoul, F. *et al.* (2002). Bioremediation of chromate: thermodynamic analysis of the effects of Cr(VI) on sulphate-reducing bacteria. *Appl Microbiol Biotechnol*, **60**, 352–60.

Chardin, B., Giudici-Orticoni, M. T., De Luca, G., Guigliarelli, B. and Bruschi, M. (2003). Hydrogenases in sulphate-reducing bacteria function as chromium reductase. *Appl Microbiol Biotechnol*, **63**, 315–21.

Choi, S. C., Chase, Jr., T. and Bartha, R. (1994). Enzymatic catalysis of mercury methylation by *Desulfovibrio desulfuricans* LS. *Appl Environ Microbiol*, **60**, 1342–6.

Czjzek, M., Guerlesquin, F., Bruschi, M. and Haser, R. (1996). Crystal structure of a dimeric octaheme cytochrome c_3 (M(r) 26,000) from *Desulfovibrio desulfuricans* Norway. *Structure*, **4**, 395–404.

Czjzek, M., ElAntak, L., Zamboni, V. *et al.* (2002). The crystal structure of the hexadeca-heme cytochrome *Hmc* and a structural model of its complex with cytochrome c_3. *Structure*, **10**, 1677–86.

De Luca, G., de Philip, P., Dermoun, Z., Rousset, M. and Vermeglio, A. (2001). Reduction of technetium (VII) by *Desulfovibrio fructosovorans* is mediated by the nickel-iron hydrogenase. *Appl Environ Microbiol*, **67**, 4583–7.

Fauque, G. D. (1994). Sulfur reductase form thiophilic sulphate-reducing bacteria. *Meth Enzymol*, **243**, 353–67.

Fauque, G., Herve, D. and LeGall, J. (1979). Structure–function relationship in hemoproteins: The role of cytochrome c_3 in the reduction of colloidal sulfur by sulphate-reducing bacteria. *Arch Microbiol*, **121**, 261–4.

Fauque, G., Peck, H. D. Jr., Moura, J. J. G. *et al.* (1988). The three classes of hydrogenases from sulphate-reducing bacteria of the genus *Desulfovibrio*. *FEMS Microbiol Rev*, **54**, 299–344.

Gilmore, C. C., Henry, E. A. and Mitchell, R. (1992). Sulphate stimulation of mercury methylation in fresh-water sediments. *Environ Sci Technol*, **26**, 2281–7.

Goulhen, F., Gloter, A., Guyot, F. and Bruschi, M. (2006). *Desulfovibrio vulgaris* strain Hildenborough: Microbe–metal interactions studies. *Appl Microbiol Biotechnol*, **71**, 892–7.

Heidelberg, J. F., Seshadri, R., Haveman, S. A. *et al.* (2004). The genome sequence of the anaerobic, sulphate-reducing bacterium *Desulfovibrio vulgaris* Hildenborough. *Nat Biotechnol*, **22**, 554–9.

Hobman, J. L., Wilson, J. R. and Brown, N. L. (2000). Microbial mercury reduction. In D. R. Lovley (ed.), *Environmental metal–microbe interactions*. Washington, DC: ASM Press. pp. 177–98.

Humphries, A. C. and Macaskie, L. E. (2002). Reduction of Cr(VI) by Desulfovibrio vulgaris and Microbacterium sp. *Biotechnol Lett*, **24**, 1261–7.

Ishimoto, M., Kondo, Y., Kameyama, T., Yagi, T. and Shirak, M. (1958). The role of cytochrome in the enzyme system of sulphate-reducing bacteria. In Science Council of Japan (ed.), *Proceedings of the International Symposium on Enzyme Chemistry*. Tokyo and Kyoto: Marüzen. pp. 229–34.

Kesen, M. A., Schicho, R. N., Kelly, R. M. and Adams, M. W. W. (1993). Hydrogenase of the hyperthermophile *Pyrococcus furiosus* is an elemental sulfur reductase or sulfurylase: Evidence for a sulfur-reducing hydrogenase ancestor. *Proc Nat Acad Sci USA*, **90**, 5341–4.

King, J. K., Kosta, J. E., Frischer, M. E. and Saunders, F. M. (2000). Sulphate-reducing bacteria methylate mercury at variable rates in pure culture and in marine sediments. *Appl Environ Microbiol*, **66**, 2430–7.

Kirk, M. F., Holm, T. R., Park, J. *et al.* (2004). Bacterial sulphate reduction limits natural arsenic contamination in groundwater. *Geol*, **32**, 953–6.

Klenk, H. P., Clayton, R. A., Tomb, J. F. *et al.* (1997). The complete genome sequence of the hyperthermophilic, sulphate-reducing archaeon *Archaeoglobus fulgidus*. *Nature*, **390**, 364–70.

Korthals, E. T. and Winfrey, M. R. (1987). Seasonal and spatial variations in mercury methylation and demethylation in an oligotrophic lake. *Appl Environ Microbiol*, **53**, 2397–404.

Lloyd, J. R., Mabbett, A. N., Williams, D. R. and Macaskie, L. E. (2001). Metal reduction by sulphate-reducing bacteria: physiological diversity and metal specificity. *Hydrometallurgy*, **59**, 327–37.

Lloyd, J. R. and Macaskie, L. E. (2000). Bioremediation of radionuclide-containing wastewaters. In D. R. Lovley (ed.), *Environmental metal–microbe interactions.* Washington, DC: ASM Press. pp. 277–329.

Lloyd, J. R., Ridley, J., Khizniak, T., Lyalikova, N. N. and Macaskie, L. E. (1999). Reduction of technetium by *Desulfovibrio desulfuricans*: biocatalyst characterization and use in a flowthrough bioreactor. *Appl Environ Microbiol,* **65**, 2691–6.

Lloyd, J. R., Yong, P. and Macaskie, L. E. (1998). Enzymatic recovery of elemental palladium by using sulphate-reducing bacteria. *Appl Environ Microbiol* **64**, 4607–9.

Lojou, E., Bianco, P. and Bruschi, M. (1998*a*). Kinetic studies on the electron transfer between bacterial *c*-type cyrochromes and metal oxides. *J Electroanal Chem,* **452**, 167–77.

Lojou, E., Bianco, P. and Bruschi, M. (1998*b*). Kinetic studies on the electron transfer between various c-type cytochromes and iron (III) using a voltametric approach. *Electrochim Acta,* **43**, 2005–13.

Lojou, E. and Bianco, P. (1999). Electrocatalytic reduction of uranium by bacterial cytochromes: biochemical factors influencing the catalytic process. *J Electroanal Chem,* **471**, 96–104.

Lovley, D. R., Giovannoni, S. J., White, D. C. *et al.* (1993*a*). *Geobacter metallireducens* gen. *nov.* sp. *nov.*, a microorganism capable of coupling the complete oxidation of organic compounds to the reduction of iron and other metals. *Arch Microbiol,* **159**, 336–44.

Lovley, D. R. and Phillips, E. J. P. (1992). Reduction of uranium by *Desulfovibrio desulfuricans. Appl Microbiol Microbiol,* **58**, 850–6.

Lovley, D. R. and Phillips, E. J. P. (1994). Reduction of chromate by *Desulfovibrio vulgaris* and its c_3 cytochrome. *Appl Environ Microbiol,* **60**, 726–8.

Lovley, D. R., Widman, P. K., Woodward, J. C. and Phillips, E. J. (1993*b*). Reduction of uranium by cytochrome c_3 of *Desulfovibrio vulgaris. Appl Environ Microbiol,* **59**, 3572–6.

Mabbett, A. N., Lloyd, J. R. and Macaskie, L. E. (2002). Effect of complexing agents on reduction of Cr(VI) by *Desulfovibrio vulgaris* ATCC 29579. *Biotechnol Bioengineering,* **79**, 389–397.

Macalady, J. L., Mack, E. E., Nelson, D. C. and Scow, K. M. (2000). Sediment microbial community structure and mercury methylation in mercury-polluted Clear Lake, California. *Appl Environ Microbiol,* **66**, 1479–88.

Macy, J. M., Santini, J. M., Pauling, B. V., O'Neill, A. H. and Sly, L. I. (2000). Two new arsenate/sulphate-reducing bacteria: mechanisms of arsenate reduction. *Arch Microbiol,* **173**, 49–57.

Marvin-DiPasquale, M., Agee, J., McGowan, C. *et al.* (2000). Methyl-mercury degradation pathways: a comparison among three mercury-impacted ecosystems. *Environ Sci Technol*, **34**, 4908–16.

Marvin-DiPasquale, M. and Agee, M. (2003). Microbial mercury cycling in sediments of the San Francisco bay-delta. *Estuaries*, **26**, 1517–28.

Michel, C., Brugna, M., Aubert, C., Bernadac, A. and Bruschi, M. (2001). Enzymatic reduction of chromate: comparative studies using sulphate-reducing bacteria. Key role of polyheme cytochromes c and hydrogenases. *Appl Microbiol Biotechnol*, **55**, 95–100.

Newman, D. K., Kennedy, E. K., Coates, J. D. *et al.* (1997). Dissimilatory arsenate and sulphate reduction in *Desulfotomaculum auripigmentum* sp. nov. *Arch Microbiol*, **168**, 380–8.

Nies, D. H. (1999). Microbial heavy-metal resistance. *Appl Microbiol Biotechnol*, **51**, 730–50.

Nies, D. H., Koch, S., Shinichiro, W., Peitzch, N. and Saier, M. H. (1998). CHR, a novel family of prokaryotic proton motive force-driven transporters probably containing chromate/sulphate antiporters. *J Bacteriol*, **180**, 5799–802.

Oremland, R. S. and Stolz, J. (2000). Dissimilatory reduction of selenate and arsenate in nature. In D. R. Lovley (ed.), *Environmental Metal–Microbe Interactions*. Washington, DC: ASM Press. pp. 199–224.

Oremland, R. S. and Stolz, J. F. (2003). The ecology of arsenic. *Science*, **300**, 939–44.

Osborn, A. M., Bruce, K. D., Strike, P. and Ritchie, D. A. (1997). Distribution, diversity and evolution of the bacterial mercury resistance (mer) operon. *FEMS Microbiol Rev*, **19**, 239–62.

Pak, K.-R. and Bartha, R. (1998*a*). Mercury methylation and demethylation in anoxic lake sediments and by strictly anaerobic bacteria. *Appl Environ Microbiol*, **64**, 1013–17.

Pak, K.-R. and Bartha, R. (1998*b*). Products of mercury demethylation of sulfidogens and methanogens. *Bull Environ Con Toxicol* **61**, 690–4.

Pak, K.-R. and Bartha, R. (1998*c*). Mercury methylation by interspecies hydrogen and acetate transfer between sulfidogens and methogens. *Appl Environ Microbiol*, **64**, 1987–90.

Payne, R. B., Casalot, L., Rivere, T. *et al.* (2004). Interaction between uranium and the cytochrome c_3 of *Desulfovibrio desulfuricans* G20. *Arch Microbiol*, **181**, 398–406.

Payne, R. B., Gentry, D. M., Rapp-Giles, B. J., Casalot, L. and Wall, J. D. (2002). Uranium reduction by *Desulfovibrio desulfuricans* strain G20 and a cytochrome c_3 mutant. *Appl Environ Microbiol*, **68**, 3129–32.

Rabus, R., Ruepp, A., Frickey, T. *et al.* (2004).The genome of *Desulfotalea psychrophila*, a sulphate-reducing bacterium from permanently cold Arctic sediments. *Environ Microbiol*, **6**, 887–902.

Saltikov, C. W. and Newman, D. K. (2003). Genetic identification of a respiratory arsenate reductase. *Proc Nat Acad Sci USA*, **100**, 10983–8.

Sani, R. K., Peyton, B. M. Smith, W. A., Apel, W. A. and Petersen, J. N. (2002). Dissimilatory reduction of Cr(VI), Fe(III), and U(VI) by *Cellulomonas* isolates. *Appl Microbiol Biotechnol*, **60**, 192–9.

Silver, S. and Phung, L. T. (2005). Genes and enzymes involved in bacterial oxidation and reduction of inorganic arsenic. *Appl Environ Microbiol*, **71**, 599–608.

Susuki, Y., Kelly, S. D., Kemner, K. M. and Banfield, J. F. (2004). Enzymatic U(VI) reduction by *Desulfosporosinus* species. *Radiochim Acta*, **92**, 11–16.

Tomei, F. A., Barton, L. L., Lemanski, C. L. *et al.* (1995). Transformation of selenate and selenite to elemental selenium by *Desulfovibrio desulfuricans*. *J Indust Microbiol*, **14**, 329–36.

Tucker, M. D., Barton, L. L. and Thomson, B. M. (1998). Reduction of Cr, Mo, Se and U by *Desulfovibrio desulfuricans* immobilized in polyacrylamide gels. *J Industr Microbiol Biotechnol*, **20**, 13–19.

Turner, R. J., Weiner, J. H. and Taylor, D. E. (1998). Selenium metabolism in *Escherichia coli*. *BioMetals*, **11**, 223–7.

Vignais, P. M., Billoud, B. and Meyer, J. (2001). Classification and phylogeny of hydrogenases. *FEMS Microbiol Rev*, **25**, 455–501.

Wang, Y.-T. (2000). Microbial reduction of chromate. In D. R. Lovley (ed.), *Environmental Metal–Microbe Interactions*. Washington, DC: ASM Press. pp. 225–6.

Warner, K. A., Roden, E. E. and Bonzongo, J. C. (2003). Microbial mercury transformation in anoxic freshwater sediments under iron-reducing and other electron-accepting conditions. *Environ Sci Technol*, **37**, 2159–65.

Yanke, L. J., Bryant, R. D. and Laishley, E. J. (1995). Hydrogenase I of *Clostridium pasteuranium* functions as a novel selenite reductase. *Anaerobe*, **1**, 61–7.

CHAPTER 16

Sulphate-reducing bacteria and their role in corrosion of ferrous materials

Iwona B. Beech and Jan A. Sunner

16.1 INTRODUCTION

In both natural environments and human-made systems, a phylogenetically diverse and heterogeneous group of anaerobic sulphate-reducing bacteria (SRB) (Barton, 1985; Odom and Singelton, 1993; Postgate, 1984) thrive as members of complex microbial communities, living on surfaces of particles, minerals and manufactured materials, i.e. within biofilms (Characklis and Marshall, 1990; Dar *et al.*, 2005 and references therein).

The presence of biofilms often results in deterioration of colonized substrata. In the case of metallic materials, undesirable change in their properties resulting from material loss under biological influence is termed microbially-influenced corrosion (MIC) or biocorrosion. A number of reviews have been published describing fundamental and practical aspects of MIC (Geesey *et al.*, 2000; Hamilton, 2000; Beech and Coutinho, 2003; Beech and Sunner, 2004).

The annual direct and derived costs of corrosion are estimated to be around 4% of the GNP of developed countries, of which 10–20% are related to biocorrosion (Geesey *et al.*, 2000). In the oil and gas industry, biocorrosion accounts for 15–30% of the corrosion cases, resulting in financial losses in a range of 100 M $ per annum in the USA alone, excluding costs of lost revenues and often necessary remediation treatments. In some cases, for example in the Gulf of Guinea, extremely high rates of pitting-corrosion related to biocorrosion have reduced the life of oil subsea lines to one year (J.-L. Crolet, personal communication). In a number of European countries, the replacement of steel piling in quays, harbours and jetties, which were designed to last for up to 60 years, often needs to be carried out after only 10 to 20 years of use, due to a MIC problem known as accelerated low water corrosion (ALWC) (Gubner and Beech, 1999; Breakell *et al.*, 2005 for technical review). Not surprisingly, many industrial sectors worldwide would greatly benefit from understanding mechanisms of biocorrosion, as it would aid in the design of efficient prevention and mitigation strategies.

The first report of biocorrosion, in which it was proposed that the interaction of bacterial metabolic products with a lead cable was a possible cause of cable corrosion, dates to 1891 (Garrett, 1891). However, it was not until 1934 that the corrosion of ferrous metal buried in anaerobic clay soil was attributed solely to microbial physiological activity, namely that of SRB (von Wolzogen Kuhr and van der Vlught, 1934). Since then, sulphate reducers has become the most investigated group of bacteria implicated in corrosion damage. SRB cause corrosion of iron and ferrous alloys in a range of aquatic and terrestrial habitats, varying in nutrient content, temperature, pressure and pH values. Several theories have been presented to explain how SRB influence corrosion of metallic materials. In this chapter, principles of corrosion are outlined to provide a background for understanding electrochemical reactions governing MIC and a brief overview of the effect biofilms can exert on corrosion is given. Proposed models of SRB-mediated corrosion of ferrous materials are described, along with the current hypothesis of biocorrosion, according to which known reactions of primary importance result in efficient coupling of electrochemical and biotic electron transfer processes (Hamilton, 2003).

16.2 FUNDAMENTAL ASPECTS OF CORROSION

Corrosion is the result of a chemical interaction between a metallic material and its environment that results in material loss. In most cases, this is an electrochemical process in which electrons are transferred from the metal, through a series of redox reactions, to an ultimate electron acceptor, located in the proximity of the metal surface. The electron acceptor is often, but not always, molecular oxygen. The reduction of the electron acceptor is coupled with oxidation of the metal, which leads to its dissolution.

For electrochemical corrosion to occur, the following components have to be present:

- an anode site
- a cathode site
- an electrolyte (a solution capable of conducting electrical flow)

- a cathodic reactant
- an anodic reactant.

The overall corrosion reaction consists of two half-reactions: an anodic reaction, involving the oxidation of the metal to metal ions, and a cathodic reaction, involving the reduction of chemical species in contact with the metal surface.

The oxidation and reduction reactions tend to occur at separate locations on the metal surface. The oxidation reaction, which releases electrons, takes place at anodic sites (or anode). It is here that metal loss occurs, as metal atoms enter the solution as metal ions. When the metal, or electrode, is electrically isolated, as is the case for, e.g., carbon steel pilings in marine habitats or stainless steel implant within the human body, electrons must flow through the metal from the anodic sites to cathodic sites (or cathode). At the cathodic sites, electrons are consumed through the reduction of surface-near chemical species.

Although the anodic and cathodic half-reactions can occur at widely separated locations on the metal surface, the anodic and cathodic sites are typically adjacent to each other, forming a "corrosion cell". A sparse and irregular distribution of anodic sites across the metallic surface causes localized attack in the form of pitting and/or crevice corrosion. When the corrosion cells are more densely packed, a more evenly distributed corrosion is observed.

The rate of corrosion, i.e. the rate of metal loss, is proportional to the magnitude of the anodic current. In order to maintain electric neutrality, the anodic current must equal the cathodic current. Generally, the rate of corrosion is limited by the rate of the cathodic reaction. In such a case, the electric potential of the electrode is low and close to the reduction potential of the metal. Enhanced corrosion rates results from the establishment of kinetically favoured reduction pathways, and this is associated with an increase in the electrode potential, i.e. cathodic depolarization. The corresponding increase in the difference between the electrode potential and the metal redox potential is directly associated with an increase in the corrosion rate.

In both aerated and oxygen-free solutions the anodic reaction is always metal dissolution:

$$Me^{O} \rightarrow Me^{n+} + ne^{-} \quad (16.1)$$

However, the cathodic reaction depends on the solution environment. In corrosion, the dominating reduction reactions are the reduction of oxygen:

$$O_2 + 2H_2O + 4e^- \rightarrow 4OH^- \quad (16.2)$$

or of protons:

$$2H^+ + 2e^- \rightarrow H_2 \quad (16.3)$$

Which of these reactions is the dominating cathodic reaction in corrosion depends on the environment in the immediate vicinity of the metal surface. The reduction of O_2 is always thermodynamically favoured, by about 1.23V. However, hydrogen evolution is generally kinetically favoured. On iron, for example, the exchange current density (current density at equilibrium) for hydrogen evolution at $pH = 0$ is approximately 10^{-6} A/cm^2 while it is 10^{-14} A/cm^2 for oxygen reduction, i.e. it is higher by about eight orders of magnitude. The exchange current density for hydrogen evolution decreases with increasing pH as it is proportional to square root of the hydrogen ion concentration. Furthermore, the redox potential for hydrogen reduction decreases with increasing pH, thus the thermodynamic driving force for the overall corrosion reaction decreases. In the case of iron, the potential difference decreases to nearly zero at $pH = 7$. The stabilization of dissolved iron ions by abiotic or biologically generated stable hydroxides or e.g. sulphides, may be significant, as the presence of stable corrosion products decreases the reduction potential for the metal. This results in an increase in the thermodynamic driving force which, therefore, would allow hydrogen production to resume.

An additional, very important factor that determines the balance between hydrogen ion and oxygen reduction in corrosion is the issue of transport of molecular oxygen. The solubility of O_2 in water is approximately 10^{-4} M. This severely limits the rate of oxygen transport to the metal surface. In combination with oxygen consumption by aerobic bacteria in biofilms, the oxygen concentration in the vicinity of a metal surface is often very low and anaerobic conditions prevail (Lee and deBeer, 1995).

In general, the above considerations mean that hydrogen evolution occurs under anaerobic conditions at any pH, while under aerobic conditions, hydrogen evolution takes place at low pH and oxygen reduction dominates at neutral and high pH.

Rusting of ferrous materials is the most familiar form of corrosion. In the presence of oxygen and water, the net chemical reaction of iron rusting can be expressed as follows:

$$Fe^O + 1/2O_2 + H_2O \rightarrow Fe^{2+} + 2OH^- \quad (16.4)$$

On surfaces of ferrous metals and their alloys, e.g. stainless steels, complex oxide and hydroxide layers are formed comprising not only Fe but also alloying elements, such as Cr, Ni and Mo. Scully (1990) provides very good description of corrosion mechanisms.

In most instances, the oxidation of solid pure metals and alloys slows to a low rate after a period of time because oxidation products tend to adhere to the metallic surface and form a deposit that slows the transport of reactants to the metal surface. Such a deposit, which is referred to as a passive layer or a passive film, forms a protective barrier that, essentially, prevents further oxidation of the underlying metal. Mechanical disruptions or chemical alteration of the passive films can lead to severe corrosion.

16.3 BIOFILMS AND CORROSION

The presence of biofilms on a metal surface often establishes new electrochemical reaction pathways, or promotes reactions which are not normally favoured in the absence of microorganisms, resulting in increased corrosion rates and deleterious effects on the performance and integrity of the material (Lewandowski *et al.*, 1997). It is well documented that bacterial metabolic products e.g. exoenzymes, organic and inorganic acids, as well as volatile compounds, such as ammonia, or in the case of SRB, hydrogen sulphide and associated metal sulphides, can alter interfacial processes between biofilm and a metallic substratum (Videla, 1996). Owing to the heterogeneous, often patchy distribution of biofilms and associated inorganic surface deposits, MIC failures are often reported as localized attack in the form of pitting.

Bacterial consortia within complex biofilms are able to catalyze electrochemical processes through cooperative metabolism in ways that single species cannot (Valencia-Cantero *et al.*, 2003; Pitonzo *et al.*, 2004) and consequently MIC has seldom been linked to a single type of bacteria. In addition to SRB, several other main groups of bacteria have been demonstrated to influence corrosion in a variety of environments, (Beech and Coutinho, 2003 for review).

The aggressive and inhibitory effects that bacterial populations exert on corrosion reactions are clearly due to very complex interactions, involving both biofilm and corrosion products on the material surface (Little and Ray, 2002). For these reasons, it has not yet been possible to assign convincingly any instance of MIC to a well-defined molecular mechanism.

It is important to realize that documenting the presence of bacterial species, including SRB, on a corroded metallic surface is not sufficient evidence for their contribution to the corrosion process, even if such species are known to produce metabolic by-products which are aggressive towards metals (Little and Wagner, 1997). Furthermore, the state of metabolic activity of biofilm microorganisms is thought to be more important for MIC than the number of viable cells that may be cultured from the biofilm.

Studies of bacterial interaction with metallic materials led to the formulation of a unifying electron-transfer hypothesis of MIC, using ferrous metals as a model system (Hamilton, 2003). According to this hypothesis, MIC is a process in which metabolic activities of microorganisms supply insoluble products, which are able to accept electrons from the base metal. This sequence of biotic and abiotic reactions produces a kinetically favoured pathway of electron flow from the metal anode to the universal electron acceptor, oxygen (Figure 16.1).

Recent years have seen significant developments in the study of SRB. Particularly noteworthy are advances in molecular microbial ecology, which have allowed detailed characterization of SRB community structure (Dar *et al.*, 2005 and references therein); considerable progress in SRB genomics and proteomics, mainly availability of *Desulfovibrio vulgaris*

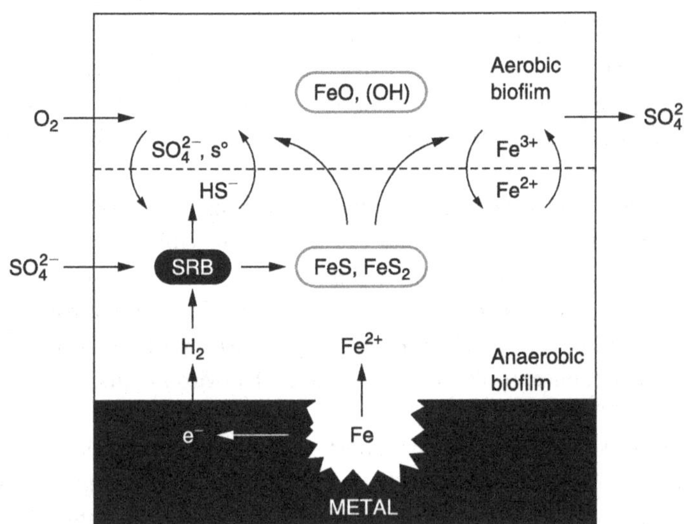

Figure 16.1. A descriptive model of the corrosion of mild steel resulting from the action of sulphate-reducing bacteria in a mixed aerobic/anaerobic system in which oxygen acts as the terminal electron acceptor. (Drawing reproduced with permission from Nielsen *et al.*, 1993 with the legend from Hamilton, 2003.)

Hildenborough genome sequence (Heidelberg *et al.*, 2004) and its proteome analysis (Fournier *et al.*, 2006); as well as the discovery of novel SRB strains which are postulated to obtain electrons directly from metallic iron (Dinh *et al.*, 2004) and the evidence of direct electron transfer between SRB hydrogenase enzyme and stainless steel surface (da Silva *et al.*, 2004). These developments have expanded the knowledge of energy transduction and electron transport processes in SRB, thus offering deeper insight into the mechanisms by which these microorganisms contribute to biocorrosion.

Indisputably, the unifying electron-transfer hypothesis of MIC accommodates most of the currently proposed models of SRB-influenced corrosion described below. Nevertheless, the evidence that cells of certain SRB species, or their enzymes alone, are able to drive the cathodic reaction by accepting electrons directly from the base metal, indicate that microbially generated insoluble products are not necessarily a prerequisite of MIC. Moreover, the emphasis solely on oxygen as an ultimate electron acceptor has also been queried and it has been advocated that the presence of alternative electron acceptors, e.g. nitrate, sulphate, ferric iron and CO_2, should not be ruled out when carrying out biocorrosion risk assessment in anoxic habitats (Lee *et al.*, 2004; 2005).

16.4 MODELS OF SRB-INFLUENCED CORROSION

The active role that SRB play in causing pitting corrosion of iron and ferrous alloys in aerated and anoxic aquatic, as well as terrestrial environments, has been long acknowledged (King and Miller, 1971). Yet, it is important to realize that, due to their phylogenetic diversity, not only different genera of SRB, but also different species of the same genus, can vary considerably in their ability to influence metal deterioration, even under identical growth conditions (Beech, 2002 for review). It is therefore not surprising that numerous investigations, in particular studies involving mixed SRB populations, describe corrosive SRB behaviour, whilst others report the lack of it (Jan-Roblero *et al.*, 2004 and references therein).

The difficulty of obtaining reproducible results when investigating MIC under rigorous experimental conditions with well-defined bacterial populations, together with a lack of experimental methods to probe directly individual electron transfer reactions, constitutes the largest obstacle to design comprehensive and cohesive models of SRB-influenced corrosion. Nevertheless, a number of mechanisms have been proposed. These include cathodic depolarizing effect of hydrogenase enzymes, cathodic depolarization due to biogenically produced iron sulphides, corrosive properties of extracellular polymeric substances (EPS) and the generation of corrosive phosphorous compounds (Beech, 2002 for review). It is now recognized that rather than one dominant mechanism, a number of processes are likely to be involved (Hamilton, 1998; Lee *et al.*, 1995; http://www.corrosionsource.com/index.htm).

A summary of biocorrosion models resulting from metabolic activities of SRB is provided below.

16.4.1 Hydrogenase enzyme as a cathodic depolarizing agent

Hydrogen metabolism plays a central role in energy-generating mechanisms of SRB. The capability of SRB to produce and consume H_2 is mediated by the enzyme hydrogenase, which catalyzes the reversible oxidation of hydrogen:

$$H_2 \leftrightarrow 2H^+ + 2e^- \qquad (16.5)$$

Hydrogenase enzymes differ in their specific activity for the evolution and consumption of hydrogen as well as in their subunit compositions, physicochemical characteristics, amino-acid sequences, immunological reactivities and catalytic properties. Furthermore, the activity of hydrogenases varies with pH, temperature, with concentration of Fe (II) and other metallic ions (Bryant *et al.*, 1993; Cheung, 1995) and, as recently demonstrated, with the presence of oxygen (Fournier *et al.*, 2004).

In *D. vulgaris* Hildenborough, the periplasmic hydrogenases are proposed to play an active role in hydrogen cycling within the cell according to the following scheme: externally produced gaseous hydrogen (H_2) diffuses to the periplasm where it is oxidized by any of four identified hydrogenases. The released electrons are captured by the *c*-type cytochrome network and are subsequently channelled through the cytoplasmic membranes, through one of several transmembrane protein conduits. Until recently, very little was known about membrane-associated proteins of SRB which could be involved in sulphate respiration. Enormous progress has been made thanks to availability of genomic information. Genome analysis of *D. vulgaris* Hildenborough, *D. desulfuricans* G20 (www.jgi.doe.gov) and *Desulfotalea psycrophila* (Rabus *et al.*, 2004) revealed the existence of two strictly conserved, i.e. essential for sulphate reduction, transmembrane redox complexes, Qmo and Dsr (Pires *et al.*, 2006). Electrons channelled through such conduits are finally used for the reduction of the terminal electron acceptors, sulphate or thiosulphate, and the production of H_2S.

Interestingly, the genome analysis of *D. vulgaris* Hildenborough revealed the presence of genes coding for, in addition to the four periplasmic hydrogenases, two membrane-bound hydrogenases with their active sites facing the cytoplasm. These hydrogenases are proposed to react with hydrogen ions generated from lactate or pyruvate oxidation to produce molecular hydrogen. The latter would diffuse to the periplasm to undergo oxidation by periplasmic hydrogenases as described above.

In 1934, von Wolzogen Kühr and van der Vlugt offered what is now referred to as the classical mechanism of anaerobic corrosion of ferrous metals in the presence of SRB, known as the "*cathodic depolarization model*" (von Wolzogen Kuhr and van der Vlugt, 1934). The essential step in this model involved the removal of hydrogen from the cathode surface (leading to cathodic depolarization) by the bacterial periplasmic hydrogenase enzyme system. The coupled reduction of sulphate to sulphide and the generation of hydroxide ions would cause the formation of iron sulphide and hydroxide products. The catalytic effect would force more iron to be dissolved at the anode.

Many studies demonstrated that SRB species with detectable hydrogenase activity were able to depolarize mild steel, i.e. an increase in the electrode potential was recorded. In contrast, depolarization did not occur in cultures of SRB where hydrogenase activity was not detected. However, other investigations focusing on the role of hydrogenase in MIC produced inconclusive or contradictory results (Beech, 2002 for review). Recently, it has been reported that hydrogenase adsorbed on the surface can influence corrosion of stainless steel through a direct electron transfer (DET) between the enzyme and the metal (Da Silva *et al.*, 2004).

Hydrogenase activity of *D. vulgaris* is regulated by the availability of dissolved Fe^{2+}, i.e. is subject to repression/derepression (Bryant *et al.*, 1993). Moreover, the Fe concentration thresholds for activating and switching off enzyme activity are SRB species-specific (Cheung, 1995). Since bacterial uptake of cathodically generated H_2, and thus the rate of the anodic reaction, is governed by the activity of hydrogenase, the rate of corrosion is expected to vary both with the ecology of SRB species within biofilms and with the local concentration of Fe^{2+} ions. Interestingly, the activity of periplasmic [Fe] hydrogenase has been shown to increase following the exposure of *D. vulgaris* Hildenborough to oxygen (Fournier et al., 2004). Proteome and transcript analysis confirmed upregulation, i.e. increases in both abundance and specific activity of this enzyme, along with other periplasmic proteins, including *c*-type cytochromes, in response to oxidative stress (Fournier *et al.*, 2006). It is tempting to speculate that such upregulation of [Fe] hydrogenase activity could have a direct effect on the sulphate reduction pathway, leading, indirectly, to increased H_2S production. Indeed, Cheung (1995) demonstrated that challenging planktonic cultures of *Desulfovibrio indonesiensis* with oxygen caused an increase in H_2S production. This enhanced H_2S release was counterintuitive, as one would expect bacterial metabolic activity to slow down as a response to stress. However, in view of the information now available from the genomic and proteomic analysis, it seems reasonable to propose that, in addition to protecting cells by removing oxygen from their vicinity, elevated H_2S levels are likely to alter the chemistry of corrosion products, e.g. rendering these products less protective. Implications of increased hydrogenase activity in the presence of oxygen, and/or varying concentrations of Fe (II) ions for the biocorrosion process has not yet been considered and/or studied experimentally.

Further expansion of SRB corrosion models involving hydrogenases has been achieved when novel corrosive types of marine SRB were isolated under conditions where metallic iron was the only electron donor (Dinh *et al.*, 2004). The high rate of sulphate reduction by these novel SRB strains could not be explained solely by consumption of chemically formed hydrogen. It was, therefore, proposed that more efficient use of metallic iron for sulphate reduction occurs by electron uptake from the metal through a cell-surface associated redox-active component. In earlier studies, Kloeke *et al.* (1995) suggested the involvement of an outer membrane cytochrome in iron corrosion according to the electron transfer scheme:

$$\begin{aligned}&\text{Fe} \rightarrow \text{cytochrome} \rightarrow \text{[Fe]hydrogenase} \rightarrow \text{H}^+/\text{H}^2\\&\quad\rightarrow \text{[Ni–Fe]hydrogenase} \rightarrow \text{electron transport system}\\&\quad\rightarrow \text{sulphate reduction enzymes.}\end{aligned} \tag{16.6}$$

The electron flow from the base metal to the sulphate reduction enzymes would involve two types of periplasmic hydrogenases with H_2 as a direct intermediate. However, based on their own findings, Dinh *et al.* (2004) suggested that H_2 is formed as a result of an imbalance between electron donation by fresh iron and electron consumption by sulphate reduction. They proposed a branched scheme of electron flow:

$$\begin{array}{c}\text{Fe} \rightarrow \text{electron transport system} \rightarrow \text{sulphate reduction enzymes}\\ \uparrow\downarrow\\ \text{hydrogenase} \leftrightarrow \text{H}_2\end{array} \tag{16.7}$$

Both schemes (16.6) and (16.7) are plausible; however, evidence for the existence of outer membrane-bound cytochrome is absent. Notwithstanding past efforts and recent progress in microbial genomic research, knowledge of topology and function of redox proteins in SRB remains incomplete, and detailed models of electron flow are lacking. To date, the role of hydrogenase in the SRB-enhanced corrosion process remains a subject of controversy. Clearly, better understanding of SRB physiology is of paramount importance in elucidating their contribution to corrosion reactions.

16.4.2 Iron sulphide as depolarizing compound

SRB carry out dissimilatory reduction of sulphur-containing compounds other than sulphate, such as thiosulphate and elemental sulphur, to sulphide (Widdel, 1988; Lovley and Philips, 1994). An intermediate product in the reduction pathway, sulphite, is reduced through a number of intermediates, to form the sulphide ion. The latter can be converted to hydrogen sulphide (H_2S) when exposed to external hydrogen (H^+) ions, or deposited as insoluble sulphides in the presence of metal ions, such as Fe^{2+}. Iron sulphide is the main corrosion product accumulating on the surface of ferrous materials in anoxic environments with metabolically active SRB. It is worth noting that H_2S production in SRB is connected with the activity of hydrogenase enzyme(s), which regulate availability of (H^+) ions in the cytoplasm.

The first evidence for the involvement of biologically generated sulphides in SRB-mediated corrosion was obtained some 40 years ago (Booth *et al.*, 1968) and prompted extensive investigations into the role of these compounds in biocorrosion. A number of models emerged which postulated that SRB-mediated corrosion of iron-based alloys proceeded under reducing conditions through electrochemical cells established between areas of unreacted metal (anode) and deposits of various biogenically generated reduced ferrous sulphide corrosion products (cathode). Moreover, it has been proposed that SRB present in biofilms on the surface of FeS continually regenerated or depolarized the FeS by the removal of hydrogen as a result of hydrogenase activity. Detailed accounts addressing the role of iron sulphides in biocorrosion are presented in earlier reviews (Hamilton, 2000; Geesey *et al.*, 2000).

16.4.3 The influence of oxygen

The presence of oxygen gradients within a biofilm matrix can be readily demonstrated, and it is accepted that anoxic niches can be found even in biofilms developed in fully oxygenated systems. Indeed, SRB are commonly isolated from natural biofilms found in aerated environments, e.g. tidal zones at low water level on steel piling structures in marine environments (Gubner and Beech, 1999; Beech *et al.*, 2001a). Although all SRB species known to date have been described as strictly anaerobic, it has been demonstrated that many SRB are able to survive for long periods in the presence of oxygen, and the existence of defence mechanisms against oxygen radicals has been confirmed, based on experimental results and genomic analysis (Cypionka, 2000 for review; see also Chapter 5, this volume). The genome of *D. vulgaris* contains pathways for the reduction of terminal electron acceptors other than sulphate, i.e. oxygen, nitrate and metal ions (Heidelberg *et al.*, 2004). *D. vulgaris* possesses genes indicating that this bacterium should have the potential to respire oxygen even under fully aerobic conditions, yet, sustainable growth of *D. vulgaris* in the presence of oxygen has yet to be demonstrated. In conclusion, the possible use of oxygen as a terminal electron acceptor by *Desulfovibrio* species has been a topic of intense debate; however, the issue remains unresolved.

Interestingly, the abundance and metabolic activity of SRB in the oxic zone of numerous biotopes are frequently determined to be higher than those in neighbouring anoxic zones. For example, measurements of bacterial sulphate reduction in hypersaline bacterial mats have revealed that sulphate reduction occurs within the oxygenated, i.e. up to 1000 uM of O_2, photosynthetic zone of the mats (Risatti *et al.*, 1994). A well-defined SRB distribution has been demonstrated within the photooxic zones of cyanobacterial mats, with *Desulfobacter* and *Desulfobacterium* restricted to the deepest levels in the mats and *Desulfococcus* and *Desulfovibrio* present in the upper photooxic zone. As already stated, it has been demonstrated that SRB species vary in their response to the presence of oxygen. For example, oxygen exposure results in an increase of *D. vulgaris* [Fe]hydrogenase activity and enzymes such as superoxide dismutases and catalases, which eliminate superoxide and hydrogen peroxide in aerobic microorganisms, are also expressed in some *Desulfovibrio* species (Fournier *et al.*, 2006 and references therein). Comparative genomic analysis of Gram-negative *D. vulgaris* and Gram-positive *Desulfitobacterium hafniense* strains revealed that while no catalase-encoding genes are present in *D. vulgaris*, the genomes of *D. hafniense* DCB2, sequenced by the US Department of Energy's Joint Genome Institute (JGI) and annotated by ORNL (www.jgi.doe.gov), and *D. hafniense* Y51 (Nonaka *et al.*, 2006) encodes three catalases, of which at least one is secretable, i.e. extracellular. These catalases are HPI (CatA_ECOLI), HPII (CatE_ECOLI) and Mn-containing catalase PMID: 8939876, gi|1752756 (Beech and Galperin, unpublished).

The involvement of oxidoreductases in the biocorrosion process has been a subject of intense research. Lai and Bergel (2000) detected catalytic activities of superoxide dismutase, catalase and peroxidase in marine biofilms formed on stainless steel under field conditions. All three enzymes were proposed to participate in the catalysis of oxygen reduction, i.e. promoting the cathodic corrosion reaction. Whether extracellular catalases, known to be encoded in some SRB species, can be expressed in biofilms and whether these catalases can contribute to oxygen-reduction reactions at the cathode, thus influencing the corrosion process, is unknown.

Laboratory observations and field studies have confirmed that the presence of dissolved oxygen plays an important part in SRB-influenced corrosion of ferrous metals (Lee *et al.*, 1993a; 1993b), and the unified electron transfer hypothesis of biocorrosion (Hamilton, 2003) accommodates these findings. The question whether oxygen is not only frequently involved in, but also required for, aggressive corrosion was posed in recent studies of carbon steel exposed to natural seawater containing SRB (Lee *et al.*, 2004). Although SRB were readily isolated from both oxygenated and anaerobic systems, sulphur-containing corrosion products were absent in the aerobic system, indicating the lack of any SRB metabolic activity. In a subsequent study, the same authors reported that upon cyclic introduction of oxygen to a carbon steel coupon, previously maintained under strictly anaerobic conditions, the corrosion rates increased significantly (Lee *et al.*, 2005). The magnitude of the instantaneous corrosion rate increase did not correspond directly with the amount of oxygen introduced at any one time. However, the lower the concentration of dissolved oxygen in the water prior to oxygen introduction, the larger was the increase in the corrosion rate. The corrosion mechanism of mild steel under alternating oxic and anoxic conditions has been described in recent reviews (Hamilton, 2000; 2003).

Analysis of cathodic depolarization models, based on the role that sulphides play, led to the complementary hypothesis that SRB promote corrosion by a mechanism of anodic depolarization, essentially dependent upon sulphide production (Crolet, 1993). Where the local sulphide concentration was low, the product formed would, in the presence of a high concentration of solvated iron, most likely be non-protective mackinawite. Where sulphide was in excess, the product would be the more protective pyrite. Although pitting corrosion of steel can theoretically be explained by the mechanism of anodic depolarization, the processes under which locally acidic conditions are maintained needs to be confirmed.

16.4.4 The role of extracellular polymeric substances

Bacterial colonization of a surface is facilitated by the production of EPS, which form the biofilm matrix. Although in the past EPS has frequently been referred to as glycocalyx or exopolysaccharide (Costerton *et al.*, 1992), it is now widely accepted that, in addition to polysaccharides, EPS constitute macromolecules, such as proteins, nucleic acids and lipids. Comprehensive reviews on bacterial exopolymers are given by Allison (1998) and Wingender *et al.* (1999). Several mechanisms have been proposed by which this macromolecular mixture can influence deterioration of metallic substrata, including:

- accumulation/entrapment of aggressive microbial metabolic products
- electronic and ionic conductivity of the EPS matrix
- the binding/sorption of metal ions.

Sequestering of metal by the biofilm matrix is a recognized phenomenon and has recently been shown to contribute to bacterial tolerance of high metal concentrations (Harrison *et al.*, 2005). Different metal ions or/and metal ions of the same element but in dissimilar oxidation states, immobilized within the biofilm matrix could, depending on their spatial distribution, participate in the electron transfer processes that drive corrosion reactions. Hence, metal binding/sorption by EPS of bacterial species thought to be involved in biocorrosion has been studied with keen interest. It has been demonstrated that EPS metal-binding/sorption capacity is both bacteria and metal species-specific (Ford *et al.*, 1990; Geesey *et al.*, 1988; Beech *et al.*, 1999). Current models of metal binding by EPS emphazise the role of polysaccharides and proteins, which possess anionic properties. EPS contain ionic groups contributing to both net negative and positive charges on the polymers at near neutral pH values. Polysaccharides owe their negative charge either to carboxyl groups of uronic acids or to non-carbohydrate substituents (Sutherland, 2001). Proteins rich in amino acids containing carboxyl groups can also contribute to the anionic properties of EPS (Dignac *et al.*, 1998). Nucleic acids are polyanionic due to the phosphate residues in the nucleotide moiety of the polymer molecule. IR studies have demonstrated that nucleic acid in EPS of *Bacillus subtilis* and *Pseudomonas aeruginosa* form monodentate complexes with Fe centres on goethite (Omoike et al., 2004). Negatively charged components of EPS are all likely to be involved in electrostatic interactions with multivalent cations (e.g. Ca^{2+}, Cu^{2+}, Mg^{2+}, Fe^{3+}).

The ability of SRB to produce EPS in batch cultures was first documented by Ochynski and Postgate (1963). Subsequent investigations demonstrated that both freely suspended and surface-associated SRB synthesise exopolymers containing a wide range of polysaccharides, including uronic acids, proteins and nucleic acids and that these EPS would vary in

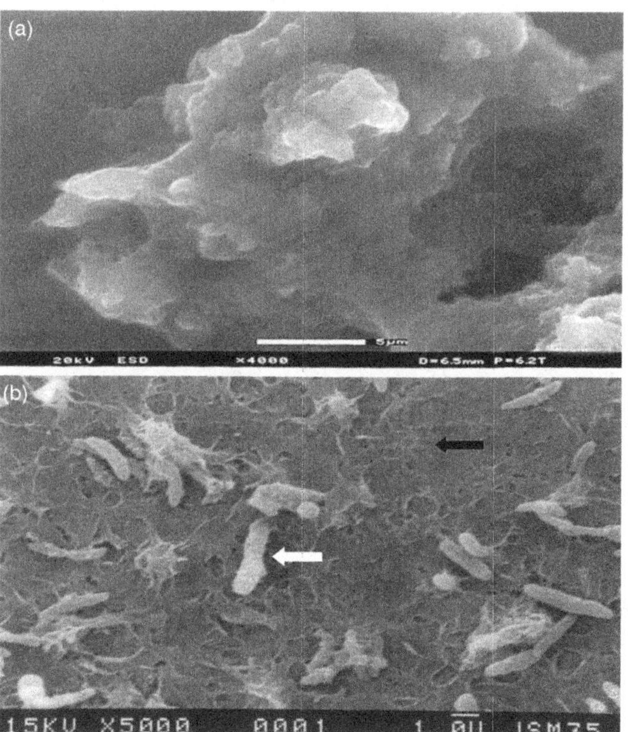

Figure 16.2. Environmental scanning electron micrograph (a) and scanning electron micrograph (b) of *D. indonesiensis* 3-week-old biofilm on the surface of carbon steel revealing (a) fully hydrated EPS matrix with bacterial cells embedded within the matrix and (b) dehydrated biofilm with EPS matrix collapsed (black arrow) and cells exposed (white arrow). (Modified from Beech and Coutinho, 2003.)

their chemical composition and metal binding abilities, depending on SRB species and growth conditions (Beech and Coutinho, 2003 for review). Electron and atomic force microscopy allowed visualization of EPS in SRB biofilms (Figure 16.2) (Beech *et al.*, 1996), and the direct involvement of a high molecular weight, thermostable protein–carbohydrate complex present in EPS of a marine SRB *Desulfovibrio indonesiensis*, in pitting corrosion of mild steel has been demonstrated (Beech *et al.*, 1998). Recently, a model has been proposed of the cathodic depolarization reaction involving EPS and cycling of (Fe^{3+}) and (Fe^{2+}) ions sequestered within the exocellular matrix (Beech and Sunner, 2004), as depicted in Figure 16.3.

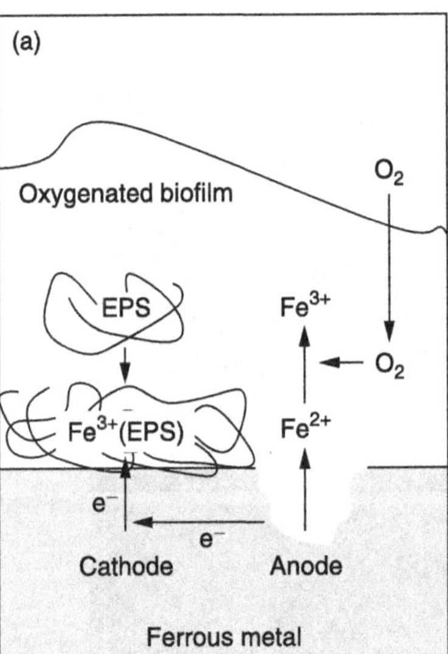

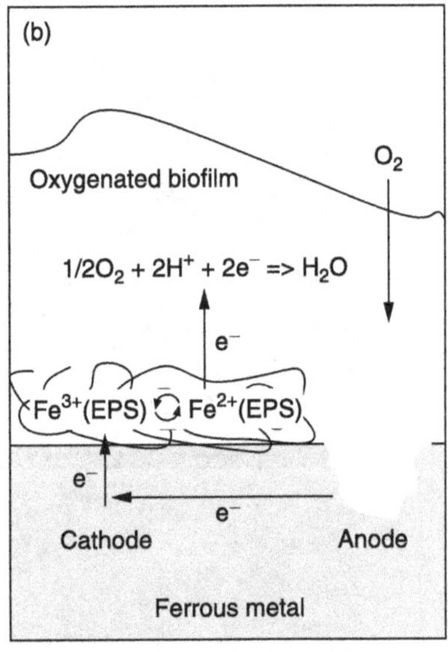

Figure 16.3. A proposed model of the involvement of SRB exopolymer in the corrosion of ferrous metal, from the perspective of the unified electron transfer hypothesis (Hamilton, 2003). (a) In the presence of oxygen, ferrous ions released from the base metal as a result of the anodic reaction are oxidized to ferric ions and are sequestered within the EPS matrix (Beech *et al.*, 1998). (b) EPS-bound ferric ions (Fe^{3+}) are reduced to ferrous ions (Fe^{2+}) through direct electron transfer from the base metal. Individual iron atoms are likely to experience widely varying coordinations while interacting with different macromolecular and inorganic components of EPS, thus display a range of redox potentials. Oxidation of the ferrous ions results in electron transfer to the ultimate electron acceptor, oxygen, through a series of redox reactions. These simultaneously occurring processes shown in (a) and (b) represent a cycling of the iron oxidation states (b), which can contribute to the biocorrosion process. (Modified from Beech and Sunner, 2004.)

A very interesting study reported that bacterial exopolymers could act as a template for the assembly of iron hydroxide pseudo-single crystals (Chan *et al.*, 2003). The observed mineralization resulted from the contact between the EPS and oxidized iron (Fe^{3+}), binding with carboxylic groups on the exopolymer. Electron and atomic force microscopy images of SRB cells coupled with elemental analysis have demonstrated close relationship between EPS and inorganic corrosion products (Beech, 1990; Beech and Sunner, 2004). Furthermore, mass spectrometric analysis revealed the presence of inorganic compounds within highly purified EPS matrix of *Desulfovibrio alaskensis* (Beech *et al.*, 2005). Whether nanoscale proximity of inorganic and organic components of SRB exopolymers could promote redox reactions essential to corrosion has to be determined. Advances in modern surface science techniques, in particular biochemical mass spectrometry and its applicability to biocorrosion research, may provide needed answers (Beech *et al.*, 2005 for review).

16.4.5 Corrosive phosphorus compounds

It has been suggested that not only iron sulphides, but also a highly active phosphorus compound produced by SRB, could cause anaerobic corrosion of ferrous alloys (Iverson and Olson, 1983). Although subsequent studies have excluded inorganic phosphate as the source of phosphorus metabolized by SRB into the corrosive phosphorus product (Iverson, 1998), it has been found that corrosive activity of hydrogenase can be augmented by availability of inorganic phosphate (Bryant and Laishley, 1993). The activity of both alkaline and acid phosphatases which use organic phosphate as a substrate to convert it into inorganic form has been demonstrated in cell extracts and in exopolymers of SRB *D. alaskensis* and *D. indonesiensis* (Beech *et al.*, 2001b), thus suggesting that, in the presence of suitable organic phosphorous sources, inorganic phosphate could accumulate. To date, however, the deleterious effect of SRB-produced exogenous phosphorous compounds on corrosion remains ambiguous.

16.5 CONCLUSIONS

It is unequivocally accepted that SRB-influenced deterioration of iron and ferrous alloys does not involve any fundamentally new mechanisms of corrosion. Instead, the observed metal deterioration in SRB-active environments is due to microbiologically influenced changes in surface conditions of the metallic substratum that promote the establishment and/or maintenance of cathodic and/or anodic reactions not normally favoured under otherwise similar conditions in the absence of these bacteria.

A hypothesis coupling electrochemical biotic and abiotic electron transfer processes from the base metal to oxygen as ultimate electron acceptor has been presented to rationalize present understanding of SRB-influenced corrosion mechanisms and to create a framework for future studies (Hamilton, 2003). The key features of this hypothesis, which emphasizes the importance of corrosion kinetics, are summarized below.

Phylogenetically and physiologically diverse anaerobic SRB, associated with mixed species microbial consortia thriving as biofilms on surfaces of metallic substrata, influence corrosion processes mainly due to the generation of sulphide ions. The extent of production of sulphide varies with SRB types and is, most likely, linked to the activity of hydrogenase enzyme(s) which are SRB species-specific. The enzymatically catalyzed reactions, which are influenced by bacterial ecology and physiological and environmental factors, facilitate oxidation of molecular hydrogen evolved at the cathode of the electrochemical corrosion cell, leading to cathodic depolarization. Metal ions, which are produced at the electrochemical anode, can combine with the biogenically generated sulphide to form metal sulphide corrosion products. These will precipitate within the anoxic zones of the biofilm, causing further stimulation of either the cathodic reaction by serving as an electron sink and/or by influencing the anodic reaction, i.e. further metal dissolution, due to local acidification at the anode. Depending on their physicochemical characteristics, the accumulating iron sulphides can be either protective or corrosive. A number of factors, e.g. traces of oxygen, levels of unbound iron ions, the concentration of soluble sulphide and the presence of other microorganisms, determine both the metabolic activity of SRB and the type of formed metal sulphide species. In the presence of SRB-harbouring biofilms, the highest corrosion rates are recorded under fluctuating aerobic–anaerobic (O_2/AnO_2) conditions. Under such conditions, sulphide ions and ferrous ions are prone to biotic/abiotic oxidation with oxygen acting as ultimate electron acceptor, which leads to the development of corrosion products such as ferric oxides/hydroxides and elemental sulphur.

Although considerable research efforts have been undertaken to elucidate the role that sulphate-reducers play in corrosion, there are still a number of ambiguities related to the contribution of biologically catalyzed reactions. These include the contribution of enzymatic activities, especially of hydrogenase but also extracellular oxidoreductases and the significance of extracellular polymeric substances to the overall corrosion reaction, as well as recent controversy querying the role of oxygen as the ultimate electron acceptor (Lee *et al.*, 2004). Undoubtedly, quantification of relationships between abiotic electrochemical corrosion reactions and spatial and temporal processes associated with metabolic activities of SRB, especially in the presence of oxygen, would greatly aid in resolving such uncertainties. Rapid development in molecular ecology of SRB coupled with growing expansion in the field of their genomics and proteomics, offers exciting perspectives of in-depth understanding of the way in which individual SRB strains or multispecies consortia can effect corrosion of ferrous metals.

REFERENCES

Allison, D. G. (1998). Exopolysaccharide production in bacterial biofilms. *Biofilm J.* 3, paper 2 (BF98002), Online Journals http://www.bdt.org.br/bioline/bf

Barton, L. L. (1985). *Sulphate-reducing bacteria.* New York: Plenum Press.

Beech, I. B. (1990). *Biofilm formation on metal surfaces.* PhD thesis, City of London Polytechnic, Council for National Academic Awards, UK.

Beech, I. B., Cheung, C. W. S., Johnson, D. B. and Smith, J. R. (1996). Comparative studies of bacterial biofilms on steel surfaces using techniques of atomic microscopy and environmental scanning electron microscopy. *Biofouling*, **10**, 65–77.

Beech, I. B., Zinkevich, V., Tapper, R. and Gubner, R. (1998). The direct involvement of extracellular compounds from a marine sulphate-reducing bacterium in deterioration of steel. *Geomicrobiol. J.*, **15**, 119–132.

Beech, I. B., Zinkevich, V., Tapper, R. and Avci, R. (1999). Study of the interaction of exopolymers produced by sulphate-reducing bacteria with iron using X-ray photoelectron spectroscopy and time-of-flight secondary ionisation mass spectrometry. *J. Microbiol. Meth.*, **36**, 3–10.

Beech, I. B., Campbell, S. A. and Walsh, F. C. (2001a). Marine microbial corrosion. In J. G. Stoecker II (ed.), *A Practical Manual on Microbially-Influenced Corrosion* volume II, Houston, Texas: NACE. pp. 11.3–11.14.

Beech, I. B., Paiva, M., Caus, M. and Coutinho, C. (2001b). Enzymatic activity and within biofilms of sulphate-reducing bacteria. In P. G. Gilbert, D. Allison, M. Brading, J. Verran and J. Walker (eds.), *Biofilm Community Interactions: chance or necessity* ? BioLine, Cardiff, UK, pp. 231–9.

Beech, I. B. (2002). Biocorrosion: role of sulphate-reducing bacteria. In G. Bitton (ed.), *Encyclopaedia of Environmental Microbiology.* John Wiley, New York, pp. 465–75.

Beech, I. B. and Coutinho, C. L. M. (2003). Biofilms on corroding materials. In P. Lens, A. P. Moran, T. Mahony, P. Stoodly and V. O'Flaherty (eds.), *Biofilms in medicine, industry and environmental biotechnology – characteristics, analysis and control.* London: IWA Publishing. pp. 115–31.

Beech, I. B. and Sunner, J. A. (2004). Biocorrosion: towards understanding interactions between biofilms and metals. *Current Opinions in Biotechnology,* **15**, 181–6.

Beech, I. B., Sunner, J. A. and Hiraoka, K. (2005). Microbe–surface interactions in biofouling and biocorrosion processes. *International Microbiology,* **8**, 157–68.

Booth, G. H., Elford, L. and Wakerley, D. S. (1968). Corrosion of mild steel by sulphate-reducing bacteria, an alternative mechanism. *Br. Corros. J.*, **3**, 242–5.

Breakell, J. E., Siegwart, M., Foster, K. *et al.* (2005). *Management of accelerated low water corrosion in steel maritime structures.* CIRIA, London, UK: Alden Press.

Bryant, R. D., Kloeke, F. V. O. and Laishley, E. J. (1993). Regulation of the periplasmic Fe hydrogenase by ferrous iron in *Desulfovibrio vulgaris* Hildenborough. *Appl. Environ. Microbiol.*, **59**, 491–5.

Bryant, R. and Laishley, E. (1993). The effect of inorganic phosphate and hydrogenase on the corrosion of mild steel. *Appl. Microbiol. Biotechnol.*, **38**, 824–7.

Chan, C. S., de Stasio, G., Welch, S. A. *et al.* (2003). Microbial polysaccharides template assembly of nanocrystal fibers. *Science,* **303**, 1656–8.

Characklis, W. G. and Marshall, K. C. (1990). *Biofilms.* John Wiley & Sons, Inc, New York.

Cheung, C. W. S. (1995). Biofilms of marine sulphate-reducing bacteria on mild steel. Unpublished PhD thesis, University of Portsmouth, UK.

Costerton, J. W., Lappin-Scott, H. M. and Cheng, K.-J. (1992). Glycocalyx bacterial. In J. Lederberg (ed.), *Encyclopedia of Microbiology,* vol. 2. San Diego: Academic Press. pp. 311–17.

Crolet, J.-L. (1993). Mechanism of uniform corrosion under corrosion deposits. *J. Mat. Sci.*, **28**, 2589–606.

Cypionka, H. (2000). Oxygen respiration by *Desulfovibrio* species. *Annu. Rev. Microbiol.*, **54**, 827–48.

Dar, S. A., Kuenen, J. G. and Muyzer, G. (2005). Nested PCR denaturing gradient gel electrophoresis approach to determine the diversity of sulphate-reducing bacteria in complex microbial communities. *Appl. Environ. Microbiol.*, **71**, 2325–30.

Da Silva, S., Basséguy, R. and Bergel, A. (2004). Electron transfer between hydrogenase and 316L stainless steel: identification of hydrogenase-catalyzed cathodic reaction in anaerobic MIC. *Journal of Electroanal. Chem.*, **561**, 93–102.

Dignac, M.-F., Urbain, V., Rybacki, D. *et al.* (1998). Chemical description of extracellular polymers: implication on activated sludge structure. *Water Sci. Technol.*, **38**, 45–53.

Dinh, H. T., Kuever, J., Mußmann, M. *et al.* (2004). Iron corrosion by novel anaerobic microorganisms. *Nature*, **427**, 829–32.

Ford, T., Black, J. P. and Mitchell, R. (1990). Relationship between bacterial exopolymers and corroding metal surfaces. In *Proceedings of the NACE Corrosion '90*, Paper No. 110. Houston, TX: NACE.

Fournier, M., Dermoun, Z., Durand, M.-C. and Dolla, A. (2004). A new function of the *Desulfovibrio vulgaris* Hildenborough Fe hydrogenase in the protection against oxidative stress. *J. Biol. Chem.*, **279**, 1787–93.

Fournier, M., Aubert, C., Dermoun, Z. *et al.* (2006). Response of the anaerobe *Desulfovibrio vulgaris* Hildenborough to oxidative conditions: proteome and transcript analysis. *Biochimie*, **88**, 85–94.

Garrett, J. H. (1891). *The Action of Water on Lead.* London: H. K. Lewis. pp. 23–9.

Geesey, G. G., Beech, I. B., Bremmer, P. J., Webster, B. J. and Wells, D. (2000). Biocorrosion. In J. Bryers (ed.), *Biofilms II: Process Analysis and Applications.* Wiley-Liss Inc, New York, pp. 281–326.

Geesey, G. G., Jang, L., Jolley, J. G. *et al.* (1988). Binding of metal ions by extracellular polymers of biofilm bacteria. *Water Sci. Technol.*, **20**, 161–5.

Gubner, R. and Beech, I. B. (1999). *Statistical assessment of the risk of biocorrosion in tidal waters.* Corrosion 99, Paper 184. Houston, TX: NACE.

Hamilton, W. A. (1998). Sulphate-reducing bacteria: physiology determines their environmental impact. *Geomicrobiol. J.*, **15**, 19–28.

Hamilton, W. A. (2000). Microbially induced corrosion in the context of metal microbe interactions. In L. V. Evans (ed.), *Biofilms: recent advances in their study and control.* Singapore: Harwood Academic Publishers. pp. 419–34.

Hamilton, W. A. (2003). Microbially influenced corrosion as a model system for the study of metal microbe interactions: a unifying electron transfer hypothesis. *Biofouling*, **19**, 65–76.

Harrison, J. J., Turner, R. and Ceri, H. (2005). Persister cells, the biofilm matrix and tolerance to metal cations in biofilm and planktonic *Pseudomonas aeruginosa. Environmental Microbiology*, **7**, 981–94.

Heidelberg, J. F., Seshadri, R., Haveman, S. A. *et al.* (2004). The genome sequence of the anaerobic sulphate-reducing bacterium *Desulfovibrio vulgaris* Hildenborough. *Nature Biotechnology*, **20**, 554–9.

Iverson, W. P. (1998). Possible source of a phosphorus compound produced by sulphate reducing bacteria that cause anaerobic corrosion of iron. *Mat. Perform.*, **37**, 46–9.

Iverson W. P. and Olson, G. J. (1983). Anaerobic corrosion by sulphate-reducing bacteria due to highly reactive volatile phosphorus compound. In *Microbial Corrosion, Metals Society*, London, 46–53.

Jan-Roblero, J., Romero, J. M. and Amaya, M. (2004). Phylogenetic characterization of a corrosive consortium isolated from a sour gas pipeline. *Appl. Microbiol. Biotechnol.*, **64**, 862–7.

King, R. A. and Miller, J. D. A. (1971). Corrosion by sulphate-reducing bacteria. *Nature*, **233**, 491–3.

Kloeke F. V., Bryant R. D. and Laishley, E. J. (1995). Localization of cytochromes in the outer membrane of *Desulfovibrio vulgaris* (Hildenborough) and their role in anaerobic biocorrosion. *Anaerobe*, **1**, 351–8.

Lai, M. E. and Bergel, A. (2000). Electrochemical reduction of oxygen on glassy carbon: catalysis by catalase. *J. Electroanal. Chem.*, **494**, 30–40.

Lee, J. S., Ray, R. I., Lemieux, E. J., Falster, A. U. and Little, B. J. (2004). An evaluation of carbon steel corrosion under stagnant seawater conditions. *Biofouling*, **20**, 237–47.

Lee, J. S., Ray, R. I., Little, B. J. and Lemieux, E. J. (2005). Evaluation of deoxygenation as a corrosion control measure for ballast tanks. *Corrosion*, **61**, 1173–88.

Lee, W.-C. and deBeer, D. (1995). Oxygen and pH microprofiles above corroding mild steel covered with a biofilm. *Biofouling*, **8**, 273–80.

Lewandowski, Z., Dickinson, W. H. and Lee, W. C. (1997). Electrochemical interactions of biofilms with metal surfaces. *Water Science and Technology*, **36**, 295–302.

Lee, W.-C., Lewandowski, Z., Morrison, M., Characklis, W. G., Avci, R. and Nielsen, P. H. (1993b). Corrosion of mild steel underneath aerobic biofilms containing sulphate-reducing bacteria – Part II. At high dissolved oxygen concentrations. *Biofouling*, **7**, 217–39.

Lee, W.-C., Lewandowski, W. Z., Nielsen, P. H. and Hamilton, W. A. (1995). Role of sulphate-reducing bacteria in corrosion of mild steel: a review. *Biofouling*, **8**, 165–94.

Lee, W.-C., Lewandowski, Z., Okabe, S., Characklis, W. G. and Avci, R. (1993a). Corrosion of mild steel underneath aerobic biofilms containing sulphate-reducing bacteria – Part I. At low dissolved oxygen concentrations. *Biofouling*, **7**, 197–216.

Little, B. and Ray, R. (2002). A perspective on corrosion inhibition by biofilms. *Corrosion*, **58**, 424–8.

Little, B. and Wagner, P. (1997). Myths related to microbiologically influenced corrosion. *Mater. Performance*, **36**, 40–4.

Lovley, D. R. and Philips, E. J. P. (1994). Novel processes for anaerobic sulfate production from elemental sulfur by sulfate-reducing bacteria. *Appl Environ Microbiol*, **60**, 2394–9.

Nielsen, P. H., Lee, W.-C., Lewandowski, Z., Morrison, M. and Characklis, W. G. (1993). Corrosion of mild steel in an alternating oxic and anoxic biofilm system. *Biofouling*, **7**, 267–284.

Nonaka, H., Keresztes, G., Shinoda, Y. *et al.* (2006). Complete genome sequence of the dehalorespiring bacterium *Desulfitobacterium hafniense* Y51 and comparison with *Dehalococcoides ethenogenes* 195. *Journal of Bacteriology*, **188**, 2262–74.

Ochynski, F. W. and Postgate, J. R. (1963). Some biological differences between fresh water and salt water strains of sulphate-reducing bacteria. In C. H. Oppenheimer and C. C. Thomas (eds.), *Marine Microbiology*, Springfield, III. pp. 426–41.

Odom, J. M. and Singleton, R. (1993). *The sulphate-reducing bacteria: contemporary perspectives.* New York: Springer-Verlag.

Omoike, A., Chorover, J., Kwon, K. D. and Kubicki, J. D. (2004). Adhesion of bacterial exopolymers to α-FeOOH. Inner-sphere complexation of phosphodiester groups. *Langmuir*, **20**, 11108–14.

Pires, R. H., Venceslau, S. S., Morais, F. *et al.* (2006). Characterization of the *Desulfovibrio desulfuricans* ATCC 27774 DsrMKJOP Complex. A membrane-bound redox complex involved in the sulphate respiratory pathway. *Biochemistry*, **45**, 249–62.

Pitonzo, B. J., Castro P., Amy, P. S. *et al.* (2004). Microbiologically influenced corrosion capability of bacteria isolated from Yucca Mountain. *Corrosion*, **60**, 64–74.

Postgate, J. R. (1984). *The Sulphate Reducing Bacteria*, 2nd edn. Cambridge, UK: Cambridge University Press.

Rabus, R., Ruepp, A., Frickey, T. *et al.* (2004). The genome of *Desulfotalea psychrophila*, a sulphate-reducing bacterium from permanently cold Arctic sediments. *Environ. Microbiol.*, **6**, 887–902.

Risatti, J. B., Capman, W. C. and Stahl, D. A. (1994). Community structure of a microbial mat: the phylogenetic dimension. *Proc. Natl. Acad. Sci. USA*, **91**, 10173–7.

Scully, J. C. (1990). *Fundamentals of Corrosion*, 3rd edn. Oxford, UK: Pergamon Press.

Sutherland, I. W. (2001). The biofilm matrix – an immobilized but dynamic microbial environment. *Trend. Microbiol.*, **9**, 222–7.

Valencia-Cantero, E., Pena-Cabriales, J. J. and Martinez-Romero, E. (2003). The corrosion effect of sulphate and ferric iron reducing bacterial consortia on steel. *Geomicrobiology Journal*, **20**, 157–69.

Videla, H. A. (1996). *Manual of Biocorrosion*. Boca Raton, FL: Lewis Publishers, CRC Press, Inc.

von Wolzogen Kuhr, C. A. H. and van der Vlugt, L. S. (1934). De grafiteering van Gietijzer als electrobiochemisch Proces in anaerobe Grunden. (Graphitization of cast iron as an electrochemical process in anaerobic soils.) *Water*, **18**, 147–51.

Widdel, F. (1988). Microbiology and ecology of sulphate and sulfur-reducing bacteria. In A. J. B. Zehnder (ed.), *Biology of Anaerobic Microorganisms*. New York: Wiley-Liss, John Wiley and Sons, Inc. pp. 469–586.

Wingender, J., Neu, T. R. and Flemming, H.-C. (1999). What are bacterial extracellular polymeric substances? In J. Wingender, T. R. Neu and H.-C. Flemming (eds.), *Microbial Extracellular Polymeric Substances: Characterization, Structure and Function*. New York: Springer-Verlag. pp. 1–15.

CHAPTER 17

Anaerobic metabolism of nitroaromatic compounds and bioremediation of explosives by sulphate-reducing bacteria

Raj Boopathy

17.1 INTRODUCTION

Many xenobiotic chemicals introduced into the environment for agricultural and industrial use are nitro-substituted aromatics. Nitro groups in the aromatic ring are often implicated as the cause of the persistence and toxicity of such compounds. Nitroaromatic compounds enter soil, water, and food by several routes, such as use of pesticides, plastics, pharmaceuticals, landfill dumping of industrial wastes, and the military use of explosives. The nitroaromatic compound, trinitrotoluene (TNT) is introduced into soil and water ecosystems mainly by military activities such as the manufacture, loading, and disposal of explosives and propellants. This contamination problem may increase in future because of the demilitarization and disposal of unwanted weapons systems.

Biotransformation of TNT and other nitroaromatics by aerobic bacteria in the laboratory has been reported frequently (Boopathy *et al.*, 1994a; 1994b; Dickel and Knackmuss, 1991; Duque *et al.*, 1993; Funk *et al.*, 1993; McCormick *et al.*, 1976; 1981; Nishino and Spain, 1993; Spain and Gibson, 1991; Zeyer and Kearney, 1984). Biodegradation of 2,4-dinitrotoluene by a *Pseudomonas* sp. has been reported to occur via 4-methyl-5-nitrocatechol in a dioxygenase-mediated reaction (Spanggord *et al.*, 1991). Duque *et al.* (1993) successfully constructed a *Pseudomonas* hybrid strain that mineralized TNT. White rot fungus has been shown to mineralize radiolabelled TNT (Fernando *et al.*, 1990). The work of Spiker *et al.* (1992) showed that *Phanerochaete chrysosporium* is not a good candidate for bioremediation of TNT contaminated sites containing high concentration of explosives because of its high sensitivity to contaminants. Michels and Gottschalk (1994) showed that the lignin peroxidase activity of *P. chrysosporium* is inhibited by the TNT intermediate hydroxylamino-dinitrotoluene. Valli *et al.* (1992) found that 2,4-dinitrotoluene is degraded completely by the white rot fungus.

Ecological observations suggest that sulphate-reducing and methanogenic bacteria might metabolize nitroaromatic compounds under anaerobic conditions if appropriate electron donors and electron acceptors are present in the environment, but this ability had not been demonstrated until recently. Under anaerobic conditions, the sulphate-reducing bacterium *Desulfovibrio* sp. (B strain) transformed TNT to toluene (Boopathy and Kulpa, 1992; Boopathy *et al.*, 1993a) by reduction. Gorontzy *et al.* (1993) reported that under anaerobic conditions, methanogenic bacteria reduced nitrophenols and nitrobenzoic acids. Preuss *et al.* (1993) demonstrated conversion of TNT to triaminotoluene by a *Desulfovibrio* sp.

17.1.1 Anaerobic transformation of nitroaromatics

The anaerobic bacterial metabolism of nitroaromatics has not been studied as extensively as of aerobic pathways, perhaps because of the difficulty in working with anaerobic cultures and perhaps the slow growth of anaerobes. Earlier studies on anaerobic metabolism of nitroaromatic compounds by McCormick *et al.* (1976) laid the foundations for such study and established the usefulness of anaerobic organisms. Successful demonstration of degradation of hexahydro-1,2,3-trinitro-1,3,5-triazine (RDX) by sewage sludge (McCormick *et al.*, 1981; Carpenter *et al.*, 1978) under anaerobic conditions further demonstrated the usefulness of anaerobes in waste treatment. RDX was reduced sequentially by the anaerobes to the nitroso derivatives, which were further converted to formaldehyde and methanol. Hallas and Alexander (1983) showed successful transformation of nitrobenzene, nitrobenzoic acid, nitrotoluene, and nitroaniline by sewage sludge under anaerobic conditions.

Methanogens are obligate anaerobes that grow in an environment with an oxidation-reduction potential of less than −300 mV. They transform various substrates to C1 products, such as CH_4 and HCOOH. The role of some novel compounds and the mechanism of single carbon flow in these bacteria remain to be formally proved, along with the arrangement of the electron transport chain. Because of the limited substrate capabilities, the metabolism of more complex molecules to methane depends on the activity of non-methanogens in association with the methanogens. Under pure culture conditions, methanogens have not been reported to degrade aromatic compounds. The studies of Gorontzy *et al.* (1993) on microbial transformation of nitroaromatic compounds by methanogenic bacteria revealed that methanogens can transform nitroaromatic compounds to corresponding amino compounds. Boopathy and Kulpa (1994) isolated a methanogen, *Methanococcus* sp., from a lake sediment which transformed TNT to 2,4-diaminonitrotoluene. This organism also transformed nitrobenzene and nitrophenol. The intermediates observed were amino derivatives of the parent compounds. According to some reports, the reductive transformation of nitroaromatic compounds leads to detoxification of the substance (Boyd *et al.*, 1983; Battersby and Wilson, 1989). The specific enzymes responsible for the reduction process in methanogens is not yet characterized. Angermeier and Simon (1983) suggested that the reduction of aromatic compounds may be catalyzed by hydrogenase and ferredoxin. The observation of sulphate reducers and methanogenic bacteria by many workers (Boopathy *et al.*, 1993a; 1993b; Boopathy and Kulpa, 1994; Boopathy, 1994; Gorontzy *et al.*, 1993; Preuss *et al.*, 1993) suggests that these organisms could be exploited for bioremediation under anaerobic conditions by supplying proper electron donors and electron acceptors.

Boopathy *et al.* (1993b) showed that TNT can be transformed under anaerobic conditions by using different electron acceptors. A soil sample collected from the Joliet Army Ammunition Plant, Joliet, IL was incubated under sulphate-reducing, nitrate-reducing and methanogenic conditions. The results showed that TNT was transformed under all three conditions. However, when no electron acceptor was supplied no TNT was transformed. The intermediates observed during the study were 4-amino-2,6-dinitrotoluene and 2-amino-4,6-dinitrotoluene. This study showed that if the appropriate electron acceptor is present in the system, anaerobic bacteria will reduce TNT to amino compounds.

17.1.2 Sulphate-reducing bacteria

Although oxygen is the most widely used electron acceptor in energy metabolism, a number of different kinds of bacteria are able to reduce other compounds and hence use them as electron acceptors. This process of anaerobic respiration is less energy efficient, but it allows these bacteria to live in environments where oxygen is absent.

Sulphate-reducing bacteria (SRB) are obligate anaerobes that are conveniently considered together because of their shared ability to perform dissimilatory sulphate reduction, a process analogous to aerobic respiration in that the sulphate ion acts as an electron acceptor, like oxygen in the aerobic process. The genera of sulphate reducers are defined on the basis of morphology rather than physiology. All sulphate reducers are gram negative, except *Desulfotomaculum*. The most frequently encountered genus is *Desulfovibrio*.

The use of various non-fermentable aromatic compounds in the absence of oxygen or nitrate is apparently of the natural roles of SRB reducing bacteria. Aromatic compounds with more than two hydroxyl groups are readily degraded by fermenting bacteria (Widdel and Hansen, 1992). Several new types of SRB have been isolated directly with aromatic compounds (Bak and Widdel, 1986; Widdel, 1988; Schnell *et al.*, 1989). Most of these isolates are extremely versatile sulphate reducers that use many aliphatic compounds. Aromatic compounds oxidized by SRB include benzoate, phenol, p-cresol, aniline, and the n-heterocyclic compounds such as nicotinate, indole, and quinoline. All the known degraders of aromatic compounds are complete oxidizers. The sulphate-reducers employ reactions like those detected in denitrifying bacteria, phototrophic bacteria, and methanogenic co-cultures using aromatic compounds (Berry *et al.*, 1987; Evans and Fuchs, 1987; Harwood and Gibson, 1986; Tschech, 1989). The SRB are capable of carrying out the following reactions: activation of benzoate to benzoyl CoA (Geissler *et al.*, 1988; Holland *et al.*, 1987), caroboxylation of phenol to p-hydroxybenzoate (Knoll and Winter, 1989; Tschech and Fuchs, 1989) or the reductive removal of hydroxyl groups (Tschech and Schink, 1986).

17.1.3 Metabolism of TNT and other nitroaromatic compounds by sulphate-reducing bacteria

Boopathy and co-workers (1993a) showed that a sulphate-reducing bacterium, *Desulfovibrio* sp. (B strain), can convert TNT to toluene. This organism, isolated from an anaerobic digester treating furfural-containing wastewater (Boopathy and Daniels, 1991), used nitrate as electron acceptor apart from using sulphate as electron acceptor. It also used nitrate as a nitrogen source. Further experiments showed that this bacterium could use the nitro group present in TNT molecules, either as an electron acceptor, or as a nitrogen source.

Some SRB can use nitrate in addition to sulphate as their terminal electron acceptor (Keith and Herbert, 1983). The reaction is coupled to electron transfer phosphorylation (LeGall and Fauque, 1988; Steenkamp and Peck, 1981) and is catalyzed by a respiratory nitrite reductase that has a molecular mass of 65 KDa and contains six *c*-type haems. This nitrite reductase, known as the hexahaem cytochrome *c*3, is widely distributed in strict and facultative aerobes (Liu and Peck, 1981; 1988). This nitrite reductase is unrelated to the regulated nitrite reductase (non-haem iron sirohaem-containing) found in many plants and bacteria (Vega and Kamin, 1977), where its function is nitrogen assimilation. According to Steenkamp and Peck (1981), nitrite reductase is closely associated with a hydrogenase and is probably a transmembrane protein. This conclusion is based on the presence of proton-releasing and nitrite-binding sites on the periplasmic aspect of the cytoplasmic membrane and a benzyl viologen-binding site on the cytoplasmic side of the membrane.

TNT (100 mg/L) was metabolized by *Desulfovibrio* sp. (B strain) within 10 days (Boopathy *et al.*, 1993a), with pyruvate as the main substrate, sulphate as the electron acceptor and TNT as the sole nitrogen source. Boopathy *et al.* (1993a) showed that, under different growth conditions, this bacterium used TNT as its sole source of nitrogen. This result indicates that the isolate has the necessary enzymes to use the nitro groups present in TNT molecules as a nitrogen source.

Apart from pyruvate, lactate served as the best substrate for TNT metabolism, followed by $H_2 + CO_2$, ethanol, and formate. Comparison of the rate of TNT biotransformation by *Desulfovibrio* sp. with that of other SRB showed that this new isolate has a unique metabolic ability to degrade TNT. *Desulfovibrio* sp. transformed 100% of TNT present in a relatively short period of time (7 days). Other *Desulfovibrio* spp. (ATCC cultures) converted 59–72% TNT within 21–23 days, whereas *Desulfobacterium indolicum* transformed 82% of TNT in 36 days of incubation (Boopathy *et al.*, 1993a).

Mass spectral analyses showed that various intermediates were produced depending upon the culture conditions of the isolate. When ammonium was the main nitrogen source, 2,4-diamino-6-nitrotoluene was the major intermediate. When TNT was the sole source of carbon and energy, it was first reduced to 4-amino-2,6-dinitrotoluene and then to 2,4-diamino-6-nitrotoluene. When TNT was the sole source of nitrogen, all the TNT in the medium was converted to 2,4-diamino-6-nitrotoluene within 10 days of incubation and traces of 2- and 4-amino compounds were identified. Later these intermediates were converted to toluene. The quantitative analysis of the aqueous and gas phases of the culture bottle by gas chromatograph showed a good mass balance of TNT to toluene (Boopathy *et al.*, 1993a).

Nitroaromatic compounds are considered resistant to microbial attack (Fewson, 1981; Haigler and Spain, 1993), partly because the reduction of electron density in the aromatic ring by the nitro groups can hinder electrophilic attack by oxygenases and thus prevent aerobic degradation of nitroaromatic compounds (Bruhn *et al.*, 1987). Under anaerobic conditions, the SRB metabolized TNT. Of all the metabolites produced, the formation of toluene from TNT seems to be very novel and significant.

TNT was reduced to diamino-nitrotoluene by the isolate through the 2-amino- and 4-amino-dinitrotoluenes when pyruvate served as the main substrate in the presence of sulphate and ammonia, in a simple reduction process carried out by the enzyme nitrite reductase. The cell free extract showed high activity of nitrite reductase. The nitroreductase activity was monitored photometrically at 325 nm by the consumption of diaminonitrotoluene. Most *Desulfovibrio* spp. have nitrite reductase enzymes that reduce nitrate to ammonia (Widdel, 1988). This isolate reduced the nitro groups present in TNT to amino groups. When TNT served as the sole source of nitrogen, toluene was formed from the TNT. McCormick *et al.* (1976) showed that TNT was reduced by H_2 in the presence of enzyme preparations of *Veillonella alkalescens* to triaminotoluene: 3 mol H_2 is required to reduce each nitro group to the amino group. Preuss *et al.* (1993) observed the formation of triaminotoluene from TNT by a sulphate-reducing bacterium isolated from sewage sludge.

Boopathy *et al.* (1993a) showed the formation of toluene from triaminotoluene and in the process, the isolate used the ammonium released from the original TNT molecule as a nitrogen source for growth. This is achieved by reduction of nitro groups followed by reductive deamination. A significant quantity of toluene concentration was observed in the culture sample (Boopathy *et al.*, 1993a), and virtually no nitrite ions were detected during TNT metabolism. The aromatic ring structure was not cleaved, and no metabolites other than toluene appeared even after six months of incubation. Reductive deamination is catalyzed by a deaminase enzyme in *Pseudomonas* sp. (Naumova *et al.*, 1986). Reductive deamination reactions were postulated first for 2-aminobenzoate degradation by methanogenic enrichment cultures (Tschech and Schink, 1988). Reductive dehydroxylation of gentisate to benzoate and acetate was demonstrated in the fermenting bacterium HQGO1 (Szewyk and Schink, 1989).

Beller *et al.* (1992) and Edwards *et al.* (1992) demonstrated the complete mineralization of toluene under sulphate-reducing conditions. These toluene-degrading sulphate-reducers could be used in combination with the *Desulfovibrio* sp. described by Boopathy *et al.* (1993a) to degrade TNT completely to CO_2.

The *Desulfovibrio* sp. (B strain) (Boopathy *et al.*, 1993a) also metabolized other nitroaromtics such as 2,4-dinitrophenol (2,4-DNP), 2,4-dinitrotoluene (2,4-DNT), 2,6-dinitrotoluene (2,6-DNT), and aniline. As shown by Boopathy and Kulpa (1993) the *Desulfovibrio* sp. used all the nitroaromatics studied as a sole source of nitrogen. It also used 2,4-DNT, 2,6-DNT and 2,4-DNP as electron acceptors in the absence of sulphate. The GC/MS analyses of the culture samples showed the presence of phenol from 2,4-DNP and benzene from aniline as intermediates. Gorontzy *et al.* (1993) showed transformation of nitrophenols and nitrobenzoic acids by the sulphate reducers *Desulfovibrio desulfuricans, D. gigas, Desulfococcus multivorans,* and *Desulfotomaculum orientis.* All of the nitroaromatics were transformed to corresponding amino compounds.

Schnell *et al.* (1989) isolated a new SRB, *Desulfobacter anilini,* which degraded aniline completely to carbon dioxide and ammonia with stoichiometric reduction of sulphate to sulphide. This is the first obligate anaerobic bacterium observed to grow in pure culture with aniline as its sole electron donor and carbon source. The organism oxidizes aniline completely to carbon dioxide and releases the aminonitrogen quantitatively as ammonia. Two metabolic pathways were suggested. Firstly, aniline could be carboxylated to 2-aminobenzoate or 4-aminobenzoate, with the aminobenzoate then reductively deaminated to benzoate and metabolized further (Zeigler *et al.*, 1987). Alternatively, aniline could be deaminated hydrolytically to phenol, which is subsequently degraded either by carboxylation to 4-hydroxybenzoate or by reductive transformation to cyclohexanol or cyclohexanone. Both pathways appear possible, because the bacterial strain used each of these intermediates as a sole source of carbon.

Schnell and Schink (1991) reported that *Desulfobacterium anilini* degraded aniline via reductive deamination of 4-aminobenzoyl CoA. The first step, the carboxylation of aniline to 4-aminobenzoate, is followed by activation of 4-aminobenzoate to 4-aminobenzoyl CoA, which is reductively deaminated to benzoyl CoA. This product enters the normal benzoate pathway leading to three acetyl CoA. Carbon monoxide dehydrogenase and formate dehydrogenase are present in *Desulfobacterium anilini,* indicating that acetyl residues are oxidized via the carbon monoxide dehydrogenase pathway (Schnell *et al.*, 1989).

Schnell and Schink (1992) isolated a sulphate-reducing bacterium that oxidized 3-aminobenzoate to carbon dioxide with concomitant reduction of sulphate to sulphide and release of ammonium. High activity of carbon monoxide dehydrogenase indicated that acetyl CoA is oxidized via the carbon monoxide dehydrogenase pathway, although 2-oxoglutarate synthase activity was found as well. Similar activity was found with pyruvate as substrate. Perhaps both synthase activities can be attributed to an enzyme needed in assimilatory metabolism. Carbon monoxide dehydrogenase and pyruvate synthase are probably also key enzymes during autotrophic growth with hydrogen and sulphate. The complete oxidation of 3-aminobenzoate yields −186 kJ per mole according to the following equation:

$$2\ C_7H_6NO_2^- + 7\ SO_4^2 + 11\ H^+ \rightarrow 14\ CO_2 + 2\ NH_4^+ + 7\ HS^- + 4\ H_2O$$
$$\times\ \Delta G'_o = -180\,kJ$$

The first step in degradation of 3-aminobenzoate by this new sulphate-reducing bacterium was found to be activation to 3-aminobenzoyl CoA (Schnell and Schink, 1992). Further reduction of 3-aminobenzoyl CoA did not yield benzoyl CoA, but rather a product tentatively described as a reduced CoA-ester. The activation of benzoyl CoA depends on the presence of the cofactors, ATP and Mg^{2+}. Acyl-CoA synthetase reactions were identified as the initial step in the degradation of benzoate by anaerobic bacteria.

17.1.4 Bioremediation of TNT under sulphate-reducing conditions

Soil and water in most US Military facilities are contaminated with explosive chemicals, mainly because of the manufacture, loading, and disposal of explosives and propellants. This contamination problem may increase in future because of demilitarization and disposal of unwanted weapon systems. Disposal of obsolete explosives is a problem for the military and the associated industries because of the polluting effect of explosives in the environment (Wyman *et al.*, 1979). TNT is the major contaminant in many US Army Ammunition facilities. TNT represents an environmental hazard because it has toxicological effects on number of organisms (Fernando *et al.*, 1990; Won *et al.*, 1974) and it is mutagenic (Kaplan and Kaplan, 1982). The disposal of large quantities of TNT in an environmentally acceptable manner poses serious difficulties. The present approach to the remediation of TNT contamination is incineration of soil, a very costly and destructive process. Bioremediation would be a safe and cost-effective method for treating TNT contamination. Biological removal of explosives from soil has been demonstrated using aerobic/anoxic soil slurry reactors (Boopathy *et al.*, 1998; Boopathy, 2000; 2001; 2002). In our lab, we isolated a well-defined sulphate-reducing consortium consists of *Desulfovibrio* spp., namely, *D. desulfuricans* strain A, *D. desulfuricans*, strain B, *D. gigas*, and *D. vulgaris* from a creek sediment (Boopathy and Manning, 1996). The ability of this consortium to degrade and remediate TNT was explored.

The consortium was grown in anaerobic serum bottles under various growth conditions including TNT as the sole carbon source, co-metabolic

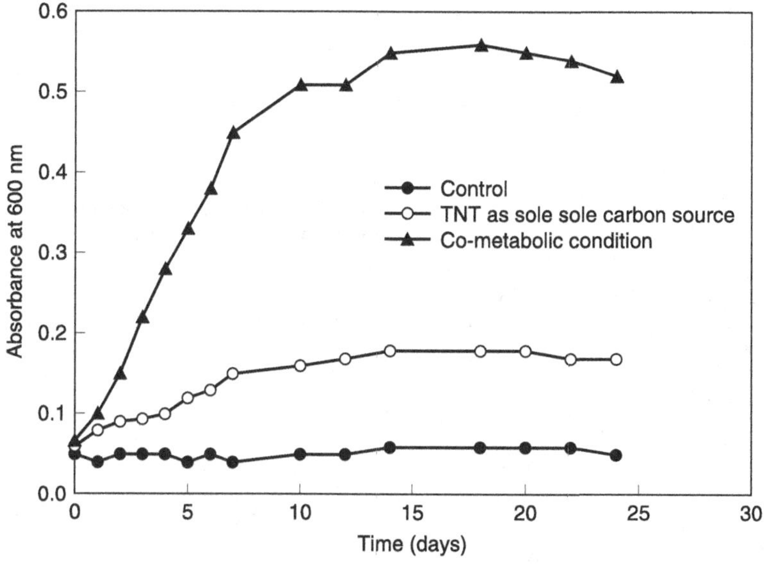

Figure 17.1. Growth of *Desulfovibrio* spp., under various conditions.

condition with pyruvate (30 mM) as co-substrate, and heat inactivated control, as shown in Boopathy and Manning (1996). Figure 17.1 shows the results of bacterial growth. Growth was observed in all conditions except in the killed control. The maximum growth was observed under co-metabolic conditions and bacteria also grew under the conditions where TNT served as the sole carbon source. Figure 17.2 shows the removal of TNT under various culture conditions. In all the cultures, the initial TNT concentration was 100 mg/L. In the killed control, the TNT concentration remained constant throughout the experiment, indicating that no physical or chemical removal of TNT occurred. TNT removal was fastest in the co-metabolic condition, where 100% of TNT was removed within 10 days of incubation. TNT removal in the culture condition where TNT served as the sole carbon source was very slow, but 100% of the TNT was still removed within 25 days. The results show that the consortium can remove TNT faster in the presence of an additional carbon source like pyruvate. This could be due to an increase in the bacterial cell numbers in the pyruvate-containing cultures.

The GC-MS (gas chromatograph-mass spectrometer) analysis of culture samples with and without pyruvate revealed the presence of various intermediates, which were identified by comparison of their GC retention times and their mass spectra with authentic standards. The first intermediates observed were 4-amino-2, 6-dinitrotolune (4-ADNT) and

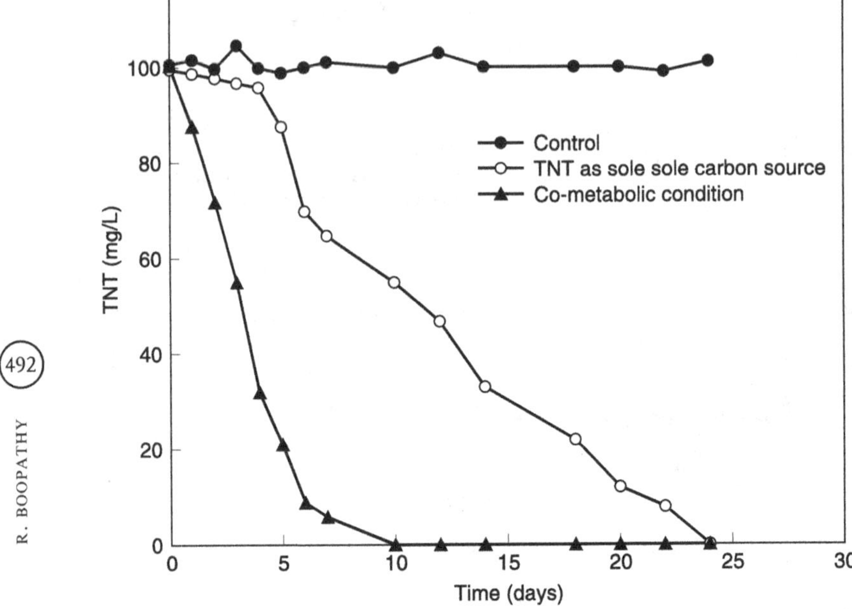

Figure 17.2. Concentration of TNT under various growth conditions.

2-amino-4,6-dinitrotoluene (2-ADNT). The ratio of 4-ADNT and 2-ADNT formed from the TNT metabolism were approximately 80:20. These products were further reduced to 2,4-diamino-6-nitrotoluene (2,4-DANT). Other compounds appearing in the culture medium in order were nitrobenzoic acid (NB), cyclohexanone, 2-methyl pentanoic acid, butyric acie, and acetic acid. All of these compounds were identified in cultures with both TNT and pyruvate as carbon sources as well as in the cultures that received TNT alone as a carbon source. These intermediates were not present in the control.

Radiolabelled study was conducted with uniformly ring labelled [^{14}C] TNT in the culture condition, where TNT served as the sole carbon source. The experimental procedure used by Boopathy and Manning (1996) was used in this study. The results of radiolabelled study are presented in Table 17.1. The data showed the production of various metabolites and biomass at the end of the experiment on day 30. TNT was not mineralized, as there was no production of CO_2. Most of the TNT was converted to acetic acid (49%) and 27% of TNT was assimilated into cell biomass. Apart from acetate, the other major intermediates present in the culture medium were nitrobenzoic acid (6%) and butyric acid (9.5%). In killed control, TNT was reduced to a smaller extent to 4-ADNT (3%), yet nearly 95% of the original TNT was

Table 17.1. *Results of radiolabelled TNT study: mass balance for TNT metabolism by* Desulfovibrio *consortium*

[14C] TNT recovered	Active culture (%)	Killed control (%)
CO_2	0	0
Biomass	27.4	0
Acetic acid	49.5	0
Nitrobenzoic acid	6.2	0
Cyclohexanone	0.01	0
Butyric acid	9.5	0
2-Methyl pentanoic acid	0.2	0
4-ADNT	0.5	3.2
2-ADNT	0.9	0
2,4-DANT	0.7	0
TNT	0	94.5
Unrecovered	5.09	2.3

recovered unaltered. Traces of cyclohexanone were observed, which accounted for 0.01% of the original [^{14}C] TNT. This radiolabelled study showed a reasonable mass balance with a recovery of 95% of [^{14}C] TNT. Since the ring carbons of TNT were uniformly labelled, conversion of TNT to acetic acid and butyric acid clearly denotes ring cleavage.

The production of various intermediates in both culture conditions (with TNT as the sole carbon source and co-metabolic condition with pyruvate) suggested that the bacterial consortium has all the necessary enzymes to degrade TNT. The anaerobic metabolic pathway as shown in Figure 17.3 was proposed for TNT metabolism by SRB. TNT was reduced to 4-ADNT and 2-ADNT, which were further reduced to 2,4-DANT. These reductions may have been accomplished by the production of sulphide from sulphate by the *Desulfovibrio* spp., as demonstrated by Preuss *et al.* (1993) and Gorontzy *et al.* (1993). The sulphide analysis showed 10.6 and 3.1 mM of sulphide on day 20 in the cultures with and without pyruvate respectively. The large difference in sulphide production in the cultures with and without pyruvate may be due to the availability of higher electron donor in pyruvate containing cultures compared to cultures with only TNT. The next metabolite identified was NB. There may be two or three intermediates between 2,4-DANT and NB, which was not identified. These compounds might be transient and thus not detected in the GC analysis. The NB was converted to cyclohexanone. This step was accomplished by ring cleavage, which under anaerobic

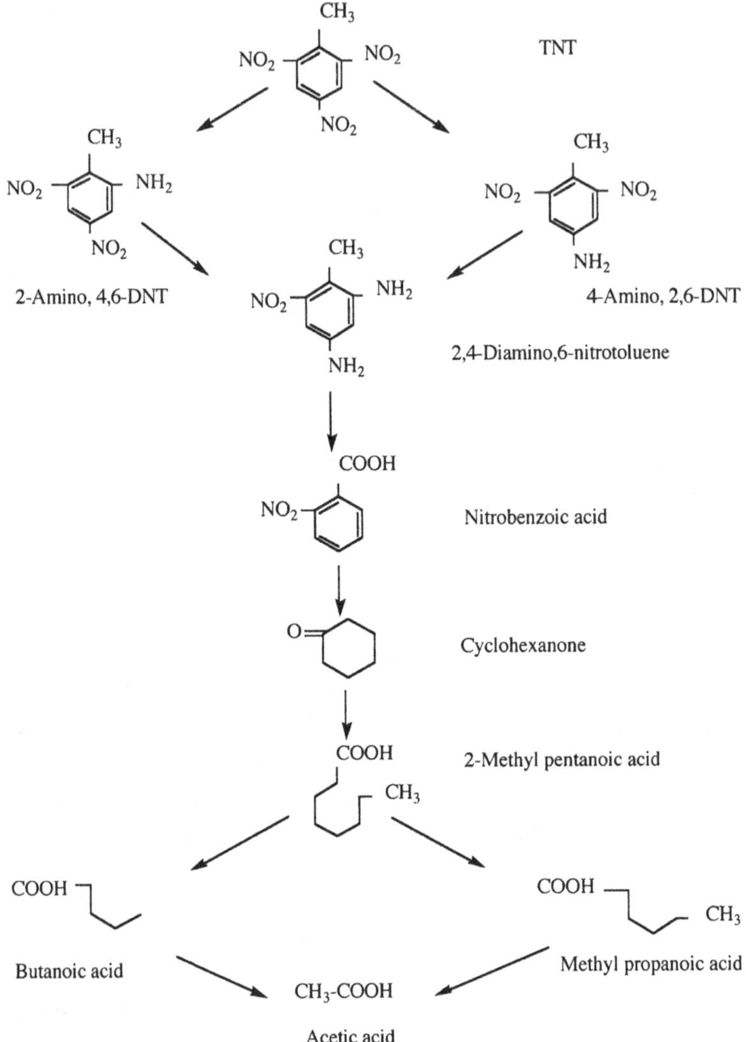

Figure 17.3. Proposed TNT metabolic pathway by *Desulfovibrio* spp.

conditions would generally be accomplished by a series of hydrogenation and dehydrogenation reactions (Harwood and Gibson, 1986), converting NB to cyclohexanone. Harwood and Gibson (1986) reported that under anaerobic conditions *Rhodopseudomonas palustris* produced pimelic acid from benzoic acidy by dehydrogenation and hydration reactions. The major intermediate observed in the study by Harwood and Gibson (1986) was cyclohexanoic acid. Cyclohexanone was further converted to 2-methyl pentanoic acid. From 2-methyl pentanoic acid, butyric acid was formed, which was further converted to acetic acid. The radiolabelled study showed no production of CO_2 from TNT metabolism and the final end product is acetic acid. This fatty acid can be easily removed under anaerobic conditions by various acetate utilizing sulphate-reducing and methanogenic bacteria.

The application of this consortium to the treatment of TNT-contaminated soil was evaluated using a TNT-contaminated soil collected from the Joliet Army Ammunition Plant (JAAP), Joliet, IL. The soil contained very high concentration of TNT of 6000 mg/kg of soil. An anaerobic soil slurry reactor was designed based on the previous study by Boopathy *et al.* (1998). The anaerobic condition in the reactor was maintained by bubbling helium gas in the headspace of the reactor. The contaminated soil was sterilized using an autoclave. A 10% soil slurry was made using sterile tapwater containing 20 mM sodium sulphate as electron acceptor, 15 mM pyruvate as co-substrate, and 5 mM ammonium chloride as nitrogen source. A 5% pre-grown inoculum of the sulphate-reducing consortium was added to the soil slurry reactor to start the bioremediation experiment. A control soil slurry reactor was maintained with similar conditions as described above except bacterial inoculum. The experiment was run for 125 days. The results shown in Figure 17.4 indicated that the sulphate-reducing bacterial

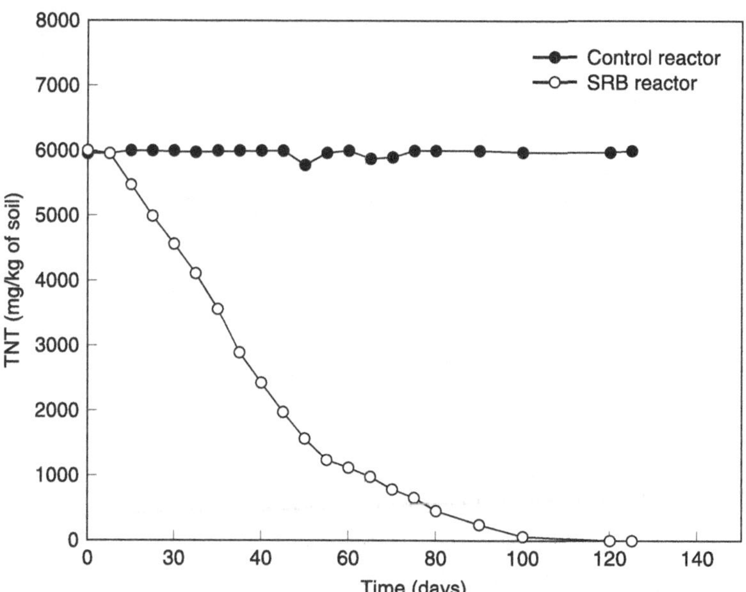

Figure 17.4. Concentration of TNT in the soil slurry reactor operated with *Desulfovibrio* spp.

consortium effectively removed TNT compared to control reactor. The TNT removal in the reactor with bacterial inoculum was almost 100% and in the control there was no TNT removal. This study showed that the sulphate-reducing bacteria can remove TNT under anaerobic conditions. This is the first report on a SRB that can remove TNT in a soil slurry condition. This report on the removal of TNT in soil by the sulphate-reducing bacterial consortium in a soil slurry reactor may have significant implications for the decontamination of TNT contaminated soil. Most munitions contamination is in the surface layer of soil, which can be excavated and treated in an anaerobic soil slurry reactor.

REFERENCES

Angermeier, L. and Simon, H. (1983). On the reduction of aliphatic and aromatic nitro compounds by Clostridia, the role of ferredoxin and its stabilization. *Hoppe Seyler's Z. Physiol Chem*, **366**, 961–75.

Bak, F. and Widdel, F. (1986). Anaerobic degradation of indolic compounds by sulphate reducing enrichment cultures and description of *Desulfobacterium indolicum* gen. nov. sp. nov. *Archiv Microbiol*, **146**, 170–6.

Battersby, N. S. and Wilson, V. (1989). Survey of the anaerobic biodegradation potential of organic chemicals in digesting sludge. *Appl Environ Microbiol*, **55**, 433–9.

Beller, H. R., Grbic-Galic, D. and Reinhard, D. (1992). Microbial degradation of toluene under sulphate reducing conditions and the influence of iron on the process. *Appl Environ Microbiol*, **58**, 786–93.

Berry, D. F., Francis, A. F. and Bellag, J. M. (1987). Microbial metabolism of homocyclic and heterocyclic aromatic compounds under anaerobic conditions. *Archiv Microbiol*, **112**, 115–17.

Boopathy, R. (1994). Transformation of nitroaromatic compounds by a methanogenic bacterium *Methanococcus* sp. (strain B). *Archiv Microbiol*, **162**, 167–72.

Boopathy, R. (2000). Bioremediation of explosives contaminated soil. *Internat Biodet Biodegrad*, **46**, 29–36.

Boopathy, R. (2001). Bioremediation of HMX-contaminated soil using soil slurry reactors. *Soil Sedi Contam*, **10**, 269–83.

Boopathy, R. (2002). Effect of food-grade surfactant on bioremediation of explosives-contaminated soil. *J Hazard Material*, **92**, 103–14.

Boopathy, R. and Daniels, L. (1991). Isolation and characterization of a furfural degrading sulphate reducing bacterium isolated from an anaerobic digester. *Curr Microbiol*, **23**, 327–32.

Boopathy, R. and Kulpa, C. F. (1992). Trinitrotoluene (TNT) as a sole nitrogen source for a sulphate reducing bacterium *Desulfovibrio* sp. (B strain) isolated from an anaerobic digester. *Curr Microbiol*, **25**, 235–41.

Boopathy, R. and Kulpa, C. F. (1993). Nitroaromatic compounds serve as nitrogen source for *Desulfovibrio* sp. (B strain). *Can J Microbiol*, **39**, 430–433.

Boopathy, R., Kulpa, C. F. and Wilson, M. (1993a). Metabolism of 2,4,6-trinitrotoluene (TNT) by *Desulfovibrio* sp. (B strain). *Appl Microbiol Biotechnol*, **39**, 270–5.

Boopathy, R., Wilson, M., Montemagno, C., Manning, J. and Kulpa, C. F. (1994b). Biological transformation of 2,4,6-trinitrotoluene (TNT) by soil bacteria isolated from TNT-contaminated soil. *Biores Technol*, **47**, 19–24.

Boopathy, R. and Kulpa, C. F. (1994). Biotransformation of 2,4,6-trinitrotoluene by a *Methanococcus* sp. (strain B) isolated from a lake sediment. *Can J Microbiol*, **40**, 273–8.

Boopathy, R. and Manning, J. F. (1996). Characterization of partial anaerobic metabolic pathway for 2,4,6-trinitrotoluene degradation by a sulphate-reducing bacterial consortium. *Can J Microbiol*, **42**, 1203–8.

Boopathy, R., Manning, J. and Kulpa, C. F. (1998). A laboratory study of the bioremediation of 2,4,6-trinitrotoluene-contaminated soil using aerobic/anoxic soil slurry reactor. *Wat Environ Res*, **70**, 80–6.

Boopathy, R., Manning, J., Montemagno, C. and Kulpa, C. F. (1994a). Metabolism of 2,4,6-trinitrotoluene by a *Pseudomonas* consortium under aerobic conditions. *Curr Microbiol*, **28**, 131–7.

Boopathy, R., Wilson, M. and Kulpa, C. F. (1993b). Anaerobic removal of 2,4,6-trinitrotoluene under different electron accepting conditions: Laboratory study. *Wat Environ Res*, **65**, 271–5.

Boyd, S. A., Shelton, D. R., Berry, D. and Tiedje, J. M. (1983). Anaerobic biodegradation of phenolic compounds in digested sludge. *Appl Environ Microbiol*, **46**, 50–4.

Bruhn, C., Lenke, H. and Knackmuss, H. J. (1987). Nitro substituted aromatic compounds as nitrogen source for bacteria. *Appl Environ Microbiol*, **53**, 208–10.

Carpenter, D. F., McCormick, N. G., Cornell, J. H. and Kaplan, A. M. (1978). Microbial transformation of ^{14}C-labeled 2,4,6-trinitrotoluene in activated sludge system. *Appl Environ Microbiol*, **35**, 949–54.

Dickel, O. and Knackmuss, H. J. (1991). Catabolism of 1,3-dinitrobenzene by *Rhodococcus* sp. QT-1. *Archiv Microbiol*, **157**, 76–9.

Duque, E., Haidour, A., Godoy, F. and Ramos, J. L. (1993). Construction of a *Pseudomonas* hybrid strain that mineralizes 2,4,6-trinitrotoluene. *J Bacter*, **175**, 2278–83.

Edwards, E. A., Wills, L. E., Reinhard, M. and Grbic-Galic, D. (1992). Anaerobic degradation of toluene and xylene by aquifer microorganisms under sulphate reducing conditions. *Appl Environ Microbiol*, **58**, 794–800.

Evans, W. C. and Fuchs, F. (1987). Anaerobic degradation of aromatic compounds. *Ann Rev Microbiol*, **42**, 289–317.

Fernando, T., Bumpus, J. A. and Aust, S. D. (1990). Biodegradation of TNT (2,4,6-trinitrotoluene) by *Phanerochaete chrysosporium*. *Appl Environ Microbiol*, **56**, 1666–71.

Fewson, C. A. (1981). Biodegradation of aromatics with industrial relevance. In T. Leisenger, A. M. Cook, R. Huttler and J. Nuesch (eds.), *Microbial degradation of xenobiotics and recalcitrant compounds*. London: Academic Press. pp. 141–79.

Funk, S. B., Roberts, D. J., Crawford, D. L. and Crawford, R. L. (1993). Initial-phase optimization for bioremediation of munitions compounds-contaminated soils. *Appl Environ Microbiol*, **59**, 2171–7.

Geissler, J. F., Harwood, C. S. and Gibson, J. (1988). Purification and properties of benzoate coenzyme. A. ligase, a *Rhodopseudomonas palustris* enzyme involved in the anaerobic degradation of benzoate. *J Bacter*, **170**, 1709–14.

Gorontzy, T., Kuver, J. and Blotevogel, K. H. (1993). Microbial transformation of nitroaromatic compounds under anaerobic conditions. *J Gen Microbiol*, **139**, 1331–6.

Haigler, B. E. and Spain, J. C. (1993). Biodegradation of 4-nitrotoluene by *Pseudomonas* sp. strain 4NT. *Appl Environ Microbiol*, **59**, 2239–43.

Hallas, L. and Alexander, M. (1983). Microbial transformation of nitroaromatic compounds in sewage effluents. *Appl Environ Microbiol*, **57**, 3156–62.

Harwood, C. S. and Gibson, J. (1986). Uptake of benzoate by *Rhodopseudomonas palustris* grown anaerobically in light. *J Bacter*, **165**, 504–9.

Holland, K. T., Knapp, J. S. and Shoesmith, J. G. (1987). *Anaerobic bacteria*. New York: Chapman and Hall.

Kaplan, D. L. and Kaplan, A. M. (1982). Mutagenicity of 2,4,6-trinitotoluene surfactant complexes. *Bulletin Environ Contam Toxicol*, **28**, 33–8.

Keith, S. M. and Herbert, R. A. (1983). Dissimilatory nitrate reduction by a strain of *Desulfovibrio desulfuricans*. *FEMS Microbiol Lett*, **18**, 55–9.

Knoll, G. and Winter, J. (1989). Degradation of phenol via carboxylation of benzoate by a defined, obligate syntrophic consortium of anaerobic bacteria. *Appl Microbiol Biotechnol*, **30**, 318–24.

LeGall, J. and Fauque, G. (1988). Dissimilatory reduction of sulfur compounds. In A. J. B. Zehnder (ed.), *Biology of anaerobic microorganisms*. New York: John Wiley and Sons. pp 587–639.

Liu, M. C. and Peck, H.D. (1981). The isolation of a hexaheme cytochrome from *Desulfovibrio desulfuricans* and its identification as a new type of nitrite reductase. *Jour Biol Chem*, **256**, 13159–64.

Liu, M. C. and Peck, H. D. (1988). Ammonia forming dissimilatory nitrite reductases as a homologous group of hexaheme-*c*-type cytochromes in metabolically diverse bacteria. In K. Kauf, K. von Dohren and H. D. Peck (eds.), *The roots of modern biochemistry*. Berlin: Walter de Gruyter and Co. pp. 685–91.

McCormick, N., Feeherry, F. E. and Levinson, H. S. (1976). Microbial transformation of 2,4,6-trinitrotoluene and other nitroaromatic compounds. *Appl Environ Microbiol*, **31**, 949–58.

McCormick, N. G., Cornell, J. H. and Kaplan, A. M. (1981). Biodegradation of hexahydro-1,3,5-trinitro-1,3,5-triazine. *Appl Environ Microbiol*, **42**, 817–23.

Michels, J. and Gottschalk, G. (1994). Inhibition of the lignin peroxidase of *Phanerochaete chrysosporium* by hydroxylamino-dinitrotoluene, an early intermediate in the degradation of 2,4,6-trinitrotoluene. *Appl Environ Microbiol*, **60**, 187–94.

Naumova, P. R., Selivanovskay, S. Y. and Mingatina, F. A. (1986) Possibility of deep bacterial destruction of 2,4,6-trinitrotoluene. *Mikrobiologiya*, **57**, 218–22.

Nishino, S. F. and Spain, J. C. (1993). Degradation of nitrobenzene by a *Pseudomonas pseudoalcaligenes*. *Appl Environ Microbiol*, **59**, 2520–5.

Preuss, A., Fimpel, J. and Diekert, G. (1993). Anaerobic transformation of 2,4,6-trinitrotoluene (TNT). *Archiv Microbiol*, **159**, 345–53.

Schnell, S., Bak, F. and Pfennig, N. (1989). Anaerobic degradation of aniline and dihydroxy-benzenes by newly isolated sulphate reducing bacteria and description of *Desulfobacterium anilini*. *Archiv Microbiol*, **152**, 556–63.

Schnell, S. and Schink, B. (1991). Anaerobic aniline degradation via reductive deamination of 4-amino-benzoyl CoA in *Desulfobacterium anilini*. *Archiv Microbiol*, **155**, 183–90.

Schnell, S. and Schink, B. (1992). Anaerobic degradation of 3-aminobenzoate by a newly isolated sulphate reducer and a methanogenic enrichment culture. *Archiv Microbiol*, **158**, 328–34.

Spain, J. C. and Gibson, D. T. (1991). Pathway for biodegradation of p-nitrophenol in a *Moraxella* sp. *Appl Environ Microbiol*, **57**, 812–19.

Spanggord, R. J., Spain, J. C., Nishino, S. F. and Mortelmans, K. E. (1991). Biodegradation of 2,4-dinitrotoluene by a *Pseudomonas* sp. *Appl Environ Microbiol*, **57**, 3200–5.

Spiker, J. K., Crawford, D. L. and Crawford, R. L. (1992). Influence of 2,4,6-trinitrotoluene (TNT) concentration on the degradation of TNT in explosive contaminated soils by the white rot fungus *Phanerochaete chrysosporium*. *Appl Environ Microbiol*, **58**, 3199–202.

Steenkamp, D. J. and Peck, H. D. (1981). On the proton translocation association with nitrite respiration in *Desulfovibrio desulfuricans*. *Jour Biol Chem*, **256**, 5450–8.

Szewzyk, U. and Schink, B. (1989). Degradation of hydroquinone, gentisate and benzoate by a fermenting bacterium in pure or defined mixed culture. *Archiv Microbiol*, **151**, 541–5.

Tschech, A. (1989). Der anaerobe annau von aromatischen verbindungen. *Forum Mikrobiol*, **12**, 251–61.

Tschech, A. and Fuch, G. (1989). Anaerobic degradation of phenol via carboxylation to 4-hydroxy benzoate: in vitro study of isotope exchange between $^{14}CO_2$ and 4-hydroxybenzoate. *Archiv Microbiol*, **152**, 594–9.

Tschech, A. and Schink, B. (1986). Fermentative degradation of monohydroxybenzoates by defined syntrophic cocultures. *Archiv Microbiol*, **145**, 396–402.

Tschech, A. and Schink, B. (1988). Methanogenic degradation of anthranilate (2-aminobenzoate). *Syst. Appl. Microbiol*, **11**, 9–12.

Valli, K., Brock, B. J., Joshi, D. K. and Gold, M. H. (1992). Degradation of 2,4-dinitrotoluene by the lignin-degrading fungus *Phanerochaete chrysosporium*. *Appl Environ Microbiol*, **58**, 221–8.

Vega, J. M. and Kamin, H. (1977). Spinach nitrite reductase. Purification and properties of a siroheme-containing iron-sulfur enzyme. *Jour Biol Chem*, **252**, 896–909.

Widdel, F. (1988). Microbiology and Ecology of sulphate and sulfur reducing bacteria. In A. J. B. Zehnder (ed.), *Biology of anaerobic microorganisms*. New York: John Wiley and Sons. pp 469–585.

Widdel, F. and Hansen, T. A. (1992). The dissimilatory sulphate and sulfur reducing bacteria. In A. Balows, H. G. Truper, M. Dworkin, W. Harder and K. H. Schleifer (eds.), *The Prokaryotes*. 2nd edn. New York: Springer Verlag. pp 583–624.

Won, D. W., Disalvo, L. H. and Ng, J. (1974). Toxicity and mutagenicity of 2,4,6-trinitrotoluene and its microbial metabolites. *Appl Environ Microbiol*, **31**, 576–80.

Wyman, J. F., Guard, H. E., Won, W. D. and Quay, J. H. (1979). Conversion of trinitrophenol to a mutagen by *Pseudomonas aeruginosa*. *Appl Environ Microbiol*, **37**, 222–6.

Zeigler, K., Braun, K., Bockler. A. and Fuchs, G. (1987). Studies on the anaerobic degradation of benzoic acid and 2-aminobenzoic acid by a denitrifying *Pseudomonas* strain. *Archiv Microbiol*, **149**, 62–9.

Zeyer. J. and Kearney, P. C. (1984). Degradation of o-nitrophenol and m-nitrophenol by a *Pseudomonas putida*. *Jour Agri Food Chem*, **32**, 238–42.

CHAPTER 18

Sulphate-reducing bacteria and the human large intestine

George T. Macfarlane, John H. Cummings and Sandra Macfarlane

18.1 THE HUMAN LARGE INTESTINAL ECOSYSTEM

The adult human colon typically contains over 200 g of digestive material (Banwell *et al.*, 1981; Cummings *et al.*, 1990; 1992), with the average daily output of faeces in Western countries being approximately 120 g (Cummings *et al.*, 1992). A large proportion of this is microbial cell mass with bacteria comprising approximately 55% of faecal solids in persons living on Western-style diets (Stephen and Cummings, 1980). The large intestine is an open system in the sense that food residues from the small intestine enter at one end, and together with bacterial cell mass, are excreted at the other end. Because of this, the colon is often viewed as being a continuous culture system, although only the proximal bowel really exhibits characteristics of a continuous culture.

The large intestine is a complex microbial ecosystem in which bacteria exist in a multiplicity of microhabitats and metabolic niches. The microbiota comprises several hundred bacterial species, subspecies and biotypes. Microbial cell counts are generally in the region of 10^{11}–10^{12} per gram of gut contents. Some organisms occur in higher numbers than others, but about 40 species make up approximately 99% of all readily culturable isolates (Finegold *et al.*, 1983). Viable counting indicates that bacteria belonging to the genera *Bacteroides*, *Bifidobacterium* and *Eubacterium*, together with a variety of anaerobic Gram-positive rods and cocci predominate in the gut (Finegold *et al.*, 1983), however, molecular methods of analysis indicate that other groups are also numerically important, including atopobium, faecalibacterium and clostridia belonging to the *C. coccoides* group (Harmsen *et al.*, 2002).

The normal gut microbiota plays an important role in host physiology and metabolism. Through fermentation and the absorption and metabolism of short chain fatty acids (SCFA), the microbiota is intimately involved in host digestive processes, enabling energy to be salvaged from unabsorbed dietary residues, as well as body tissues and secretions. Intestinal micro-organisms therefore play a major role in health and disease, and affect human physiology in a variety of ways, through, for example, obligate host requirements for SCFA, maintenance of colonization resistance to microbial pathogens, activation or destruction of mutagenic substances, and interactions with the host immune system.

18.2 SULPHATE-REDUCING BACTERIA IN THE HUMAN COLON

Sulphate-reducing bacteria (SRB) are normal inhabitants of the intestine in humans and animals. Early reports of SRB in human faeces were made by Moore *et al.* (1976) who isolated a species named *Desulfomonas pigra*, which has since been reclassified as *Desulfovibrio piger* (Loubinoux *et al.*, 2002). Bacteria identified as desulfomonas and desulfovibrio were subsequently isolated by Beerens and Romond (1977). Gibson *et al.* (1988c) reported that the predominant SRB in human faeces were desulfovibrios, but using chemostat enrichments, species belonging to the genera *Desulfobacter, Desulfotomaculum* and *Desulfobulbus* were also isolated (Gibson and Macfarlane, 1988). Although a minor component of the culturable SRB community in faecal material, desulfotomaculum rRNA was later reported to account for 1.5–3.3% of total RNA in human faeces (Hristova *et al.*, 2000). Carriage rates of SRB in different human populations vary in that these bacteria were detected in 70% of stools from UK faecal donors, but only from 15% of South African blacks (Gibson *et al.*, 1988c). A number of studies have shown that numbers of SRB in stools vary considerably. Leclerc *et al.* (1979) studied 143 stools and found that 83% had SRB, with counts ranging from $<10^2–10^{11}$ per gram of faeces. Ten percent of individuals had SRB counts in excess of 10^9 per gram of stool. Gibson *et al.* (1988c) reported that 35% of stools from people in the UK had SRB counts in the range $10^9–10^{10}$ per gram of faecal material. These bacteria can therefore be a major component of the colonic microbiota in some individuals, and as will be discussed later, this may have considerable metabolic and health implications for the host. A later study, involving 87 healthy volunteers, confirmed that SRB occurred in high numbers, ranging from 10^7 to 10^{11} per gram wet weight, in faeces of non-methanogenic persons (Gibson *et al.*, 1993). Table 18.1 shows that three

Table 18.1. *Dissimilatory sulphate reduction and methanogenesis in human gut contents*[a]

	Volunteer group		
	Group 1 ($n = 21$)	Group 2 ($n = 9$)	Group 3 ($n = 57$)
Sulphate reduction rates[b]	Trace	Trace	82.0 ± 34.1
SRB counts[c]	n.d.[f]	4.1 ± 1.2	9.3 ± 3.3
Breath methane[d]	24.7 ± 10.2	23.5 ± 6.2	n.d.
Methane production rates[e]	120 ± 97	101 ± 49.6	n.d.

[a] Adapted from Gibson *et al.* (1993). Stools were obtained from 87 donors, values are means ± SEM.
[b] nmol sulphate reduced (gram stool)$^{-1}$.
[c] Log_{10} (gram stool)$^{-1}$.
[d] ppm.
[e] nmol methane formed (gram stool)$^{-1}$.
[f] n.d.

distinct population groupings were evident in these human volunteers: Group 1 consisted of 21 individuals who were strong methane producers: in their stools SRB were completely absent. In Group 2, methanogenesis occurred and low numbers of SRB were detected, although their metabolic activities were negligible. The final group comprised 57 volunteers who had high counts of faecal SRB, and complete absence of methane production. The numerically predominant SRB were desulfovibrios, which accounted for 67–91% of total SRB counts. Species belonging to the genera *Desulfobacter* (9–16%), *Desulfobulbus* (5–8%) and *Desulfotomaculum* (2%) were recovered in considerably lower numbers from the stools. In the same investigation, the activities of SRB were studied in fresh intestinal contents obtained from human sudden death victims, and were found to be most active in the distal bowel. This probably reflects the fact that the proximal colon is an acidic environment (pH typically <5.5), whereas pH is neutral in distal colon contents (Gibson *et al.*, 1990).

Using real-time PCR, Fite *et al.* (2004) found that SRB were not only present in faecal material, but that they also colonized the gut wall. Cell counts of these bacteria in rectal biopsies were in the region 10^6–10^7 per gram, with no differences found in males and females. This longitudinal study lasted for 12 months, and it was shown that mucosal desulfovibrio numbers varied by several orders of magnitude in some individuals during the experimental period. Other recent work has shown that SRB are acquired very early in life. Using northern hybridizations and real-time ploymerase chain reaction (PCR), Hopkins *et al.* (2005) detected relatively high levels of faecal desulsulfovibrios in children younger than six months of age, in both bottle-fed and breast-fed infants. Northern hybridizations showed that in children up to six months, desulfovibrios accounted for about 2% of extractable RNA, while real-time PCR indicated that cell counts were approximately $\log_{10}$ 3.7 and 4.5 per gram of stool, in breast-fed and bottle-fed babies, respectively.

18.3 SUBSTRATES FOR GROWTH OF SRB IN THE GUT

Host tissues and other substrates of endogenous origin (sloughed epithelial cells, mucins, pancreatic and other secretions) are continually being broken down and recycled by bacteria growing in the large gut, however, the species composition and metabolic activities of the colonic microbiota are primarily determined by diet. In quantitative terms, starches and non-starch polysaccharides (dietary fibre) are the principal sources of carbon and energy for bacteria that inhabit the large intestine, although significant amounts of oligosaccharides, proteins and peptides are also available (Macfarlane and McBain, 1999). The main products of fermentation in the large intestine are the SCFA acetate, propionate and butyrate, and H_2 and CO_2. Other fermentation products include lactate, succinate and ethanol, as well as CH_4 in some individuals, together with branched chain SCFA, amines, phenols, indoles, H_2S and thiols produced from amino acid fermentation (Cummings and Macfarlane, 1991). Many of these fermentation products are further metabolized by cross-feeding species, such as SRB (Gibson, 1990). Studies on substrate utilization by human faecal SRB showed that these organisms employed a wide range of electron donors, as shown in Figure 18.1, but that lactate, pyruvate, acetate and ethanol were the most important (Gibson and Macfarlane, 1988).

18.3.1 Sulphate

Sulphate is poorly absorbed in the human gut, and studies with ileostomists show that approximately 2–9 mmol dietary sulphate reaches the large bowel every day. Most of this sulphate appears to be

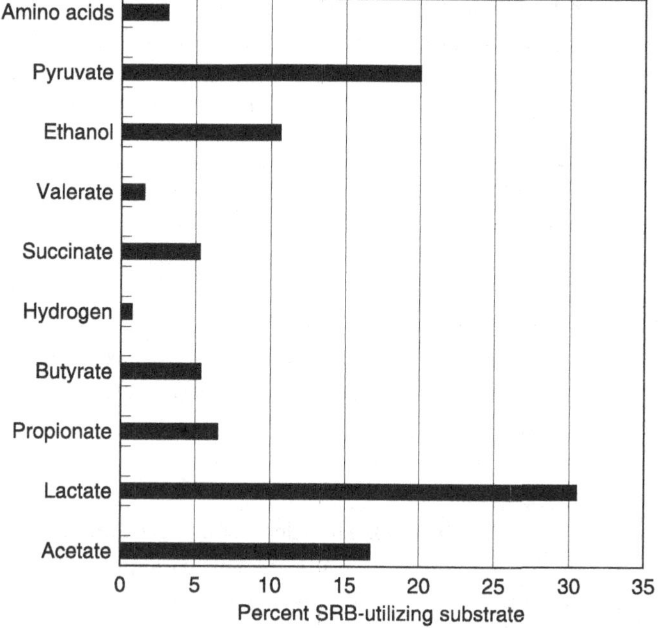

Figure 18.1. Utilization of different electron donors by human faecal sulphate-reducing bacteria. Results are adapted from Gibson and Macfarlane (1988).

metabolized in the colon, because faecal excretion is less than 0.5 mmol per day (Florin *et al.*, 1991). Sulphate can be present in high levels in some drinking water, and in some vegetables, but in addition, sulphur dioxide, sulphite, bisulphite, metabisulphite and sulphate are widely used as food additives. Collectively known as S(iv) these substances can be found in many foodstuffs (beer, cider, wine, bread, preserved meats and fruits, pickled foods) where they function as preservatives, antioxidants and bleaching agents. In vitro studies have demonstrated that intestinal bacteria can also acquire sulphate by depolymerization and desulphation of host glycoproteins that have a high sulphate content, such as salivary, gastric, hepatic, small bowel and colonic mucins, which are secreted by goblet cells lining the gastrointestinal tract (Gibson *et al.*, 1988a, b). Another potential source of sulphate is chondroitin sulphate, which is an acidic mucopolysaccharide that is widespread in mammalian tissues, and is thought to be a significant source of carbon and energy in the colon due to sloughing of epithelial cells lining the gut (Salyers and O'Brien, 1980). This polymer has also been shown to stimulate sulphide production in faecal material containing SRB (Table 18.2). Chondroitin sulphate and mucins

Table 18.2. *Stimulation of sulphide production in the human faecal microbiota by mucin and chondroitin sulphate*[a]

Test substrate[a]	Rate of sulphide production (nmol h^{-1} ml $slurry^{-1}$)
Starch (non-sulphated control)	60.0 ± 8.7
Chondroitin sulphate	117.5 ± 10.8
Mucin	162.5 ± 11.2

[a]Adapted from Gibson *et al.* (1988a).
The polysaccharides (0.2% w/v) were added to faecal slurries (5% w/v) which were incubated at 37 °C. Results are means of triplicate experiments ± SEM.

are not directly digested by SRB. They depend on saccharolytic species to do this and make free sulphate available. This is supported by studies with mice (Deplancke *et al.*, 2003) which show that SRB were most numerous in parts of the gut containing goblet cells that specialize in secreting sulphomucins.

18.3.2 Lactate

Lactate is a major fermentation product formed by many gut micro-organisms, particularly bifidobacteria and lactobacilli. In healthy people, concentrations of this metabolite seldom exceed more than a few mmol kg^{-1} of faeces, however, measurements of digestive contents taken directly from the intestine at autopsy show that it is mainly produced in the proximal bowel (Macfarlane *et al.*, 1992). Lactate is well absorbed by the colon, and under normal circumstances, concentrations are also kept low by microbial cross-feeding. Lactate production is associated with the breakdown of readily fermentable carbohydrates, such as starches (Etterlin *et al.*, 1992), while little is formed during bacterial digestion of non-starch polysaccharides. Lactate would appear to be an important electron donor for human intestinal SRB since organisms utilizing this compound are numerically predominant in the gut (Gibson *et al.*, 1988c). Measurements of lactate and residual carbohydrate in intestinal material taken from sudden death victims show a positive correlation between lactate concentrations and starch availability (Macfarlane *et al.*, 1994), suggesting that starchy diets may select for growth of SRB in the gut, if sufficient sulphate is available.

18.3.3 Hydrogen

Hydrogen gas is an major fermentation product in the large bowel. Colonic bacteria use protons as electron sinks in catabolic reactions involving both sugars and amino acids, but carbohydrates, particularly low molecular mass, rapidly fermentable molecules, are quantitatively more important sources of gas production. Theoretically, daily production of H_2 in the human colon can be in excess of 1 litre from a dietary input of 40–50 grams of carbohydrate (Levitt *et al.*, 1995), but total flatus volume in healthy people seldom exceeds this value (Kirk, 1949; Levitt *et al.*, 1969; 1971). Human volunteer studies using whole body calorimeters show that H_2 excretion is only about 2.5–14% of predicted production from fermentation stoichiometries (Christl *et al.*, 1992). This apparent discrepancy between theoretical and actual levels of H_2 excretion results from the activities of H_2 consuming microbial communities in the gut.

In the UK, about 30% of people harbour significant numbers of methanogens in the large bowel, these individuals have either non-detectable, or low levels of SRB (Gibson *et al.*, 1988c; 1993). Hydrogen is the sole electron donor for the principal gut methanogen *Methanobrevibacter smithii*, and these studies showed that SRB and methanogens competed for the mutual growth substrate hydrogen, and that if sufficient sulphate was available, dissimilatory sulphate production became predominant, and methanogenesis was abolished.

Subsequent feeding studies with human volunteers demonstrated that the amount of sulphate in the diet can have a significant effect on competition between SRB and methanogens in the large bowel. In an investigation involving six methane-excreting volunteers, Christl *et al.* (1992) showed that when 15 mmol sulphate per day was added to their diet, breath methane decreased in three individuals, while faecal sulphate reduction rates increased threefold. In these volunteers, SRB which were not detected in faeces during the control dietary period increased markedly, while methanogen counts were reduced by up to 1000-fold. It was concluded from these observations that methanogenesis can be regulated by sulphate availability, if SRB are present in the gut, even in very low numbers.

18.4 EFFECTS OF SRB ON FERMENTATION PROCESSES IN THE GUT

Due to their abilities to utilize H_2 as an electron donor, SRB can have a significant influence on fermentation processes in the colonic ecosystem.

This was demonstrated when faecal material was incubated with either sulphate, to stimulate SRB activities, or molybdate, to inhibit dissimilatory sulphate reduction. Acetate and propionate formation was stimulated by sulphate, while butyrate production was reduced and lactate did not accumulate in the incubations. However, in the presence of molybdate, SCFA formation declined markedly, while fermentation intermediates such as pyruvate and lactate accumulated in the cultures (Macfarlane *et al.*, 1992). Further studies using chemostats were done to investigate the way in which *Desulfovibrio desulfuricans*, one of the predominant SRB species in the gut, interacted with simplified human colonic microbiotas (Newton *et al.*, 1998). The results showed that an extensive multispecies biofilm developed on chemostat walls after introduction of the desulfovibrio, which profoundly affected metabolic processes and carbon flow in the community, as evidenced by marked reductions in overall SCFA production, and increased acetate formation, which is a characteristic of other habitats in which SRB occur (Gibson, 1990). Lactate disappeared in the chemostats, indicating that it was being used as an electron donor by the desulfovibrio. However, it was evident that these organisms were also scavenging H_2, since apart from the increase in acetate, butyrate production was reduced by a factor of three, despite the presence of substantial numbers of butyrate-producing species, such as *Clostridium butyricum*. These experiments highlighted the occurrence of syntrophic interactions between *Dsv. desulfuricans* and saccharolytic and amino acid-fermenting bacteria that occur in the large intestine. The extent to which this was due to biofilm creation by the SRB is unclear, but extensive microbial biofilms form on digestive residues in the gut lumen, and on mucosal surfaces. Therefore, through their ecological and physiological effects on butyrate production, the occurrence and activities of SRB in the large bowel have the potential to play a significant role in host metabolism.

18.5 SULPHATE-REDUCING BACTERIA AND HUMAN DISEASE

Although not generally considered to be important human pathogens, SRB have been implicated as aetiologic agents in a number of disease states, ranging from cholecystitis, and brain and abdominal abscesses, involving a variety of *Desulfovibrio* spp., to bacteraemia caused by *Dsv. fairfieldensis*. This organism has been suggested to have greater pathogenic potential than other species belonging to the genus *Desulfovibrio* (Goldstein *et al.*, 2003; Loubinoux *et al.*, 2000). Most of these mono- and polymicrobial infections have been sequelae to gastrointestinal surgery. Subsequent studies by

Loubinoux *et al.* (2003), using multiplex PCR, found that 12 out of 100 abdominal and pleural pus specimens contained *Dsv. piger, Dsv. fairfieldensis* or *Dsv. desulfuricans*. Bacteraemia was also caused by an unspeciated *desulfovibrio*, which had 97% sequence identity with *Dsv. desulfuricans*, and was believed to have resulted from bleeding colonic polyps (McDougall *et al.*, 1997). Together, these investigations suggest that the colon is the principal reservoir of infection by SRB in humans. Although growth of SRB in the mouth is thought to be limited by sulphate availability, these bacteria are also found in periodontal lesions (Boopathy *et al.*, 2002; Loubinoux *et al.*, 2002; Van der Hoeven *et al.*, 1995), where like methanogens, their syntrophic abilities are thought to contribute towards general disease processes (Lepp *et al.*, 2004).

Intestinal SRB may also be linked to rheumatic disease, because patients with ankylosing spondylitis have been found to have a higher prevalence of these organisms than healthy controls (Stebbings *et al.*, 2002). It has also been suggested that SRB are involved indirectly in the aetiology of some forms of colorectal cancer, through sulphide formation. Studies have shown that this reducing agent activates a number of biochemical pathways believed to be involved in initiation of the disease (Huycke and Gaskins, 2004). Moreover, at physiological concentrations, sulphide has been shown to protect colon cancer cells from drugs such as β-phenylethyl isocyanate that induce apoptosis (Rose *et al.*, 2005), thereby providing a mechanism for the promotion of tumorigenesis.

18.6 SULPHATE-REDUCING BACTERIA AND INFLAMMATORY BOWEL DISEASE

Microorganisms cover the surface of the large bowel mucosa, in the mucus layer, and on underlying epithelial cell surfaces, where bacterial cell densities reach about $10^8 cm^2$ (Macfarlane *et al.*, 2004). As discussed previously, microbial communities in adjacent lumenal contents are several orders of magnitude higher (Finegold *et al.*, 1983). Most people tolerate this complex metabolically active and antigenic biota, but in approximately 2 per 1000 adults living in Europe and the United States (Loftus *et al.*, 2000; Mayberry *et al.*, 1989; Montgomery *et al.*, 1998) an intense inflammation develops in the mucosa. Symptoms vary between individuals, but in general, the disease is associated with bloody diarrhoea, abdominal pain, weight loss, urgency to defecate, arthritic effects, increased gut permeability and general malaise (Head and Jurenka, 2003). Despite many studies, ulcerative colitis (UC) has not been shown to be directly associated with any known microbial pathogens. UC is one of the two major forms of idiopathic inflammatory bowel disease (IBD). It is an acute and chronic illness that affects the colon and rectum. The disease can be a highly disabling condition, and is incurable. Current maintenance therapies rely on anti-inflammatory drugs and steroids. If these strategies are unsuccessful, the last resort is partial or complete removal of the bowel by surgery.

Convincing evidence from research using animal models of gut inflammation, as well as the study of UC patients, shows that intestinal bacteria are essential in the initiation and maintenance of inflammatory processes in this condition. In humans, UC is non-infectious, and is generally believed to occur as the result of a genetically mediated abnormal immune response to elements of the normal commensal microbiota, and a number of lines of evidence point to bacterial sulphate metabolism having some involvement in the disease. Firstly, in experimental animals, several sulphated polymers, such as partially degraded carrageenan, sulphated amylopectin and sodium lignosulphonate, when fed orally in drinking water, induce an acute attack of colitis (Marcus and Watt, 1969; Marcus *et al.*, 1983; Watt and Marcus, 1973). While the clinical and pathological features closely resemble human UC, the lesions extend distally from the caecum, whereas human disease always initiates in the distal large bowel. Disease severity is related to the amount of sulphate present in the polymer.

Although the mechanisms are unclear, it would appear that in these animals, the processing of sulphate by intestinal microorganisms forms a product that causes immune activation, followed by mucosal damage. The link with SRB is that these organisms reduce sulphate to sulphide in the large bowel, which is toxic to colonic epithelial cells (Pitcher *et al.*, 1998), indeed studies with animal models have indicated that sulphide is as toxic as cyanide (Levitt *et al.*, 1999). Sulphide has a multiplicity of effects on the gut mucosa; it inhibits energy generation in colonocytes (Roediger *et al.*, 1993a, b), inhibits phagocytosis and bacterial killing (Gardiner *et al.*, 1996), and induces hyperproliferation and metabolic abnormalities in epithelial cells, which are similar to those seen in UC (Christl *et al.*, 1996). Untreated UC patients have been reported to have significantly higher levels of faecal sulphide excretion than healthy controls (Pitcher *et al.*, 2000), while 5-ASA, the main drug used to treat the disease, is known to inhibit sulphide production (Dzierzewicz *et al.*, 2004; Pitcher *et al.*, 2000). However, contrary to the work of Pitcher and co-workers, Moore *et al.* (1998) reported that there were no significant differences in stool sulphide levels in UC patients.

Despite this, support for a role for sulphide being involved in inflammatory processes in the gut comes from a recent study by Ohge *et al.* (2005)

which found that production of this metabolite, and SRB numbers, were greatly increased during inflammation of the ileo-anal pouch (pouchitis) in 45 patients who had previously received surgery for UC. These individuals responded well to antibiotic therapy (metronidazole or ciprofloxacin), which was associated with reduced SRB counts and sulphide production.

Pitcher *et al.* (2000) using 39 UC patients, showed that while 95% carried SRB in the active phase of the illness, only 55% did so in quiescent disease. The mean counts of viable SRB were three orders of magnitude higher in the active disease group than those in remission. Investigations in France, involving 151 subjects (Loubinoux *et al.*, 2002) found a lower prevalence of SRB in faeces in healthy subjects (12%), while the incidence of desulfovibrios was significantly higher in IBD (UC and Crohn's disease) patients (55%). Gibson *et al.* (1991) observed that although SRB were present in lower numbers in UC faeces, compared with stools from healthy people, their metabolic activities, as assessed by specific rates of sulphate reduction, were considerably higher than in the controls. It was also shown that SRB isolated from UC patients were particularly adapted to grow under low sulphate concentrations, at high specific growth rates, which was related to ecological selection by environmental conditions in the colitic bowel.

Little is known of the way in which SRB interact with the gut epithelium, however, using culturing methods, the existence of mucosal-associated SRB was detected in 92% of UC samples compared with 52% in controls (Zinkevich and Beech, 2000). In these studies, PCR analysis further indicated that SRB were ubiquitous, and were present in all of these biopsies studied. Interest in a role for mucosal SRB in IBD was stimulated by the discovery of an intracellular Gram-negative pathogen genetically related to desulfovibrios, known as ileal symbiont intracellularis (ISI). This organism has been shown to cause bloody diarrhoea, anorexia and weight loss in animals such as pigs and rodents (Fox *et al.*, 1994; Gebhardt *et al.*, 1993). Histological examination in this infectious condition shows epithelial hyperplasia, a profusion of crypt abscesses, goblet cell depletion and inflammatory cell infiltrates similar to those observed in human IBD. While this was thought to be a possible link with SRB and UC in humans, studies subsequently showed that ISI does not play a similar role in human inflammatory disease. None of these bacteria were detected when rectal biopsies from 19 UC patients were investigated using an immunofluorescent assay, involving mouse monoclonal antibody IG4 targeted to ISI (Pitcher, 1996). This work further showed that there was no reactivity against a variety of SRB isolates from UC patient stools.

18.6.1 Effect of sulphide on energy metabolism in colonocytes

Is there a mechanism whereby SRB could potentially induce inflammatory symptoms in UC? Butyrate produced in bacterial fermentation reactions in the gut lumen is the principal fuel for colonic epithelial cells, especially in the distal large bowel. Studies with rats have shown inhibiting butyrate metabolism with 2-bromo octanoate rapidily induces UC-like lesions, and death of the animal (Roediger and Nance, 1986). Oxidation of this fatty acid is also inhibited by a variety of reducing sulphur-containing substances (Roediger *et al.*, 1993a, b). In particular, sulphide, mercaptoacetate and methanethiol, at low millimolar concentrations, reduced fatty acid oxidation by between 30–50% without significantly affecting glucose metabolism, reflecting metabolic changes that are seen in UC. Selective impairment of the uptake of butyrate by physiological concentrations of sulphide (0.1–0.5 mM) was also found to occur in rat colonocytes (Babidge *et al.*, 1998; Roediger *et al.*, 1993a; Stein *et al.*, 1995). Experiments aimed at elucidating how sulphide inhibits butyrate oxidation in gut epithelial cells showed that it causes a metabolic block at the level of FAD-linked oxidation by butyryl-CoA dehydrogenase (Roediger *et al.*, 1993a, b). It is clear from these experiments that sulphide is detrimental to energy generation in colonocytes, and that these mechanisms point to reduced-sulphur compounds generated in the lumen of the large bowel being good candidates for mediators of disease in UC.

However, it is likely that the story is much more complicated than UC simply being caused by inhibition of butyrate metabolism. Fite *et al.* (2004) showed that large numbers of desulfovibrios (ca. 10^6–10^7 per gram) were present in rectal biopsies taken from both colitis patients and healthy people, and that no disease-related differences were apparent. While it is possible that the disease might only be linked to certain bacterial strains, in view of the fact that both healthy and diseased mucosae were heavily colonized by SRB (these bacteria play a role in colitis) some host defect, possibly in sulphide detoxication pathways, or in bacterial antigen handling, would be required for manifestations of pathogenicity to become apparent. There is some experimental support for UC patients having impaired sulphide detoxication pathways. Although the exact pathway of sulphide disposal in the human body is unclear (see Figure 18.2 for an outline of bacterial S-metabolism in the colon), it is thought to involve the enzymes rhodanese, mercaptopyruvate sulphur transferase (MST) and sulphite oxidase (SO). Recent studies by Kong *et al.* (2004; 2005) showed that rhodanese and MST activities were significantly lower in gut tissues taken from UC patients as compared with

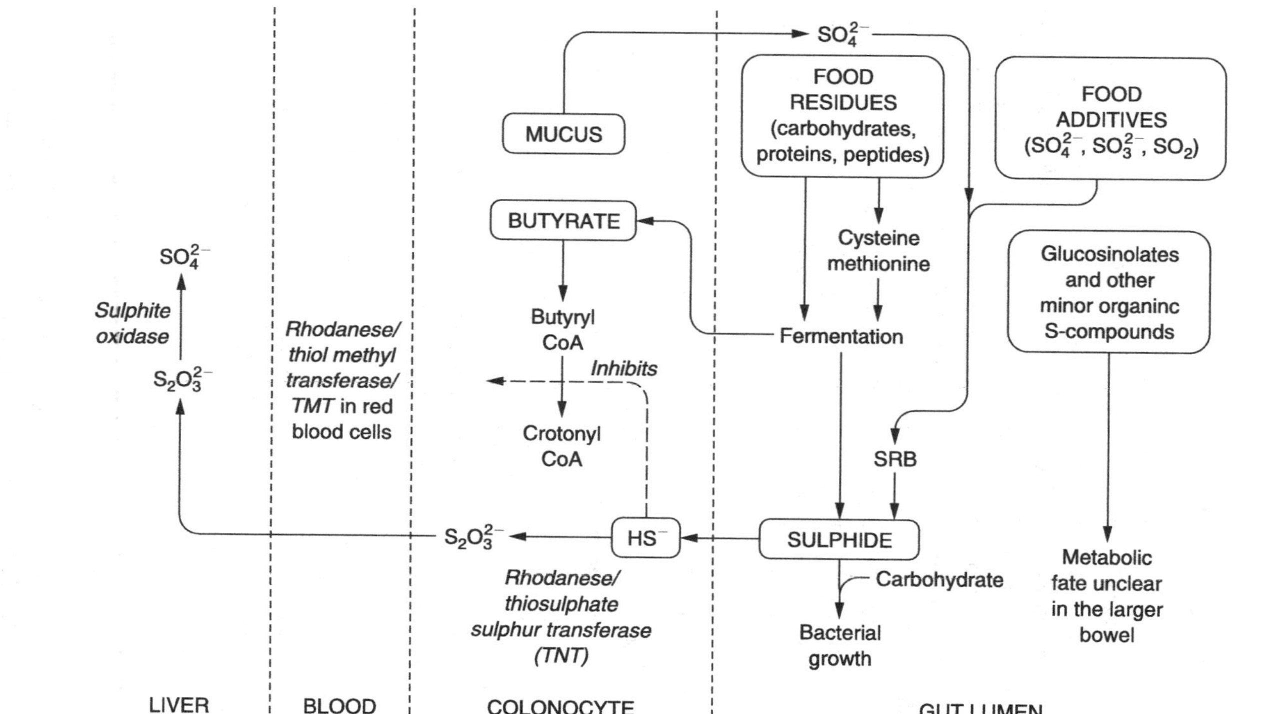

Figure 18.2. Involvement of sulphate-reducing bacteria in the metabolism of sulphur-containing substances in the human large bowel, and the fate of sulphide metabolites in host tissues.

healthy controls, particularly in the distal bowel, where the disease always initiates, although there was no difference with SO.

18.7 CONCLUSIONS

In conclusion, SRB are normal inhabitants of the human large intestine in health and disease, irrespective of age and sex. However, while a number of investigations have linked SRB to human inflammatory bowel disease, the evidence is circumstantial, and the case has not been made as to whether these organisms are true aetiologic agents, or that they are simply taking advantage of the dramatic environmental changes in the gut ecosystem that result from mucosal inflammation, tissue destruction and diarrhoea. Sulphide is also produced by some amino acid fermenting anaerobes in the human large bowel, so even if this metabolite is an important mucosal toxin in UC, the presence or absence of SRB, as well as their metabolic propensities, does not necessarily explain the prevalence or severity of the disease.

REFERENCES

Babidge, W., Millard, S. and Roediger, W. E. W. (1998). Sulfides impair short chain fatty acid beta-oxidation at acyl-CoA dehydrogenase level in colonocytes: implications for ulcerative colitis. *Mol. Cell. Biochem.*, **18**, 117–24.

Banwell, J. G., Branch, W. J. and Cummings, J. H. (1981). The microbial mass in the human large intestine. *Gastroenterology*, **80**, 1104.

Beerens, H. and Romond, C. (1977). Sulphate-reducing anaerobic bacteria in human feces. *Am. J. Clin. Nutr.*, **30**, 1770–6.

Boopathy, R., Robichaux, M., LaFont, D. and Howell, M. (2002). Activity of sulphate-reducing bacteria in human periodontal pocket. *Can. J. Microbiol.*, **48**, 1099–103.

Christl, S., Eisner, H. D., Kasper, H. and Scheppach, W. (1996). Antagonistic effects of sulfide and butyrate on proliferation of colonic mucosa: a potential role for these agents in the pathogenesis of ulcerative colitis. *Dig. Dis. Sci.* **41**, 2477–81.

Christl, S. U., Murgatroyd, P. R., Gibson, G. R. and Cummings, J. H. (1992). Production, metabolism and excretion of hydrogen in the large intestine. *Gastroenterology*, **102**, 1269–77.

Cummings, J. H., Banwell, J. G., Englyst, H. N., Coleman, N., Segal, I. and Bersohn, D. (1990). The amount and composition of large bowel contents. *Gastroenterology*, **98**, A408.

Cummings, J. H., Bingham, S. A., Heaton, K. W. and Eastwood, M. A. (1992). Fecal weight, colon cancer risk, and dietary intake of nonstarch polysaccharides (dietary fiber). *Gastroenterology*, **103**, 1783–9.

Cummings, J. H. and Macfarlane, G. T. (1991). The control and consequences of bacterial fermentation in the human colon. *J. Appl. Bacteriol.*, **70**, 443–59.

Deplancke, B., Finster, K., Graham, W. V. *et al.* (2003). Gastrointestinal and microbial responses to sulphate-supplemented drinking water in mice. *Exp. Biol. Med.*, **228**, 424–33.

Dzierzewicz, Z., Cwalina, B., Weglarz, L., Wisniowska, B. and Szczerba, J. (2004). *Med. Sci. Mon.*, **10**, 185–90.

Etterlin, C., McKeowen, A., Bingham, S. A. *et al.* (1992). D-lactate and acetate as markers of fermentation in man. *Gastroenterology*, **102**, A551.

Finegold, S. M., Sutter, V. L. and Mathisen, G. E. (1983). Normal indigenous intestinal flora. In D. J. Hentges (ed.), *Human intestinal microflora in health and disease*. London: Academic Press. pp. 3–31.

Fite, A., Macfarlane, G. T., Cummings, J. H. *et al.* (2004). Identification and quantitation of mucosal and faecal desulfovibrios using real-time PCR. *Gut*, **53**, 523–9.

Florin, T. H. J., Neale, G., Gibson, G. R., Christl, S. U. and Cummings, J. H. (1991). Metabolism of dietary sulphate: absorption and excretion in humans. *Gut*, **32**, 766–73.

Fox, J. G., Dewhirst, F. E., Fraser, G. J. *et al.* (1994). Intracellular *Campylobacter*-like organism from ferrets and hamsters with proliferative bowel disease is a *Desulfovibrio* sp. *J. Clin. Microbiol.*, **32**, 1229–37.

Gardiner, K. R., Halliday, M. I., Barclay, G. R. *et al.* (1996). Significance of systemic endotoxaemia in inflammatory bowel disease. *Gut*, **36**, 897–901.

Gebhart, C. J., Barns, S. M., McOrist, S., Lin, G.-F. and Lawson, G. H. K. (1993). Ileal symbiont intracellularis, an obligate intracellular bacterium of porcine intestines showing a relationship to *Desulfovibrio* species. *Int. J. Syst. Bacteriol.*, **43**, 533–8.

Gibson, G. R. (1990). A review: physiology and ecology of the sulphate-reducing bacteria. *J. Appl. Bacteriol.*, **69**, 769–97.

Gibson, G. R., Cummings, J. H. and Macfarlane, G. T. (1988a). Competition for hydrogen between sulphate-reducing bacteria and methanogenic bacteria from the human large intestine. *J. Appl. Bacteriol.*, **65**, 241–7.

Gibson, G. R., Cummings, J. H. and Macfarlane, G. T. (1988b) Use of a three-stage continuous culture system to study the effect of mucin on dissimilatory sulphate reduction and methanogenesis by mixed populations of human gut bacteria. *Appl. Environ. Microbiol.*, **54**, 2750–5.

Gibson, G. R., Cummings, J. H. and Macfarlane, G. T. (1991). Growth and activities of sulphate-reducing bacteria in gut contents from healthy subjects and patients with ulcerative colitis. *FEMS Microbiol. Ecol.*, **86**, 103–12.

Gibson, G. R., Cummings, J. H., Macfarlane, G. T. *et al.* (1990). Alternative pathways for hydrogen disposal during fermentation in the human colon. *Gut*, **31**, 679–83.

Gibson, G. R. and Macfarlane, G. T. (1988). Chemostat enrichment of sulphate-reducing bacteria from the large gut. *Lett. Appl. Microbiol.*, **7**, 127–33.

Gibson, G. R., Macfarlane, G. T. and Cummings, J. H. (1988c). Occurrence of sulphate-reducing bacteria in human faeces and the relationship of dissimilatory sulphate reduction to methanogenesis in the large gut. *J. Appl. Bacteriol.*, **65**, 103–11.

Gibson, G. R., Macfarlane, S. and Macfarlane, G. T. (1993). Metabolic interactions involving sulphate-reducing and methanogenic bacteria in the human large intestine. *FEMS Microbiol. Ecol.*, **12**, 117–25.

Goldstein, E. J. C., Citron, D. M., Peraino, V. A. and Cross, S. A. (2003). *Desulfovibrio desulfuricans* bacteremia and review of human *Desulfovibrio* infections. *J. Clin. Microbiol.*, **41**, 2752–4.

Harmsen, H. J. M., Raangs, G. C., He, T., Degener, J. E. and Welling, G. W. (2002). Extensive set of 16S rRNA-based probes for detection of bacteria in human feces. *Appl. Environ. Microbiol.*, **68**, 2982–90.

Head, K. A. and Jurenka, J. S. (2003). Inflammatory bowel disease part 1: Ulcerative colitis – pathophysiology and coventional and alternative treatment options. *Alt. Med. Rev.*, **8**, 247–83.

Hopkins, M. J., Macfarlane, G. T., Furrie, E., Fite, A. and Macfarlane, S. (2005). Characterisation of intestinal bacteria in infant stools using real-time PCR and northern hybridisation analyses. *FEMS Microbiol. Ecol.*, **54**, 77–85.

Hristova, K. R., Mau, M., Zheng, D. *et al.* (2000). *Desulfotomaculum* genus- and subgenus-specific 16S rRNA hybridization probes for environmental studies. *Environ. Microbiol.*, **2**, 143–59.

Huycke, M. M. and Gaskins, H. R. (2004). Commensal bacteria, redox stress, and colorectal cancer: mechanisms and models. *Exp. Biol. Med.*, **229**, 586–97.

Kirk, E. (1949). The quantity and composition of human colonic flatus. *Gastroenterology*, **12**, 782–94.

Kong, S. C., Furrie, E., Macfarlane, G. T. and Cummings, J. H. (2005). Regional variation of mRNA of hydrogen sulphide detoxification enzymes in the colon may predispose to ulcerative colitis. *Gut*, **54**, Suppl. 2, 352.

Kong, S. C., Furrie, E., Madden, J. *et al.* (2004). Comparison of hydrogen sulphide detoxification enzyme mRNA expression in normal and ulcerative colitis rectal mucosae. *Gut*, **53**, Suppl. 3, 142.

Leclerc, H., Oger, C., Beerens, H. and Mossel, D. A. (1979). Occurrence of sulphate-reducing bacteria in the human intestinal flora and in the water environment. *Water Res.*, **14**, 253–6.

Lepp, P. W., Brinig, M. M., Ouverney, C. C. *et al.* (2004). Methanogenic *Archaea* and human periodontal disease. *Proc. Nat. Acad. Sci.*, **101**, 6176–81.

Levitt, M. D. (1969). Production and excretion of hydrogen gas in man. *New Eng. J. Med.*, **281**, 122–7.

Levitt, M. D. (1971). Volume and composition of human intestinal gas determined by means of an intestinal washout technique. *New Eng. J. Med.*, **284**, 1394–8.

Levitt, M. D., Furne, J., Springfield, J., Suarez, F. and DeMaster, E. (1999). Detoxification of hydrogen sulfide and methanethiol in the cecal mucosa. *J. Clin. Invest.*, **104**, 1107–14.

Levitt, M. D., Gibson, G. R. and Christl, S. U. (1995). Gas metabolism in the large intestine. In G. R. Gibson and G. T. Macfarlane (eds.), *Human colonic bacteria: role in nutrition, physiology and health.* Boca Raton, FL: CRC Press. pp. 131–54.

Loftus, E. V., Silverstein, M. D., Sandborn, W. J. *et al.* (2000). Ulcerative colitis in Olmsted County, Minnesota, 1940–1993: incidence, prevalence, and survival. *Gut*, **46**, 336–43.

Loubinoux, J., Bisson-Boutelliez, C., Miller, N. and Le Faou, A. E. (2002). Isolation of the provisionally named *Desulfovibrio fairfieldensis* from human periodontal pockets. *Oral Microbiol. Immunol.*, **17**, 321–3.

Loubinoux, J., Bronowicji, J.-P., Pereira, I. A. C., Moungenel, J.-L. and Faou, A. E. (2002). Sulphate-reducing bacteria in human feces and their association with inflammatory diseases. *FEMS Microbiol. Ecol.*, **40**, 107–12.

Loubinoux, J., Jaulhac, B., Piemont, Y., Monteil, H. and Le Faou, A. E. (2003). Isolation of sulphate-reducing bacteria from human thoracoabdominal pus. *J. Clin. Microbiol.*, **41**, 1304–6.

Loubinoux, J., Mory, F., Pereira, I. A. C. and Le Faou, A. E. (2000). Bacteremia caused by a strain of *Desulfovibrio* related to the provisionally named *Desulfovibrio fairfieldensis. J. Clin. Microbiol.*, **38**, 931–4.

Loubinoux, J., Valente, F. M. A., Pereira, A. C. *et al.* (2002). Reclassification of the only species of the genus *Desulfomonas, Desulfomonas pigra*, as *Desulfovibrio piger* comb. nov. *Int. J. Syst. Evol. Microbiol.*, **52**, 1305–8.

Macfarlane, G. T., Gibson, G. R. and Cummings, J. H. (1992). Comparison of fermentation reactions in different regions of the human colon. *J. Appl. Bacteriol.*, **72**, 57–64.

Macfarlane, G. T., Gibson, G. R. and Macfarlane, S. (1994). Short chain fatty acid and lactate production by human intestinal bacteria grown in batch and continuous culture. In H. J. Binder, J. H. Cummings and K. H. Soergel (eds.), *Short chain fatty acids.* Lancaster: Kluwer Academic Publishers. pp. 44–60.

Macfarlane, G. T. and McBain, A. J. (1999). The human colonic microbiota. In G. R. Gibson and M. Roberfroid (eds.), *Colonic microflora, nutrition and health.* London: Chapman & Hall. pp. 1–25.

Macfarlane, S., Furrie, E., Cummings, J. H. and Macfarlane, G. T. (2004). Chemotaxonomic analysis of bacterial populations colonizing the rectal mucosa in patients with ulcerative colitis. *Clin. Infect. Dis.*, **38**, 1690–9.

Marcus, R., Marcus, A. J. and Watt, J. (1983). Chronic ulcerative disease of the colon in rabbits fed native carrageenans. *Proc. Nutr. Soc.*, **42**, 155A.

Marcus, R. and Watt, J. (1969). Seaweeds and ulcerative colitis in laboratory animals. *Lancet*, **2**, 489–90.

Mayberry, J. F., Ballantyne, K. C., Hardcastle, J. D., Mangham, C. and Pye, G. (1989). Epidemiological study of asymptomatic inflammatory bowel disease: the identification of cases during a screening programme for colorectal cancer. *Gut*, **30**, 481–3.

McDougall, R., Robson, J., Paterson, D. and Tee, W. (1997). Bacteremia caused by a recently described novel *Desulfovibrio* species. *J. Clin. Microbiol.*, **35**, 1805–8.

Montgomery, S. M., Morris, D. L., Thompson, N. P. *et al.* (1998). Prevalence of inflammatory bowel disease in British 26 year olds: national longitudinal birth cohort. *Brit. Med. J.*, **316**, 1058–9.

Moore, J., Babidge, W., Millard, S. and Roediger, W. E. W. (1998). Colonic luminal hydrogen sulfide is not elevated in ulcerative colitis. *Dig. Dis. Sci.*, **43**, 162–5.

Moore, W. E. C., Johnson, J. L. and Holdeman, L. V. (1976). Emendation of *Bacteroidaceae* and *Butyrivibrio* and descriptions of *Desulfomonas* gen. nov. and ten new species of the genera *Desulfomonas, Butyrivibrio, Eubacterium, Clostridium* and *Ruminococcus. Int. J. Syst. Bact.*, **26**, 238–52.

Newton, D. F., Cummings, J. H., Macfarlane, S. and Macfarlane, G. T. (1998). Growth of a human intestinal *Desulfovibrio desulfuricans* in continuous cultures containing defined populations of saccharolytic and amino acid fermenting bacteria. *J. Appl. Microbiol.*, **85**, 372–80.

Oghe, H., Furne, J. K., Springfield, J. *et al.* (2005). Association between fecal hydrogen sulfide production and pouchitis. *Dis. Col. Rect.*, **48**, 469–75.

Pitcher, M. C. L. (1996). Sulphate-reducing bacteria, sulphur metabolism and ulcerative colitis. MD Thesis, University of Cambridge.

Pitcher, M. C. L., Beatty, E. R. and Cummings, J. H. (2000). The contribution of sulphate reducing bacteria and 5-aminosalicylic acid to faecal sulphide in patients with ulcerative colitis. *Gut*, **46**, 64–72.

Pitcher, M. C. L., Beatty, E. R., Harris, R. M., Waring, R. H. and Cummings, J. H. (1998). Sulfur metabolism in ulcerative colitis. Investigation of detoxification enzymes in peripheral blood. *Dig. Dis. Sci.*, **43**, 2080–5.

Roediger, W. E. W., Duncan, A., Kapaniris, O. and Millard, S. (1993a). Sulphide impairment of substrate oxidation in rat colonocytes: a biochemical basis for ulcerative colitis? *Clin. Sci.*, **85**, 1–5.

Roediger, W. E. W., Duncan, A., Kapaniris, O. and Millard, S. (1993b). Reducing sulfur compounds of the colon impair colonocyte nutrition: implications for ulcerative colitis. *Gastroenterology*, **104**, 802–9.

Roediger, W. E. W. and Nance, S. (1986). Metabolic induction of experimental ulcerative colitis by inhibition of fatty acid oxidation. *Brit. J. Exp. Pathol.*, **67**, 773–82.

Rose, P., Moore, P. K., Ming, S. H. *et al.* (2005). Hydrogen sulphide protects colon cancer cells from chemopreventative agent β-phenylethyl isocyanate induced apotosis. *World J. Gastroenterol.*, **11**, 3990–7.

Salyers, A. A. and O'Brien, M. (1980). Cellular location of enzymes involved in chondroitin sulphate breakdown by *Bacteroides thetaiotaomicron*. *J. Bacteriol.*, **143**, 772–80.

Stebbings, S., Munro, K., Simon, M. A. *et al.* (2002). Comparison of the faecal microflora of patients with ankylosing spondylitis and controls using molecular methods of analysis. *Rheumatology*, **41**, 1395–401.

Stein, J., Schroder, O., Milovic, V. and Caspary, W. F. (1995). Mercaptopropionate inhibits butyrate uptake in isolated apical membrane vesicles of the rat distal colon. *Gastroenterology*, **108**, 673–9.

Stephen, A. M. and Cummings, J. H. (1980). The microbial contribution to human faecal mass. *J. Med. Microbiol.*, **13**, 45–56.

Van der Hoeven, J. S., Van den Lieboom, C. W. A. and Schaeken, M. J. M. (1995). Sulphate-reducing bacteria in the periodontal pocket. *Oral Microbiol. Immunol.*, **10**, 288–90.

Watt, J. and Marcus, R. (1973). Experimental ulcerative disease of the colon in animals. *Gut*, **14**, 506–10.

Zinkevich, V. and Beech, I. B. (2000). Screening of sulphate-reducing bacteria in colonoscopy samples from healthy and colitic gut mucosa. *FEMS Microbiol. Ecol.*, **34**, 147–55.

Index

ABC permease 147
Acetate kinase 148
Acetivibrio FISH probe 60
Acetobacterium 388
Acetyl-CoA 11, 12, 14–16, 489
Acetyl-CoA synthase 16, 450
Acetyl-phosphate 11, 14
Acetylserine sulfhydrylase 160
Acid mine waters 412, 419, 421
Acid pH stress 145, 148, 156, 158
Acidic fens 41
Adenosine phosphosulphate 5, 7
Aerotaxis 172
Aggregation 171
Agmatinase 158
Alcaligenes eutrophus 246
Alcohol dehydrogenase 129, 219
Alkaline pH stress 145, 148, 159
Alkane utilization 281–3, 308
Alkene utilization 287
Alkyl hydroperoxide reductase 148
Alkylbenzene utilization 281
Allochromatium vinosum 101, 228
Alvinella pompejana symbiont 320
Aminobenzoate 490
Ammonification 242, 256
Anaerofilum FISH probe 60
Analysis of natural systems 89
Aniline 488, 489
Anoxic layers 168
Antibiotic inhibition 386
APS reductase 7, 92, 96, 97, 101, 128, 133, 148, 197, 227, 229
ApsA 97, 101
ArcAB regulon 204
Archaea 24, 39, 274
Archaea probe 42
Archaeoblobus lithotrophicus 316
Archaeoglobus 1, 22, 24, 39, 99, 101, 306, 309, 316, 321, 350
Archaeoglobus fulgidus 6, 16, 22, 102, 123, 124, 131, 201, 227, 228, 310, 316, 352, 440, 447, 450
Archaeoglobus probe 42
Archaeoglobus profundus 6, 22, 310, 316, 321
Archaeoglobus veneficus 316, 321, 348
Arginine decarboxylase 158
Aromatic hydrocarbons oxidized 289
Arsenate 14
Arsenate reductase 441, 447
Arsenate reduction 447, 448
Arsenic reduction 447
Arsenic transporter 441, 447
Arsenical sulphides 410, 416
Arsenite pump 441, 447
ATP sulphurylase 7, 96, 133
ATP synthase 10, 147, 148, 154, 215, 243
ATPase 10, 157–9, 441
a-type haem 192

Bacteria 39, 306, 364
Bacteria FISH probe 60
Bacteria probe 42
Bacterioferritin 187, 188, 202
Bacteroides fragilis 170
bc1 complex 193
bd- type oxygen reductase 187, 191, 192
Bdellovibrio bacteriovorus 132
Beggiatoa 171
Benzoyl CoA 490
Bilophila wadsworthia 23, 99
Biofilm 159, 169, 360–2, 367, 373, 391, 406, 417, 459, 463, 469, 472, 510
Biogeochemical cycling 405, 410
Bioinformatical studies 131, 186–8
Biophila wadsworthia FISH probe 60
Bioreactors 386, 387
Bioremediation 405, 408, 452, 490
Bisulphite disproportionation 5
Bisulphite reductase 7, 9, 41, 92, 96, 99, 101, 320
Bisulphite reduction 7
Bradyrhizobium japonicum 152
BTEX utilization 282
b-type haem 191, 192
Butyrate metabolism 514
Butytyl CoA dehydrogenase 514

C_1-carrier 16
caa_3 oxygen reductase 192
Caldivirga 39, 306
Caldivirga maquilingensis 103
Carbon mineralization 372
Carbon monoxide 15, 16
Carbon monoxide dehydrogenase 15, 16, 117, 124, 129, 219, 489
Carboxydothermus hydroformans 99
Catalase 169, 188, 196, 198, 200, 202, 470
Cathodic depolarising and corrosion 465–7
Caulobacter crescentus 152
cbb3 oxidases 191
Cellulomonas 446
Chaperones 152, 153
Chemoclines 168
Chloroflexi 349
Chondroitin sulphate 507, 508
Chromate detoxification 445
Chromate reduction 414, 438, 440
Chromate stress 145, 148, 195, 440
Chromate transport 440, 441
Chromosome size 124
Citric acid cycle 16
Clostridium FISH probe 60
CO cycling 215
Cobalamin 450
Cobalt sorbents 394
COD/sulphate ratio 387
Coenzyme B 9, 279
Coenzyme M 277, 278
Cold stress 145, 148
Colonizing the gut wall 505
Community analysis 96
Community structure 360–2
Consortium 274, 490, 493
CooMKLXUHF complex 11, 14, 128
Copper inhibition 386
Copper sulphide 419
Corrosion 307, 360, 459, 460, 463, 469, 472
Corrosion models 465
Corrosion of alloys
Corrosion of cables 460
Corrosion rates 461
Corrosion reaction 461
Corrosive phosphorus 475
Cr(III) precipitates 445
Crenarchaeote 39
Crohn's disease 513
Cryptanaerobacter FISH probe 60
c-type haem 191, 233, 440
Cu(II) toxicity 131
Cysteine desulfurase 441, 448
Cysteine synthase 441
Cytochrome – decahaem 224
Cytochrome – octahaem 224
Cytochrome bd 170
Cytochrome bd oxygen reductase 188
Cytochrome bd quinol:oxygen oxidoreductase, 192
Cytochrome c 8, 9, 14, 128, 148, 155, 170, 192, 195, 203, 218, 219, 222, 227, 232, 448, 468

Cytochrome c oxidase 187
Cytochrome c_3 222–4, 232, 414, 415, 425, 437, 439, 446, 448, 486
Cytochrome c_{553} 192, 193, 222, 224
Cytochrome c_{554} 224
Cytochrome c_7 425, 437, 439
Cytochrome in outer membrane 468

Denaturing gradient gel electrophoresis (DGGE) 41
Desulfacinum 24, 309, 318
Desulfacinum hydrothermale 310, 320
Desulfacinum infernum 313, 320
Desulfacium subterraneum 310
Desulfarculus 24
Desulfarculus baarsii FISH probe 60
Desulfatibacillum 24
Desulfatibacillum aliphaticivorans 284, 287
Desulfatibacillum alkenivorans 284, 287
Desulfitobacterium 99
Desulfitobacterium hafniense 470
Desulfoacinum infernum 313
Desulfoacinum subterraneum 313
Desulfoarculus baarsii 97
Desulfobacca 24
Desulfobacca acetoxidans probe 42
Desulfobacter 1, 24, 168, 185, 320, 345, 450, 470, 504, 505
Desulfobacter anilini 489
Desulfobacter FISH probe 60
Desulfobacter postagatei 16, 117
Desulfobacter probe 42
Desulfobacter vibroformis 310
Desulfobacterace 24, 364
Desulfobacteraceae FISH probe 60
Desulfobacteraceae probe 42
Desulfobacterales probe 42
Desulfobacteriaceae 22, 41, 97, 117, 124, 134
Desulfobacterium 1, 24, 185, 309, 320, 362, 450, 470
Desulfobacterium anilini 97, 99, 348, 489
Desulfobacterium autotrophicum 117, 124, 131, 320
Desulfobacterium autotrophicum FISH probe 60
Desulfobacterium autotrophicum probe 42
Desulfobacterium catecholicum 242
Desulfobacterium catecholicum FISH probe 60
Desulfobacterium cetonicum 283, 310, 313
Desulfobacterium indolicum 487
Desulfobacterium niacini FISH probe 60
Desulfobacterium niacini probe 42
Desulfobacterium phenolicum 282
Desulfobacterium vacuolatum FISH probe 60
Desulfobacterium vacuolatum probe 42
Desulfobacula 24
Desulfobacula FISH probe 60
Desulfobacula phenolica 282
Desulfobacula toluolica 283, 284
Desulfobibrionaceae 23
Desulfobotulus 24
Desulfobotulus sapovorans FISH probe 60
Desulfobulbaceae 24, 97, 124
Desulfobulbaceae FISH probe 60
Desulfobulbaceae probe 42
Desulfobulbus 1, 24, 41, 168, 274, 309, 362, 364, 450, 504, 505
Desulfobulbus FISH probe 60
Desulfobulbus mediterraneus 348
Desulfobulbus probe 366
Desulfobulbus propionicus 12, 16, 242, 243, 363
Desulfobulbus rhabdoformis 310, 313
Desulfocapsa 24
Desulfocapsa thiozymogenes FISH probe 60
Desulfocella 24
Desulfocella FISH probe 60
Desulfococcus 24, 40, 41, 90, 91, 134, 168, 185, 274, 450, 470
Desulfococcus FISH probe 60
Desulfococcus multivorans FISH probe 60, 489
Desulfococcus probe 42
Desulfofaba 24
Desulfofaba FISH probe 60
Desulfofaba gelida FISH probe 60
Desulfofaba probe 42
Desulfoferrodoxin 199
Desulfofrigus 24, 40, 345
Desulfofrigus FISH probe 60

Desulfofrigus probe 42
Desulfofustis 24
Desulfofustis FISH probe 60
Desulfohalobiaceae 24, 320
Desulfohalobiaceae FISH probe 60
Desulfohalobium 24
Desulfomic cavernae 346
Desulfomic macestense 346
Desulfomicrobiaceae 24
Desulfomicrobioaceae FISH probe 60
Desulfomicrobium (strain Ben-RB) 448
Desulfomicrobium 24, 314, 345, 362, 416, 439
Desulfomicrobium aspheronum 310
Desulfomicrobium baculatum 437
Desulfomicrobium probe 42
Desulfomonas hansenii FISH probe 60
Desulfomonas pigra 504
Desulfomonile 24, 41, 95
Desulfomonile FISH probe 60
Desulfomonile probe 42
Desulfomonile tiedjei 97
Desulfomonile tiedjei FISH probe 60
Desulfomusa 24
Desulfomusa FISH probe 60
Desulfonatronovibrio 24
Desulfonatronovibrio hydrogenovorans 320
Desulfonatronum 24
Desulfonatronum lacustre 348
Desulfonatronumaceae 24
Desulfonauticus 24, 318
Desulfonauticus submarinus 310, 320
Desulfonema 90, 168, 172, 345, 362–4, 366
Desulfonema FISH probe 60
Desulfonema magnum 348
Desulfonema probe 42
Desulforegula 24
Desulforegula FISH probe 60, 362
Desulforhabdus 24
Desulforhabdus amnigena 320
Desulforhopalus 24
Desulforhopalus FISH probe 60
Desulforhopalus singoporenssi 242
Desulfosarcina 24, 40, 41, 90, 134, 274, 345
Desulfosarcina cetonica 282, 284
Desulfosarcina FISH probe 60
Desulfosarcina probe 42
Desulfosarcina variabilis FISH probe 60
Desulfospira 24
Desulfospira FISH probe 60
Desulfosporomuse 39
Desulfosporosinus 24, 39, 306, 314, 345, 351
Desulfosporosinus 446, 447
Desulfosporosinus probe 42
Desulfotalea 24, 345
Desulfotalea FISH probe 60
Desulfotalea psychrophila 6, 123, 124, 127, 131, 203, 215, 222, 225, 226, 440, 447, 450
Desulfothermus 24
Desulfotignum 24
Desulfotignum FISH probe 60
Desulfotomaculum (strain B 10) 350
Desulfotomaculum (strain B2T) 346
Desulfotomaculum 1, 22, 24, 39, 101, 168, 306, 314, 345, 350, 351, 504, 505
Desulfotomaculum FISH probe 60
Desulfotomaculum halophilum 310, 314, 315
Desulfotomaculum kuznetsovii 310, 314, 346
Desulfotomaculum nigrificans 310, 314
Desulfotomaculum norvegicum 224, 437, 439
Desulfotomaculum orientis 489
Desulfotomaculum probe 42
Desulfotomaculum reducens MI-1 103
Desulfotomaculum thermobenzoicum 242
Desulfotomaculum thermocisternum 310, 314, 315
Desulfotomaculum thermosapovorans 348
Desulfovibrio (strain B) 484
Desulfovibrio (strain Ben-RA) 448
Desulfovibrio 24, 40, 41, 91, 96, 168, 185, 306, 309, 314, 318, 345, 364, 388, 414, 439, 450, 470
Desulfovibrio acrylicus FISH probe 60
Desulfovibrio aestuarii FISH probe 60
Desulfovibrio africanus 223
Desulfovibrio alaskensis 123, 216, 310, 313, 475

Desulfovibrio bastinii 310, 313
Desulfovibrio capillatus 310, 313
Desulfovibrio desulfuricans (strain ATCC 27774) 202, 223, 227, 228, 242, 244, 248, 250, 251
Desulfovibrio desulfuricans (strain CSN) 243
Desulfovibrio desulfuricans (strain Dv01) 185, 243
Desulfovibrio desulfuricans (strain Essex 6) 243
Desulfovibrio desulfuricans (strain G-20) 102, 123, 124, 131, 188, 215, 224, 225, 437, 440, 446, 447, 450
Desulfovibrio desulfuricans 6, 21, 170, 198, 200, 242, 313, 415, 446, 449, 489, 510, 511
Desulfovibrio fairfieldensis 510
Desulfovibrio fairfieldensis FISH probe 60
Desulfovibrio fructosovorans 12, 216, 437
Desulfovibrio furfuralis 242
Desulfovibrio gabonensis 310, 313
Desulfovibrio geothermicum 346
Desulfovibrio gigas 170, 192, 194, 198, 216, 223, 437, 439, 489
Desulfovibrio gracilis 310, 313
Desulfovibrio halophilus probe 42
Desulfovibrio hydrothermalis 310, 318
Desulfovibrio indonesiensis 346, 468, 474, 475
Desulfovibrio infernum 310
Desulfovibrio latus 320
Desulfovibrio longreachii 313
Desulfovibrio longus 310, 313
Desulfovibrio magneticus 132
Desulfovibrio multispirans 242
Desulfovibrio oxamicus 242
Desulfovibrio oxyclinae probe 42, 185, 313
Desulfovibrio piger 504, 511
Desulfovibrio profundus 242, 310, 318, 346, 350, 351
Desulfovibrio putealis 351, 352
Desulfovibrio salexigens 170
Desulfovibrio simplex 242
Desulfovibrio termitidis 170, 242
Desulfovibrio thermophilus 315
Desulfovibrio vietnamensis 310, 313
Desulfovibrio vulgaris (strain DP4) 103
Desulfovibrio vulgaris 5, 6, 8–12, 15, 102, 123, 124, 131, 141, 170, 172, 186, 188, 200, 215, 223, 224, 257, 320, 414, 415, 425, 437, 439, 445, 446, 450, 466, 484
Desulfovibrio vulgaris Myazaki 192, 198, 437
Desulfovibrio zoesterae 318
Desulfovibrioaceae 22
Desulfovibrionaceae 24, 41, 93, 117, 124
Desulfovibrionaceae FISH probe 60
Desulfovibrionaceae probe 42
Desulfovibrionales probe 42
Desulfovirga 24
Desulfurellales probe 42
Desulfuromonadaceae 23
Desulfuromonales FISH probe 60
Desulfuromonales probe 42
Desulfuromonas 345
Desulfuromonas acetoxidans 425, 437, 439
DhcARnfCDGEAB 224, 226
Diffusion chamber 172
Dimethyl mercury 450
Dimethyl sulfoxide reductase 155
Dimethyl sulfoxide reduction 1
Dissimilatory sulphate reduction 6, 12, 15
Diurnal changes 168, 172, 185
DMSO reductase 244, 246
dsrAB/DsrAB 99, 101, 154, 228, 232
DsrMKJOP complex 128, 154, 227, 228, 233, 466
d-type haem 191

EchABCDEF complex 11, 14, 127
Electron transfer and corrosion 464, 471
Electron transport 9
Energy coupling/energetics 5, 266, 267, 270, 279
Engineered systems 90
Environmental diversity 96
Environmental genomics 102
Environmental technology 383, 384
ERGO 121
Ethylbenzene utilized 290, 292

Eubacterium FISH probe 60
Euryarchaeota 1, 124
Evolution 22, 96, 132, 292
Extracellular substances 472

Fatty acid oxidation 308, 389, 390, 409, 504, 506, 514
Fe(II) transporter 148
Fe(III) reduction 1, 19, 351
Fecal sulphate reduction 509
Fe-hydrogenase 8, 188, 195, 437, 439, 447, 467, 468
Fermentation 20, 23, 307, 506, 508
Ferredoxin 11, 14, 227, 247, 439, 485
Ferredoxin-thioredoxin reductase 441
Ferritin 188, 202
Fibrobacter 88
Fingerprint 40, 41, 58, 102, 123
Firmicutes 24, 39, 99
Firmicutes FISH probe 60
Firmicutes probe 42
FISH technique 60, 361
Flavodoxin 148
Flavoproteins 194
Flavorubredoxin 194
Florescence in situ hybridization (FISH) 91
FNR regulon 204
Formate cycling 15, 222, 231
Formate dehydrogenase 12, 15, 128, 155, 215, 218, 219, 222, 224, 227, 231, 246
Fumarate reduction 1
Fumarate:quinone oxidoreductase 192
Fur regulon 161, 200

GAP4 STADEN package 120
Gas hydrate formations 333, 335
GenDB 121
Gene organization 129
Gene targeted primers 42
Genes 96
Genome 9, 12, 15, 103, 117, 118, 123, 186, 187, 195, 215
Genome annotation 120
Genome sequencing 118
Geobacter 131, 446
Geobacter metallireducens 132
Geobacter sulfurreducens 128, 132
Global regulators 145, 160, 162, 186
Glutaredoxin 187, 188
Glycine betaine 146
Glycolate oxidase 148, 230
Glycolysis 194
Glycoprotease 153
Glycoproteins 507
Goethite 473
Greigite 407
Growth in the gut 506

Haem a 192
Haem b 219, 222, 228
Haem c 253
Haem-copper oxygen reductase 188, 192, 193
Heam d 191
Heam o 192
Heat shock response 131, 145, 148, 153, 162
Hemerythrin 199
Hg(II) toxicity 131
Hmc complex 9, 128, 223, 225–8, 232
hmc genes 9, 148
Hme complex 9, 227
Hme genes 9
Horizontal gene transfer 103, 133
HrcA 152
Human colon 504
Human disease 510
Hydrocarbon oxidation 266, 267, 270, 279, 280, 284
Hydrogen cycling 14, 20, 215, 216, 222
Hydrogen evolution 462, 509
Hydrogen peroxide 196
Hydrogen peroxide reductase 197–9
Hydrogenase 8, 10–12, 14, 41, 127, 128, 170, 190, 195, 215, 216, 218, 219, 222, 223, 227, 231, 440, 445, 465–7, 475, 485
Hydroperoxidase I 188, 198
Hydrothermal vents 241, 310, 316
Hydroxyl radicals 269
Hydroxylamine 256
HynBAC3 224
HysBA 226

Inflammatory bowel disease 511
Influent dilution 386
Inhibition of SRB 386
Interspecies hydrogen transfer 17, 20, 274, 282, 451
Intestinal SRB 511
Intraspecies formate transfer 16
Intraspecies hydrogen transfer 12, 14, 15, 20, 21
Iron oxidation
fixed-bed reactor 411
Iron storage enzymes 202
Iron sulphide 407, 421
Iron sulphide and corrosion 469
Isoprenoid biosynthesis 133
Isothermal microcalorimetry 440
Isotope array approach 96
Isotope ratios for carbon 272, 282
Isotopes of sulphur 22

KEGG database 121

Lactate dehydrogenase 14, 148, 170, 230
Lactate metabolism 508
Lactate oxidation 11, 12
Lactate permease 148, 230
Lawsoni intracellularis 23
Lawsonia intracellularis FISH probe 60
Lead precipitation 424
Low temperature stress 144

Mackinawite 407
MAGPIE 121
Malonomonas rubra 23, 170
Marine arctic sediment 41
Marine sediments 343, 345, 350
Marine subsurface 344, 348
Mats 168, 171–3, 185, 241
Membrane fluidity 145
Membrane hybridization 59
Membrane reactor 396
Menaquinone 10, 12
Mercaptopyruvate sulphur transferase 514
Mercuric lyase 450
Mercuric reductase 441, 450
Mercuric transport 441
Mercury reduction 450
Mercury resistance 441, 449, 450
Metabolic island 99
Metal hydroxides 412
Metal precipitation 393, 412
Metal recovery 393, 408
Metal reductases 445
Metal reduction 332, 414, 437, 451
Metal removal 416, 417
Metal sorption 393
Metal sulphides 405–8, 419
Metalloid reduction 414
Metalloid sulphides 409, 411
Methane 2, 307, 335
Methane monooxygenase 269
Methane oxidation 269, 330, 335
Methane presence 272
Methanobacterium foricicum 249
Methanogenesis in gut 505
Methanogens 88, 90, 225, 227, 274, 385, 387, 389, 410, 484
Methanosaeta 372
Methanosarcina acetivorans 132
Methanosarcina thermophila 132
Methionine sulfoxide reductase 153, 188
Methyl mercury 450, 451
Methylmalonyl-CoA pathway 16, 129
Methylnaphthalene utilized 290, 292
Methyltetrahydrofolate 450
Microarry 93, 95, 103, 121, 131, 142, 154, 159, 187
Microautoradiography 361
Microbial communities 361, 451, 459
Microbial Community Analysis (MiCA) 58
Microelectrodes 361, 369, 391
Migration 172
Mn(IV) reduction 1, 19, 340, 351
Mn-containing catalase 471
MnS 407
Mo(V) species 248, 249
Modeling 332
Molybdate inhibition 372, 386, 451
Molybdate transport 157
Mössbauer data 251, 254

Mucin 508
Myxococcales probe 42

NAD(P)H hydrogen peroxide oxidoreductase 201
NAD(P)H peroxidase 188, 197
NAD(P)H superoxide oxidoreductase 201
NADH dehydrogenase 128, 231
NADH oxidase 169, 170
NADH:rubredoxin oxidoreductase 170
NADH:quinone oxidoreductase 128, 193
NADP-reducing hydrogenase 216
Nanometer metal precipitates 445
Nanowires 275
nap genes 246
Naphthalene oxidization 277, 282
Neelaredoxin 199
NhcB 228
Nickel sorbents 394
NiFe-hydrogenase 8, 96, 195, 216, 222, 224, 437, 439, 468
NiFeSe-hydrogenase 195, 224, 437
Nigerythrin 188, 200
Nitrate as electron acceptor 1, 18, 171, 233, 242, 257, 308, 332, 351, 373, 423
Nitrate reductase 154, 244, 248, 249, 441
Nitrate stress 145, 148, 155, 243, 371
Nitric oxide 194, 256
Nitric oxide reductase 194
Nitrite inhibition 131, 233, 243
Nitrite reductase 154, 224, 233, 242, 244, 250, 251, 253, 257, 486
Nitrite stress 145, 148, 152, 157
Nitroaniline 484
Nitroaromatics 484, 486
Nitrobenzene 484
Nitrogenase 175, 226, 266
Nitrospira 39
Nitrospira FISH probe 60
Nitrospiraceae FISH probe 60
NrfA 255
Numbers of SRB in stools 504

Oceanic crust 342
Off-gas treatment 384
OhcABC 226
Oil fields 241, 243, 309, 310
Oil reservoirs 281, 307, 308, 320
Olavius algarvensis symbiont FISH probe 60
Olavius crassitunicatus symbiont FISH probe 60
Organomercury lyase 450
Organosulphonate reduction 99, 102
Osmotic stress 145, 146, 148
o-type haem 192
Oxic/anoxic interface 17, 18, 168, 169, 363, 408, 471, 476
Oxoglutarate synthase 489
Oxygen and corrosion
Oxygen detoxification 169
Oxygen gradients 171
Oxygen reduction 1, 170, 187, 188, 191, 194–7
Oxygen stress 131, 145, 148, 186, 194–6, 198
Oxygen tolerance 169, 172, 173, 185, 187, 196
Oxygen toxicity 169
Oxygenases 266
Oxygen-sulphide gradients 173
OxyR 204

PAPS reductase 148
Paracoccus pantotrophus 244, 246, 249
Pb(II) immobilization 424, 446
PEDANT 121
Pelobacter 23
Pelobacter carbinolicus 132, 348
Pelotomaculum 99, 101
Pelotomaculum FISH probe 60
Pelotomaculum probe 42
Pelotomaculum thermopropionicum FISH probe 60
Peptococcaceae FISH probe 60
Periplasm 15
Peroxidase 197, 199, 201, 202
Peroxiredoxin 198
PerR regulon 161, 200
pH homeostasis 158
Phosphate acetyltransferase 148, 230
PHRED program 120

Phylochip 93, 96
Phylogenic Assignment Tool (PAT) 58
Phylogenic diversity 21
Plasmids 124, 198, 440
Pollution control 383
Polyamine biosynthesis 159
Polyglucose reserves 194
Probe characterization 88
Probe labeling 59
Probes for 16S rRNA 364, 391
Propionate oxidation 16
Propionly-CoA 16
Proteases 152, 153
Proteobacteria FISH probe 60
Proteobacteria probe 42
Proteomic analysis 121, 190, 446
Proteomics 131, 186, 187, 467
Protohaem IX farnesyl transferase 192
Proton stoichiometries 10
Proton translocation 10
Protonophores 15
Pyrite 407
Pyrobaculum 101
Pyrobaculum islandicum 22
Pyrococcus 317
Pyrophosphatase 7, 148, 229
Pyruvate synthase 489
Pyruvate:ferredoxin oxidoreductase 11, 14, 148, 230
Pyruvate-formate lyase 15, 230

Qmo complex 9, 128, 225, 227, 233, 466
Quinol oxidase 191
Quinol:cytochrome c oxidoreductase 193

Reactive oxygen species 186, 187, 196, 197, 202
Redox cycling 19
Redox potentials 3, 13
Reverse sample genome probing (RSGP) 92
Rhodanese 514
Rhodobacter sphaeroides 244, 246
Ribonucleotide reductase 173
Rice roots 41
RnfCDGEAB 226
rRNA sequencing 40
Rubredoxin 195, 199, 201
Rubredoxin:oxygen oxidoreductase 169, 188, 194, 199
Rubrerythrin 148, 186, 188, 198, 199
Ruminococcus FISH probe 60

Saccharopine dehydrogenase 158
Saline subsurface environment 346
Salt stress 146
Salt stress response 131
Salt tolerance 350
Sapropels 340
Selenate reduction 415, 448
Selenide,water dikinase 441
Selenium removal 414
Selenocysteine synthetase 448
Selenophosphate synthase 448
Sensing system for nitrate 247
Seryl-tRNA selenium transferase 441
Shewanella oneidensis 425
Sigma factors 147, 153
Sodium/proton transporter 157, 159, 160
Soil treatment 384
Solid waste treatment 384
Spatial distribution 368
Spermidine decarboxylase 158
Sporobacter FISH probe 60
Sporotomaculum 99
Sporotomaculum FISH probe 60
Sporotomaculum probe 42
Stress response 145, 160
Structure of ccNir 251
Structure of nitrate reductase 245
Subsurface metabolism 351
Succinate:quinone oxidoreductase 192
Succinyl-CoA 11
Sulfide concentration 386
Sulfospirillum deleyianum 251, 253
Sulphate activation 6, 7, 167, 215
Sulphate adenylyltransferase 147, 148
Sulphate in the gut 506
Sulphate permease 148
Sulphate reduction 41, 96, 168, 215, 218, 229, 330, 332, 335, 340, 343, 360, 385, 389

Sulphate reduction in gut 505
Sulphate supply 335
Sulphate transport 10
Sulphate:methane interface 337
Sulphide effects gut mucosa 512
Sulphide oxidation 243, 273, 308, 332, 333, 337, 360, 371
Sulphide production in gut 508
Sulphidogenesis 19
Sulphidogenic reactors 388
Sulphidogenic vs methanogenis 386
Sulphite oxidase 514
Sulphite reductase 9, 22, 101, 128, 133, 148, 154, 227–9, 233, 243
Sulphur cycle 241, 374, 375, 383, 384
Sulphur oxidation 363
Sulphur reduction 1, 317
Superoxide 196
Superoxide dismutase 169, 188, 196, 199, 201, 446, 470
Superoxide reductase 186, 198
Synechococcus 249
Syntrophic interaction 275
Syntrophobacter 95
Syntrophobacteraceae 41, 97
Syntrophobacteraceae FISH probe 60
Syntrophobacteraceae probe 42
Syntrophobacterales probe 42
Syntrophus FISH probe 60

Taxonomic differences 24
Tc(VII) reduction 446
Technecium precipitation 409, 413, 415
Terminal restriction fragment length polymorphism (T-RFLP) 58
Tetrahydrofolate 16
Tetrahydromethanopterin 16
Thermoacetogenium 314
Thermoacetogenium phaeum 129
Thermocladium 39
Thermococcus 317
Thermodesulfatator 306, 318, 321
Thermodesulfatator indicus 310, 321
Thermodesulfobacterium 24, 39, 97, 99, 306, 309, 315, 318
Thermodesulfobacterium commune 103, 132, 310, 315
Thermodesulfobacterium hydrogenophilum 310, 315, 321
Thermodesulfobacterium probe 42
Thermodesulfobacterium thermophilum 310, 315
Thermodesulfobiaceae 1
Thermodesulfobium 23, 24, 309
Thermodesulfobium narugense 39, 242
Thermodesulforhabdus 24, 309, 314
Thermodesulforhabdus norvegicum 310, 313, 320
Thermodesulfotomaculum orientis 6
Thermodesulfovibrio 1, 23, 24, 39, 97, 306, 309, 350
Thermodesulfovibrio FISH probe 60
Thermodesulfovibrio islandicus 99, 242
Thermodesulfovibrio yellowstonii 103, 132
Thermodesulfovibrionaceae 24
Thermodynamics 2
Thermoproteales 39
Thio peroxidase 197, 198, 202
Thiobacillus denitrificans 97, 396
Thiobacillus ferroxidans 396
Thiobacillus thioparus 396
Thiol-peroxidase 187, 188
Thioploca 171
Thioredoxin 188, 441
Thioredoxin disulphide reductase 441
Thioredoxin reductase 153, 188, 441, 448
Thiosulphate reductase 8, 425
Thiothrix 363
Thiovirga sulfuroxydans 363
Thiovulvum 171
Tiosphaera pantotropha 246
TmcCBA 226, 232
Toluene oxidation 283, 290
TpI-c3 complex 8, 222
TpII-c3 complex 9, 128
Transcriptomic analysis 121
Transcriptomics 102, 187
Transporters 157
Transposase 124, 127, 155, 157

TRF-CUT 58
tRFLP fragment sorter 58
Trinitrotoluene 483, 493
Trophic interactions 17
Tryptophanase 160
Twin Arginine Translocator system 246

U(VI) reduction 415, 446
Ubiquinone pool 248
Uranium reduction 413, 415, 446

Vectoral distributions 366
Veillonella 446
Vibrio alginolyticus 128

Wastewater treatment 384
Wautersia eutropha 246
Wetland systems 419
Wolinella succinogenes 251, 253

Xylene utilized 290

YACOP 121

Znic sulphide 419